Risk Governance

Urbano Fra.Paleo
Editor

Risk Governance

The Articulation of Hazard, Politics and Ecology

Springer

Editor
Urbano Fra.Paleo
School of Social Sciences and Humanities
University of Extremadura
Cáceres, Spain

ISBN 978-94-024-0372-5 ISBN 978-94-017-9328-5 (eBook)
DOI 10.1007/978-94-017-9328-5
Springer Dordrecht Heidelberg New York London

Softcover reprint of the hardcover 1st edition 2015

Printed on acid-free paper

Springer is part of Springer Science+Business Media (www.springer.com)

To Encarnación, whose plans have shaped the present
To Belén, who has seen plans happen or vanish and is able to anticipate the risks
To Valeria, whose indecisive plans are shaping an uncertain time
Only after understanding can we take action.

About the Editor

Urbano Fra.Paleo is Professor in Human Geography at the University of Extremadura, Spain. He is also Visiting Professor at the UN-mandated University for Peace (UPEACE), Costa Rica. He worked at the United States Geological Survey in Denver (1995) and Hawaii (1999) and was Research Associate at the Environment Institute of the University of Denver (1996). Fra.Paleo was Fellow of the American Geographical Society Library, University of Wisconsin-Milwaukee, in 2005. In 2009, he received the Award for Sustainable Actions in Social Entrepreneurship and the Innovation Award by the University of Santiago de Compostela. Since 2011, he is a member of the Scientific Committee of the Integrated Risk Governance (IRG) Project, an initiative within the International Human Dimensions Programme on Global Environmental Change (IHDP), and of the Disaster Risk Reduction Thematic Group within the International Union for Conservation of Nature (IUCN). Urbano Fra.Paleo has edited the book *Building Safer Communities. Risk Governance, Spatial Planning and Responses to Natural Hazards* in 2009, a volume that addresses the adoption of spatial planning as a proactive approach to deal with risk exposure. His interests lie in the collaborative evaluation of risk governance and the development of proactive strategies of mitigation and adaptation to risk from natural hazards.

About the Editor

Acknowledgements

I gratefully acknowledge those scholars who generously contributed their time and expertise to review the manuscripts of the book chapters and provided valuable feedback to the authors. Thoughtful comments and constructive input after careful reading helped to sharpen analysis and rethink many issues. I am grateful to anonymous reviewers and to the scholars listed below.

Abdul-Akeem Sadiq, Assistant Professor at Indiana University-Purdue University Indianapolis (IUPUI), United States of America
Dr. Anne Marie O'Hagan, Researcher at the Hydraulics and Maritime Research Centre, University College Cork, Ireland
Claudia Baldwin, Senior Lecturer at the University of the Sunshine Coast, Australia
Professor David Sanderson from Oxford Brookes University, United Kingdom
Professor Lance Gunderson from Emory University, United States of America
Professor John Fitzgibbon from the University of Guelph, Canada
Nicole Gurran, Associate Professor at the University of Sydney, Australia
Dr. Michael Brett-Crowther, Editor of the International Journal of Environmental Studies
Michael Palenchar from the University of Tennessee, United States of America
Murray Lee, Senior Lecturer at the University of Sydney, Australia
R.B. Singh, Associate Professor at the University of Delhi, India
Dr. Russel Reichelt, Chairman and Chief Executive of the Great Barrier Reef Marine Park Authority, Australia
Annett Steinführer, Researcher at the Johann Heinrich von Thünen-Institut (vTI), Germany
Professor Tim Doyle from the University of Adelaide, Australia
Professor William Waugh from Georgia State University, United States of America
Professor Yvonne Rydin from University College London, United Kingdom

However, the final responsibility for the editing and for any errors is mine entirely.

Contents

Contributors

Daniel P. Aldrich Department of Political Science, Purdue University, West Lafayette, IN, USA

Helen Elizabeth Allison School of Veterinary and Life Sciences, Murdoch University, Murdoch, WA, Australia

Rhoda Ballinger School of Earth and Ocean Sciences, Cardiff University, Cardiff, UK

Timothy Beatley School of Architecture, University of Virginia, Charlottesville, VA, USA

Frederic Bouder Assistant Professor, Department of Technology and Society Studies, Maastricht University, Maastricht, The Netherlands

David J. Brunckhorst UNESCO Centre for Bioregional Resource Management, University of New England, Armidale, NSW, Australia

Ian Burton School of the Environment, University of Toronto, Toronto, ON, Canada

Andrew E. Collins Department of Geography, Northumbria University, Newcastle upon Tyne, UK

Ana Maria Cruz Disaster Prevention Research Institute, Kyoto University, Uji, Kyoto, Japan

Simin Davoudi Environmental Policy and Planning, School of Architecture, Planning and Landscape, Newcastle University, Newcastle upon Tyne, UK

Bruna De Marchi Centre for the Study of the Sciences and the Humanities (SVT), University of Bergen, Bergen, Norway

Gert de Roo Department of Spatial Planning and Environment, Faculty of Spatial Sciences, University of Groningen, Groningen, The Netherlands

Thea Dickinson Department of Physical and Environmental Sciences, University of Toronto, Toronto, ON, Canada

Bob Evans School of Social Science, University of Waikato, Hamilton, New Zealand

Urbano Fra.Paleo School of Social Sciences and Humanities, University of Extremadura, Cáceres, Spain

Georg Frerks Conflict Prevention and Conflict Management, Centre for Conflict Studies, Utrecht University, Utrecht, The Netherlands

International Security Studies, Faculty of Military Sciences, Netherlands Defense Academy, Utrecht, The Netherlands

Disaster Studies, Sociology of Development and Change, Social Sciences Group, Wageningen University, Wageningen, The Netherlands

James S. Gardner Natural Resources Institute, University of Manitoba, Winnipeg, MB, Canada

Armin Haas Global Contract for Sustainability, Institute for Advanced Sustainability Studies, Potsdam, Germany

Robert L. Heath School of Communication, University of Houston, Houston, TX, USA

Michael Heazle School of Government and International Relations, Griffith University, Brisbane, QLD, Australia

Carlo C. Jaeger State Key Laboratory of Earth Surface Processes and Resource Ecology, Beijing Normal University, Beijing, China

Yoshio Kajitani Central Research Institute of Electric Power Industry, Chiyodaku, Tokyo, Japan

Naim Kapucu School of Public Administration, University of Central Florida, Orlando, FL, USA

Roger E. Kasperson George Perkins Marsh Institute, Clark University, Worcester, MA, USA

Andreas Klinke Department of Political Science and Environmental Policy Institute, Memorial University of Newfoundland, Canada

Katherine McComas Department of Communication, Cornell University, Ithaca, NY, USA

David A. McEntire Emergency Administration and Planning, Department of Public Administration, University of North Texas, Denton, TX, USA

Bruce Mitchell Department of Geography and Environmental Management, University of Waterloo, Waterloo, ON, Canada

James K. (Ken) Mitchell Department of Geography, Rutgers University, Piscataway, NJ, USA

Gabe Mythen Department of Sociology, University of Liverpool, Liverpool, UK

Geoff O'Brien Department of Geography, Faculty of Engineering and Environment, School of Built and Natural Environment, Northumbria University, Newcastle upon Tyne, UK

Marg O'Brien Coastal & Freshwater Group, Cawthron Institute, Nelson, New Zealand

Phil O'Keefe Department of Geography, Faculty of Engineering and Environment, School of Built and Natural Environment, Northumbria University, Newcastle upon Tyne, UK

Ortwin Renn Stuttgart Research Center for Interdisciplinary Risk and Innovation Studies ZIRIUS, University of Stuttgart, Stuttgart, Germany

John B. Rundle Department of Physics, University of California, Davis, CA, USA

Rajib Shaw Graduate School of Global Environmental Studies, Kyoto University, Kyoto, Japan

Peijun Shi State Key Laboratory of Earth Surface Processes and Resource Ecology, Beijing Normal University, Beijing, China

Hirokazu Tatano Disaster Prevention Research Institute, Kyoto University, Uji, Kyoto, Japan

Graham A. Tobin School of Geosciences, University of South Florida, Tampa, FL, USA

Donald L. Turcotte Department of Geology, University of California, Davis, CA, USA

Fiona Tweed Geography and Environment, Science Centre, Staffordshire University, Stoke-on-Trent, UK

Gordon Walker Lancaster Environment Centre, Lancaster University, Lancaster, UK

Geoffrey Wescott School of Life and Environmental Sciences, Deakin University, Burwood, VIC, Australia

Qian Ye State Key Laboratory of Earth Surface Processes and Resource Ecology, Beijing Normal University, Beijing, China

Christian Zuidema Department of Spatial Planning and Environment, Faculty of Spatial Sciences, University of Groningen, Groningen, The Netherlands

List of Figures

List of Tables

Chapter 1
Introduction

Urbano Fra.Paleo

> The future is not a pre-existing land towards which we are all moving, and which it is our task to discern through the mist and prepare for, but something which is created and shaped through all the decisions we make.
>
> Szerszynski et al. (1996)

> In characterizing the authoritarian technics that has begun to dominate us, I have not forgotten the great lesson of history: Prepare for the unexpected!.
>
> Lewis Mumford (1964)

> The challenge for social-ecological systems is to enhance the adaptive capacity to deal with disturbance and to build preparedness for living with change and uncertainty.
>
> Berkes et al. (2003)

> I believe most citizens do not want to confront the need for major social changes on any issues except those that seem directly to threaten them.
>
> Anthony Downs (1972)

> There is little place for self-governing in risk issues and because I see complexity first, dynamics second and diversity last, a mix of hierarchical and co-governance seems to me the appropriate mix to govern risk situations in modern societies.
>
> Jan Kooiman (2003)

Not long before the writing of this book, the 2011 Tōhoku disaster struck Japan. Its occurrence helped make the paradigm of contemporary risk and disaster in the anthropocene era globally visible and has synthesized most of the attributes of risk society and global environmental change. At the same time, this event, also known as the Great East Japan disaster, showed how natural processes—earthquake and tsunami—interact, and provided clear evidence of the increasingly complex interference between technology and natural processes resulting in nuclear disaster in

U. Fra.Paleo (✉)
School of Social Sciences and Humanities, University of Extremadura,
Campus universitario, 10071 Cáceres, Spain
e-mail: upaleo@unex.es

U. Fra.Paleo (ed.), *Risk Governance*, DOI 10.1007/978-94-017-9328-5_1

this instance. The disaster also exhibited how its effects and loss extend globally, well beyond the local area, with the release of radiation through air and seawater, the dispersion of disaster debris throughout the North Pacific Ocean and by the simultaneous impact on the national economy and global production.

1.1 *Homo risicus*

According to Alexander (2008: 82), "One key aspect of the reductionism that has characterized modernist planning and design, at least in North America, is the tendency to see people as atomized egos, as isolated monads." However, any person has multiple, compatible, conflicting or complementary dimensions of the self that lead to "applying different preferences in different contexts", or "hold multiple preference orderings" (Nyborg 2000: 306) that engender complexity in the environmental knowledge-attitude-behaviour sequence. Nyborg has studied how individual preferences fluctuate between a maximization of personal well-being as *Homo economicus* and social welfare as *Homo politicus*, agreeing with Faber et al. (1997), who deem they are not mutually exclusive, making political processes the result of their mutual interaction. In other words, as Montani (2007: 51) points out: "The economic *Homo* does not exist, but we can observe individuals who operate on the market. The political *Homo* does not exist, but we can observe individuals who fight to conquer power within the State." Any person, at times, acts as a consumer or a producer and others as a citizen (Sagoff 1988) who seeks political justice (Faber et al. 2002). These authors state that "arguments of utility maximisation are of importance for all decisions of political actors, although they are neither always decisive nor are they the only ones." (p. 325). Numerous models of human behaviour (*Homo*) have been proposed to depict the multiple non-exclusory dimensions of human behaviour. Zawojska (2011) and Faber et al. (2002) have pulled together and interpreted a range of *Homines*. A number of these are alternatives to *Homo economicus*, others an integration of various human aspects, as in the case of *Homo agricola*, which Zawojska identifies as an integration of *Homo economicus* and *Homo politicus*.

The landscape is not complete without considering the intervention and role of *Homo faber*, the human being that develops and uses technology. Hanna Arendt has been a principal proponent of this concept. "The role of *Homo faber* for Arendt is thus central for human life" (Hickman 1988: 161) because, according to Hickman, it allows humans to both escape from the subjectivity of the private (*Homo sapiens*) and to stabilize human life against the variations of nature. Arendt is reluctant to acknowledge the natural condition of human beings and supports a strong anthropocentric stance (Szerszynski 2003) aligned with the economic thought of Malthus who sees nature as a restriction on human activity (Becker et al. 2004). "According to Malthus, human mind and nature remain—as already for Francis Bacon (1561–1626) and Rene Descartes (1596–1650)—two disparate and opposed entities, whereby it is the human mind which represents the higher principle" (Becker et al. 2004: 6).

Consistent with the paradigm of risk society, the intensification of the activity of *Homo faber* in modern society brought the emergence of new technological hazards, and at the same time science and technology were seen as political instruments to control nature and limit risk—a vicious circle that gave birth to *Homo risicus* (hérédité, nécessité et non-connaissance). According to the TNK (transfer, needs, knowledge) framework (Fra.Paleo 2009), risks are either inherited (hérédité) as the result of the history of the location, the result of the search for the satisfaction of human needs (nécessité) (*Homo biologicus*, as dependent on nature (Baumgärtner and Quaas 2009)), or the result of a lack of knowledge (non-connaissance) or risk awareness, or any combination of the three. "Risks are created and selected by human actors" (Renn, Chap. 2, this volume), and again, a dual character is found. Sometimes *Homo risicus* acts as a risk taker and at other times a risk avoider. In the first case, the behaviour responds to the nature of the *Homo economicus* alone, or in combination with *Homo faber*, while the second type of action is the result of the intervention of *Homo politicus*, with the occasional cooperation of *Homo faber*. Politics is the instrument that controls the intervention of factors and balances the role of actors influencing decisions in the risk field, while science and technology play a double role as controller and promoter. Rather than describing modes of human behaviour, both *Homo ecologicus* (Dryzek 1996) and *Homo sustinens* (Siebenhüner 2000) represent the ideals of the human–nature relationship that fit the paradigm of sustainable development. However Faber et al. (2002: 328) judge that the behaviour of *Homo politicus* serves the purposes of ecological economics "since justice denotes the generally approved ordering of a political community, it implies that the natural foundations of existence have to be secured in the long run, i.e., sustainability. Hence, sustainability is implied by political justice."

The notion of risk society (Beck 1992), a prevalent theoretical paradigm in the social sciences, represents an "alternative sociological imagination" (Ekberg 2007) to address and manage the emergent challenge of technological risk—in parallel to pervasive natural hazards—associated with the reflexive second modernity. The paradigm serves as an analytical framework for research and policymaking in late modernity, in competition and cooperation with paradigms of contemporary society, including post-industrial society (Daniel Bell), liquid modernity (Zygmunt Bauman), or network society (Manuel Castells), that altogether build an era of multiple modernities (Lee 2006). Turner and Boyns (2001) ask for the return of the Grand Theory in sociology to advance the explanation of big processes, and to fill the theoretical gap between macro and micro levels of reality. However, Gomá (2013) argues that contemporary philosophy—and perhaps theory in general—is not producing ideals but instead hands out descriptive schemes of how society is. He contends that this makes these present theories inoperative because there is a need, not just for explanation, but also for prescription, for clues of how society might be, for recipes for political advancement.

Not surprisingly the identification of the attributes of risk and governance alike is discussed in this book, in relation with the features of contemporary society, and a high level of agreement can be found. While Renn (Chap. 2, this volume) believes risk is a mixture of complexity, scientific uncertainty and socio-political ambiguity,

Mitchell (Chap. 16, this volume) contends it is a combination of change, complexity and uncertainty, and Allison (Chap. 6, this volume) sees a complex future characterised by connectivity, uncertainty and ambiguity. A descriptive triad can also be found in Kooiman (2003: 151), when he maintains that, "in a governance perspective … complexity comes first, followed by dynamics and then diversity." After all, Kooiman does not draw attention to uncertainty, for his primary focus is society at large and not specifically risk.

1.2 Social, Political and Ecological Subsystems

States and sub-national administrations have been progressively acting and placing greater emphasis on the role of institutional regulation and on the almost unlimited capacity of the public sector to design and successfully implement policies, including disaster risk reduction, ecosystem protection and the control of the impact of human activity on the environment. However, in recent decades, of equal prominence has been a neo-liberal discourse stressing the efficiency of the market and claiming the replacement of the state in the provision of public services (with government downsize), by turning citizens into consumers, and ecosystem components and services into assets with measurable economic value. Both discourses have been following and overlapping each other, although lately the latter seems to become more dominant when the legitimacy of the state to govern is being challenged. This has been particularly visible in liberal and transition economies during the financial crisis of the early twenty-first century.

The dominant approach to political decision-making and management, formulated by John Dewey in his 1910 book *How we think* and based on perfect rationality, follows three stages: (i) what is the problem, (ii) what are the alternatives, and (iii) which alternative is best? In contrast, Pomerol (2009) believes this perspective is simplistic and hardly operational, and perhaps one of the main causes for failure in policy implementation is the poor understanding of decision-making as a process of permanent adaptation, learning and negotiation (Kay et al. 1999; Mitchell 2002; McCarthy et al. 2006;). Pomerol (2009) argues against each of the three stages, stating that a (i) clear definition of the problem is not feasible but dependent on multiple agent interpretations, (ii) the alternatives are not readily available—owing to bounded rationality—but still have to be constructed, (iii) and agreement has to be reached regarding the criteria to evaluate the best option. Bounded rationality provides support to an alternative pathway for decision-making that looks more at the process than at the outcomes because all the policy options will not be accessible, the only way to expand them and to gain diversity is to incorporate multiple viewpoints and societal actors in political decision-making and management.

However, little attention has been paid to the multiple societal and economic actors, operating more or less autonomously, and to their roles—either through action or inaction—as creators and managers (Mythen, Chap. 3, this volume), or as bearers of risk. Instead of considering them as major political players with specific interests, who plan and act according to the satisfaction of their needs and wants, supporting, opposing or deflecting public policies, or as actors who continuously interfere with and modify the extant risk conditions, their action is considered marginal, merely a behavioural issue that increases the complexity of governance but devoid of political dimensions and legitimacy. The influence of multiple, unconnected decisions at the ground level is not considered an element of policy-making but merely a challenge to be addressed during policy implementation. Civil society organizations and citizens do not seem solely passive subjects whose role it is to inactively accept government decisions and regulations, who suffer the impact of disasters and receive external assistance, but key actors who continuously intervene in the environment, using natural resources or benefiting from environmental services, determining the evolution of political processes. The social–political system is as complex, dynamic and diverse (Kooiman 2003) as the ecological system, and possibly increasingly so. Thus, the task of governing can neither be entirely assigned to government for it does not have the full capacity to successfully implement the designed policies, nor to the market for it does not have the legitimacy or a shared understanding of societal values. As decision-making and governing are neither equivalent nor separate processes, the extant gap can be bridged with effective public participation.

The encounter between ecology and risk science—or risk management—that may drive theoretical cross-fertilization (Dogan and Pahre 1990), and ultimately the conceptual drift of the latter, has not been hitherto very fruitful, perhaps limited by the conventionally strong emphasis of risk management in the adoption of structural measures that seek to increase resistance of the built environment. The construction of major coastal protection structures, watercourse dams and channels, with the purpose of reducing the dimensions of hazards, particularly frequency, has taken priority over the long-term impact of hazards in societal and ecological systems.

In recent years, the most prevalent strategy adopted in risk policy, resistance—the search for the stability of environmental conditions—seems to have weakened, albeit slowly. Starting in 2006, the Netherlands is implementing a 10-year plan, Room for the Water (*Ruimte voor de Rivier*), following a strong push for flood control infrastructure (particularly with the Zuiderzee and Delta works) in the twentieth century. The new plan exemplifies a change of focus from resistance to resilience, in a process of destructuralisation of flood risk. The resilience paradigm is based on the perspective of living with floods (Liao 2012) rather than living with no floods. As Liao (2012) reasons, "cities that are dependent on flood-control infrastructure are highly resistant—but not resilient—to floods because they have physically adapted to the artificially expanded dry-and-stable conditions to become intolerant of wet conditions."

The adoption of resilience as a policy approach in risk governance represents a turning point in the interaction between both fields of study. First introduced by Holling (1973) in ecology, ecological resilience has been adopted by social sciences as a theory, policy strategy and metaphor (Norris et al. 2008), in a sort of theoretical cross-fertilization. Dogan and Pahre (1990) claim that term borrowing is common in multiple disciplines, although some seem to be more averse to this practice than others, and find that disciplines follow a life cycle with successive phases of specialization, fragmentation and hybridization. Risk science has too short a history as a discipline—Gilbert F. White published his seminal work *Human adjustment to floods* in 1942—to have exhibited extreme internal diversity or weakness in the generation of theory; And remains to some degree impermeable to the concept of ecology, although not to the earth and social sciences' ideas. The study of disasters is a notable example of the latter. Pitirim Sorokin's pioneer work, *Man and society in calamity* (1942), is no less important than White's in integrating earth sciences, although a more contemporary example is the work in two volumes, *What is a disaster? Perspectives on the question* (Quarantelli 1998) and *What is a disaster? New answers to old questions* (Perry and Quarantelli 2005). Rosa (1998) argues that risk science continues to mature as a field of study as the progressive consensus over several conceptual foundations shows. The currently dominant discourse of resilience is the result of both evidences of the failure of a past strong confidence on structural measures—New Orleans (2005) and Eastern Japan (2011) are prominent cases—and of the inevitability of environmental change. However, the approach of resilience apparently brings the perspective of an inexorable course of action, as if those acts of nature were acts of god. The finding that some degraded ecosystems exhibit forms of resilience and resist ecological restoration that, through human intervention, seeks to take them back to an earlier steady state, brings new avenues of understanding for the management of risk. According to Suding et al. (2004: 4) "degraded communities often do not respond predictably to management efforts, producing inconsistent and sometimes unexpected results" while new "feedbacks can make a degraded system resilient to restorative change". Similarly, disaster areas where recovery measures are in progress may exhibit attributes of resilience to restorative action and prior vulnerability is frequently replicated.

What is required now is more than a hybridisation between ecology and risk science, but rather a unifying theory that combines the knowledge gained about human–nature (directional) interactions within the natural sciences and the nature–human (directional as well) exchanges recognized by risk science. Beyond borrowing concepts from each other—although certainly fruitful—it may look at integrating the existing paradigms to elaborate a complete view of the problem of risk, in a kind of co-evolution—bridging disciplinary divides—because we "cannot effectively parse the patterns and dynamics of today's environmental systems into 'natural' and 'anthropogenic' components" (Kinzig 2001). An upstream integration (Kinzig 2001) would be preferable to a downstream assimilation. After all, as Mythen (Chap. 3, this volume) contends, "explicitly risk based strategies to governance first emerged in most nations in response to environmental problems and were embedded in policies of ecological protection".

Where Cohen (1997) identifies occasions for convergence through a unified framework between Ulrich Beck's risk society theory and ecological modernization, Buttel (2000) finds key theoretical obstacles. This author contends various views are shared by both theories, particularly the notion that the problems caused by modernization can be solved with more modernization. However, Buttel (2000) distinguishes various inconsistencies that prevent ecological modernization from resting on the first theory, and they concern policy and decision-making. Within his own framework, Beck (1992) recognizes the role and importance of social movements in driving radical political processes, pushing structural changes in state and society. Thus, social action might drive ecological modernization too, with environmental problems as the arena for reflections on modern society (Szerszynski et al. 1996). Buttel (2000) argues that ecological modernization theorists such as Arthur Mol and Gert Spaargaren may be adopting an opportunistic stance to help mainstream ecological modernization within sociological thought and to gain legitimacy as a theory: "ecological modernization has become attractive as a concept because it provides alternatives to the pessimistic connotations of frameworks such as the treadmill of production and the growth machine" (Buttel 2000: 63). Notwithstanding, he also identifies weaknesses. In particular, Buttel highlights its reactive stance against radical environmentalism with a strong political perspective and a focused conceptualisation based on the success of eco efficiency and pollution control in certain Northern European countries. Barry (2006) points to a stagnation in the path to ecological sustainability attributable to the co-optation—and possible neutralization—of environmental citizenship by the market and the state. Thus, corporative, state-based or state-backed support a limited environmental progress that does not challenge injustice or ecological unsustainability. "neglecting the economic, political, and cultural dimensions of sustainability and sustainable development" (Barry 2006: 21) "in a manner that hollows out its transformative, oppositional, and radical political dimensions" (Barry 2006: 24). To agitate the academic debate and motivate political action, Barry advocates active sustainability citizenship for its transformative potential, based on the civic republican tradition of citizenship, to attain a different type of society. He contends "citizenship is something that has to be learned rather than something that comes naturally to members of society. Just as citizenship can be learned and therefore needs to be taught or encouraged, it can also be forgotten" (Barry 2006: 27). The idea of sustainability citizenship stands as a pathway to *Homo sustentans* by strengthening and deepening the *Homo politicus*, without rejecting *Homo economicus*: "The point is not to reject consumption and the consumer identity in favor of a citizen one" (Barry 2006: 38). Sustainability citizenship turns into a prospective political approach to temper *Homo risicus*, to reduce disaster risk in a wider context, and challenging contemporary modes of justice, participation, social and economic development and human–environment relations.

However, risk is not the single and absolute object of concern in risk governance—as exemplified in this book—for the reason that risk policies are oriented or shaped by specific approaches to human–nature interactions and to political systems. There are four critical pillars of risk governance: sustainability, as a response to environmental degradation, resource depletion and ecosystem service deterioration; governance, as a reaction to government inability to address key societal and environmental problems; mitigation, as a means to avoid exposure and reduce vulnerability; and adaptation, as a response to expected and unexpected changes in environmental conditions. Although both sustainability and governance may have become buzzwords for some—De Marchi (Chap. 9, this volume) believes many do not know the meaning of the latter—they also represent valid paradigms to orient policy-making. So while the notion of a social–ecological system (Berkes and Folke 1998) helps to explain the processes of change in the context of natural science, Szerszynski et al. (1996) claim that social sciences have made little progress to adopt the concept—and what has been achieved has been done with caution and critically—as a framework for theoretical and empirical analysis, perhaps with the exception of political ecology.

1.3 Organization of the Book

Drawing on various disciplines, this book provides a comprehensive examination of the complexity of both risk and environmental policy-making and of their multiple—and not always visible—interactions in the context of social–ecological systems, considering the intricate relations between society and nature that lead to environmental change and the emergence of risk. No less importantly, it also addresses the unseen and neglected complementarities between regulatory policy-making and ordinary individual decision-making through the actions of nongovernmental actors—individuals, private sector and community-based organizations—and their effect on the development of inclusive governance. As Mythen (Chap. 3, this volume) indicates, "Citizens [should not be] seen exclusively as subjects of risk who require institutional protection against harm, but also as both creators and managers of risk."

A range of distinguished scholars from a diverse set of disciplines have contributed to the book with their expertise in many areas, including disaster studies, emergency planning and management, ecology, sustainability, environmental planning and management, climate change, geography, spatial planning, development studies, economy, political sciences, public administration, communication, as well as physics and geology.

The book consists of three parts. The book opens with chapters addressing the concept of risk governance and its unfolding in the current conditions of increasing complexity, diversity and change, with different perspectives from various disciplines. As the thread of the discourse is the exploration of a harmonized mode of governance of the social–political and ecological systems, the second part deals with how governance is organised and operates in specific regions of risk. These include mountain environments, coastal or oceanic areas, landscapes, and water basins, megacities (on a region level) and climate (at a global scale). There is also a focus on natech disasters, as an example where natural processes and technology interact at a higher level of complexity. To conclude, the third part consists of short notes that provide glimpses of key concepts, ideas and frameworks, and give directions for future research, whose integration in the framework of risk governance requires further investigation.

As a general introduction, the first chapter by Ortwin Renn focuses on the debate on the relationships between resilience and risk governance. Based on the characteristics of risk—complexity, uncertainty and ambiguity—he calls for the reform of governance structures and procedures to allow for the assimilation of not just knowledge and political legitimacy, but also of public values and preferences. Governance is thus interpreted a process of the negotiation and construction of landscapes of risk. The author distinguishes the various interpretations of risk and highlights its ambiguity, attributable to the varied selection of information and to perceptions on how to act, which leads to various forms of risk management. Resilience is examined and its emergence is ascribed to the capacity of individuals to learn and negotiate.

In the second chapter, Gabe Mythen retakes the thread of the interplay between knowing and not-knowing, or *Nichtwissen* in Beck's terms of discourse. Mythen draws attention to the increasing levels of societal risk aversion and distrust, both in decision-makers, as responsible for action/inaction, and in experts, as producers of knowledge. The limits of risk knowledge have been put manifest through public attitude, so that it is not an issue of increasing the understanding of processes but rather of finding the appropriate political instruments to deal with the unmanageability of contemporary risks. Furthermore, and interestingly, he establishes a foundation connection between ecology and risk in the response to environmental problems that influence risk assessment.

In an increasingly urbanised world, the chapter authored by Christian Zuidema and Gert de Roo is devoted to the examination of the implications of the process of power decentralization. They contend local governments are in a better position to address complexity because they can gain a better understanding of the local conditions, and can more successfully access actors and get them involved in collaborative processes. According to the authors, the allocation and integration of environmental policies in the Netherlands followed loops of centralisation and decentralisation, assisted by area specific policies to help integration. Change and cycles of political learning were followed by governance renewal. In addition, the authors identify limiting factors, whose intervention may reduce local base quality; however, these can be counteracted with the central state retaining enabling and support functions.

Soft infrastructure is presented as a resource to foster civic engagement in public processes in the next chapter, entitled *Local governance and soft infrastructure for sustainability and resilience*, by Bob Evans and Marg O'Brien. They argue that the challenges faced by societies cannot be addressed by both local and central government alone. Collaborative governance may drive a more efficient environmental policy-making and management, favouring social learning that may secure behaviour change if matched with self-interest. Resilience and sustainable development are greatly dependent on ecolacy, and not simply on knowledge but also on awareness of the consequences of individual action. Soft infrastructure, with its components of institutional capital, social capital and social learning, together with augmented government–civil society interaction, may propel new forms of governance.

The fifth chapter, written by Helen Allison, adopts an ontological perspective, supported by the understanding that our decisions are ultimately built on how we think of the world around us. Allison deals with the concept of an ontological-driven risk governance system, and argues for a paradigmatic change as a replacement for incremental and adaptive change, to deal with increasingly complex social–ecological systems. New models based on systems thinking and resilience leave behind assumptions of equilibrium and stability to help better define problems, identify solutions and implement actions.

In their chapter, Robert L. Heath and Katherine McComas seek pathways for a more fully functioning society based on a novel paradigm of risk analysis and communication, as they understand that the identification and assessment of risk is a central component of risk governance. Again, reality only becomes meaningful if multiple interpretations are integrated through dialogue within the context of risk democracy. Legitimacy, responsibility and control for risk identification and distribution should shift and be shared to correct loss–benefit imbalances because “organizations and individuals bend reality to their interests and their interests to reality”. Thus, risk communication is understood as a dialogue where multiple interests are balanced to manage risk with more efficacy.

The emergence of civil society is the aim of the chapter written by Daniel P. Aldrich. He contends that strong social ties are both the outcome of and a vital condition in disasters because they help increase resilience. The case of the Tōhoku disaster serves well to illustrate his assumption. Aldrich identifies various types of social capital, their contribution and mechanisms of action. Social capital is a key component of risk governance in disasters when the state cannot assist response and recovery, particularly in weak states where the government is mostly absent, or when governmental action ignores or weakens social networks.

The integration of uncertainty and of the various types of knowledge in risk governance is discussed in the following chapter by Bruna De Marchi. She warns against the mirage of complete calculable risk through standard scientific knowledge and management, because uncertainty is expected to increase rather than decrease owing to the amplified capacity of human intervention in nature. De Marchi criticizes the attitude of normal science and decision-makers of ignoring lay knowledge, monopolizing legitimacy and policy input with the illusion of objective facts and rational modelling. Such actions have led to hard engineering mitigation, and De Marchi calls for the inclusion of other forms of knowledge and the accommodation of uncertainty in risk management to improve disaster prevention and response.

In the following chapter, Rajib Shaw examines the role education can play in the reduction of disaster risk, by integrating disaster risk reduction into school curricula and making schools more resistant and resilient. The learning process dealing with safety should continue throughout the time the student remains at school and extended to the interactions between education institutions and society at large. The implications of the Hyogo Framework for Actions in education and advances in three national initiatives serve to measure progress. Finally, the author observes that its impact should be identified based on the adoption of both non-structural and structural measures.

The disaster epidemic is the topic of the chapter authored by Thea Dickinson and Ian Burton, and leads to propose the Forensic Investigation of Disasters as a new research approach to better identify and understand disasters. For the authors it is not just the issue of the unnatural nature of natural disasters—with the intervention of human choice and action—but the recognition that the human action behind its triggering is changing the perspective to address the problem. The processes associated with globalization and today's information society have increased the level of connectedness and dependence within the social–political system, and thereby adds extra complexity to risk management.

The following two chapters deal with the transition of risk and emergency management to address the failures of focus and separate operation. David McEntire explores the conceptual bridges between proactive risk management and reactive emergency management by way of the notion of comprehensive vulnerability management; however, in some cases it is difficult to tell the difference between preparedness and response. This approach facilitates to overcome the limitations of emergency management as regards to inclusiveness and the neglect of mitigation, and the narrowness of risk management assuming that all disasters can be eliminated with sufficient knowledge. Comprehensive vulnerability management is proactive in that it focuses on viable vulnerability reduction and reactive in considering resilience in the phases of response and recovery.

Naim Kapucu focuses on the role of actors during disasters and the failures of emergency management in his chapter *Leadership and collaborative governance in managing emergencies and crises*, while proposing collaborative emergency management as an alternative approach. Kapucu advances the requirements for a more effective management of response and recovery: social capital and crisis leadership to induce collaboration among groups with different views. He also contends risk communication plays a key role in disasters, not just to provide accurate information about the event but to scrutinize government performance and to promote accountability. Finally, Kapucu analyses the operation of the cluster approach by the international humanitarian system in weak states such as Haiti.

The last chapter of this section by Urbano Fra.Paleo is devoted to determining a theoretical framework to support and advance a system to evaluate risk governance with a participatory approach. The author argues evaluation does not just serve the purpose of measuring performance but can be extended to social learning and steady advancement in disaster risk reduction if other societal groups are involved in the process of dialogue, and consensus is promoted. Evaluation as deliberation facilitates reaching consensus over a more equitable distribution of risks and a negotiation of the preferred level of risk. Finally, an evaluation framework based on a mixed hierarchical and networked structure of criteria, components and dimensions of risk governance is advanced to complete the evaluation system.

The chapter written by Geoff O'Brien and Phil O'Keefe opens the second part of the book, where governance in different contexts is examined. The authors argue that government inaction to address climate change responds to the lack of perception and interest of citizens, who only react when large-scale changes in environmental conditions occur. Mitigation and adaptation are complementary approaches that tackle two sides of the problem: changes that have occurred and those that will occur in the future. However, those governments that are acting to mitigate seem to rely heavily upon the position of civil society and not just upon law instruments that, by themselves, are insufficient.

The metaphor of the Iraq invasion is used by Michael Heazle in his chapter to illustrate how uncertainty and science are used in politics, whether for action or inaction—particularly in dealing with climate change—through what has been called the scientisation of politics. Values and interests are behind political stances, and decisions are argued to be supported by objective rationality. Science is thus a screen of legitimacy administered by governments and political elites to avoid or control public debate and downplay competing interests. Heazle proposes to adopt post-normal science to address uncertainty and policy-making, favouring early competition between multiple antagonistic priorities and goals to be resolved through dialogue and consensus.

The next chapter by Bruce Mitchell concerns the implementation of integrated water risk management and the inability to adequately deal with the needs, interests and priorities of multiple stakeholders, and the lack of success in increasing global efficiency by way of optimising it locally. Water governance requires the reorganization of relationships in different directions, horizontally and vertically, as well as upstream–downstream cooperation, reassigning tasks and competencies. However implementation gaps reduce system efficiency with failures in stakeholder participation, and in the integration of land and water uses or between economic development and ecosystem management.

According to David Brunckhorst, landscapes are socially constructed spaces that "synthesize and reflect human and ecological interactions", revealed as spatial patterns that reflect past and present processes, and where decisions are reflected as changes in pattern. Perceptions of pattern and process engender the sense of community, and landscapes turn into an appropriate context for the design and implementation of both natural resource and natural hazard management. The idea of landscape governance is put forward in Brunckhorst's chapter as an avenue for the translation of the principles and notion of landscape ecology through co-management, which may assist "the design of institutional arrangements towards increased adaptive capacity".

In the chapter titled *Risk complexity and governance in mountain environments*, James Gardner describes the complexity of mountain environments in terms of extreme spatial and temporal geo-ecological variability with changing human action, which has evolved from local subsistence to a globally-linked and diversified economy that does not take into account the local environment as a reference. Growth in population by in-migration, an increased economic base and the expansion of transportation infrastructures are increasingly disconnected from local knowledge and traditional avoidance of hazard prone areas. Finally, the author argues for a different mode of governance in mountain areas, with increased horizontal and vertical cooperation, to manage the combined character as a source of both natural resources and large-scale cascading hazards.

Rhoda Ballinger discusses in the following chapter the problematical and multifaceted challenges of coastal risk governance, which has been chiefly built on piecemeal reactive decision-making and the regulation of individual hazards. Furthermore, other factors come into play such as the largely hard-engineering approach used in response to local demands and spatial planning by planners who suffer from poor knowledge and low awareness of risk. Moreover, coastal zones have a high level of complexity owing to multiple property rights, arbitrary division of responsibilities and access to resources. Ballinger maintains that while conventional governance is being questioned in a context of economic slowdown and budget strain, integrated forms of governance such as integrated coastal management find it difficult to advance with the same factors.

The next chapter, written by Geoffrey Wescott, takes a closer look at the governance of oceans and speculates whether risk management is an option for ocean governance. According to Wescott, the process of the removal of large areas of ocean from the commons has been decisive for the increased pressure to develop further forms of governance. However, difficulties in implementing an ecosystem based management (EBM) approach soon arise because change is perceived as a threat by various policy sectors. EBM transcends traditional approaches (monospecific, sectoral) by delivering shared enforcement, regional cooperation and marine spatial planning (however, because delays in the implementation are expected, costly learning processes will be required).

According to James Mitchell, the study of the governance of megacity disaster risk is still at an early stage. Megacities are a relatively recent phenomenon, and represent the highest level of community complexity, particularly in the sprawl form. This social–political system needs to operate with a large volume and variety of information, whereby both solutions and institutions need to be designed and implemented to address problems that worsen at a great speed. Novel approaches based on decentralisation, diffuse forms of risk management and combined forms of knowledge are required. However, a one-size-fits-all approach to megacities will not accommodate the extant heterogeneity, and yet, the various forms of participatory governance will not become a panacea for managing risks either.

The last chapter of the second part, by Ana Maria Cruz, Yoshio Kajitani and Hirokazu Tatano, focuses on natech (as combined natural and technological disaster risks), a manifestation of the prospect of widespread disasters with the intensification of technology. Natech disasters exhibit some of the most salient attributes of risk society, such as the delocalisation of impacts and cascading effects, and—as happened with megacities—new methods and approaches are needed to properly address their complexity. The governance of natech requires a change in the regulatory frameworks to consider the multiple interactions among natural and technological factors, a better integration of natural and technological risk management and, concurrently, a separation between the authorities that regulate and promote a certain technology.

The chapters in this book seek to shed light on the underlying structural factors, processes, players and interactions in the risk scenario, all of which influence decision-making that both increases and reduces disaster risk. Action cannot be confined to palliation but should operate within the context of the social–political–ecological system to satisfactorily address complexity, diversity and change within the subsystems and in their interactions. Still, risk governance is far from being systemic and comprehensive, far from being both proactive and reactive, far from becoming effectively integrated with other sectoral policies and far from being inclusive and equitable. The construction of the concept of risk governance is a work in progress that, although partly in the context of disciplinary boundaries, is supported by a dynamic debate and fluent exchange as a manifestation of its lively nature and young history. However, neglecting the social–ecological dimension of risk and that of the adaptive character of human action may lead to disciplinary failure. In his work *Risk*, John Adams (1995) makes clear his aim to bridge the hard and soft approaches to risk. Similarly, it is vital to cross the conceptual divides. The purpose of this book is to explore the common language of politics, ecology and risk.

Acknowledgement Copy editing of this chapter has been made possible with the financial support of the Government of Extremadura and the European Regional Development Fund (ERDF) through the grant GR10071 awarded to the research group on Geospatial Engineering and Urban Heritage of the University of Extremadura.

References

Adams, J. 1995. *Risk*. London: UCL Press.

Alexander, D. 2008. Physical determinism, modernism and mental health. *Environments: A Journal of Interdisciplinary Studies/Revue d'études interdisciplinaire* 35(3): 81–86.

Barry, J. 2006. Resistance is fertile: From environmental to sustainability citizenship. In *Environmental citizenship*, ed. A. Dobson and D. Bell, 21–48. Cambridge: MIT Press.

Baumgärtner, S., and M. Quaas. 2009. *What is sustainability economics?* Working paper series in economics, no. 138. Lüneburg: University of Lüneburg.

Beck, U. 1992. *Risk society: Towards a new modernity*. London: Sage.

Becker, C., M. Faber, K. Hertel, and R. Manstetten. 2004. *Malthus vs Wordsworth: Perspectives on humankind, nature and economy*, Discussion paper series no. 404. Heidelberg: University of Heidelberg, Department of Economics.

Berkes, F., and C. Folke (eds.). 1998. *Linking social and ecological systems: Management practices and social mechanisms for building resilience*. Cambridge: Cambridge University Press.

Berkes, F., J. Colding, and C. Folke (eds.). 2003. *Navigating social-ecological systems: Building resilience for complexity and change*. Cambridge: Cambridge University Press.

Buttel, F.H. 2000. Ecological modernization as social theory. *Geoforum* 31: 57–65.

Cohen, M.J. 1997. Risk society and ecological modernisation alternative visions for post-industrial nations. *Futures* 29(2): 105–119.

Dogan, M., and R. Pahre. 1990. *Creative marginality: Innovation at the intersections of social sciences*. Boulder: Westview.

Downs, A. 1972. Up and down with ecology: The issue attention cycle. *The Public Interest* 28: 38–50.

Dryzek, J.S. 1996. Foundations for environmental political economy: The search for homo ecologicus? *New Political Economy* 1(1): 27–40.

Ekberg, M. 2007. The parameters of the risk society: A review and exploration. *Current Sociology* 55(3): 343–366.

Faber, M., R. Manstetten, and T. Petersen. 1997. Homo oeconomicus and Homo politicus. Political economy, constitutional interest and ecological interest. *Kyklos* 50: 457–483.

Faber, M., T. Petersen, and J. Schiller. 2002. Homo oeconomicus and homo politicus in ecological economics. *Ecological Economics* 40: 323–333.

Fra.Paleo, U. 2009. On exposure to natural hazards: Revisiting a neglected primal action. In *Building safer communities. Risk governance, spatial planning, and responses to natural hazards*, ed. U. FraPaleo, 61–78. Amsterdam: IOS Press.

Gomá, J. 2013. ¿Dónde está la gran filosofía? *El País*. 14/03/2013. Available at: http://cultura.elpais.com/cultura/2013/03/14/actualidad/1363283723_240070.html. Accessed 1 Mar 2013.
Hickman, L. 1988. The phenomenology of the quotidian artifact. In *Technology and contemporary life*, ed. P. Durbin, 161–176. Dordrecht: Reidel.
Holling, C. 1973. Resilience and stability of ecological systems. *Annual Review of Ecology and Systematics* 4: 1–23.
Kay, J.J., H. Regier, M. Boyle, and G.R. Francis. 1999. An ecosystem approach for sustainability: Addressing the challenge of complexity. *Futures* 31(7): 721–742.
Kinzig, A.P. 2001. Bridging disciplinary divides to address environmental and intellectual challenges. *Ecosystems* 4: 709–715.
Kooiman, J. 2003. *Governing as governance*. London: Sage.
Lee, R.L.M. 2006. Reinventing modernity. Reflexive modernization vs liquid modernity vs multiple modernities. *European Journal of Social Theory* 9(3): 355–368.
Liao, K.H. 2012. A theory on urban resilience to floods: A basis for alternative planning practices. *Ecology and Society* 17(4): 48.
McCarthy, D., G. Whitelaw, P. Jongerden, and B. Craig. 2006. Sustainability, social learning and the Long Point World Biosphere Reserve. *Environments Journal* 34(2): 1–15.
Mitchell, B. 2002. *Resource and environmental management*, 2nd ed. Harlow: Prentice Hall.
Montani, G. 2007. The ecocentric approach to sustainable development. Ecology, economics and politics. *The Federalist: A Political Review* XLIX(1): 25–60.
Mumford, L. 1964. Authoritarian and democratic technics. *Technology and Culture* 5(1): 1–8.
Norris, F.H., S.P. Stevens, B. Pfefferbaum, K.F. Wyche, and R.L. Pfefferbaum. 2008. Community resilience as a metaphor, theory, set of capacities, and strategy for disaster readiness. *American Journal of Community Psychology* 41(1–2): 127–150.
Nyborg, K. 2000. Homo economicus and Homo politicus: Interpretation and aggregation of environmental values. *Journal of Economic Behavior & Organization* 42: 305–322.
Perry, R.W., and E.L. Quarantelli (eds.). 2005. *What is a disaster? New answers to old questions*. Philadelphia: Xlibris.
Pomerol, J.C. 2009. Human decision: Recognition plus reasoning. In *Decision making process: Concepts and methods*, ed. D. Bouyssou, D. Dubois, H. Prade, and M. Pirlot, 157–197. London: Wiley.
Quarantelli, E. 1998. *What is a disaster? Perspectives on the question*. London: Routledge.
Rosa, E.A. 1998. Metatheoretical foundations for post-normal risk. *Journal of Risk Research* 1(1): 15–44.
Sagoff, M. 1988. *The economy of the earth. Philosophy, law, and the environment*. Cambridge: Cambridge University Press.
Siebenhüner, B. 2000. Homo sustinens. Towards a new conception of humans for the science of sustainability. *Ecological Economics* 32: 15–25.
Suding, K.N., K.L. Gross, and G. Houseman. 2004. Alternative states and positive feedbacks in restoration ecology. *Trends in Ecology and Evolution* 19: 46–53.
Szerszynski, B. 2003. Technology, performance and life itself: Hannah Arendt and the fate of nature. *The Sociological Review* 51(Supplement s2): 203–218.
Szerszynski, B., S. Lash, and B. Wynne. 1996. Introduction: Ecology, realism and the social sciences. In *Risk, environment and modernity: Towards a new ecology*, ed. S. Lash, B. Szerszynski, and B. Wynne, 1–26. London: Sage.
Turner, J.H., and D.E. Boyns. 2001. The return of grand theory. In *Handbook of sociological theory*, ed. J.H. Turner, 353–378. New York: Plenum Press.
Zawojska, A. 2011. Economic, social and political lives of Homo agricola. *Journal of US-China Public Administration* 8(3): 275–287.

Urbano Fra.Paleo is Professor in Human Geography at the University of Extremadura, Spain. He is also Visiting Professor at the UN-mandated University for Peace (UPEACE), Costa Rica. He worked at the United States Geological Survey in Denver (1995) and Hawaii (1999) and was

Research Associate at the Environment Institute of the University of Denver (1996). Fra.Paleo was Fellow of the American Geographical Society Library, University of Wisconsin-Milwaukee, in 2005. In 2009, he received the Award for Sustainable Actions in Social Entrepreneurship and the Innovation Award by the University of Santiago de Compostela. Since 2011, he is a member of the Scientific Committee of the Integrated Risk Governance (IRG) Project, an initiative within the International Human Dimensions Programme on Global Environmental Change (IHDP), and of the Disaster Risk Reduction Thematic Group within the International Union for Conservation of Nature (IUCN). Urbano Fra.Paleo has edited the book *Building Safer Communities. Risk Governance, Spatial Planning and Responses to Natural Hazards* in 2009, a volume that addresses the adoption of spatial planning as a proactive approach to deal with risk exposure. His interests lie in the collaborative evaluation of risk governance and the development of proactive strategies of mitigation and adaptation to risk from natural hazards.

Part I
Risk Governance Under Conditions of Increasing Complexity, Diversity and Change

Chapter 2
Risk Governance and Resilience: New Approaches to Cope with Uncertainty and Ambiguity

Ortwin Renn and Andreas Klinke

2.1 Introduction

Deciding on suitable locations for hazardous facilities, setting standards for chemicals, making decisions about cleaning up contaminated land, regulating food and drugs, or designing and enforcing safety limits all have one element in common: these activities are collective endeavours to understand, assess and handle risks to human health and the environment. These attempts are based on two requirements. Firstly, risk managers need sufficient knowledge about the potential impacts of the risk sources under investigation and the likely consequences of the different decision options for controlling these risks. Secondly, they need criteria to judge the desirability or undesirability of these consequences for the people affected and the public at large (Klinke and Renn 2012; Renn and Schweizer 2009; Horlick-Jones et al. 2007; Rowe and Frewer 2000). Criteria concerning desirability are reflections of social values such as good health, equity, or efficient use of scarce resources. Both components –knowledge and values– are necessary for any decision-making process independent of the issue and the problem context.

Anticipating the consequences of human actions or events (knowledge) and evaluating the desirability and moral quality of these consequences (values) are the core elements of risk analysis. Anticipating future events and judging their desirability poses particular problems if the consequences are complex and uncertain and the values contested and controversial (in our terms ambiguous). Dealing with

O. Renn (✉)
Stuttgart Research Center for Interdisciplinary Risk and Innovation Studies ZIRIUS,
University of Stuttgart, Seidenstr. 36, 70174 Stuttgart, Germany
e-mail: ortwin.renn@sowi.uni-stuttgart.de

A. Klinke
Department of Political Science and Environmental Policy Institute,
Memorial University of Newfoundland, Canada
e-mail: aklinke@grenfell.mun.ca

U. Fra.Paleo (ed.), *Risk Governance*, DOI 10.1007/978-94-017-9328-5_2

complex, uncertain and ambiguous outcomes often leads to the emergence of social conflict relating to both epistemological as well as moral issues (van Asselt and Renn 2011). Questions of how to deal with complex, uncertain and controversial risks demand procedures for dealing with risks that go beyond the conventional risk management routines. Numerous strategies to cope with this challenge have evolved over time. They include technocratic decision-making through the explicit involvement of expert committees, muddling through in a pluralist society, negotiated rule-making via stakeholder involvement, deliberative democracy or ignoring probabilistic information altogether (see reviews in Nelkin and Pollak 1979, 1980; Brooks 1984; Renn 2008: 290ff.) The main thesis of this chapter is that risk management institutions need more adequate governance structures and procedures that enable them to integrate professional assessments (systematic knowledge), adequate institutional process (political legitimacy), responsible handling of public resources (efficiency) and public knowledge and perceptions (reflection on public values and preferences). These various inputs are necessary to increase the resilience of our risk management and regulatory institutions to deal with complex, uncertain and ambiguous risks.

The way in which actors negotiate and construct 'landscapes' of risk can be subsumed under the term risk governance (Renn 2008: 8; IRGC 2005). Hutter (2006: 215) characterizes the move from governmental regulation to governance in the following manner:

> This decentring of the state involves a move from the public ownership and centralized control to privatized institutions and the encouragement of market competition. It also involves a move to a state reliance on new forms of fragmented regulation, involving the existing specialist regulatory agencies of state but increasingly self-regulating organizations, regimes of enforced self-regulation … and American-style independent regulatory agencies.

Risk governance involves the 'translation' of the substance and core principles of governance to the context of risk and risk-related decision-making. Based on our previous work on risk governance and risk evaluation (Klinke and Renn 2001, 2002, 2010, 2012; Klinke et al. 2006; Renn 2008; Renn et al. 2011), it is the objective of this chapter to explore the connections between risk governance and resilience.

The concept of resilience has been adopted from ecological research and denotes the resistance of natural ecosystems to cope with stressors (Holling 1973; Walker et al. 2004). Resilience is focused on the ability and capacity of systems to resist shocks and to have the capability to deal and recover from threatening events (Carpenter et al. 2001; Rose 2007). This idea of resistance and recovery can also be applied to social systems (Review in Norris et al. 2008; Adger et al. 2005). The main emphasis here is on organizational learning and institutional preparedness to cope with stress and disaster. The US Department of Homeland Security (DHS) uses this definition: "Resilience is the ability of systems, infrastructures, government, business, and citizenry to resist, absorb, and recover from or adapt to an adverse occurrence that may cause harm, destruction, or loss [that is] of national significance (cited after Longstaff et al. 2010: 19). Hutter (2011) added to this analysis the ability of systems to respond flexibly and effectively when a system is under high stress from unexpected crisis. Pulling from an interdisciplinary body of theoretical and policy-oriented literature, Longstaff et al. (2010) regard resilience as a function of resource robustness and adaptive capacity. The governance framework suggested by the International Risk Governance Council (IRGC 2005) depicts resilience as a normative goal for risk management systems to deal with highly uncertain events or processes (surprises). It is seen as a property of risk-absorbing systems to withstand stress (objective resilience) but also the confidence of risk management actors to be able to master crisis situations (subjective resilience).

In this chapter we like to make the link between different risk management strategies and the need to enhance resilience, understood here as the capability of a socio-technical system to cope with events that are uncertain and ambiguous (Klinke and Renn 2012). This approach has been inspired by Lorenz (2010), who distinguishes adaptive, coping and participative aspects of resilience. I will use this classification to discern between three management styles which correspond to these three aspects of resilience. I have called them: risk-informed (corresponding to adaptive capability); precaution-based (corresponding to coping capability) and discourse-based (corresponding to participative capability).

In this chapter I will first explain the major characteristics of risk knowledge, and then address major functions of the risk governance process: pre-estimation, interdisciplinary risk estimation (including scientific risk assessment and concern assessment), risk characterization and risk evaluation, and risk management, including decision-making and implementation (based on the IRGC risk governance model). Each of these stages is described in the light of the three aspects of resilience: adaptive management capability, coping capability, and participative capability. Furthermore, the article expands this perspective by suggesting four different forms of public and stakeholder involvement for coping with the three aspects of resilience, including a strategy where all none of the three aspects matter. Finally, the chapter concludes with some general remarks about the relationship between governance and resilience.

2.2 Three Characteristics of Risk Knowledge

Integrative risk governance is expected to address the challenges raised by three risk characteristics that result from a lack of knowledge and/or competing knowledge claims about the risk problem. Transboundary and collectively relevant risk problems, such as global environmental threats (climate change, loss of biological diversity, chemical pollution, etc.), new and/or large-scale technologies (nanotechnology, biotechnology, offshore oil production, etc.), food security or pandemics are all characterized by limited and sometimes controversial knowledge with respect to their risk properties and their implications (Horlick-Jones and Sime 2004). The three characteristics are complexity, scientific uncertainty and socio-political ambiguity (Klinke and Renn 2002, 2010, 2012; Renn 2008; Aven and Renn 2009).

2.2.1 Complexity

Complexity refers to the difficulty of identifying and quantifying causal links between a multitude of potential candidates and specific adverse effects (see Lewin 1992; Underdal 2009). A crucial aspect in this regard concerns the applicability of probabilistic risk assessment techniques. If the chain of events between cause and effect follows a linear relationship (as for example in car accidents, or an overdose of pharmaceutical products), simple statistical models are sufficient to calculate the probabilities of harm. But even such simple relationships may be associated with a high degree of uncertainty, for example when very few data are available, or the effect is stochastic by nature. Sophisticated models of probabilistic reasoning are required if the relationship between cause and effect becomes more complex (Renn and Walker 2008). The nature of this difficulty may be traced back to interactive effects among these candidates (synergisms and antagonisms, positive and negative feedback loops), long delay periods between cause and effect, inter-individual variation, intervening variables, and others. It is precisely these complexities that make sophisticated scientific investigations necessary, since the cause–effect relationship is neither obvious nor directly observable.

2.2.2 Scientific Uncertainty

Scientific uncertainty relates to the limitedness or even absence of scientific knowledge (data, information) that makes it difficult to exactly assess the probability and possible outcomes of undesired effects (Rosa 1997; Aven and Renn 2009; Filar and Haurie 2010). It most often results from an incomplete or inadequate reduction of complexity in modelling cause-effect chains (see Marti et al. 2010). Whether the world is inherently uncertain is a philosophical question that is not pursued here. It is essential to acknowledge in the context of risk assessment that human knowledge is always incomplete and selective, and, thus, contingent upon uncertain assumptions, assertions and predictions (Functowicz and Ravetz 1992; Laudan 1996; Renn 2008: 75). It is obvious that the modeled probability distributions within a numerical relational system can only represent an approximation of the empirical relational system that helps elucidate and predict uncertain events. It therefore seems prudent to include additional aspects of uncertainty (van Asselt 2000: 93–138). Although there is no consensus in the literature on the best means of disaggregating uncertainties, the following categories appear to be an appropriate means of distinguishing between the key components of uncertainty (Renn 2008: 76):

- *Variability* refers to different vulnerability of targets such as the divergence of individual responses to identical stimuli among individual targets within a relevant population such as humans, animals, plants, landscapes, etc.;
- *Inferential effects* relate to systematic and random errors in modeling, including problems of projecting inferences from small statistical samples, from animal

data or experimental data onto humans or from large doses to small doses, etc. All of these are usually expressed as statistical confidence intervals;

- *Indeterminacy* results from a genuine stochastic relationship between cause and effects, apparently non-causal or non-cyclical random events, or badly understood non-linear, chaotic relationships;
- *System boundaries* allude to uncertainties stemming from restricted models and the need for focusing on a limited number of variables and parameters;
- *Ignorance* means a lack of knowledge about the probability of occurrence of a damaging event and about its possible consequences.

The first two components of uncertainty qualify as statistically quantifiable uncertainty and can be reduced by improving existing knowledge, applying standard statistical instruments such as Monte Carlo simulation and estimating random errors within an empirically proven distribution. The last three components represent genuine uncertainty components and can be characterized to some extent by using scientific approaches, but cannot be completely resolved. Risk assessment and management agencies require additional information and input, such as a subjective confidence level in risk estimates, potential alternative pathways of cause–effect relationships, ranges of reasonable estimates, maximum loss scenarios, and others. Examples of high uncertainty include many natural disasters, such as earthquakes, possible health effects of pandemics, and long-term effects of introducing genetically modified species into the natural environment.

2.2.3 Socio-political Ambiguity

While more and better data and information may reduce scientific uncertainty and provide the prerequisites for a common understanding of the respective risk, more knowledge does not necessarily reduce ambiguity. Ambiguity thus indicates a situation of ambivalence in which different and sometimes divergent streams of thinking and interpretation about the same risk phenomena and their circumstances are apparent (see Feldman 1989; Zahariadis 2003; Klinke and Renn 2012). Our risk governance frameworks distinguishes between interpretative and normative ambiguity which both relate to divergent or contested views regarding the justification, severity or wider 'meanings' associated with a given threat (Stirling 2003; Renn 2008: 77; Renn and Walker 2008). Entering the realm of ambiguity opens the dimensions of space towards a whole set of subjective interpretations and meanings.

Interpretative ambiguity denotes the variability of (legitimate) interpretations based on identical observations or data assessment results, e.g. an adverse or non-adverse effect. Variability of interpretation, however, is not restricted to expert dissent. Lay people's perception of risk often differs from expert judgements because it is related to qualitative risk characteristics such as familiarity, personal or institutional control, assignment of blame, and others. Moreover, in contemporary pluralist societies diversity of risk perspectives within and between social groups is generally fostered by divergent experiences of space and location.

This leads us to the aspect of *normative ambiguity*. It alludes to different concepts of what can be regarded as tolerable, referring for instance to ethics, quality of life parameters, distribution of risks and benefits, etc. Ambiguity emerges where the problem is agreeing on the appropriate values, priorities, assumptions or boundaries to be applied to the definition of possible outcomes. Normative ambiguities can be associated, for example, with exposure to noise, aquaculture in sensitive areas, pre-natal genetic screening, or genetically modified food. A good example of this is the use of phthalates in toys. All analysts are aware that this substance is potentially carcinogenic, but given the known exposure and the dose-response functions there is hardly any possibility of young children being negatively affected (Wilkinson and Lamb 1999). Yet the mere idea of having a carcinogenic substance in children's toys has incited a fierce debate about the tolerability of such an ingredient in rubber toys.

Most risks are characterized by a mixture of complexity, uncertainty and ambiguity. Passive smoking may be a good example of low complexity and uncertainty, but high ambiguity. Nuclear energy may be a good candidate for high complexity and high ambiguity, but relatively little uncertainty. The massive emission of aerosols into the atmosphere to combat the effects of greenhouse gases might be cited as an example of high complexity, uncertainty and ambiguity.

2.3 Inclusive Risk Governance: Structure and Processes

The ability of risk governance institutions to cope with complex, uncertain and ambiguous consequences and implications has become a central concern of scientists and practitioners alike. In 2005, the International Risk Governance Council suggested a process model of risk governance (IRGC 2005; Renn 2008; Renn and Walker 2008). This framework structures the risk governance process in four phases: pre-assessment, appraisal, characterization and evaluation, and risk management. Communication is conceptualized as a constant companion to all four phases of the risk governance cycle. The framework's risk process, or risk handling chain, is illustrated in Fig. 2.1.

Since its publication in 2005, the IRGC Risk Governance Framework has been applied to diverse risk governance issues in various case studies. Publications of these case studies are available on IRGC's homepage (www.irgc.org/Publications.html). The case studies deal with emerging risks such as air quality, bioenergy, carbon capture and storage, critical infrastructure, nanotechnology, pollination services, and synthetic biology. Furthermore, the IRGC has commissioned several case studies as tests of the applicability, efficacy and practicability of the Risk Governance Framework (Renn and Walker 2008). The applications have shown that the Framework can be used as broad conceptual guidance on the critical elements of the risk governance process. To date, the IRGC risk framework has been discussed and partially applied to a number of institutions and organizations, including most

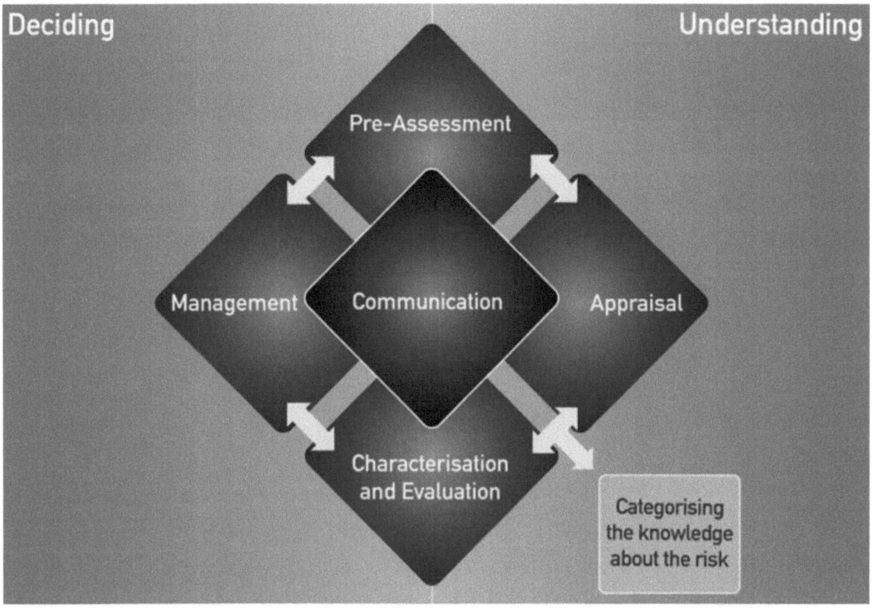

Fig. 2.1 The Risk Governance Framework (Adapted from IRGC 2005)

prominently the European Food Safety Authority (Vos and Wendler 2009) and the Health Council of the Netherlands (Health Council 2006). Reports using the Framework have been given by the German Occupational Health and Safety Committee (Bender 2008), the International Occupational Safety Association (Radandt et al. 2008), the UK Treasury (2005), the US EPA (2009), and several private organizations. In addition, the Framework was applied to strategic risk management by the US Joint Chiefs of Staff (Rouse 2011). The model has been used for major military operations and has, according to the source, improved the risk management process considerably.

The framework was primarily developed to deal with technological risks. It has been criticized as overstating the demarcation line between assessment and management, as being too rigid in its phasing of the governance process and in being not specific enough on stakeholder involvement and participation (see articles in Renn and Walker 2008; van Asselt 2005). For the purpose of developing a more adaptive and inclusive version of the IRGC framework, Klinke and Renn (2012) and Renn et al. (2011) suggest a slightly modified version as illustrated in Fig. 2.2.

The modified framework consists of the steps: pre-estimation, interdisciplinary risk estimation, risk characterization, risk evaluation and risk management. This is all related to the ability and capacity of risk governance institutions to use resources effectively (see Fig. 2.2). Appropriate resources include institutional and financial means as well as social capital (e.g. strong institutional mechanisms and configurations, transparent decision-making, allocation of decision making authority, formal and informal networks that promote collective risk handling, education), technical

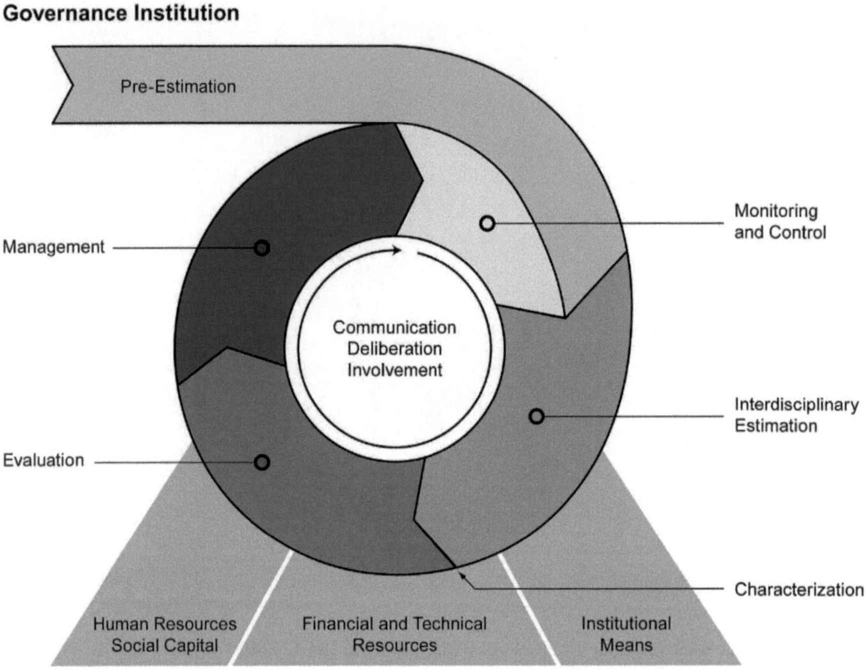

Fig. 2.2 Adaptive and integrative risk governance model (Adapted from Renn et al. 2011)

resources (e.g. databases, computer soft- and hardware, etc.), and human resources (e.g. skills, knowledge, expertise, epistemic communities, etc.). Hence, the adequate involvement of experts, stakeholders and the public in the risk governance process is a crucial dimension to produce and convey adaptive and integrative capacity in risk governance institutions (Pelling et al. 2008). The following sections will explore the significance of each step for risk governance and resilience.

2.3.1 *Pre-estimation*

Risks are not straightforwardly objective phenomena. They are based on the observation of hazards, i.e. the inherent potential for causing harm. Risks are mental constructions that reflect how people perceive uncertain phenomena and the ways in which their interpretations and responses are determined by social, political, economic and cultural contexts and judgments (see Luhmann 1993; OECD 2003; IRGC 2005). In this sense, risks have both an objective (some property that could act as a hazard) and a subjective component (a mental model that links these properties with possibilities of experiencing harm). The conceptualization of risk as a mental construct has major implications for how risk is considered. Risks are created and selected by human actors. What counts as a risk for one person may be seen by another as a destiny explained by religion, or even as an opportunity by a third party. Although societies have over time gained experience and collective knowledge of the potential impacts of events and activities, one cannot anticipate all potential scenarios and be worried about all the many potential consequences of a proposed activity or an expected event. By the same token, it is impossible to include all possible options for intervention. Therefore, societies have been *selective* in what they have chosen to consider worth addressing and what to ignore. One of the most significant selection rules is related to vulnerability and resilience. What is being rated as high risk depends very much on the confidence that one links with the personal or organizational capability to deal with its consequences. Therefore, resilience is also a mental construct that demonstrates the degree of self-confidence in one's coping capacity.

It is important to explore what major political and societal actors such as governments, companies, epistemic communities, nongovernmental organizations and the general public identify as risks and what types of problems they label as problems associated with risks and their ability to cope with risks. This is called *framing* and it specifies how society and politics rely on schemes of selection and interpretation to understand and respond to those phenomena which are socially constructed as relevant risk topics (Kahneman and Tversky 2000; Reese et al. 2003). Interpretations of risk experience depend on the frames of reference (Daft and Weick 1984). The process of framing corresponds to resilience. For example, Merad et al. (2008) were able to prove that managers of hazardous sites were more often convinced that stringent risk management actions were necessary the more they felt that a disaster could come as a surprise to them – regardless whether they believed they had sufficient control over risk management outcome. Conversely, those who held the conviction that accidents in their facilities could be anticipated and planed for had little doubt that the risks were well manageable.

Another issue is variety among the actors. What counts as a serious risk may vary among different actor groups. Whether an overlapping consensus evolves about what requires consideration as a relevant risk depends on the legitimacy of the selection rule. For example, the risks and benefits of biomass conversion for energy purposes can be seen under the frame of energy security, national independence, climate protection or economic development opportunities for rural areas (IRGC 2008). Depending on the frame, different types of risks and benefits may emerge; furthermore some benefits under one frame (for example national independence) may be a risk for another frame (economic opportunities for developing countries).

2.3.2 Interdisciplinary Risk Estimation

For politics and society to make reasonable decisions about risks in the public interest, it is not enough to consider only the results of (scientific) risk assessment. In order to understand the concerns of the people affected and various stakeholders, information about both risk perceptions and the further implications of the direct consequences of a risk is needed and should be taken into account by risk management (Renn 2008: 72).

Interdisciplinary risk estimation thus includes scientific assessment of risks to human health and the environment and assessment of related concerns, as well as social and economic implications (IRGC 2005; Renn and Walker 2008). The interdisciplinary estimation process should be clearly dominated by scientific analyses, but, in contrast to traditional risk regulation models, the scientific process includes the natural sciences and engineering as well as the social sciences, including economics.

The interdisciplinary risk estimation comprises two stages (Klinke and Renn 2012):

1. *Risk assessment*: Experts from the natural and technical sciences produce the best estimate of the physical harm that a risk source may cause;
2. *Concern assessment*: Experts from the social sciences, including economics, identify and analyze the issues that individuals or society as a whole link to a certain risk. For this purpose, the repertoire of the social sciences, such as survey methods, focus groups, econometric analysis, macro-economic modeling, or structured hearings with stakeholders may be used.

There are different approaches and proposals regarding how best to address the issue of interdisciplinary risk estimation. The German Advisory Council on Global Change (WBGU) has developed a set of eight criteria to characterize risks beyond the established assessment criteria (Klinke and Renn 2002; WBGU 2000). Some of the criteria have been used by different risk agencies or risk estimation processes (for example HSE 2001). These criteria include:

- *Extent of damage*: Adverse effects in natural units, e.g. death, injury, production loss, etc.
- *Probability of occurrence*: Estimate of relative frequency, which can be discrete or continuous.
- *Incertitude*: How do we take account of uncertainty in knowledge, in modeling of complex systems or in predictability in assessing a risk?
- *Ubiquity*: Geographical dispersion of damage (space dimensions).
- *Persistence*: How long will the damage last (time dimension)?
- *Reversibility*: Can the damage be reversed?
- *Delay effects*: Latency between initial event and actual damage.
- *Potential for mobilization*: The broad social impact. Will the risk generate social conflict, outrage, etc.?
  - *Inequity and injustice* associated with the distribution of risks and benefits over time, space and social status;
  - *Psychological stress and discomfort* associated with the risk or the risk source (as measured by psychometric scales);
  - *Potential for social conflict and mobilization* (degree of political or public pressure on risk regulatory agencies);

– *Spill-over effects* that can be expected when highly symbolic losses have repercussions on other fields, such as financial markets or loss of credibility in management institutions.

These four sub-criteria in the last category reflect many factors that have been proven to influence risk perception. The 'appraisal guidance' published by the UK Treasury Department in 2005 recommends a risk estimation procedure that is similar to this proposal and includes both the results of risk assessment and the direct input from data on public perception and the assessment of social concerns (HM Treasury 2005).

2.3.3 *Risk Evaluation*

A heavily disputed task in the risk governance process relates to the procedure for judging a given risk and justifying an evaluation of its societal acceptability or tolerability (see Fig. 2.2). In many approaches, risks are ranked and prioritized on the basis of a combination of probability (how likely is it that the risk will occur) and impact (what are the consequences, if the risk does occur). In the so-called traffic-light model (see Fig. 2.3), risks are located in the diagram of probability versus expected consequences, and three areas are identified: green, yellow and red (Klinke and Renn 2002; Renn 2008: 149ff.). A risk falls in the green area if the occurrence is highly unlikely and the impact is negligible. No further formal intervention is necessary. A risk is seen as tolerable when serious impacts might occur occasionally (yellow area). The benefits are worth the risk, but risk reduction

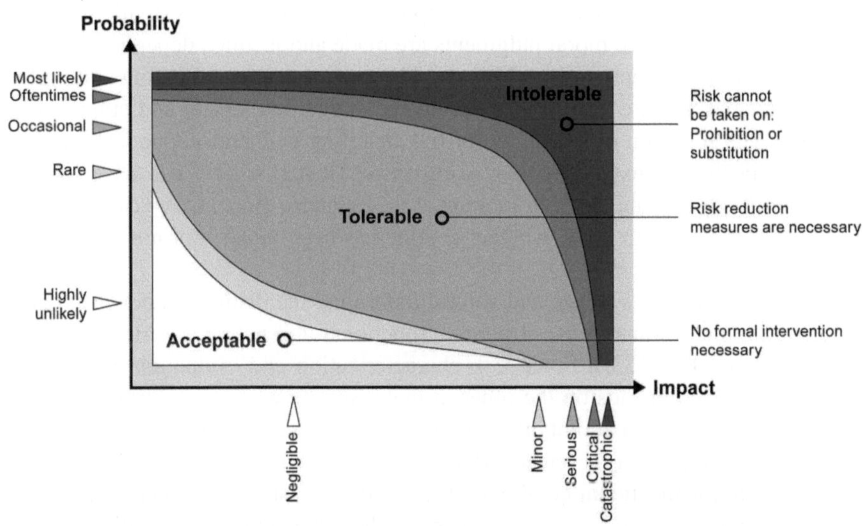

Fig. 2.3 Risk areas (Adapted from Renn 2008: 150)

measures are necessary. Finally, a risk is viewed as intolerable when the occurrence of catastrophic impacts is very likely (red area). Possible negative consequences of the risk are so catastrophic that in spite of potential benefits it cannot be tolerated.

Drawing a line between 'acceptable', 'tolerable' and 'intolerable' risk is one of the most controversial tasks in the risk governance process. The UK Health and Safety Executive has developed a procedure for chemical risks based on risk-risk comparisons (Löfstedt 1997). Some Swiss cantons such as Basel County have experimented with Round Tables as a means to reach consensus on drawing the two demarcation lines, where participants in the Round Table represented industry, administrators, county officials, environmentalists, and neighborhood groups. Irrespective of the means selected to support this task, the judgment concerning acceptability or tolerability is contingent on making use of a variety of different knowledge sources. It is important to include all data and insights resulting from risk assessment activities, and additional data from the concern assessment.

Risk evaluations in general rely on factual knowledge about cause-effect relationships and moral judgments about the desirability of anticipated consequences (see Goldstein and Keohane 1993). Factual knowledge refers to the scientific evidence from risk assessment, whether, how and to what extent the hazard potential causes harm to the environment or to human health. This dimension emphasizes cause-effect relations and provides factual guidance as to which strategy is appropriate to achieve the goal of risk avoidance or reduction. However, the question of what is safe enough also implies a moral judgment about the acceptability of risk and the tolerable burden that risk producers can impose on others (Klinke and Renn 2012). The results of the concern assessment can provide insights into what kind of associations are present and which moral judgments people would prefer in a choice situation. Of major importance is the perception of just or unjust distribution of risks and benefits. How these moral judgments are made and justified depends to a large degree on cultural values and worldviews. They affect personal thinking and evaluation strategies and are shaped by collectively shared ontological and ethical convictions. The selection of strategies for risk handling is, therefore, understandable only within the context of broader worldviews. Hence, society can never derive acceptability or tolerability from looking at the evidence alone. Likewise, evidence is essential if we are to know whether a value has been violated or not (or to what degree).

The separation of evidence and values underlying the distinction between characterization and evaluation is, of course, functional and not necessarily organizational (Renn 2008: 153f). Since risk characterization and evaluation are closely linked, and each depends on the other, it may even be wise to perform these two steps simultaneously in a joint effort by both risk assessment experts and risk management decision makers (Frewer and Salter 2007). The US regulatory system tends to favour an organizational combination of characterization and evaluation, while European risk management tends to maintain the organizational separation, e.g. in the food area (Löfstedt and Vogel 2001).

2.3.4 Risk Management

Risk management starts by reviewing all relevant data and information generated during the previous steps of interdisciplinary risk estimation, characterization and risk evaluation. The systematic analysis of risk management options focuses on tolerable risks (yellow area) and those where tolerability is disputed (light green and orange transition zones). The other cases (green and red areas) are fairly easy to deal with. Intolerable risks demand prevention and prohibition strategies aimed at replacing the hazardous activity with another activity leading to identical or similar benefits. The management of acceptable risks is left to private actors (civil society and economy). They may initiate additional and voluntary risk reduction measures, or seek insurance to cover for possible minor losses. If risks are classified as tolerable, or if there is a dispute as to whether they are in the transition zones of tolerability, public risk management needs to design and implement actions that make these risks either acceptable or at least tolerable by introducing reduction strategies. This task can be described in terms of classical decision theory (see Aven and Vinnem 2007; Klinke and Renn 2010):

- Identification and generation of generic risk management options;
- Assessment of risk management options with respect to predefined criteria;
- Evaluation of risk management options;
- Selection of appropriate risk management options;
- Implementation of risk management options, and
- Monitoring and control of option performance.

This is the place where considerations of resilience play a major role. As outlined in the introduction, our approach distinguishes different strategies of risk management that relate to the three resilience aspects by Lorenz (2010). These three aspects can be described in the language of the three risk characteristics: complexity, uncertainty, and ambiguity.

The first category refers to linear risk problems: they are characterized as having low scores regarding the dimensions of complexity, uncertainty and ambiguity. They also do not require major efforts to deal with the three aspects of resilience: adaptive, coping, and participative capacities. They can be addressed by *linear risk management* because they are normally easy to assess and quantify. Routine risk handling within risk assessment agencies and regulatory institutions is appropriate for this category, since the risk problems are well known, sufficient knowledge of key parameters is available and there are no major controversies about causes and effects or conflicting values. The management includes risk-benefit analysis, risk-risk comparisons or other instruments for balancing pros and cons. An example might be requiring that cyclists should wear helmets.

If risks are ranked high for complexity but low for uncertainty and ambiguity, they require the systematic involvement and deliberation of experts representing the relevant epistemic communities in order to produce an accurate estimate of the complex relationships. This situation requires professional adaptive management skills (Norris et al. 2008; Kuhlicke et al. 2011). It does not make much sense to integrate public concerns, perceptions or any other social aspects in the resolving of complexity, unless specific knowledge from the concern assessment helps to untangle complexity. Complex risk problems therefore demand *risk-informed management* that can be offered by scientists and experts applying methods of expanded risk assessment, determining quantitative safety goals, consistently using cost-effectiveness methods, and monitoring and evaluating outcomes. It is therefore essential to investigate perceptions and positions with respect to the risk source and initiate either risk communication or risk modification programs to address these concerns.

Risk problems that are characterized by high uncertainty but low ambiguity require strategies that can be labeled as *precaution-based management*. This corresponds with the coping capacity of resilience. Since sufficient scientific certainty is currently either not available or unattainable, expanded knowledge acquisition may help to reduce uncertainty and thus move the risk problem back to first stage of handling complexity. If, however, uncertainty cannot be reduced by additional knowledge, risk management should foster and enhance precautionary strategies that enhance coping capacities, decrease vulnerabilities and provide sufficient emergency preparedness in order to avoid irreversible effects (Renn 2007). Appropriate instruments include containment, diversification, monitoring and substitution.

Finally, if risk problems are ranked high for ambiguity (regardless of whether they are low or high for uncertainty), *discourse-based management* is required, demanding participative processing. This again corresponds with the third aspect of resilience: participative capacity (Lorenz 2010). This includes the need to involve major stakeholders as well as the affected public. The goal of discourse-based risk management is to produce a collective understanding among all stakeholders and the concerned public on interpretative ambiguity or to find legitimate procedures of justifying collectively binding decisions on acceptability and tolerability. It is important that a consensus or a compromise is achieved between those who believe that the risk is worth taking (perhaps because of self-interest) and those who believe that the impending consequences do not justify the potential benefits of the risky activity or technology.

2.4 Communication, Deliberation and Involvement of Non-governmental Actors

The effectiveness and legitimacy of the risk governance process depends on the capability of management agencies to resolve complexity, characterize uncertainty and handle ambiguity by means of communication and deliberation.

2.4.1 Instrumental Processing Involving Governmental Actors

Dealing with linear risk issues, which are associated with low scores for complexity, scientific uncertainty and socio-political ambiguity, requires hardly any changes to conventional public policy-making. The data and information regarding such linear (routine) risk problems are provided by statistical analysis; law or statutory requirements determine the general and specific objectives; and the role of public policy is to ensure that all necessary safety and control measures are implemented and enforced (Klinke and Renn 2012). Traditional cost-benefit analyses including effectiveness and efficiency criteria are the instruments of political choice for finding the right balance between under- and over-regulation of risk-related activities and goods. In addition, monitoring the area is important to help prevent unexpected consequences. For this reason, linear risk issues can well be handled by departmental and agency staff and enforcement personnel of state-run governance institutions. The aim is to find the most cost-effective method for a desired regulation level. If necessary, stakeholders may be included in the deliberations as they have information and know-how that may help to make the measures more efficient.

2.4.2 Epistemic Processing Involving Experts

Resolving complex risk problems requires dialogue and deliberation among experts. The main goal is to scan and review existing knowledge about the causal connections between an agent and potential consequences, to characterize the uncertainty of this relationship and to explore the evidence that supports these inferences. Involving members of various epistemic communities which demonstrate expertise and competence is the most promising step for producing more reliable and valid judgments about the complex nature of a given risk. Epistemic discourse is the instrument for discussing the conclusiveness and validity of cause-effect chains relying on available probative facts, uncertain knowledge and experience that can be tested for empirical traceability and consistency. The objective of such a deliberation is to find the most cogent description and explanation of the phenomenological complexity in question as well as a clarification of dissenting views (for example, by addressing the question which environmental and socio-economic impacts are to be expected in which areas and in what time frame). The deliberation among experts might generate a profile of the complexity of the given risk issue on selected inter-subjectively chosen criteria. The deliberation may also reveal that there is more uncertainty and ambiguity hidden in the case than the initial appraisers had anticipated (Birkmann 2011; Bovenkerk 2012). It is advisable to include natural as well as social scientists in the epistemic discourse so that potential problems with risk perception and risk frames can be anticipated. Controversies would then be less of a surprise than is currently the case. Such epistemic discourse is meant to lead to adaptive management procedures that monitor the state of knowledge and proficiency in the field and adjust management responses according to the various levels of knowledge available at each time period (Wiering and Arts 2006; Klinke and Renn 2012).

2.4.3 *Reflective Processing Involving Stakeholders*

Characterizing and evaluating risks as well as developing and selecting appropriate management options for risk reduction and control in situations of high uncertainty poses particular challenges. How can risk managers characterize and evaluate the severity of a risk problem when the potential damage and its probability are unknown or highly uncertain? Scientific input is, therefore, only the first step in a series of steps constituting a more sophisticated evaluation process. It is crucial to compile the relevant data and information about the different types of uncertainties to inform the process of risk characterization. The outcome of the risk characterization process provides the foundation for a broader deliberative arena, in which not only policy makers and scientists, but also directly affected stakeholders and public interest groups ought to be involved in order to discuss and ponder the 'right' balances and trade-offs between over- and under-protection (Renn and Schweizer 2009). This reflective involvement of stakeholders and interest groups pursues the purpose of finding a consensus on the extra margin of safety that potential victims would be willing to tolerate and potential beneficiaries of the risk would be willing to invest in to avoid potentially critical and catastrophic consequences. If too much precaution is applied, innovations may be impeded or even eliminated; if too little precaution is applied, society may experience the occurrence of undesired consequences. The crucial question here is how much uncertainty and ignorance the main stakeholders and public interest groups are willing to accept or tolerate in exchange for some potential benefit.

This issue has direct implications for resilience. As this concept reflects the confidence of all actors to deal with even uncertain outcomes, it provides a mental guideline for the negotiations between beneficiaries and potential victims of risks (IRGC 2005). Furthermore, it includes a discourse about coping capacity and compensation schemes if the worst were to happen. The boundary between subjective and objective resilience is, however, fuzzy under the condition of effect uncertainty (Brown and Kulig 1996/97; Norris et al. 2008). In cases of known risks past experience can demonstrate whether the degree of self-confidence was accurate and justified. Over long time spans one would expect an emerging congruence between objective and subjective resilience (learning by trial and error). However, for extremely rare events or highly uncertain outcomes, one necessarily relies on models of anticipation and expectations that will widely vary among different stakeholder groups, in particular those who benefit and those who will bear the risks (Berkes 2007). Furthermore, there will be lots of debates about the potential distribution of effects over time and space. The degree of coping capacity that is regarded as sufficient or justified for approving a new risk agent or a disaster management plan to become enacted depends therefore on a discourse between the directly affected groups of the population. Such a reflective involvement of policy makers, scientists, stakeholders and public interest groups can be accomplished through a spectrum of different procedures such as negotiated rule-making, mediation, round-table or open forums, advisory committees, and others (see Beierle and Cayford 2002; Klinke 2006; Rowe and Frewer 2000; Stoll-Kleemann and Welp 2006).

2.4.4 Participative Processing Involving the Wider Public

If risk problems are associated with high ambiguity, it is not enough to demonstrate that risk regulation addresses the public concerns of those directly affected by the impacts of the risk source. In these cases, the process of evaluation and management needs to be open to public input and new forms of deliberation. This corresponds with the participative aspect of resilience (Lorenz 2010). Such discursive activities should start with revisiting the question of proper framing. Is the issue really a risk problem or is it an issue of lifestyle or future vision? Often the benefits are contested as well as the risks. The debate about 'designer babies' may illustrate the point that observers may be concerned not only about the social risks of intervening in the genetic code of humans but also about the acceptability of the desired goal to improve the performance of individuals (Hudson 2006). Thus the controversy is often much broader than dealing with the direct risks only. The aim here is to find an overlapping consensus on the dimensions of ambiguity that need to be addressed in comparing risks and benefits, and balancing pros and cons. High ambiguity would require the most inclusive strategy for involvement because not only directly affected groups but also those indirectly affected should have an opportunity to contribute to this debate.

Resolving ambiguities in risk debates necessitates the participatory involvement of the public to openly discuss competing arguments, beliefs and values. Participatory involvement offers opportunities to resolve conflicting expectations through a process of identifying overarching common values, and to define options that will allow a desirable lifestyle without compromising the vision of others. Critical to success here is the establishment of equitable and just distribution rules when it comes to common resources and a common understanding of the scope, size and range of the problem, as well as the options for dealing with the problem (Renn and Schweizer 2009). Unless there is some agreement on the boundaries of what is included, there is hardly any chance for a common solution. Such a common agreement will touch upon the coping capacity of systems to deal with different frames of risks and not only with the physical impacts of risks. There are various social constructions of resilience that the participants associate with the management options. The set of possible procedures for involving the public includes citizen panels or juries, citizen forums, consensus conferences, public advisory committees and similar approaches (see Rowe and Frewer 2000; Beierle and Cayford 2002; Hagendijk and Irwin 2006; Klinke 2006; Abels 2007; Renn 2008: 284ff.).

An overview of the different participation and stakeholder involvement requirements with respect to linear, complex, uncertain and ambiguous risks is displayed in Fig. 2.4 (Renn 2008: 280). As is the case with all classifications, this scheme shows a simplified picture of the involvement process and it has been criticized for being too rigid in its linking of risk characteristics (complexity, uncertainty, and ambiguity) and specific forms of discourse and dialogue (Löftstedt and van Asselt 2008). In addition to the generic distinctions shown in Fig. 2.4, it may, for instance, be wise to distinguish between different types of risks and different types of regulatory cultures or styles (Löfstedt and Vogel 2001; Renn 2008: 358ff.). To conclude these caveats, the purpose of this scheme is to provide a general orientation and to make a generic distinction between ideal cases rather than to offer a strict recipe for participation.

The classification in Fig. 2.4 offers a taxonomy of requirements for stakeholder and public inclusion based on the characteristics of risk knowledge. These general guidelines can be further specified by looking into each phase of the risk governance cycle (Renn and Walker 2008: 356ff; Renn 2008).

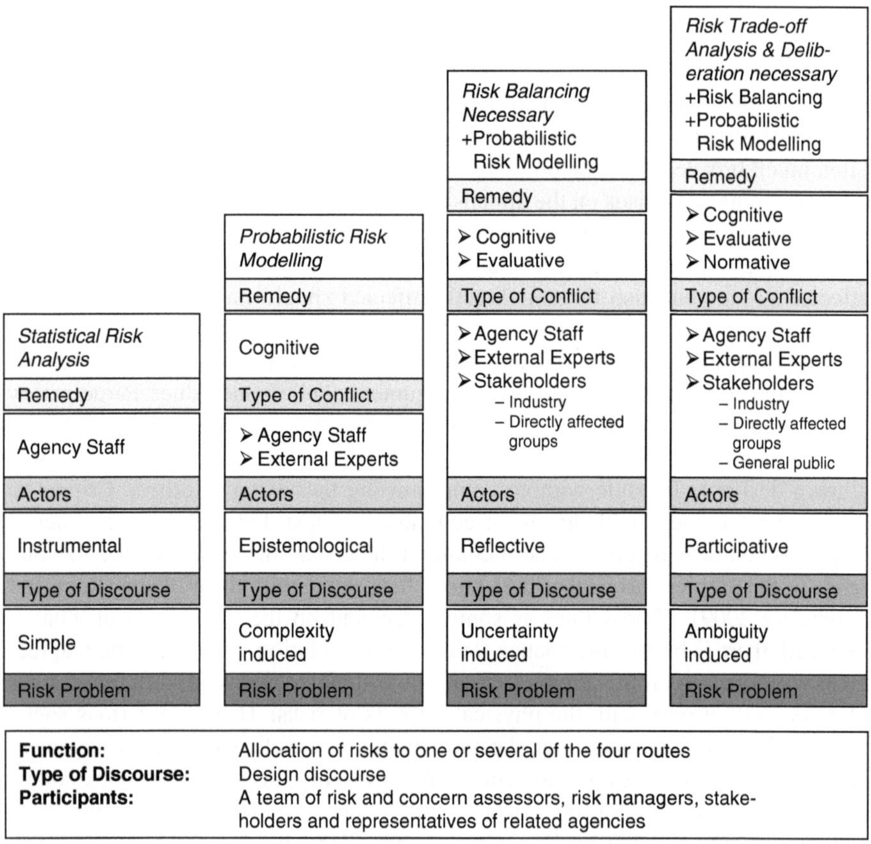

Fig. 2.4 The risk escalator: a guide to inclusive risk governance (Adapted from Renn 2008: 280)

2.5 Conclusions

The goal of this paper has been to illustrate the significance of resilience for risk governance, including all stages from pre-assessment to management and communication. For this purpose, the resilience concept by Lorenz was applied to link risk governance strategies with the three major aspects of resilience: adaptive management capacity, coping capacity, and participative capacity. The three risk characteristics –complexity, uncertainty and ambiguity– were linked to these three aspects of resilience. Furthermore, the three aspects were used to develop four major risk management and discourse strategies; beginning with simple risk management in which none of these characteristics and capacity requirements were involved to discourse-based management in which all three characteristics and capacity requirements were combined.

Whereas the analysis of simple and –to some degree– complex problems is better served by relying on the physical understanding of experienced resilience, uncertain and ambiguous problems demand the integration of social constructions and mental models of resilience, operationalized as confidence in one's coping capacity, for both understanding and managing these problems. The distinction of risks according to risk characteristics not only highlights deficits in our knowledge concerning a risk issue, but also points the way forward for the selection of the appropriate management options. Thus, the risk governance framework attributes an important function to public and stakeholder participation, as well as risk communication, in the risk governance process. The framework suggests efficient and adequate public or stakeholder participation procedures. The concerns of stakeholders and/or the public are integrated in the risk appraisal phase via concern assessment. Furthermore, stakeholder and public participation are an established part of risk management. The optimum participation method depends on the characteristics of the risk issue. In this respect, the three aspects of resilience are gradually included into the various discourses. The need for finding an agreement on what constitutes an adaptive, coping and participative response to ensuring resilience underlines the necessity to understand and comprehend the objective and subjective nature of resilience.

References

Abels, G. 2007. Citizen involvement in public policymaking: Does it improve democratic legitimacy and accountability? The case of pTA. *Interdisciplinary Information Science* 13(1): 103–116.

Adger, N., T. Hughes, C. Folke, S. Carpenter, and J. Rockstrom. 2005. Social-ecological resilience to coastal disasters. *Science* 309: 1036.

Aven, T., and O. Renn. 2009. The role of quantitative risk assessments for characterizing risk and uncertainty and delineating appropriate risk management options, with special emphasis on terrorism. *Risk Analysis* 29(4): 587–600.

Aven, T., and J.E. Vinnem. 2007. *Risk management: With applications from the offshore petroleum industry*. Heidelberg/London: Springer.

Beierle, T.C., and J. Cayford. 2002. *Democracy in practice. Public participation in environmental decisions*. Washington, DC: Resources for the Future.

Bender, H.F. 2008. Ergebnisse der Projektgruppe Risikoakzeptanz des AGS. *Gefahrstoffe Reinhaltung der Luft* 68(7/8): 287–288.

Berkes, F. 2007. Understanding uncertainty and reducing vulnerability: Lessons from resilience thinking. *Natural Hazards* 41(2): 283–295.

Birkmann, J. 2011. First- and second-order adaptation to natural hazards and extreme events in the context of climate change. *Natural Hazards* 58(2): 811–840.

Bovenkerk, B. 2012. Consensus conferences: The influence of contexts. In *The biotechnology debate*, ed. B. Bovenkerk. Library of Ethics and Applied Philosophy 29(3): 189–232. New York: Springer.

Brooks, H. 1984. The resolution of technically intensive public policy disputes. *Science, Technology, and Human Values* 9: 39–50.

Brown, D., and J. Kulig. 1996/97. The concept of resiliency: Theoretical lessons from community research. *Health and Canadian Society* 4: 29–52.

Carpenter, S., B. Walker, J.M. Anderies, and N. Abel. 2001. From metaphor to measurement: Resilience of what to what? *Ecosystems* 4(8): 765–781.

Daft, R.L., and K.E. Weick. 1984. Toward a model of organizations as interpretation systems. *Academy of Management Review* 9(2): 284–295.

Feldman, M.S. 1989. *Order without design: Information production and policy making*. Stanford: Stanford University Press.

Filar, J.A., and A. Haurie (eds.). 2010. *Uncertainty and environmental decision making*. New York: Springer.

Frewer, L.J., and B. Salter. 2007. Societal trust in risk analysis: Implications for the interface of risk assessment and risk management. In *Trust in cooperative risk management: Uncertainty in scepticism in the public mind*, ed. M. Siegrist, T.C. Earle, and H. Gutscher, 143–158. London: Earthscan.

Functowicz, S.O., and J.R. Ravetz. 1992. Three types of risk assessment and the emergence of post-normal science. In *Social theories of risk*, ed. S. Krimsky and D. GoldingD, 251–273. Westport/London: Praeger.

Goldstein, J., and R.O. Keohane. 1993. Ideas and foreign policy. An analytical framework. In *Ideas and foreign policy. Beliefsi Institutions, and political change*, ed. J. Goldstein and R.O. Keohane, 3–30. Ithaca: Cornell University Press.

Hagendijk, R., and A. Irwin. 2006. Public deliberation and governance: Engaging with science and technology in contemporary Europe. *Minerva* 44: 167–184.

Health Council of the Netherlands. 2006. *Health significance of nanotechnologies*, Publication No. 2006/06E. The Hague: Health Council of the Netherlands.

HM Treasury. 2005. *Managing risks to the public: Appraisal guidance*. London: HM Treasure. http://www.hm-treasury.gov.uk/media/0/B/Managing_risks_to_the_public.pdf. Accessed May 2011.

Holling, C.S. 1973. Resilience and stability of ecological systems. *Annual Review of Ecology and Systematics* 4(1): 1–23.

Horlick-Jones, T., and J. Sime. 2004. Living on the border: Knowledge, risk and transdiciplinarity. *Futures* 36: 441–456.

Horlick-Jones, T., G. Rowe, and J. Walls. 2007. Citizen engagement processes as information systems: The role of knowledge and the concept of translation quality. *Public Understanding of Science* 16: 259–278.

HSE (Health and Safety Executive). 2001. *Reducing risk – Protecting people*. London: Health and Safety Executive.

Hudson, K.L. 2006. Preimplantation diagnosis: Public policy and public attitudes. *Fertility & Sterility* 58(6): 1638–1645.

Hutter, B.M. 2006. Risk, regulation, and management. In *Risk in social science*, ed. P. Taylor-Gooby and J. Zinn, 202–227. Oxford: Oxford University Press.

Hutter, G. 2011. Organizing social resilience in the context of natural hazards: A research note. *Natural Hazards*. Special Volume: Resilience in hazards research and planning – A promising concept, eds. G. Hutter, C. Kuhlicke, T. Glade, and C. Felgentreff, 67(1): 1–14.

IRGC (International Risk Governance Council). 2005. *Risk governance: Towards an integrative approach*. White paper no. 1, O. Renn with an Annex by P. Graham. Geneva: IRGC.

IRGC (International Risk Governance Council). 2008. *Risk governance guidelines for bioenergy politics. Policy brief.* Geneva: IRGC.

Kahneman, D., and A. Tversky (eds.). 2000. *Choices, values, and frames*. Cambridge: Cambridge University Press.

Klinke, A. 2006. *Demokratisches Regieren jenseits des Staates. Deliberative Politik im nordamerikanischen Große Seen-Regime*. Opladen: Barbara Budrich Publication.

Klinke, A., and O. Renn. 2001. Precautionary principle and discursive strategies: Classifying and managing risks. *Journal of Risk Research* 4(2): 159–173.

Klinke, A., and O. Renn. 2002. A new approach to risk evaluation and management: Risk-based, precaution-based, and discourse-based strategies. *Risk Analysis* 22(6): 1071–1094.

Klinke, A., and O. Renn. 2010. Risk governance: Contemporary and future challenges. In *Regulating chemical risks: European and global perspectives*, ed. J. Eriksson, M. Gilek, and C. Ruden, 9–28. Berlin: Springer.

Klinke, A., and O. Renn. 2012. Adaptive and integrative governance on risk and uncertainty. *Journal of Risk Research* 1(1): 3–20.

Klinke, A., M. Dreyer, O. Renn, A. Stirling, and P. van Zwanenberg. 2006. Precautionary risk regulation in European governance. *Journal of Risk Research* 9(4): 373–392.

Kuhlicke, C., A. Steinführer, C. Begg, C. Bianchizza, M. Bründl, M. Buchecker, B. De Marchi, M. di Masso Tarditti, C. Höppner, B. Komac, L. Lemkow, J. Luther, S. McCarthy, L. Pellizzoni, O. Renn, A. Scolobig, M. Supramaniam, S. Tapsell, G. Wachinger, G. Walker, R. Whittle, M. Zorn, and H. Faulkner. 2011. Perspectives on social capacity building for natural hazards: Outlining an emerging field of research and practice in Europe. *Environmental Science and Policy* 2: 23–56.

Laudan, L. 1996. The pseudo-science of science? The demise of the demarcation problem. In *Beyond positivism and relativism. Theory, method and evidence*, ed. L. Laudan, 166–192. Boulder: Westview Press.

Lewin, R. 1992. *Complexity: Life at the edge of chaos*. New York: Macmillan.

Löfstedt, R.E. 1997. *Risk evaluation in the United Kingdom: Legal requirements, conceptual foundations, and practical experiences with special emphasis on energy systems*, Working paper no. 92. Stuttgart: Center of Technology Assessment.

Löfstedt, R., and D. Vogel. 2001. The changing character of regulation: A comparison of Europe and the United States. *Risk Analysis* 21(3): 393–402.

Löftstedt, R., and M. van Asselt. 2008. A framework for risk governance revisited. In *Global risk governance. Concept and practice using the IRGC framework*, ed. O. Renn and K. Walker, 77–86. Dordrecht/Heidelberg: Springer.

Longstaff, P.H., N.J. Armstrong, K. Perrin, W.M. Parker, and M.A. Hidek. 2010. Building resilient communities: A preliminary framework for assessment. *Homeland Security Affairs* VI(3): 1–23.

Lorenz, D.F. 2010. The diversity of resilience: Contributions from a social science perspective. *Natural Hazards*. Special Volume: Resilience in hazards research and planning – A promising concept?, eds. G. Hutter, C. Kuhlicke, T. Glade, and C. Felgentreff, 67(1): 1–18.

Luhmann, N. 1993. *Risk: A sociological theory*. Berlin: de Gruyter.

Marti, K., Y. Ermoliev, and M. Makowski (eds.). 2010. *Coping with uncertainty. Robust solutions*. Berlin/Heidelberg: Springer.

Merad, M., N. Rodrigues, and O. Salvi. 2008. Urbanisation control around industrial Seveso sites: The French context. *International Journal of Risk Assessment and Management* 8(1, 2): 158–167.

Nelkin, D., and M. Pollak. 1979. Public participation in the technological decisions: Reality or grand illusion? *Technology Review* 6: 55–64.

Nelkin, D., and M. Pollak. 1980. Problems and procedures in the regulation of technological risk. In *Making bureaucracies work*, ed. C.H. Weiss and A.F. Burton, 233–253. Beverly Hills: Sage.

Norris, F.H., S.P. Stevens, B. Pfefferbaum, K.E. Wyche, and R.L. Pfefferbaum. 2008. Community resilience as a metaphor, theory, set of capabilities, and strategy for disaster readiness. *American Journal of Community Psychology* 41: 127–150.

OECD (Organisation for Economic Co-operation and Development). 2003. *Emerging systemic risks: Final report to the OECD Futures Project*. Paris: OECD Press.

Pelling, M., C. High, J. Dearing, and D. Smith. 2008. Shadow spaces for social learning: A relational understanding of adaptive capacity to climate change within organisations. *Environment and Planning A* 40: 867–884.

Radandt, S., J. Rantanen, and O. Renn. 2008. Governance of occupational safety and health and environmental risks. In *Risks in modern society*, ed. H.-J. Bischoff, 127–258. Heiderlberg/Berlin: Springer.

Reese, S.D., O.H. Gandy Jr., and A.E. Grant (eds.). 2003. *Framing public life: Perspectives on media and our understanding of the social world*. Mahwah: Lawrence Erlbaum Associates.

Renn, O. 2007. Precaution and analysis: Two sides of the same coin? *EMBO Reports* 8: 303–305.

Renn, O. 2008. *Risk governance. Coping with uncertainty in a complex world*. London: Earthscan.

Renn, O., and P. Schweizer. 2009. Inclusive risk governance: Concepts and application to environmental policy making. *Environmental Policy and Governance* 19: 174–185.

Renn, O., and K. Walker. 2008. Lessons learned: A re-assessment of the IRGC Framework on Risk Governance. In *The IRGC risk governance framework: Concepts and practice*, ed. O. Renn and K. Walker, 331–367. Dordrecht/Heidelberg: Springer.

Renn, O., A. Klinke, and M. van Asselt. 2011. Coping with complexity, uncertainty and ambiguity in risk governance: A synthesis. *AMBIO* 40(2): 231–246.

Rosa, E. 1997. Metatheoretical foundations for post-normal risk. *Journal of Risk Research* 1(1): 15–44.

Rose, A. 2007. Economic resilience to natural and man-made disasters: Multidisciplinary origins and contextual dimensions. *Environmental Hazards* 7: 383–398.

Rouse, J. 2011. *The chairman of the Joint Chiefs of Staff Risk assessment system. Incorporation of the international risk governance council framework*. Paper at the annual meeting of the Society for Risk Analysis in Salt Lake City, December 6. Washington, DC: Arete Associates.

Rowe, G., and L.J. Frewer. 2000. Public participation methods: A framework for evaluation. *Science, Technology & Human Values* 25(1): 3–29.

Stirling, A. 2003. Risk, uncertainty and precaution: Some instrumental implications from the social sciences. In *Negotiating change*, ed. F. Berkhout, M. Leach, and I. Scoones, 33–76. London: Edward Elgar.

Stoll-Kleemann, S., and M. Welp (eds.). 2006. *Stakeholder dialogues in natural resources management: Theory and practice*. Heidelberg/Berlin: Springer.

UK Treasury Department. 2005. *Managing risks to the public: Appraisal guidance*. Draft for Consultation. London: HM Treasury Press. http://www.hm-treasury.gov.uk. Accessed 11 June 2011.

Underdal, A. 2009. Complexity and challenges of long-term environmental governance. *Global Environmental Change* 20: 386–393.

US Environmental Protection Agency (EPA). 2009. *Potential nano-enabled environmental applications for radionuclides*, EPA-402-R-06-002. Washington, DC: EPA.

van Asselt, M.B.A. 2000. *Perspectives on uncertainty and risk*. Dordrecht/Boston: Kluwer.

van Asselt, M.B.A. 2005. The complex significance of uncertainty in a risk area. *International Journal of Risk Assessment and Management* 5(2, 3 & 4): 125–158.

van Asselt, M.B.A., and O. Renn. 2011. Risk governance. *Journal of Risk Research* 1(4): 431–449.

Vos, E., and F. Wendler. 2009. Legal and institutional aspects of the general framework. In *Food safety governance. Integrating science, precaution and public involvement*, ed. M. Dreyer and O. Renn, 83–109. Berlin: Springer.

Walker, B., C.S. Holling, S.R. Carpenter, and A. Kinzig. 2004. Resilience, adaptability and transformability in social–ecological systems. *Ecology and Society* 9(2): 5. http://www.ecologyandsociety.org/vol9/iss2/art5/

WBGU (German Advisory Council on Global Change). 2000. *World in transition: Strategies for managing global environmental risks*. Heidelberg/New York: Springer.

Wiering, M.A., and B.J.M. Arts. 2006. Discursive shifts in Dutch river management: 'Deep' institutional change or adaptation strategy? *Hydrobiologia* 565(1): 327–338.

Wilkinson, C.F., and J.C. Lamb. 1999. The potential health effects of phthalate esters in children's toys. A review and risk assessment. *Regulatory Toxicology and Pharmacology* 30(2): 140–155.

Zahariadis, N. 2003. *Ambiguity and choice in public policy. Political decision making in modern democracies*. Washington, DC: Georgetown University Press.

Ortwin Renn serves as Full Professor and Chair of Environmental Sociology and Technology Assessment at Stuttgart University (Germany). Among his many political advisory activities the chairmanship of the State Commission for Sustainable Development (German State of Baden-Württemberg) is most prominent. His honours include an honorary doctorate from the Swiss Institute of Technology (ETH Zurich), an honorary affiliate professorship at the Technical University Munich and the Distinguished Achievement Award of the Society for Risk Analysis (SRA). Renn is primarily interested in risk governance, political participation and technology assessment. His has published more than 30 books and 250 articles, most prominently the monograph *Risk Governance* (2008).

Chapter 3
The Problem of Governance in the Risk Society: Envisaging Strategies, Managing Not-knowing

Gabe Mythen

3.1 Introduction

A great deal of academic work in the social sciences in the last two decades has focused on the subject of risk (see *inter alia*, Beck 1992, 2009; Boyne 2003; Mythen 2004, 2008; Strydom 2002). The primary focus within such work has varied from issues of risk assessment and calibration, to the representation and communication of risk and, most prolifically, the issue of risk perception. In more recent times, a growing body of academic research has homed in on risk governance and the problems and issues faced by institutions deploying risk based strategies of regulation (see Amoore and de Goede 2008; Hutter 2005; Mythen and Walklate 2008). The rationale underpinning risk based approaches to governance is not necessarily to secure an environment of perfect safety in which deleterious outcomes are rendered obsolete, but rather to seek to reduce levels of risk via probability assessment, regulation and effective communication to the public and other stakeholders. If successful, risk based approaches are seen to be not only the most ideationally rational means of managing limited State resources but also the most cost effective. In a world characterized by threats that emerge as unintended consequences or 'side effects' of capitalist development, the problem of risk management becomes ever more critical (Beck 1992). While risk has long been used as a technology of categorization and classification (see Ewald 1991; O'Malley 2008), in the last two decades risk governance has become an accepted mode of administration, regulation and management across a range of institutional domains (see Denney 2005; Kemshall 2003; Rothstein 2003). Explicitly risk based strategies to governance first emerged in most nations in response to environmental problems and were embedded in policies of ecological protection and hazard management. However, practices and

G. Mythen (✉)
Department of Sociology, University of Liverpool, Liverpool L69 7ZA, UK
e-mail: g.mythen@liverpool.ac.uk

U. Fra.Paleo (ed.), *Risk Governance*, DOI 10.1007/978-94-017-9328-5_3

processes of risk governance have spread rapidly over the last three decades and are now entrenched in regulatory strategy in areas as diverse as criminal justice, policing, education, health, food safety and child welfare. Not only have techniques of governing with and through risk proliferated, they have also become routine aspects of institutional practice in many Anglo-Saxon countries (Rothstein et al. 2006). Alongside various projects and programs initiated by nation-state agencies, risk based approaches have also been deployed by global supranational organizations such as the WTO, Oxfam, the EU and the OECD.

In the course of this chapter I will address the issue of risk governance from a critical sociological position. To this end, the various strategies used by regulatory institutions in seeking to manage manufactured risks are identified and the dilemmas and contradictions that emerge in this process are subjected to analysis. It is my intention to consider some of the key problems and issues that arise for politicians, policy makers, the State and other regulatory agencies in managing and governing through risk in contemporary society. The broader theoretical context within which this discussion will be situated is that of the risk society, as proposed by Beck (1992, 1999, 2009) and endorsed by Giddens (1998). Capturing problems of governance in the risk society is a potentially expansive endeavor and it is necessary to impose some limits and raise some caveats. The discussion developed here is both schematic and conceptual as opposed to empirically grounded. Governing through, with and by risk is such a situated and diverse institutional practice that attempting to empirically validate 'truths' about the right and wrong ways of using risk as a driver for governance is futile (see O'Malley 2008). I wish instead to embark on an exploratory discussion which will connect together the changing nature of threats in society to preferred modes of institutional governance and the problems that arise from deploying risk as a technology for making judgments and validating interventions. The chapter falls into three interconnected sections. In Sect. 3.2, in order to contextualise the discussion, the key elements of the 'turn to risk' that has occurred in advanced Western nations over the last three decades are catalogued. This provides an appropriate entree in Sect. 3.3 for a more focused appraisal of the problem of non-knowing, or *nichtwissen*, and its prescience in the risk society. In Sect.3.4, using the management of crime and national security as an exemplar, I examine salient institutional strategies that have emerged as a response to endemic uncertainties in the risk society.

3.2 The Turn to Risk

It has become commonplace for social science thinkers to observe that risk has become something of a signature of the age (Arnoldi 2009; Beck 2009; Giddens 1998). Assessing, identifying and managing risk are no longer assumed to be technical exercises presided over by engineers, medics, scientists and business analysts, but have become taken for granted features of everyday life in the modern world. Citizens themselves are not seen exclusively as subjects of risk who require institutional protection against harm, but also as both creators and managers of risk. Many mundane decisions we make are effectively about evaluating and balancing risk, from career choices, to personal relationships, mortgage lending and pension options. While individuals set about managing risk as it arises in everyday life, social institutions bear greater collective responsibility for public safety. State agencies have a binding duty to communicate risk swiftly and effectively to the public such that people can make informed decisions about appropriate behaviour and actions. Insofar as the aspirations and objectives of risk regulating institutions are diffuse, scientific, technological and informational advances mean that a greater range of risks are detectable than was possible in earlier epochs. In many respects, technological and scientific developments can be conceived as something of a double-edged sword. On one blade, institutional capacity to identify and assess a greater range of social risks has been enhanced. On the other, the increasing sophistication of scientific and technological tools enabling institutions to identify risks that might previously have remained unknown means that we have, as a society, to deal with the consequences of our knowledge. In such a climate of indeterminacy it is unsurprising that empirical work has demonstrated that the general public have become more skeptical of expert systems in general and more willing to challenge expert opinion on risk issues (see Wynne 2002). As a result, risk practitioners are now more aware that they must accurately assess harm while also understanding the subjective processes by which people make sense of risk. Although a more nuanced appreciation of the role of the individual in risk perception is to be welcomed, critical thinkers such as Beck (1992, 2009) and Strydom (2002) have been bound up with recording the deleterious effects of environmental risks produced by institutions, companies and agencies on the natural habitat and its inhabiting populations. As global warming, air pollution and nuclear reactor leaks show, capitalist techno-scientific development produces its own risks. In as far as the positive aspects of globalization in stimulating mobility and speeding up the movement of people, information, products and services have been well documented, these very processes also aid the fluidity and rapidity with which risks travel across populations and continents.

Although it might be argued that Western citizens have generally become inured to living with risk (see Mythen and Walklate 2006; Taylor-Gooby and Zinn 2006), it needs to be acknowledged that the 'turn to risk' -in society and academia- is a relatively recent phenomenon. Given that the contours of risk have become pronounced in society, politics and the media, it is unsurprising that sociologists have been busy trying to decipher what all of this might mean for social structures, identity and interpersonal relationships. However, all was not ever thus. Risk has become a recognized subject of research in the social sciences only over the last two decades. Prior to the 1980s, risk was traditionally approached from the point of view of the natural sciences and predominantly seen as a method of calculation through which harms could be identified. As such, the developed language of risk was a distinctly technical one, oriented toward objectively quantifying harm and developing countermeasures to alleviate threats. Following the technical natural science model, once risks are identified through assessment -be they environmental, biological or crimogenic- regulations and procedures can be introduced and/or reinforced to improve safety. Historically speaking, the language of risk and the evolution of tools of risk assessment have been advanced primarily in the natural sciences, medicine and engineering as a way of objectively measuring hazards. Meanwhile, business management and insurance approaches to risk have tended to deploy risk as a way of accounting for the balance between possible acquisition and potential loss. However, from the early 1980s onwards, various contributions from within the social sciences sought to draw attention to the comparatively undeveloped social dimensions of risk. These formative contributions to debates about risk were initially relatively diffuse and scattered across the disciplines of psychology (Slovic 1987), geography (Adams 1985) and anthropology (Douglas and Wildavsky 1982). In the early 1990s, the uptake of risk in sociology moved on apace, largely due to the input of heavyweight thinkers such as Ulrich Beck and Anthony Giddens. Arguing in similar grooves, these two authors sought to throw light on the social salience of risk and advocated risk as a theoretical prism through which macro social shifts could be viewed. Since the publication of Beck's best-selling book, *Risk society: Towards a new modernity* (1992), a burgeoning group of scholars have become engrossed in the subject of risk. Alongside sociologists inspired by Beck's landmark text, followers of Mary Douglas's work in anthropology and a collection of political science and criminological scholars deploying Michel Foucault's work have contributed toward the development of a large body of literature about risk. Today, understanding the evolution, construction, assessment and management of risks has become a central preoccupation in the social sciences. I wish here to draw attention in particular to some of the contradictions and dilemmas that emerge for institutions and agencies involved in risk governance in contemporary society. As we shall see, one of the fundamental ironies of risk governance is that it is, in and of itself, a risky venture. Decision-making through risk inevitably involves deciding on courses of (in)action, determining the appropriate use of resources and engaging stakeholders in dialogue around potentially controversial issues. Thus, while national governments may resolve to spend billions of pounds stocking up on Tamiflu vaccinations in anticipation of a flu pandemic, the eventuation of a single terrorist attack raises thorny questions about the security reach of the State and its capacity to ensure public safety at a time of cost cutting and reduced expenditure in policing and the intelligence services. In a future oriented and risk averse society, in which situations of intense uncertainty around threats is routine, the pressure on institutions and regulators to foresee hazards and to intervene pre-emptively becomes pronounced. Before we broach these *problematiques* head on, it is first necessary to situate our analysis within the context of the uncertainties and ambiguities of the risk society.

3.3 Nichtwissen in the Risk Society: The Problem of Not-knowing

Regarded by some commentators as most impactful theory of risk to emerge from within the social sciences (see Boyne 2003; Elliott 2009: 284), the risk society perspective is systematically laid out by Beck (1992) in his magnum opus *Risk society: Towards a new modernity*. For Beck, and fellow risk society travellers such as Giddens (1998, 1999) and Strydom (2002), the shadow of risk is ubiquitous, hovering over the environment, family, education, employment and politics. Not only has living with risk become a stock feature of everyday experience, the combined effects of the 'social explosion' of risks by the media (Beck 1995) and the 'staging' of potential future threats (Beck 2009) has not only raised awareness of risk but also encouraged lay actors to challenge the authority of expert institutions. In the original formulation of the risk society thesis, Beck (1992) was keen to stress the magnitude of the threats faced by contemporary society and to urge recognition of the human-made nature of contemporary hazards. Although these features remain important in his most recent iteration of the thesis, the problem of how to assess, communicate and manage manufactured risks has become a central pre-occupation for Beck. Responding to critiques of his work, Beck (2009) now wishes to distinguish between harms with known large scale effects -which he now classifies as 'catastrophes'- and risks which are defined as hazards with indeterminate effects and consequences. It is this latter category of *risks* -delineated as uncertainties that evade calibration- that he wishes to spotlight. In situations of non-knowing -or, in Beck's (2009: 47) terms, *nichtwissen*- the State and social institutions must act, or, indeed, choose to remain intransigent, on the basis of partial, incomplete or flawed information. Beck (2009: 115) avers that risk society is a 'non-knowledge' society as not knowing is often a direct result of the application of scientific and technological principles. For him, the more that we know, the less we understand. Whether one agrees with this proposition or not, it is evident that 'not-knowing' and the pressures to acknowledge the uncertainties present has impacted heavily on a range of social institutions from law, criminal justice, policing and the intelligence services to medical health, politics and finance. *Nichtwissen* not only presents micro challenges for the individual. At a macro level, radical uncertainties threaten both the legitimacy and the control of institutional systems (Beck 2009: 53). As Beck (1999: 78) puts it: "institutional power holders are rendered accountable for making decisions in a miasma of imperfect information and incomplete knowledge". While the strategies used to manage, regulate and communicate not-knowing vary, the task of risk governance has acquired another layer -and one which is potentially hazardous and volatile. Moreover, the catastrophic potential of risks that may appear as 'worst imaginable accidents' (WIAs) mean that the stakes cannot be higher for regulatory institutions. Whether it be the location of terrorist attacks, appropriate means of managing the global financial crisis or limiting the occurrence of Severe Acute Respiratory Syndrome (SARS), the evidence on which risk assessments are based is often imperfect and partial. What is more, somewhat ironically, the positive dreams of progress and development which fuelled the Enlightenment dream bring with them modern nightmares: "living in risk society means living with ineradicable non-knowing (nichtwissen) or, to be precise, with the simultaneity of threats and non-knowing and the resulting political, social and moral paradoxes and dilemmas. In contrast to the modern era, it cannot be overcome by more and better knowledge, more and better science; rather precisely the opposite holds: it is the *product* of more and better science" (Beck 2009: 115).

I have noted elsewhere that Beck's original formulation of the risk society thesis rests on three conceptual assumptions -or 'pillars of risk'- on which his assertions regarding the passing from 'industrial modernity' into the contemporary 'risk society' are founded (Mythen 2004: 17). First, Beck believes that the relationship between risk, time and space has fundamentally altered. Where the natural hazards of pre-industrial times and the accidents that characterized industrial modernity were relatively localized and tended to affect specific populations and groups, manufactured uncertainties are global and universal affecting both rich and poor nations and peoples. Second, Beck believes that the quality of harms is fundamentally different in the risk society. While natural hazards such as earthquakes, tornadoes and floods blighted pre-modern cultures, the magnitude of present threats far surpasses those that occurred in previous epochs. Manufactured risks challenge the whole fabric and structure of society. Third, the volatility and inherent unmanageability of manufactured risks means that institutional mechanisms and systems of protection and insurance are unable to effectively regulate to maintain public safety. As Beck (2009: 28) puts it:

> Large scale threats are abolishing the three pillars of the risk calculus. They involve, first, often irreparable harms that cannot be limited, so that the concept of monetary compensation fails. Second precautionary aftercare [*vorsorgende nachsorge*] for the worst imaginable accident is out of the question because it is impossible to gauge outcomes in advance. Third, the 'accident' has no limits in time and space, it becomes an event without a beginning and an end, an open ended festival of creeping, galloping and overlapping waves of destruction.

While the substance and presence of each of these aspects of Beck's thesis have been widely debated in the social sciences (Ekberg 2007; Elliott 2002; Mythen 2008), in his most recent work, Beck elucidates the problem of 'non-knowledge'. Insofar as situations of risk are always marked by uncertainty, Beck believes that the quality of that uncertainty has become more intense and pronounced at the same time as the magnitude of harm has increased. Thus 'old risks' are being replaced by 'new risks' -such as catastrophic climate change and the global economic crisis- which threaten not only to derail the system but to render it extinct (Beck 2009: 19). The uncertainty that surrounds new risks that have emerged in recent years in terms of their impacts and effects are not solely attributable to the physical properties of the threat. Rather, the world in which risks appear is one in which contests and debates are waged by scientists, politicians, campaign groups and the public over what constitutes 'proof' of danger. In the current phase of risk society the production and representation of knowledge about threats is a contested and contestable process. Furthermore, in comparison with other historical phases, public tolerance of harm is low and a culture of risk aversity has taken hold in the West. All of this makes for a heady cocktail, so far as questions of risk communication and risk governance are concerned. While technological and scientific advances are enabling us to detect a greater range of harms -often at earlier points of gestation- the residual uncertainties about the scale, geography and effects of risk, mean that the task of risk regulation is an arduous and testing one. Importantly for our analysis, these uncertainties that characterize the 'new risks' of world risk society encourage the 'anticipation of global catastrophes' (Beck 2009: 52). For Beck, this means that State institutions are obliged to tread a very fine line, whereby overstepping on either side will produce accusations of fear-mongering on the one hand or recklessness on the other. In such a climate, questions of whether, how and when to intervene become vital. The difficulties of responding to these questions are magnified by the juxtaposition between information/knowledge and action (Beck 1999: 78): "the ultimate deadlock of risk society ... resides in the gap between knowledge and decision: there is no one who really knows the global outcome -at the level of positive knowledge, the situation is radically 'undecideable'- but we none the less have to decide ... so risk society is provoking an obscene gamble, a kind of ironic reversal of predestination: I am accountable for decisions which I was forced to make without proper knowledge of the situation". Insofar as states of not-knowing present challenges for individuals, in terms of personal decision-making, fear management and safety strategies, it is the implications for invested governments -and the associated agencies of the State- that is our focus here. As we shall see, non-knowing places the institutions responsible for the safety, security and health of citizens under considerable strain. While one might argue that risks have always presented themselves as potential thorns in the side of government, a concatenation of factors have raised both the stakes involved in decision making and the consequences of failing to act appropriately. The globalization of the media and its enshrinement in everyday culture in Western nations means that the general public are exposed to and sensitized to a panoply of threats -which may or may not materialize as catastrophes in Beck's terms- from flesh eating bugs to asteroids set to collide with the earth. At the same time, challenges to the authority of risk regulating institutions emerging out of high-profile imbroglios, serve as indicators that public trust in expert systems is contingent rather than unconditional. Further, in a risk sensitive culture in which public tolerance for harm is low, the pressures on the State to act become greater. All of these underlying factors situate risk regulating institutions in a difficult, compromising and potentially litigious position.

3.4 Risk Governance in a Climate of Not-knowing: The Case of Counter-Terrorism Policy in the UK

It is important to note that not all risks are marked by endemic uncertainty or evade institutional regimes of regulation. The tendency in the academic literature has been to concentrate on what Taleb (2010) refers to as 'Black Swan' events, where an unexpected and highly unpredictable occurrence -such as the Chernobyl disaster, the 9/11 attacks and the Aceh tsunami- causes harm, death and destruction on a massive scale. While scholars such as Beck and Giddens have been keen to alert us to potentially catastrophic dangers that lurk in the ether, it needs to be recognized that advances in science, technology and medicine have increased our capacity to 'know' about the causes and frequency of a range of risks, from sexually transmitted diseases to road traffic accidents. Despite the development of systems and databases that can provide indicators of the frequency and effects of such risks, specific threats about which institutional knowledge is more limited have been found to produce public anxieties. A well-established finding of studies in risk research in the psychometric tradition has been that the risks that concern most people are those which have dreadful consequences and those that are largely unknown at the level of detection and impacts (see Slovic 1987, 2000). Given the penchant within the mass media for reporting potential catastrophes and bringing into vision spectacular one-off incidents that cause harm -plus sporadic cultivation of a culture of fear by politicians and those in the security industries- it is perhaps little wonder that this is so. But how have institutions responded to the specter of risks with dreadful consequences, in which the frequency and magnitude of harm remains unknown or unknowable? Of course, this is a huge question and one which extends well beyond the ambit of this chapter. The range of risks that exist 'out there' is overwhelmingly large and institutional responses to managing situations in which uncertainty is high will vary according to many factors including region, country, political party, history, culture and religion. In order to begin to respond and to bring to the surface some of the problems involved in risk governance, I wish to narrow the frame of reference and examine certain aspects of the management of national security against the threat of terrorism in the UK since 9/11. In so doing, I will be drawing out some of the dilemmas that regulatory institutions face in situations of high uncertainty and some of the negative consequences that risk based policies engender for those at the sharp end of their application.

In order that we do not get drawn into a facile mud-slinging exercise, in which all attempts at risk governance by State institutions are seen as no more than acts of social control, we should acknowledge some of the intrinsic problems that arise when risk based interventions occur in a climate of indeterminacy, such as that surrounding the protection of national security. First, the evidence on which risk assessments have to be made may be limited, patchy or contradictory. Second, and partly as a consequence of this, decision-making is complex and contradictory views can be expected from different stakeholders involved in the process. Third, in such states of indeterminacy, communication with the public and other agencies is a complicated activity and may produce unexpected consequences.

While we will return to examine how State institutions have responded to these inherent difficulties shortly, it is necessary to grasp the context in which the institutional strategies analyzed here need to be understood. As has been frequently observed, 9/11 served as something of an iconic moment so far as policies of (inter) national security are concerned. Following on from the shock of these attacks, the United States and Governments in many other nations not only re-evaluated their national security strategies but also set in train processes of securitization characterized by the introduction of new counter-terrorism legislation, increased surveillance by intelligence services, intensified scrutiny at borders and airports and expanded policing powers. In one sense, it was the fact that an attack such as 9/11 was not foreseen -it was, in the words of the 9/11 Commission (2004), "a failure of the imagination"- that drove forward new modes of horizon scanning and systems of datavaillance (see Amoore and de Goede 2008; Mythen and Walklate 2008).

Although Donald Rumsfeld's somewhat convoluted summation of the problems faced by the intelligence services was much maligned at the time, in retrospect it can be seen as a harbinger for risk based forms of governance that have materialized since: "As we know, there are known knowns. These are the things we know we know. We also know there are known unknowns. That is to say we know there are some things we do not know. But there are also unknown unknowns, the ones we don't know we don't know". Arguably, it is the unknown unknowns -and fear about their manifestation- that has served as a catalyst for the revising and revamping of security policies. In the UK, the subsequent 7/7 attacks -undertaken by four British citizens of Pakistani heritage and one naturalized Briton born in Jamaica- led to the then Labour Government introducing sweeping and draconian counter-terrorism measures, buttressed by the belief that the scale of the terrorist threat merited a response that was wide ranging and dramatic. Prime Minister Tony Blair's observation at that juncture that the "rules of the game are changing" were later reinforced in his testimony to the Iraq War inquiry where he referred to the establishment of a 'new calculus of risk'. So what does measuring danger along this new calculus of risk involve? How has the State responded to the known unknowns and, moreover, the unknown unknowns that surround national security at this particular historical moment?

Granting the liberty of broad brush strokes -and on the basis of an impressionistic as opposed to an empirical vista- it is possible to identify a series of institutional responses to the problem of not knowing. Here I shall focus on just four strategies that have frequently emerged: *staging, responsibilization, targeting and pre-emption*. In order to ground these responses, I wish to tether them to a particular example, that of counter terrorism measures and security policies designed to diminish the threat of a terrorist attack. Some of these responses are dramaturgical acts of managing risk, while others involve invoking hard-edged policies that make deep impressions on groups that are adversely affected by them. As Donald Rumsfeld's tongue-twister infers, the communication of strategies of security post 9/11 has involved the concession that public safety cannot be assured as the complexity of terrorist plots means that they may be impenetrable to intelligence services. The admission that national security cannot be guaranteed has -as Beck recognizes- encouraged the *staging of risk*, whereby possible future attacks are hypothetically played out. Such practices of imagining the future -which critical security studies thinkers have called 'pre-mediations' (Salter 2008; Grusin 2004)- have ranged in the UK from emergency management drills in mock situations of a dirty bomb being exploded to politicians raising public sensitivity to future attacks by describing types of attack that terrorists may use in the future. In addition to being constructed by State institutions, pre-mediations are often performed by the media and cultural agencies -for instance, through television dramas such as *24* and *Spooks*. Via the staging of risk, politicians are able to use the media and other cultural products to bring into view dystopic futures. Further, upcoming counter measures and strategies can be envisaged in the present. At its heart, the staging of risk realities involves something of a guess. Pre-mediative practices of bringing future attacks into sight are attempts to imagine the unknown. Such attempts to engage with threats which stretch the boundaries of calculability have reversed the use of probability estimates based on past incidents and advanced speculation about future unknowns (Salter 2008). Through the lens of 'vigilant visualities' responsibility for guarding against future harms very much depends on assertive action in the present (Amoore 2007). Although methods and techniques of risk assessment are inherently attempts to estimate future outcomes, the intensification of horizon scanning practices post the 2001 attacks has been marked and is embedded in extensive changes in legislation, policing and criminal justice (Aradau and van Munster 2008; Mythen and Walklate 2010). The institutional naval gazing that arose in response to the intelligence services failure to predict the attacks on the United States produced tangible pressures on securocrats to pose more volubly the 'What if?' question, as immortalized in Rumsfeld's 'unknown unknowns'. Pre-mediations act as harbingers of harm that do not so much present what *has* happened, but speculate about what *may* happen next. Broadcast the day before the announcement of the 2008 Counter-Terrorism and Security Bill, former Security Minister Tony McNulty's (2008) provocation acts as a case in point: "Imagine three 9/11s. Imagine two 7/7s. Given the evidence we've got such scenarios aren't fanciful". In such circumstances, the ideological possibilities of the staging of risk realities is stripped bare. The governmental rationale for staging risk in advance -notwithstanding whether they eventuate or not- is that the public can be exposed in advance to possible modes of attack, risk consciousness may be heightened and the public can be primed for the introduction of new measures. Thus, the staging of risk realities invokes a second strategy of *responsibilization* through which citizens are encouraged to partake in security seeking behaviour for themselves and on behalf of the State. Hence, the emphasis on 'resilience' in more recent policy making around counter-terrorism. If individuals, groups and communities can be encouraged to prepare and brace themselves for attack then the security burden of the State is ideationally, materially and economically reduced. Responsibilization around terrorism takes many forms and is embedded in government funded projects based amongst communities in which young people 'at risk' of radicalization live, advertisements and posters that invite citizens to look out for suspicious behaviour of neighbors and the issuing of emergency advice booklets so the public can learn how to prepare for critical incidents (Kearon et al. 2007). Each of these ventures in vigilance serves as attempt to recruit citizens and to involve them in both their own security and the safety of the nation. For some, encouraging citizen responsibility may be interpreted as common sense and sensible risk governance. For others a more deep seated migration of responsibility for risk is taking place from the State to the individual. Further, strategies of responsibilization around terrorism have been seen by some as deliberate attempts by the State to manage populations through fear. As both Burkitt (2005) and Vertigans (2010) argue, the staging of dystopic futures frequently emerges during moments at which governments seek public support for the implementation of stringent law and order measures.

While the staging of risk and responsibilization act as warnings and prompts, a further strategy of *targeting* identifies specific people, places and communities that are dangerous or threatening. Again, there is something of an appeal to common sense rationale which courses through policies of targeting. Often deployed with recourse to statistical or factual data, the logic of targeting is that discrete populations can be classified as more risky than others based around past propensity to commit attacks. This logic was elucidated in a base, but nonetheless revealing way by the Head of British Transport Police, Sir Ian Johnston after the 7/7 attacks: "Intelligence-led stop and searches have got to be the way … we should not waste time searching old white ladies. It is going to be disproportionate. It is going to be young men, not exclusively, but it may be disproportionate when it comes to ethnic groups" (cited in Dodd 2005).

Considered in the context of the regulation of terrorism, targeting brings practical and moral dilemmas. Intelligence may be erroneous, the assessments on which individual and communities are rendered suspect may be faulty and there is no guarantee that targeting will yield successful results (see Biglino 2002). A number of high profile cases bear out these concerns, including the mistaken execution of Jean Charles de Menezes by the British security services and the bungled Forest Gate police raid in which Mohammed Koyar was shot in the chest (see Mythen and Walklate 2006). The use of risk profiling and the targeting of Black and Asian ethnic minorities in the UK serves as a cautionary tale of generalized risk-based intervention, with stop and search legislation permitting apprehension and questioning without cause for suspicion producing inordinately few arrests in proportion to its application (see Travis 2010). The assumptions made in the formative media coverage of the Norwegian terrorist attacks serve as an important lesson regarding the impacts of labeling. Despite reports about the physical characteristics of the perpetrator, initial media coverage of the terrorist attacks in Oslo and Utoya that claimed 77 lives suggested that Islamic Fundamentalists were to blame. When it emerged that the attacker was not a Muslim of Asian descent but Norwegian, white, Middle Class and Christian, the risk heuristics and biases not of the general public but of institutions such as the media, police and intelligence services tumbled into the spotlight. This example serves as a neat example of the limits of targeting and a reminder of the unknown unknowns that defy institutional risk governance. It is, of course, no coincidence that those labeled as risky and targeted by counter-terrorism measures are identified as outsiders. The strategy of labeling is, of course, more pronounced in some areas that others. In the case of counter-terrorism in particular, applying 'imagination' in situations of not knowing has resulted in the cementation of a 'culture of suspicion'. The ramifications of surveillance for those labeled as outsiders are not simply external but reproduce internal forms of evaluation and reflexivity. For instance, young British Pakistanis defined in security discourses as dangerous are not simply subjected to the surveillant gaze of the State. Rather, this gaze invokes subject introspection in terms of the modification of everyday practices, from political and linguistic censorship to choices of clothing and modifying one's appearance in order to present a 'safe' identity (see Mythen et al. 2009).

Individuals defined as risky in the broader sweep of forms of risk based governance such as profiling can find themselves subjected to a fourth strategy of *pre-emption*. As a result of pressures to act and a lack of public acceptance of risk more generally, in areas in which the manifestation of harm is seen as critical -such as counter-terrorism and national security- various pre-emptive strategies have been developed in recent years. Pre-emption involves taking early action in order to prevent an anticipated event happening in the future. Situations where potential risks of a high magnitude emerge, but in which knowledge is limited or contingent are often those in which the State intervenes by introducing pre-emptive policies. In the case of counter-terrorism, various pre-emptive responses have been enshrined in law. It would seem that the condition of non-knowing about future terrorist attacks has engendered sweeping attempts to manage the future through forms of pre-criminalization. Echoing the activities of State agents in Phillip K. Dick's *Minority Report* (1956), the use of control orders, stop and search laws and the extension of detention without charge in the UK are all examples of pre-criminalisation processes formally designed to reduce the possibility of tail end risks (see McCulloch and Pickering 2010; Mythen and Walklate 2010). To add a further layer of intricacy to matters, it is probable that such misguided forays in risk governance may actually end up reproducing the law of inverse consequences, whereby the very actions and behaviours that institutional intervention invokes are actually invoked rather than attenuated. Community relations between the police and Muslim minority communities have been damaged via risk interventions such as stop and search and in conditions in which trust is low the reporting of suspicious behaviour to the police is less likely to occur (see Thiel 2009; Travis 2010). Poignantly, the overuse of stop searches was frequently cited by those from ethnic minorities involved in the urban unrest across the UK in 2011 as a major contributory factor in their behaviour (LSE/Guardian Report 2011).

3.5 Conclusion

Using post 9/11 counter terrorism policing and security policies in the UK as a vignette, I have outlined a range of institutional strategies of risk governance that emerge as responses to situations of not-knowing. The impacts and effects of these risk-based strategies of governance have been considered and the limits to their efficacy discussed. There are several important qualifications to make before we conclude. First, although I have traced the contours of prominent institutional strategies designed to counter the uncertainties of risk, a broader range of responses than has been documented here can be evidenced. Establishing limits to risk and acceptable modes of risk governance are ventures that will differ across space and place and will be linked to moral, political and religious views. Second, processes of managing and communicating threats will change according to the potential risk in question. To this end, in as much as tiers of legislation have been introduced through counter-terrorism law, one might provocatively ask why mechanisms of financial risk governance to regulate bankers and mortgage lenders have changed so little since 2008 given the dramatic toll that the global financial crisis has taken around the world. Similarly, the worst case scenario for environment despoliation described in the Stern Review (2006) does not appear to have led to radical changes in production or consumption nor serious commitments to reduce carbon emissions amongst major polluting countries such as the United States and China. Third, while I have focused my attention on some of the problems that have arisen in one particular case, it is important to note that application of the very strategies critically scrutinized here -staging, responsibilization, targeting and pre-emption- may, in other instances, lead to positive outcomes in terms of public protection and safety management. As always, context is everything. In situations in which potentially losses are marginal, where there is a strong consensus amongst stakeholders about interventions, where the gains are demonstrable and the losses of intransigence are high, it is likely that some of the strategies of risk governance discussed here may have greater purchase and appeal.

This chapter began by sketching out the key facets of the turn to risk that has materialized in Western nations over the last three decades. In documenting the processes and practices that characterize the turn to risk, the problem of *nichtwissen* was highlighted as a fundamental problematic for institutions seeking to govern with, through and by risk. In situations of endemic uncertainty, Beck's (2009: 117) reflections regarding the brittle line that social institutions have to negotiate between being prudent and being accused of inciting public hysteria are resonant. The relative risk aversity of the public places the State in something of an invidious position in terms of balancing interventions that are premature against accusations of neglect in cases of inaction. Thus, the lens of academic attention in the contemporary age ought to be trained not only on adverse consequences in situations where the State is intransigence but also the potential perils of proactive policies of risk governance. In order to bring out some of the specific tensions and issues around risk governance we illuminated the case of domestic security in the UK in the light of the terrorist threat. In considering the perils and pitfalls of risk based strategies of governance attention was drawn to the use of worst imaginable assumptions, the limits to risk profiling and the danger of producing the law of inverse consequences. In conclusion, I have argued here that there has been a tendency since 9/11 for practitioners and professionals involved in security policy-making to overlook the question of what it is that we do know in preference for what it is that we don't know. Notwithstanding the positive possibilities of preparedness in efforts to heighten risk consciousness around terrorism, the staging of worst imaginable scenarios has largely raised rather than reduce public anxieties. In many regards, what Beck (2009: 40) calls "the threat imagination of world risk society" is as expansive as it is troubling. Thinking the unthinkable and predicting tail events of low probability but high potential consequences has become an increasingly prevalent aspect of the architecture of the modern nation modern State and this facet of risk governance is worthy of further academic investigation.

References

9/11 Commission. 2004. *Final report of the National Commission on terrorist attacks upon the United States*. Washington: National Commission on Terrorist Attacks.

Adams, J. 1985. *Risk*. London: UCL Press.

Amoore, L. 2007. Vigilant visualities: The watchful politics of the war on terror. *Security Dialogue* 38: 215–232.

Amoore, L., and M. de Goede. 2008. *Risk and the war on terror*. London: Routledge.
Aradau, C., and R. Van Munster. 2008. Governing terrorism through risk: Taking precautions, (un) knowing the future. *European Journal of International Relations* 13(1): 89–115.
Arnoldi, J. 2009. *Risk*. Cambridge: Polity Press.
Beck, U. 1992. *Risk society: Towards a new modernity*. London: Sage.
Beck, U. 1995. *Ecological politics in an age of risk*. Cambridge: Polity Press.
Beck, U. 1999. *World risk society*. London: Sage.
Beck, U. 2009. *World at risk*. Cambridge: Polity Press.
Biglino, I. 2002. *Anti-terrorism legislation in Britain*. London: Liberty.
Boyne, R. 2003. *Risk*. London: Sage.
Burkitt, I. 2005. Powerful emotions: Power, government and opposition in the war on terror. *Sociology* 39: 679–695.
Denney, D. 2005. *Risk and society*. London: Sage.
Dick, P.K. 1956. *Minority report*. New York: Basic Books.
Dodd, V. 2005. Asian men targeted in stop and search. *The Guardian*, August 17.
Douglas, M., and A. Wildavsky. 1982. *Risk and culture: An essay on the selection of technical and environmental dangers*. Berkeley: University of California Press.
Ekberg, M. 2007. The parameters of the risk society: A review and exploration. *Current Sociology* 55: 342–366.
Elliott, A. 2002. Beck's sociology of risk: A critical assessment. *Sociology* 36(2): 293–315.
Elliott, A. 2009. Series editors foreward. In *Risk, vulnerability and everyday life*, ed. I. Wilkinson. London: Routledge.
Ewald, F. 1991. Insurance and risk. In *The Foucault effect: Studies in governmentality*, ed. G. Burchell, C. Gordon, and P. Miller. London: Harvester.
Giddens, A. 1998. Risk society and the context of British politics. In *The politics of risk society*, ed. J. Franklin. Cambridge: Polity Press.
Giddens, A. 1999. Risk and responsibility. *The Modern Law Review* 62(1): 1–10.
Grusin, R. 2004. Premediation. *Criticism* 46: 17–39.
Hutter, B. 2005. *The attractions of risk-based regulation: Accounting for the emergence of risk ideas in regulation*, CARR discussion paper no. 33. London: LSE.
Kearon, T., G. Mythen, and S. Walklate. 2007. Making sense of the terrorist risk: Public perceptions of emergency advice. *Security Journal* 20(2): 77–95.
Kemshall, H. 2003. *Understanding risk in criminal justice*. Buckingham: OUP.
LSE/Guardian Report. 2011. *Reading the riots*. London: Guardian.
McCulloch, J., and S. Pickering. 2010. Pre-crime, state terror, and dystopia in the twenty-first century. *Criminal Justice Matters* 81(1): 32–33.
McNulty, T. 2008. Prepare for three 9/11s in one day. *Daily Mirror*, January 23, p. 1.
Mythen, G. 2004. *Ulrich Beck: A critical introduction to the Risk Society*. London: Pluto.
Mythen, G. 2008. Sociology and the art of risk. *Sociology Compass* 2(1): 299–316.
Mythen, G., and S. Walklate. 2006. Communicating the terrorist risk: Harnessing a culture of fear? *Crime, Media, Culture: An International Journal* 2: 123–142.
Mythen, G., and S. Walklate. 2008. Terrorism, risk and international security: The perils of asking what if? *Security Dialogue* 39(2–3): 221–242.
Mythen, G., and S. Walklate. 2010. Pre-crime, regulation and counter-terrorism: Interrogating anticipatory risk. *Criminal Justice Matters* 81(1): 34–36.
Mythen, G., S. Walklate, and F. Khan. 2009. I'm a Muslim but I'm not a terrorist: Victimisation, risky identities and the performance of safety. *British Journal of Criminology* 49(6): 736–754.
O'Malley, P. 2008. Experiments in risk and criminal justice. *Theoretical Criminology* 12(4): 451–471.
Rothstein, H. 2003. Neglected risk regulation: The institutional attenuation phenomenon. *Health, Risk and Society* 5(1): 85–103.
Rothstein, H., M. Huber, and G. Gaskell. 2006. A theory of risk colonisation: The spiralling regulatory logics of societal and institutional risk. *Economy and Society* 35(1): 91–112.

Salter, M. 2008. Conclusion: Risk and imagination in the war on terror. In *Risk and the war on terror*, ed. L. Amoore and M. de Goede. London: Routledge.
Slovic, P. 1987. Perception of risk. *Science* 236(4799): 280–285.
Slovic, P. 2000. *Perception of risk*. London: Earthscan.
Stern, N. 2006. *The economics of climate change*. London: HM Treasury.
Strydom, P. 2002. *Risk, environment and modernity*. Buckingham: Open University Press.
Taleb, N. 2010. *The black swan*. London: Penguin.
Taylor-Gooby, P., and J.O. Zinn. 2006. *Risk in social science*. Oxford: Oxford University Press.
Thiel, D. 2009. *Policing terrorism: A review of the evidence*. London: The Police Foundation.
Travis, A. 2010. Stop and search powers illegal, European Court rules. *The Guardian*, January 12.
Vertigans, S. 2010. British Muslims and the UK government's 'war on terror' within: Evidence of a clash of civilizations or emergent de-civilizing processes? *British Journal of Sociology* 61(1): 26–44.
Wynne, B. 2002. Risk and environment as legitimatory discourses of technology: Reflexivity inside out. *Current Sociology* 50(3): 459–477.

Gabe Mythen is Reader in Sociology in the Department of Sociology, Criminology and Social Policy at the University of Liverpool. His current research interests revolve around risk and regulation, specifically in the context of counter-terrorism policy, security management and the hegemonic objectives of the state. Mythen's teaching interests lie in the areas of sociological criminology and social theory. He wrote the monograph *Ulrich Beck: A critical introduction to the Risk Society* (2004), which provides a critique of the risk society thesis. Alongside Sandra Walklate, he edited *Beyond the Risk Society: Critical reflections on risk and human security* (2006).

Chapter 4
Making Sense of Decentralization: Coping with the Complexities of the Urban Environment

Christian Zuidema and Gert de Roo

4.1 Introduction

During most of the twentieth century, the exercise of governance was left largely to the discretion of formal governments, most notably the central state. Reliance on government control was supported by the long-held assumptions that (central) governments are able to exercise a high degree of control over social processes and, while doing so, are also best equipped to represent the 'public good'. Although long left relatively undisputed, the past few decades are characterized by much greater skepticism towards these assumptions. This skepticism is fuelled by the works of scholars such as Castells (2000), Beck (1992) and Harvey (1989), which have led to the rather mainstream acceptance that our societies are far more complex and fragmented than many assumed in the 1950s and 1960s. In response, we have also increasingly come to accept that relying on central government control is often incompatible with the societal conditions we face (e.g. Kooiman 1993; Hooghe and Marks 2001; Pierre and Peters 2000). It has resulted in several prime 'shifts in governance' that mean to improve our societies' governance capacity to respond to the complexities of our twenty-first century societies.

Since the 1980s in particular, extensive governance renewal operations can be witnessed in many countries. As Jessop (1994) explains, these changes result in the 'hollowing out' of the nation state. This implies that power and responsibility are reallocated from the central state 'upward' to supranational bodies, 'sideways' to non-government, market and civil organizations and 'downwards' to lower tiers of government. In doing so, central governments hope and expect to increase the societal capacity to govern by involving a wide diversity of societal and market parties

C. Zuidema (✉) • G. de Roo
Department of Spatial Planning and Environment, Faculty of Spatial Sciences, University of Groningen, PO Box 800, 9700 AV Groningen, The Netherlands
e-mail: c.zuidema@rug.nl; g.de.roo@rug.nl

U. Fra.Paleo (ed.), *Risk Governance*, DOI 10.1007/978-94-017-9328-5_4

and by making policies more tailored to different circumstances. Among the dominant means of pursuing governance renewal is the decentralization of power and authority from the central government to the local level (also De Vries 2000). The idea is that local authorities are in a better position to engage in bargaining or collaborative processes with local stakeholders and civil organizations. In addition, familiarity with local circumstances and interests helps local parties developing more integrated policies that are tailored to the local situation. Hence, local authorities are assumed important players in helping governance respond to the complexities of our twenty-first century societies.

Decentralization has in recent decades also begun to affect environmental policies in countries such as Sweden (Bergström and Dobers 2000), Norway (Hovik and Reitan 2004), the United Kingdom (Gibbs and Jonas 1999) and the Netherlands (Zuidema 2011). In the meantime, various authors point out that decentralization is often pursued without a keen understanding of its possible or likely consequences (e.g. De Vries 2000; Flynn 2000; Prud'homme 1994; Walberg et al. 2000). Lower levels of local performance as compared to the situation before decentralization are therefore among the possible outcomes of decentralization. This is clearly a risk, especially when it comes to environmental policy and its focus on protecting public health and safety. In this chapter we will therefore reflect on the increasing role of the local level in environmental policy.

In this chapter we take inspiration from a recent study into the experiences in the Netherlands. This study was issued by the Dutch Ministry of Housing, Spatial Planning and the Environment. Both authors participated in the research, which largely took place in 2008 and was followed by various discussion meetings with Ministry experts on interpreting and explaining the outcomes during 2009 and 2010 (see Spreeuwers et al. 2008; Zuidema 2011). The study's objective was to explain the consequences of decentralization in Dutch environmental policies. In this chapter, we go beyond such an explanation. Instead, we use the findings of our study to explain that central policies and regulations are important conditioning factors for the outcomes of decentralization, as they can provide a robust set of central policies and regulations that stimulate, enable and guarantee levels of local performance. That is, we argue that even if we pursue decentralization to develop more dynamic policy approaches that are tailored to complex local circumstances, central policies and regulations remain to play a key role.

The study was based on four related sub-studies. The first is a desk-study of policy reports and legal documents to better understand the decentralization measures in five environmental policy fields (soil remediation, energy, noise nuisance, air pollution, and odor). The second study was a desk-study of recent research reports on the performance of Dutch municipalities in environmental management. They were based on research conducted by provincial and national institutions that are responsible for supervising municipal performance. The third study was based on a series of interviews with over 40 experts from various governmental and non-governmental organizations, each involved in supervision of municipal performance or advising municipalities. The fourth study consisted of surveys in 28 Dutch municipalities. The surveys focused both on levels of municipal performance and on explaining the causes for success or failure. More details on these studies are presented in the empirical sections of this chapter.

Section 4.2 contains a short introduction of the initial development of Dutch environmental policies and its dependence on central government control. We subsequently discuss the arguments in favor of decentralization in Sect. 4.3 and the main doubts and risks associated with decentralization in Sect. 4.4. The decentralization operations as pursued in the Netherlands are discussed in Sect. 4.5, while the prime consequences we uncovered in our research are discussed in Sect. 4.6. In Sect. 4.7 we will use our discussion on the benefits, doubts and risks surrounding decentralization to explain the consequences and reflect on the Dutch approach. Based on that we will come to our main conclusions in Sect. 4.8, where we suggest how the Dutch experiences indicate the importance of central policies and regulations to support decentralization.

4.2 The Netherlands, an International Frontrunner?

Modern day Dutch environmental policies find their origins in the 'environmental revolution' of the late 1960s and early 1970s that hit much of the western world. Just as many other countries, the Dutch were at this time confronted with the environmental consequences of rapid economic growth following World War II. The response was fast and very well structured. It made the Netherlands an internationally renowned 'pioneer' in environmental policies (e.g. Liefferink and Van der Zouwen 2004; Weale 1992).

The Dutch approach might have been fast and well structured; it was not that dissimilar to that in other western countries (e.g. Andersen and Liefferink 1997). Just as many other nations during the early 1970s, the Dutch were inspired by a confidence in the coordinative capacity of the central state to command and control the delivery of societal objectives. It resulted in a strong reliance on centrally issued policy objectives (environmental standards), hierarchical control (sanctions) and the specialization of policies and bureaucracies in distinct topics such as noise abatement, air quality, soil remediation, water quality, etc. Developing and implementing these policies relied rather solely on the regulatory capacity of government bureaucracies and the technical solutions available for those implementing them. It is also clearly visible in the early developments of Dutch environmental policies (also De Roo 2003).

During the 1970s environmental policies in the Netherlands focused on reducing the effect of environmental stress and on cleaning up the most urgent environmental pollution. It called for a fast response with reliable outcomes. This is easier to accommodate by relying on a hierarchical organization than on more collaborative or horizontal organizations, as it is possible to rely on a direct top-down exercise of control. In the *Emergency memorandum on Environmental Hygiene* (Urgentienota Milieuhygiëne) of 1972, the Dutch laid down the foundations for the development of environmental legislation and policies in the 1970s. The Dutch indeed chose to install a series of environmental quality standards that dictated maximum levels of environmental stress tolerated. Such environmental quality standards would for example dictate the maximum amounts of pollutants tolerated in air, water and soil, the height of safety risks tolerated in areas where people would live of work or the maximum amount of noise tolerated in residential areas. These standards were not just meant to be 'strict but realistic' (De Roo 2003), but given their reliance on health and safety, they were also generic. Within this 'command-and-control' tradition, the state set ambitious environmental targets which local authorities had to meet. Local authorities only had a role to play in implementing national policies and ensuring that local environmental quality levels were within the legal limits.

At the end of the 1980s, there was a well-developed body of environmental legislation and policies. Most environmental priority themes were well covered by national policies and regulations. During the 1980s the Dutch national government had furthermore worked hard to ensure these thematic policies and regulations would be well coordinated. In doing so, the Dutch had successfully evaded one of the most serious drawbacks of relying on central government control and the associated thematic specialization: incoherent policies and coordination deficits between separate policy themes (also see Andersen and Liefferink 1997). In 1989 the first *Dutch National Environmental Policy Plan* was therefore published in the face of a more or less mature and well-coordinated body of environmental policies (TK 1989). The plan provided a strategic vision upon the future of Dutch environmental planning. Based on strong societal support and political commitment, it first of all suggested even more ambitious and strict environmental standards. Secondly though, the Dutch aimed to move away from a focus on 'damage control' and cleaning up. Instead of only providing 'checks' on developments, environmental standards were to be achieved within integrated policy strategies; i.e. where environmental targets were pursued alongside other policy objectives, most notably spatial planning.

For improving the relation between environmental policies and other policies the Dutch national government promoted what it called an 'area specific approach'. Based on the desired or existing qualities of a specific area, environmental priorities were expected to be successfully combined with other priorities in integrated policy strategies. In practice, the integration of environmental policies and ambitions in an area specific strategy proved more difficult than expected (e.g. De Roo 2003). After all, while trying to integrate environmental targets with other policy objectives, compliance with strict environmental standards had to be maintained. Local authorities were therefore faced with environmental goals that had to be met a priori to any balancing or integrative effort. Both environmental standards and the framework for meeting them locally remained imposed by the national government. This is not necessarily a problem, if these environmental standards can be realistically combined with other local policy priorities. This, however, proved problematic.

In practice, it was not always realistic to meet noise levels in inner-city areas, while there were many localities where meeting environmental standards for air, water or soil would be impossible, at least in the short term. In more extreme cases, meeting environmental standards was simply impossible within reasonable technological, financial or time frames; i.e. it would imply consequences such as shutting down large industrial estates, closing main transport routes or even demolishing large parts of cities (e.g. Borst et al. 1995). To illustrate, much of the inner-city of Dordrecht would have to be demolished or, as alternative, the larger part of its industry should be closed. Similarly, soil remediation would, according to the standards of the time, require a full remediation of most Dutch city centers. But even in other cases, fully prioritizing environmental standards was considered undesirable, or at least controversial, due to its social, economic, financial, etc. consequences. It could for example result in the need for sound screens in the middle of historic city centers or seriously downsizing Schiphol airport in Amsterdam (e.g. Borst et al. 1995).

Given the problems encountered, a rethinking of the status of strong regulatory instruments and state intervention occurred in Dutch environmental policies during the early and mid-1990s (e.g. De Roo 2003; Liefferink and Van der Zouwen 2004). Setting strict a priori targets was increasingly considered to 'suffocate' local authorities in their attempts to produce area specific policies and to pursue the desired external integration. In response, the Dutch national government chose to increase the room to maneuver by reducing the impact of a priority targets, especially with regards to noise policies, soil remediation, odour nuisance and later also safety risks (e.g. De Roo 2004; Zuidema 2011). This fitted in neatly with the established Dutch confidence in decentralization and deregulation for improving the efficiency of most governmental policies (e.g. De Roo 2004). This confidence in decentralization as a means for governance renewal had, in the meantime, also manifested itself in both international theoretical debates and practice on planning and administration during the 1980s and 1990s.

4.3 Decentralization in Environmental Governance

Decentralization is pursued for a variety of reasons, few of which are actually undisputed (see also De Vries 2000; Prud'homme 1994). To begin with the main arguments supporting decentralization, De Vries points towards "the possibility of tailor-made policies, short lines between the allocating agency and the receivers thereof, service delivery based on greater knowledge of the actors at the local level, with regard for local circumstances, greater possibilities for civil participation and, in general, more effective and efficient allocation of public goods and services" (2000, p. 493). Each of these arguments means that decentralization is often considered among the dominant means to renew governance for dealing with more complex policy issues (e.g. De Vries 2000; Fleurke and Hulst 2006). Indeed, this is also the main motive used for decentralization in Dutch environmental policy (e.g. De Roo 2004).

In recent decades there has been an increased acceptance that traditional policies and regulations associated with central government control are often incompatible with the societal complexity we face (e.g. Kooiman 1993; Hooghe and Marks 2001; Pierre and Peters 2000). On the one hand, complexity can be related to an increased fragmentation and diversity in societies, its processes and activities (e.g. Healey 1997; Martens 2007; Sassen 2002; Stoker and Mossberger 2001). Social fragmentation has fuelled an increase of different interpretations regarding the challenges that policies mean to address and the various ambitions that groups and individuals hold on to. In the meantime, various societal and market parties are claiming their place in the process of governance and also have the resources to exercise influence (e.g. Kooiman 1993; Pierre and Peters 2000; Van Tatenhove et al. 2000). Social fragmentation and power dispersal combined therefore challenge a monistic approach to governance whereby the central state controls policy development and delivery. In the face of the resulting socio-cultural plurality, support for the supremacy of central state control is undermined, while the central state is also increasingly confined in its capacity to dominate societal or market parties. On the other hand, we also face an increased acceptance of the interrelation between problems, their causes and effects (e.g. De Roo 2003; Martens 2007; Van Tatenhove et al. 2000). Not only does this imply that it becomes increasingly cumbersome to model or map the exact consequences of policy actions. There are also usually multiple, interrelated and potentially conflicting objectives to which policies should respond, begging for more integrated and cross-sectoral policy approaches that engage and respect different stakeholders and their interests in their unique local manifestations (also De Roo 2003; Rydin 1998). Doing so can be hard for central governments, as they often fail to have sufficient knowledge of the exact local circumstances and stakeholder interests.

In contrast, decentralization brings decision making closer to the local level and can thus facilitate local parties and people to influence policy development and delivery. It allows political constituents to exert influence over the issues they face in their own areas based on representative democracy and increases the possibility for direct democratic involvement. In addition, decentralization can increase the capacity of governance to respond to local circumstances. Decentralized units are closer to these local circumstances and are considered well located for translating the interrelatedness of issues and interests *in situ* into integrated strategies (see De Roo 2004; Liefferink et al. 2002). Indeed, a key purpose of decentralization is to help "local government to act pragmatically and develop locally contingent solutions to problems rather than feeling compelled to fit with guidance" (Coaffee and Headlam 2008), p. 1595). Summarizing then, bringing policy making closer to local circumstances and related stakeholder interests can allow for a more dynamic development of policy approaches in relation to the local circumstances. But while this means that decentralization can be seen as a response to deal with more complex policy issues, there are also some important doubts and risks associated with decentralization.

4.4 Doubts About Decentralization

To begin with, decentralization implies that top-down policies and guidance are increasingly replaced by specifically designed policy approaches and local responsibilities. This might be relevant and beneficial when dealing with more complex local issues. However, not all issues are strongly embedded in their local contexts and are similarly *complex*. There are also circumstances when we face much agreement on a set of objectives, are confident which causes and effects need to be addressed to meet these objectives, and therefore have a unified perspective on dealing with these circumstances. For coping with such relatively simple and straightforward issues, there is little reason to invest time and resources in developing specific policy approaches that are tailored to the local circumstances. Rather, there are now important economies of scale involved in organizing policy development and delivery at higher levels of authority. Central guidance can benefit the repetitiveness of implementing, at a central level, common policy formats that can be applied at all lower levels of authority. These formats include environmental standards, licensing and permit systems and common solution strategies, such as the size of sound screens, spatial buffer zones or air filters. Local authorities no longer need to 'reinvent the wheel', but can routinely implement these common formats. It is only when they face more complex circumstances that specifically designed policy approaches that are tailored to the local circumstances are needed.

Secondly, decentralization means that the outcomes of governance become increasingly dependent on local performance and therefore, of the available local willingness and ability to perform decentralized tasks and responsibilities. Willingness and ability are, however, not self-evident in a local realm. Especially, we argue, when it comes to environmental interests (e.g. Burström and Korhonen 2001; Zuidema 2011). When it comes to *local ability*, time, funding, competent staff and legal instruments are among the immediate resources needed to do so. Research findings suggest that the availability of these resources is certainly not evident at the local level, also when it comes to environmental policies (e.g. Burström and Korhonen 2001; Flynn 2000; Walberg et al. 2000). As Prud'homme (1994) explains, central governments can more easily invest in research and development, innovative projects and in attracting people with many different competences and forms of expertise. This is more problematic for smaller units such as municipalities. When decentralizing tasks there is thus a need to assess whether local units can realistically be assumed to acquire the required resources to perform them. In our case, it is therefore important to assess whether municipalities host or can attract the kind of creative, innovative and visionary persons needed for developing and implementing specific and integrated policy approaches.

Local willingness to pursue environmental objectives is also not self-evident. Local characteristics such as the local political color, the balance of power between stakeholders and local urgencies will influence local willingness pursue environmental objectives. In the meantime, there are also two general constraints on local willingness to pursue environmental objectives that provide arguments for some degree of central government influence. On the one hand, environmental priorities have a relatively 'weak profile' as compared to more development-oriented priorities. Environmental benefits can, for example, be hard to express in financial terms (such as noise nuisance), are often invisible (as with safety risks), diffuse (as with air pollution), are highly subjective (odor) or focus on a long time horizon (e.g. sustainability). Economic growth, social development or, for example, financial costs are easier to envision and often relate to the short term. Decentralization allows local authorities to balance environmental objectives with other local objectives; i.e. they become more 'tradable'. Their 'weak profile', however, means that environmental objectives might be easily 'overruled' by other more powerful economic and social objectives (e.g. Eckersley 1992; Jordan 1999; Walberg et al. 2000).

On the other hand, many environmental issues manifest themselves as 'social dilemmas' (e.g. Lemos and Agrawal 2006; Wätli 2004). Typically, these are issues that manifest themselves on scales incompatible with the local such as air pollution, the depletion of the ozone layer, global warming, river pollution, etc. They are issues where the benefits of local investments in improved environmental conditions 'spill over' to adjacent jurisdictions, while the costs are confined to the local realm. This is hardly an incentive for local action. Decentralization is now risky. Without sufficient incentives local authorities might not be inclined to take action. They might even be inclined to accept the 'costs' of additional environmental pollution of which they only experience a modest share, in the face of the 'benefits' of additional growth which they can experience to the fullest. Instead, coordination between local authorities is desirable, where central governments can play a key role in organizing coordination, installing incentives and avoiding 'free riders'.

We here argue that, despite its potential benefits, decentralization should be approached with some degree of skepticism. This is especially true when it comes to environmental policy. We have just seen that decentralization results in a decreased certainty regarding the kind of ambitions that are pursued locally. In the face of possible limits to local willingness and ability to pursue environmental objectives, this can have serious consequences. Some localities might choose for ambition levels that are quite lower than previously installed by central governments. Also, different localities might choose for different ambition levels, inspired by different circumstances and social or political preference. Decentralization can therefore easily conflict with a desire to maintain a generically enforced minimal degree of performance, such as inspired by the protection of humans and ecosystems to environmental stress or by the desire to avoid strong inequalities between localities. That is not to say that decentralization is no longer possible. Rather, it is to say that there are good motives to at least hold on to some central control over local performance.

4.5 Decentralization and Deregulation in the Netherlands

Decentralization in Dutch environmental policies started somewhere in the mid-1990s. In the face of the 'suffocating' effect of strict generic standards, the national government argued that local government "must be afforded greater freedom and as much integrated responsibility for the local living environment as possible" (VROM

2001, p. 68). In response, traditional generic targets in for example noise, odor and soil policies were replaced by less ambitious and more flexible standards, while changes in financial structures also gave local authorities more flexibility. The Dutch national government now only considers itself responsible for guaranteeing what it calls a minimal 'base quality'. This 'base quality' is no longer dominated by the ambitious 'strict but realistic' targets of the early 1990s. Rather it is based on more 'modest' targets that only mean to protect a less ambitious 'minimum quality level'.

Decentralization in Dutch environmental policies is clearly inspired by the assumption that local authorities are in the best place to cope with local circumstances and stakeholders. This is not just meant to enable local authorities to cope with more complex local circumstances. The Dutch national government also assumed decentralization would "result in improvements to the quality of the living environment" (VROM 2001, p. 68). The idea was that in addition to protecting a minimal base quality, local authorities would actively pursue an environmental quality that goes 'beyond the minimum' (VROM 2001). Inspired by knowledge of local circumstances and their proximity to local stakeholders, the national government assumed local authorities were best equipped to pursue these higher quality ambitions. Therefore, the national government also explicitly called upon local authorities to do so in its policy documents.

Initially, the Dutch national government aimed for national policies to provide local authorities with fairly strong incentives to go 'beyond the minimum'. Described as the 'Quality of the Living Environment' approach (MILO), the plan was to install indicative environmental standards that were more ambitious than the base quality. Based on a contract between local authorities and the national government, the exact legal status of these indicative standards would be decided. MILO has, however, turned out as a fully voluntary instrument without binding contracts or indicative standards. It merely provides municipalities with guidance upon how to link environmental interests and ambitions with spatial planning processes. As no genuine alternative for MILO has been developed, local authorities face no obligatory national guidelines or incentives to 'go beyond the minimum' or to develop and deliver strategic and integrated policies. Instead, they merely face persuasive policies. Consequently, it is up to local authorities whether or not they want to follow up on the national call to go 'beyond the minimum'. In the meantime, it is important to recognize that minimal requirements have in recent years become less ambitious and more flexible. Failure to go 'beyond the minimum' will thus imply that environmental objectives are pursued that are lower than before decentralization and deregulation operations began.

Clearly then, the Dutch case is an example where the strong dependence on central government control has been replaced by a fairly strong dependence on local performance. It offers a good case to reflect on the possible benefits and risks of decentralization as discussed in the previous sections. We will also do so here, based on our research into identifying and explaining the consequences of decentralization in Dutch environmental policies (see Spreeuwers et al. 2008; Zuidema 2011).

4.6 Reporting on the Dutch Experience

We began our research with a quick scan of five environmental policy themes that have seen some important changes in the past 15 years. The fields studied were soil remediation, energy, noise nuisance, air pollution, and odor and was based on policy reports and legal documents. The objective was to identify how the role of the national and local government had changed during the last decades. Therefore, we aimed to assess how minimal and higher ambitions were formulated, whether and how deviations were possible and to assess the allocation of responsibilities between various levels of authority.

While conducting our quick scan, we soon discovered all policy fields studied are organized in their own specific ways, often accompanied by a multitude of regulations, directives and case law. It means that large differences exist regarding the legal status of environmental standards, the possibilities to deviate from them and the available stimuli to 'go beyond the minimum'. Some policy fields even have no definitions of a minimal base quality (e.g. odor, energy), are ambiguous regarding this definition (e.g. noise, air pollution), while neither has clear stimuli to 'go beyond the minimum'. In addition, each policy field relies on a different allocation of responsibilities between levels of authority and between governmental and semi-governmental agencies. Clarity on the current role of national and local government was hardly to be found in the midst of this complex and fragmented body of policies.

During the early 1990s, Dutch environmental policies were considered fairly coherent, clear and easy to implement in all but really complex cases. Following the recent decentralization and associated deregulation operations, we now found a policy field that has become hard to oversee. The survey of 29 municipalities also revealed that many professionals no longer recognize a single environmental policy field. Rather, they consider environmental policies as a combination of sectoral and thematic policies, such as air, water, soil, noise, etc., each requiring highly specialist knowledge. Furthermore, respondents complained about incoherent and often swiftly and unexpectedly changing regulations. This has not been without consequences for implementing these policies. This was most clearly revealed during our second study; focusing on a series of nationwide and government issued studies on the performance of local authorities in environmental governance (also see Spreeuwers et al. 2008).

The first series of reports was produced by the Interprovincial Consultative Body (IPO), which is an agency developed by all twelve Dutch regional authorities (provinces) to communicate, collaborate and issue studies such as those discussed here. These reports (IPO et al. 2003) reveal that "… none of the competent authorities complies with all requirements, whilst about half the authorities has problems to meet even half these requirements" (Huberts and Verberk 2005, p. 54). Many municipalities did subsequently improve their performance, but the follow up study (IPO et al. 2005) shows that 2/3 of all municipalities still did not meet all legal obligations. Another series of reports produced by the Dutch Ministry of Housing,

Spatial Planning and the Environment (VROM Inspection 2006, 2007) confirms the picture, although taking a somewhat different approach. While again considering legal responsibilities, these reports also make a judgment on local performance. Taking into account locally specific circumstances and potential complex environmental problems, they are less harsh than the IPO reports. Still, the conclusion is that "... one on every four or five municipalities scores inadequate on surveillance and control in environmental management" (VROM Inspection 2006, p. 29). Hence, both the reports made by the IPO and the VROM Inspection indicate that quite a large number of Dutch municipalities fails to meet even legal requirements.

The third step in our study was assessing the degree in which municipalities 'go beyond the minimum'. To do so, we conducted interviews with almost 60 experts from professional environmental organizations that have a clear overview of the performance of Dutch municipalities in environmental management. We also collected surveys in 29 Dutch municipalities, which were chosen to present a cross section of Dutch municipalities. They varied in geographic location and population size and on the levels of performance in meeting minimal legal requirements based on the research reports studied. The sample is not a representative sample, but was meant to reflect on the other data sources and, mostly, to gain more understanding of the causes for success or failure to 'go beyond the minimum'. Finally, we also conducted eight case studies to gain a more 'in-depth' understanding of these causes. More details on the data collection, results and analysis, see Spreeuwers et al. (2008).

As the nationwide studies revealed that many municipalities struggle with meeting minimal requirements, we expected that 'going beyond the minimum' would not necessarily be a priority for many municipalities. This was clearly confirmed by the interviews conducted with Dutch environmental experts. From the 11 experts of the Environmental Inspection of the Ministry of Housing, Spatial Planning and the Environment, only four were moderately positive regarding municipalities' success in 'going beyond the minimum'. Instead, most recognize that environmental requirements are being considered as limiting conditions for development. Some argue how they are even considered as a 'hindrance' to municipalities, rather than as indicators of health and quality. This was confirmed by experts interviewed working for regional health departments. Out of the 13 respondents, 10 indicate that improving public health is hardly or not at all taken into account by municipalities in planning and policy making in their regions. They do have to comply with legal environmental standards for many environmental stressors that are installed for the protection of human health. These are however mere *minimal* requirements and do not consider the cumulative environmental load of several stressors. Meeting minimal requirements is therefore not supposed to guarantee 'good' quality, but rather to prevent excessive pollution on one specific stressor. Still, as the health experts note, "... most municipalities only focus on what is legally required" (Spreeuwers et al. 2008, p. 62). The thirteen experts of the non-governmental Provincial Environmental Federations and the Society for Nature and Environment finally confirmed the message. They were unanimous in concluding that "... by far most municipalities limit themselves to what is strictly necessary" (Spreeuwers et al. 2008, p. 64).

The surveys conducted subsequently allowed for municipal staff members to react themselves. It revealed that pursuing higher environmental ambitions is indeed far from evident in the survey municipalities. In 11 of the 29 municipalities, there was no intent at all to do more than meet environmental standards; whilst in 8 others explained they would ‘possibly do more than meeting minimal requirements’. Clearly, only a small minority of municipalities actually aimed for more ambitious policies on a more or less structural basis. In the meantime, decentralization has resulted in less ambitious and more flexible environmental standards. Especially in noise, odor and soil remediation policies there are now possibilities to deviate from the environmental standards that were in the 1990s still considered minimal requirements for protecting human health. As was illustrated by our case studies (Spreeuwers et al. 2008), municipalities are indeed using these possibilities. They are however not necessarily doing so in the face of highly complex local circumstances where meeting environmental standards would result in excessive financial, economic or social costs. They are also doing so to allow for additional development such as building houses in environmentally strained areas or constructing additional road infrastructure. It is an indication that decentralization in Dutch environmental policy has not just failed to push municipalities to pursue environmental ambitions that ‘go beyond the minimum’. It has even pushed levels of performance below levels previously considered necessary to protect human health. So instead of the intended transition in environmental management towards a higher local environmental quality, we are seeing a transition in the opposite direction (Fig. 4.1).

The fourth and final stage of our research was a study to explain the main consequences encountered. The interviews with experts, surveys collected and case studies were also instrumental to this study. It also allows us to use the Dutch experience to reflect on the possible benefits and risks of decentralization as discussed earlier.

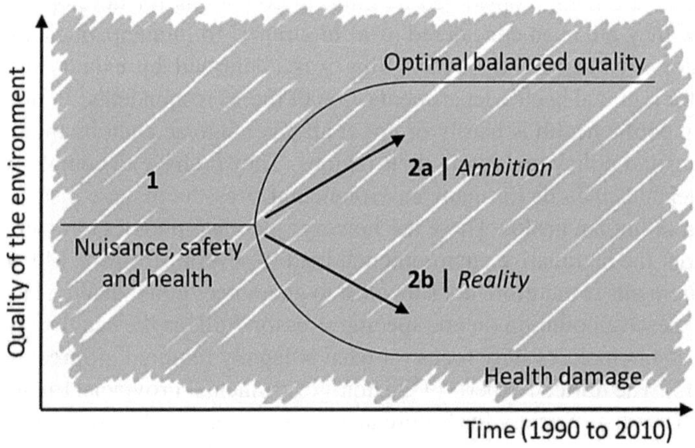

Fig. 4.1 Transitions in Dutch environmental management

4.7 Explaining Causes: A Reflection

We begin with reflecting on the argument that decentralization mostly makes sense when dealing with more complex issues. It is clear that dealing with such more complex issues was a clear motivation for decentralization in Dutch environmental policy. Decentralization, also accompanied by deregulation, has resulted in less ambitious and more flexible environmental standards and indeed allows local authorities to pursue their own strategies to deal with their local environmental challenges. In doing so, they have however had an important side-effect.

Before decentralization operations were instigated, most local environmental issues could be well dealt with by municipalities based on a routine implementation of the common policy formats (see also Borst et al. 1995; De Roo 2003). This has significantly changed. In the midst of the current complex and voluminous set of environmental policies and regulations, municipalities constantly have to choose their own strategy with regards to how deal with national policies and regulations. They not only have to interpret these policies; they also have to try and comply with the many rules regarding whether they are allowed to deviate from indicative environmental standards. Our research now shows that most municipalities are engaged in a constant struggle to understand and implement national policies and regulations. Many fail to do so, increasing the risk of exposure of humans and ecosystems to levels of environmental stress beyond what legal requirements would allow.

Clearly, the organization of national environmental policies puts serious pressure on local time and resources. The research reports studied also suggest that a lack of local resources contribute to problems to implement national policies and regulations. The IPO for example found that many municipalities do not meet criteria such as 'sufficient expertise', 'financial means to hire specialists' and 'systems for controlling on responsibilities' (IPO 2005; also see VROM Inspection 2007). This is also hardly a surprise, as many municipalities have to work with only two to five employees who have to cover all environmental policies, varying from soil to air quality and energy. It is not realistic to suppose each of these employees can be an expert on all themes. In the meantime, decentralization now even puts extra strains on local resources.

Decentralization means that municipalities are now responsible for developing their own environmental policies and to strategically position environmental interests in integrated local policies. Such strategic and cross-sectoral working requires competences such as visionary thinking, communicative skills and strategic planning. As the interviews and cases revealed, Dutch environmental personnel is however traditionally more technically oriented and, as one respondent convincingly explained, 'simply not educated' for their new responsibilities in environmental policy. In response, some municipalities divided municipal environmental departments into a division of specialists working on permits and control and a division of more strategic generalists. The idea is that strategic staff members can function as a bridge between the environmental specialists and spatial planners and urban designers. However, doing so is especially problematic for smaller departments and not common in most Dutch municipalities.

In addition, the difficulties to meet minimal requirements mean that pursuing voluntary integrated strategic environmental policies has little priority. One municipality in the case study was very clear about this: "ambitions and targets which go beyond the minimal legal requirements will only be pursued when, next to the resources and capacities invested in implementing legal regulations, any capacity remains" (Geldermalsen 2006, p. 5). In the meantime, there are also other important constraints on local ability. Background levels of pollution can prevent more ambitious policies in quite some urban municipalities, while lack of resources and time were also mentioned as important bottlenecks during the expert interviews and cases. As one case respondent noted, "…the thought that municipalities should produce so called 'tailor-made' outcomes is very nice and all, but it remains the question whether municipalities are actually up to it" (Spreeuwers et al. 2008, p. 93). In asking this respondent about his municipality the reaction was simple: 'we are not ready yet'.

Clearly, willingness to 'go beyond the minimum' becomes especially relevant when faced with limited time and resources and high levels of background pollution. During the interviews with experts and surveys, the importance of political support and administrative enthusiasm where addressed as essential conditions to 'go beyond the minimum'. Most respondents, however, noted that these conditions are exceptions in Dutch municipalities. A large majority of experts stated that "… the environment is not an important issue in most municipalities" (Zuidema 2011, p. 214), illustrating the 'weak profile' of environmental interests. The survey provided further confirmation of the relative 'weak profile' of environmental interests. To most respondents, the financial or spatial design principles are respected as leading criteria for plans and policies. The environment is instead often surrounded by a limited sense of urgency and awareness, undermining support for environmental ambitions other than within the legal framework. During the case studies this was also illustrated. Even when there are more ambitious environmental objectives, these are usually amongst the first to be dropped in the face of budgetary challenges or high demands for housing.

Finally then, we argued that national policies can be crucial conditions for creating sufficient levels of local willingness and ability. To help and persuade local authorities to pursue proactive and integrated policies, or to 'go beyond the minimum', the Dutch national government uses only policy recommendations and minor financial support as stimuli. There are no 'checks and balances' such as the need to account for local results, legal or contractual obligations or financial sanctions, which urge municipalities to do so. Those 'checks and balances' that are in place have to do with fulfilling minimum legal requirements (e.g. air quality, safety risks) or with deciding environmental standards locally (e.g. noise, odor). The result is that minimum legal requirements continue to dominate municipal environmental management. After all, as a respondent in Rheden stated, "…it is these criteria that the Ministry of VROM uses to judge you".

In the meantime, supportive and persuasive policies make only a limited impact in a local realm. Respondents indicate that many policy recommendations end up on a shelf or a desk, as there is simply no time and urgency to read them. The survey also confirmed their limited impact. During the survey questions were asked regarding the familiarity of respondents with some key concepts and instruments the national government uses. Extensively used in national policies and advices are the concepts 'optimal living quality' and 'quality of the living environment'. However, more than one third (16) of the 42 respondents indicated they did not know or recognize these concepts. In addition, these concepts proved hard to make operational by those who did know them. In asking respondents to explain what these concepts meant, a wide variety of answers was generated. For some respondents, environmental aspects were at the heart of these concepts, while others considered them not part of the concepts, apart from legal requirements. It shows that these concepts are not convincing in showing what 'going beyond the minimum' implies. The same story goes for the main instrument the national government designed to persuade and help municipalities to achieve the desired 'optimal living quality': MILO. Only half of the respondents were familiar with the MILO approach, while only 9 of the 28 municipalities surveyed actually used any of the national instruments that meant to help them achieve the desired 'optimal living quality', including MILO. Clearly, national policies to 'go beyond the minimum' are hardly reaching their target population, let alone that they have a strong impact in encouraging municipalities to do so.

4.8 Reflections and Conclusions

About 15 years ago Gershberg concluded that "despite the great attention paid to decentralization in the past two decades, we still know too little about the impact various decentralizing reforms have had on service outcomes in the social and urban sectors" (1998, p. 405). Looking back at the last decade in environmental management in the Netherlands, it is clear that these outcomes are not always as positive as expected at the beginning. The Dutch national government hoped to increase the development of dynamic policy approaches that were tailored to local circumstances though decentralization. Decentralization in Dutch environmental policies indeed did allow for more flexibility to develop local policy solutions, tailored to local circumstances. Many municipalities also used this flexibility to develop their own new plans and policies, seemingly suggesting that decentralization has been successful. The ambition was, however, that decentralization would result in improvements to the quality of the living environment (VROM 2001). That has, clearly, not occurred.

Decentralization was to help local authorities deal with a minority of highly complex circumstances, but also to encourage more ambitious environmental policies in the large majority of other cases. In practice, only a minority of municipalities aims for environmental quality ambitions that go 'beyond the minimum'. Many even aim for environmental quality ambitions that are below levels previously considered unacceptable, and not just when faced with highly complex local circumstances. Instead, they are also doing so to allow for additional development such as building houses in environmentally strained areas or constructing additional road infrastructure. While possibly attractive to many local stakeholders and local governments, this does strongly contrast with the ambitions of decentralization in Dutch environmental policy.

The results of this study teach us that decentralization should not be pursued without assessing whether those local authorities that have to deal with new responsibilities are up for it. We however argue that it is just as important to look at how national policies have enabled, stimulated and supported local performance. To begin with, decentralization and deregulation have caused an 'implosion' of national policies into a complex, fragmented, voluminous and sometimes conflicting body of policies. This gives municipalities a hard time getting them implemented. Ideally, meeting legal regulations in environmental management should be part of the normal routine of municipalities, where common national policy formats can be implemented routinely in all but the more extreme cases. Currently, implementing minimum quality levels proves to be a serious challenge for much Dutch municipalities, even in the face of fairly straightforward policy issues. Instead of supporting a routine implementation, the current organization of Dutch national environmental policies makes it hard to be 'able' to deal even with issues of limited complexity.

Secondly then, while municipal employees try to maneuver through the complexities of the environmental regulations, they are only modestly supported and stimulated by the national government to 'go beyond the minimum'. The national call to do so is communicated with policy advises and manuals. In the midst of their struggle to meet minimal requirements, many municipalities don't hear this call. Those that do are merely 'asked' to take on new responsibilities. Most municipalities consider this request as a merely additional and hardly interesting pursuit for which they have little time and resources. Faced with no rewards, they are also not stimulated. Consequently, success in going 'beyond the minimum' depends on highly skilled, enthusiastic and hard working professionals in combination with political will. And as political will proved to be limited and 'super employees' are an exception, so is success.

The Dutch case exemplifies that we should not deny the conditioning role of national policies for the outcomes of decentralization. We do accept that decentralization *can* help to accommodate the development of more dynamic policy approaches that are tailored to the local circumstances. As such, decentralization can indeed be a means of renewing governance so as to cope more effectively with the challenges with which our complex twenty-first century societies are confronting us with. But when we do so, we also need a robust basis upon which to build the kind of dynamic approaches that can also respond to these challenges. That is, we argue that the support and stimuli provided by central policies and regulations are a crucial precondition for decentralization to result in its envisioned outcomes. They provide the necessary guarantees for meeting minimum outcomes and stimulus for local parties in developing and delivering the desired flexible dynamic and tailor-made approaches. It is in doing so that these policies can help environmental policies in its pursuit of the desired 'livable' future for our spatial environment. It is a lesson the Dutch national government hopefully learned, at least if they hope to regain their traditional position as an international frontrunner in environmental management…

References

Andersen, M.S., and D. Liefferink. 1997. *The innovation of EU environmental policy*. Oslo: Scandinavian University Press.

Beck, U. 1992. *Risk society: Towards a new modernity*. New Delhi: Sage.

Bergström, O., and P. Dobers. 2000. Organising sustainble development: From diffusion to translation. *Sustainable Development* 8: 167–179.

Borst, H., G. de Roo, H. Voogd, and H. van der Werf. 1995. *Milieuzones in beweging*. Alphen aan den Rijn: Samsom H.D. Tjeenk Willink.

Burström, F., and I. Korhonen. 2001. Municipalities and industrial ecology: Reconsidering municipal environmental management. *Sustainable Development* 9: 36–46.

Castells, M. 2000. *The rise of the network society*. Oxford: Blackwell.

Coaffee, J., and N. Headlam. 2008. Pragmatic localism uncovered: Experiences from English urban policy reform. *Geoforum* 39: 1585–1599.

De Roo, G. 2003. *Environmental planning in the Netherlands: Too good to be true: From command-and-control planning towards shared governance*. Aldershot: Ashgate.

De Roo, G. 2004. *Toekomst van het Milieubeleid: Over de regels en het spel van decentralisatie – een bestuurskundige beschouwing*. Assen: Koninklijke Van Gorcum BV.

De Vries, M.S. 2000. The rise and fall of decentralization: A comparative analysis of arguments and practices in European countries. *European Journal of Political Research* 38(2): 193–224.

Eckersley, R. 1992. *Environmentalism and political theory*. London: UCL Press.

Fleurke, F., and R. Hulst. 2006. A contingency approach to decentralization. *Public Organization Review: A Global Journal* 6(1): 39–58.

Flynn, B. 2000. Is local truly better? Some reflections on sharing environmental policy between local governments and the EU. *European Environment* 10: 75–84.

Gemeente Geldermalsen. 2006. *Milieuprogramma 2007–2010*. Geldermalsen: Gemeente Geldermalsen.

Gershberg, A.I. 1998. Decentralisation, recentralisation and performance accountability: Building an operationally useful framework for analysis. *Development Policy Review* 16: 405–431.

Gibbs, D., and A.E.G. Jonas. 1999. Governance and regulation in local environmental policy: The utility of a regime approach. *Geoforum* 31: 299–313.

Harvey, D. 1989. *The condition of postmodernity: An enquiry into the origins of cultural change*. Cambridge: Blackwell.

Healey, P. 1997. *Collaborative planning: Shaping places in fragmented societies*. London: MacMillan Press.

Hooghe, L., and G. Marks. 2001. *Multilevel governance and European integration*. Lanham: Rowman and Littlefield.

Hovik, S., and M. Reitan. 2004. National environmental goals in search of local institutions. *Environment and Planning C: Government and Policy* 22: 687–699.

Huberts, L.W.J.C., and S. Verberk. 2005. *Overtredende overheden, Op zoek naar de omvang aan regelovertreding door overheden*. Den Haag: Boom Juridische Uitgevers.

IPO, VROM, VNG, UvW, and V&W. 2003. *Nulmeting Milieuhandhaving*. Den Haag: Ministerie van VROM.

IPO (Interprovinciaal Overlegorgaan). 2005. *Eindmeting Professionalisering Milieuhandhaving*. Gebaseerd op cijfers per 12 mei 2005 in samenwerking met Ministeries van VROM en V&W, de Unie van Waterschappen en de Vereniging Nederlandse Gemeenten. Den haag: Ministerie van Vrom.

Jessop, B. 1994. Post-fordism and the state. In *Post-Fordism: A reader*, ed. A. Amin. Oxford: Blackwell.

Jordan, A.J. 1999. *Subsidiarity and environmental policy: Which level of government should do what in the European Union?* CSERGE working paper. Norwich: Centre for Social and Economic Research on the Global Environment.

Kooiman, J. (ed.). 1993. *Modern governance: New government-society interactions*. London: Sage.

Lemos, M.C., and A. Agrawal. 2006. Environmental governance. *Annual Review of Environment and Natural Resources* 31(3): 297–325.

Liefferink, D., and M. Van der Zouwen. 2004. The Netherlands, the advantages of being 'Mr Average'. In *Environmental policy in Europe: The Europeanization of national environmental policy*, ed. A. Jordan and D. Liefferink. London: Routledge.
Liefferink, D., M.A. Hajer, and J.P.M. van Tatenhove. 2002. The dynamics of European nature policy: The interplay of front stage and back stage by 2030. In *Bestuurlijke trends en het natuurbeleid*, Planbureaustudies, no. 3, ed. W. Kuindersma. Wageningen: Natuurplanbureau.
Martens, K. 2007. Actors in a fuzzy governance environment. In *Fuzzy planning: The role of actors in a fuzzy governance environment*, ed. G. de Roo and G. Porter. Aldershot: Ashgate.
Pierre, J., and B.G. Peters. 2000. *Governance, politics and the state*. London: McMillan.
Prud'homme, R. 1994. *On the dangers of decentralization*, Policy research working paper series, TWUTD, no. 31005, February. Washington, DC: World Bank.
Rydin, Y. 1998. The enabling local state and urban development: Resources, rhetoric and planning in East London. *Urban Studies* 35: 175–191.
Sassen, S. (ed.). 2002. *Global networks, linked cities*. New York: Routledge.
Spreeuwers, W.J., C. Zuidema, and G. de Roo. 2008. *De Basiskwaliteit voorbij...: een zoektocht naar milieukwaliteit op lokaal niveau*. Groningen: URSI.
Stoker, G., and K. Mossberger. 2001. The evolution of urban regime theory: The challenge of conceptualization. *Urban Affairs Review* 36(6): 810–835.
TK (Tweede Kamer). 1989. *Nationaal Milieubeleidsplan 1989–1993, Kiezen of verliezen*. Den Haag: SdUUitgevers.
Van Tatenhove, J., B. Arts, and P. Leroy (eds.). 2000. *Political modernisation and the environment: The renewal of environmental policy arrangements*. Dordrecht: Kluwer Academic.
VI (VROM-inspectie). 2006. *Landelijke Rapportage VROM-brede gemeenteonderzoeken 2005*. Den Haag: Ministerie van VROM.
VI (VROM-inspectie). 2007. *Landelijke Rapportage VROM-brede gemeenteonderzoeken 2006*. Den Haag: Ministerie van VROM.
VROM (Ministerie van Volkshuisvesting, Ruimtelijke Ordening en Milieubeheer). 2001. *Nationaal Milieubeleidsplan 4, Een wereld en een wil: werken aan Duurzaamheid*. Den Haag: Ministerie van VROM.
Walberg, H.J., S.J. Paik, A. Komukai, and K. Freeman. 2000. Decentralization: An international perspective. *Educational Horizons* 78(3): 153–164.
Wätli, S. 2004. How multilevel structures affect environmental policy in industrialized countries. *European Journal of Political Research* 43: 599–634.
Weale, A. 1992. *The new politics of pollution*. Manchester: Manchester University Press.
Zuidema, C. 2011. *Stimulating local environmental policy: Making sense of decentralization in environmental governance*. Zutphen: Wöhrmann.

Christian Zuidema is Assistant Professor in Spatial Planning at the Department of Spatial Planning and Environment, Faculty of Spatial Sciences, University of Groningen. Zuidema has done extensive research on topics such as sustainable urban management, sustainable area developments and governance renewal in environmental planning. Following his research, he has published various books, such as *Stimulating Local Environmental Policy* (2011), *Towards Liveable Cities in Europe* (forthcoming) and, as co-author, *Smart Methods for Environmental Externalities* (2012). Furthermore, he is board-member of the International Urban Planning and Environment Association.

Gert de Roo is Full Professor in Spatial Planning and heads the Department of Spatial Planning and Environment, Faculty of Spatial Sciences, University of Groningen. He is President of the Association of European Schools of Planning for the period 2011–2015 and is editor of the Ashgate Publishing Series on Urban Planning and Environment and on Planning Theory. De Roo edited and published various books on environmental planning, including *Urban environmental planning: Policies, instruments and methods in an international perspective (2005), Environmental planning in the Netherlands; Too good to be true (2010), and Smart methods for environmental externalities* (2012).

Chapter 5
Local Governance and Soft Infrastructure for Sustainability and Resilience

Bob Evans and Marg O'Brien

5.1 Introduction

There is widespread agreement that there is a global resource and environmental crisis. This has been variously expressed in terms of the limits to growth; the impact of climate change and global warming; resource depletion and 'peak oil'; ecosystem damage; oceanic pollution; the threat to food production in the context of both a rising global population and increasing consumer demands; and so on.

The response at local, national and international levels to these issues has been mixed. Whilst there have been a number of high profile attempts to both secure international commitment to climate change adaptation and mitigation and to promote more sustainable styles of living, it is apparent that the 'business as usual' pattern of economic growth still dominates any attempts to change to more eco-efficient and low carbon ways of life.

It is often argued that one key reason underpinning the reluctance of the governments of the more prosperous and high consuming nations to challenge the 'business as usual' growth model is that there is a well-entrenched public unwillingness to see such alternative policies as realistic or necessary. As a consequence, politicians are unlikely, it is argued, to promote policies which are perceived as being counter to the self-interest of the electorates of the world's most prosperous nations.

In this chapter we explore the possibility that public attitudes, values and behaviour are amenable to influence and possible change through a process of nurturing what we refer to as soft infrastructure. Moreover, as Berkes (2009) has argued, the

B. Evans (Retired)
School of Social Science, University of Waikato,
Private Bag 3105, Hamilton 3240, New Zealand
e-mail: bob.evans100@gmail.com

M. O'Brien (✉)
Coastal & Freshwater Group, Cawthron Institute, Private Bag 2, Nelson 7042, New Zealand
e-mail: margobrien@clear.net.nz

U. Fra.Paleo (ed.), *Risk Governance*, DOI 10.1007/978-94-017-9328-5_5

issues around resource use are becoming too complex to be governed effectively by a single agency. Increased complexity and rapid change has created a heightened awareness amongst agencies of the limits of hierarchical control and of formal/scientific expertise as decision makers find that they are failing to solve problems unilaterally. In short, the challenges facing governments – central and local – are so great that they are unlikely to be adequately addressed by governments acting alone. The knowledge, skills, resources and ingenuity of civil society will need to be an increasingly important element of the policy process.

Drawing upon research conducted in both New Zealand and Europe, we examine how recent work on governance, civic engagement and sustainability might assist local governments wishing to promote more sustainable patterns of living within their communities through greater understanding of the mechanisms and processes of soft infrastructure. Equally, such processes can serve to enable local communities to change entrenched attitudes within governments and associated agencies. We also reflect upon how insights from social psychology might further assist in understanding how individuals' self-belief can play a part in behaviour change through processes of social learning.

The central theme of this chapter is that soft infrastructure – the institutions, networks, relationships and social and psychological processes which foster civic engagement in public life – is inevitably emerging as a key part of public policy making and implementation as societies grapple with the complex challenges of risk and uncertainty. Furthermore, we reflect on the processes of collaborative governance as a device for promoting effective civic engagement in environmental management and policy making and, as a part of this, the role of governance in nurturing behavioral change supportive of sustainability.

5.2 Sustainability, Resilience and Behaviour Change

During the years since the publication of the *Brundtland Report* in 1987, sustainable development, or some variant of this concept, has become a generally accepted policy principle at many levels of government worldwide, albeit often honored more in rhetoric than in the reality of action. In major part, this is a response to the substantial evidence indicating that there are clear environmental limits to economic growth, but it is also a response to the projected levels of material and resource consumption expected in countries such as China, India and Brazil which are anticipated to be placing unsustainable demands upon the earth's resources within the near future, compounding the impact that the prosperous countries of the 'North' have historically placed upon global resources.

More recently, the policy discourse around sustainability has been augmented – some would argue replaced – by the concept of resilience. Whilst sustainable development has global ecological constraints and poverty as central themes, resilience prioritises self-interest and survivability over altruism (Adger 2000; Dale and Newman 2006). Although both the sustainability and resilience discourses are fundamentally concerned with resource depletion and the limits to growth, resilience, with its implied emphasis upon self-interest, may resonate more easily with publics in the more prosperous nations in times of austerity and high energy and food prices. It is argued that the focus within resilience upon *transition* as a strategy – exemplified through the global Transition Towns movement and energy descent plans – is more tangible and politically achievable than the call to simply be more sustainable (Hopkins 2008). Such movements represent an important shift from 'out of our control' attributions to 'under our control' attributions. The sense of self-efficacy (to which we return below) becomes one of 'we can choose to manage our own resources and futures'.

In response to both the need to implement international agreements, such as those on climate change, and the wider concern with issues such as energy security, governments of some of the world's prosperous nations, such as New Zealand or those of Europe are increasingly committing to policies which encourage lower energy and resource usage; promote climate change mitigation and adaptation; and protect environmental and ecological resources in various ways. Such policies are often poorly understood by the general public, and will frequently be in conflict with individual immediate self-interest. As a consequence, there is an understandable reluctance on the part of politicians and policy makers to pursue such policies in a full-blooded or radical manner.

Policy instruments designed to secure behaviour change, such as taxation or regulation are likely to meet with resistance from both organized interests and the citizen body unless such instruments are linked to a process of widening public understanding and engagement with the issues. Moreover, such behaviour change is more possible when personal beliefs coincide with self-interest, for example when householders see that there are real economic benefits arising from energy saving insulation. However, local communities may also be driven by a wider sense of collective interest. It is this knowledge and commitment which can contribute to, or even drive a process of collaboration between governmental agencies and local communities.

It is important to differentiate the concept of civic engagement from that of public participation or consultation. Whilst public 'participation' (and the elusive 'empowerment' of Arnstein's (1969) much quoted *Ladder of Participation*) is widely touted as a desirable element of public policy making, most evaluations of the process conclude that it is, at best, a form of consultation and at worst, window dressing and legitimation for decisions that have already been taken. Civic engagement, on the other hand, should be understood as involving elements of participation and consultation, but it is more than this in that the concept implies a process of social learning, the co-generation of knowledge, and an on-going civil society–government partnership of dialogue.

5.3 Soft Infrastructure

The concept of soft infrastructure has no universally agreed definition. As a general definition it is often taken to refer to all the institutions which are required to maintain the health, cultural and social standards of a country, state or sometimes even a company. In other words, it is everything that 'hard infrastructure' is not. In a societal context the concept might therefore include public education, health systems, libraries and social welfare. In contrast, within the computer industry, the term relates to networking, communications technology, and other internet-enabled systems.

We have chosen to use the term 'soft infrastructure for sustainable development' to refer to those institutions, networks, and social and psychological processes which foster civic engagement in public life which in turn helps nurture a resilient and sustainable society. We see this concept as a heuristic device to enable us to examine the potential for behaviour change at the local level.

We wish to argue that soft infrastructure may be understood as comprising four key variables which are likely to determine levels of local civic engagement.[1] There is, of course considerable overlap between these four variables and, as will be seen below, it is the interaction between these variables which together contribute to greater levels of social learning and heightened capacity to secure more resilient and sustainable outcomes. In summary, these variables are:

- The levels of **institutional capital** in local government;
- The levels of local **social capital** expressed as 'civic tone' or awareness via the activities of community organizations;
- The existence of **learning processes** that nurture self- and collective-efficacy or nurture the individual and group transformation that are fundamental to institutional and social capacity building;
- The existence of processes of **governance** that can through learning build on, yet be determined, by levels of both institutional and social capital.

By using the term **institutional capital** we are referring to the internal patterns of behaviour and ways of working, as well as the collective values, knowledge and relationships that exist within any organized group in society. It incorporates formal, institutionalised procedures and relationships as well as more informal networks. Institutional capital can be created, nurtured, developed or, conversely, eroded and dismantled. In the particular case of local government, research suggests that institutional capital may be built or increased as a consequence of relationships established with civil society actors. It is this process of **governance**, defined as the sphere of public debate, partnership, dialogue, collaboration and conflict entered into by local citizens and organizations and local government – which enables a dialogue to be established between civil society actors and governments and in turn contributes to '*institutional capacity*'– that we will return to later.

Social capital (a term which we use interchangeably with '*social capacity*') is an over-exposed concept which nevertheless has considerable significance for this discussion. The term has secured wide usage in social science since being popularized in the 1980s by Bourdieu (1986), Coleman (1988) and others and then further developed by Putnam (2000). As a broad generalisation, social capital may be understood

[1] Although it should be emphasised that these may in turn be conditioned by the social context of class, ethnicity and life-cycle stage.

as incorporating the level of trust; extent of networks; density of relationships; knowledge of relationships; obligations and reciprocity; forms of local knowledge; operating norms and sanctions to punish free riders present in any particular locality (Rydin and Pennington 2000). Collier (1998) differentiates between government social capital (e.g. enforceability of societal contracts, rule of law, and the extent of civil liberties) similar to our institutional capital, and civil social capital (e.g. common values, shared traditions, norms, informal networks and associational membership). In societies where government social capital is limited, a large proportion of contracts may depend on civil social capital and trust. Higher levels of social capital imply social stability and social cohesion, lower levels, the reverse. Putnam has argued that the existence of social capital is closely tied to the effective functioning of a democracy, although there is no clear association between, for example, levels of social capital and traditional political participation. Moreover, although social movements and social capital are closely connected, it is not clear which of these is cause and which is effect.

Earlier, Putnam et al. (1993) also argued that social capital in Italy is a legacy of long periods of historical development, and therefore it cannot be added to in the short-run, but this prognosis has been challenged by a number of empirical studies (e.g. Schneider et al. 1997) that indicate that it is possible to create social capital, although the process may be incremental. For instance, Falk and Kilpatrick (2000) argue that social capital can be created and the accumulation is the outcome of the process of learning interactions, and these we anticipate, would impact on the cognitive social capital – the norms, values, attitudes and beliefs (Uphoff 1999) amenable to change. To date, the literature on social capital has tended to focus on the collective rather than the individual. However, we suggest that insights from social psychology are also helpful here in explaining how change and social learning might occur through self-efficacy and the learning process – our third variable – to which we now turn.

5.4 Self-Efficacy and the Learning Process

At an individual level one of the most potent ingredients for behavioural change is that of self-efficacy, a belief in one's own effectiveness to create change (Bandura 1977). As Bandura argues; "Self-efficacy beliefs contribute to motivation in several ways. They determine the goals people set for themselves, how much effort they expend, how long they persevere in the face of difficulties, and their resilience to failures." (1993, p. 117) A key point is that self-efficacy, the belief that is so fundamental to change, while shaped by our ability to observe and imitate models and through our susceptibility to verbal persuasion is primarily shaped by mastery – through positive success experiences or the learning-by-doing process. Success here is not only the sense of control gained by the accomplishment or 'doing' of a task but also in the reflection on the 'doing' and the rich relationships that develop from sharing the experience with others (O'Brien 1990, 1995). Success in fact can come through multiple avenues – learning that one has choices, sharing concerns with others, tackling a problem successfully or understanding that one could do that in the future.

Beliefs about self-efficacy can influence not only individual but also collective capacity, "...People change their lives for the better not only through self-development but by acting together to alter adverse institutional practices. If the practices of social systems impede or undermine the personal development of some sectors of society, then a large part of the solution lies in changing the adverse practices of social systems through the exercise of collective efficacy. To shape their social future, people must believe themselves capable of accomplishing significant social change..." (Bandura 1997, p. 33). What we find is that in trying to change individual or collective efficacy – in changing people's beliefs – we are also changing the way they think, their attitudes, their mind-sets. The precursor to developing the capacity to progress change is changing the way people think so it becomes important to understand more about the learning process so as to facilitate and maximize the potential for these changes to occur.

The centrality of learning from experience like this is a core concept of educational theorists, going back to Dewey (1938), Hahn (1949), Freire (1972), Kolb (1984) and more recently Orr (1992). All have been explicit in their encouragement of success experience in the form of action and reflection upon that action – a change of behaviour occurs as a result of the process – the medium is the message.

But can it be the right message? Can it be that those who experience mastery through an experiential process are more ecologically minded and more caring of the environment? In 1978 Hardin distinguished between numeracy, literacy and ecolacy, a third level of education in which a person develops a critical and more accurate understanding of the complexity of the world. Such a person would be able to think broadly (*reflecting*) and clearly regarding the consequences of his/her behaviour (*actions*). They would no longer seek to understand the world by dissecting it bit by bit but rather consider how the parts interact to sustain life. Ecolacy would reflect this change in perspective – to develop the 'and then what?' frame of mind.

One could say of course that this 'habit of mind' is both the means and the outcome of any informed or deliberative community, and yet, a similar proposition has been put forward quite independently by one of the key scholars of environmental education in the UK with *sustainability being seen as a social learning process* (Sterling 2007, p. 63) and later, "...the process of sustainable development is seen as essentially one of systemic learning and change. At the same time, the context of learning is seen as sustainability, with emphasis on developing resilience, integrity, capacity in individuals, groups, communities, organizations and human systems." (Sterling 2004, p. 70).

Again, change is seen as a result of an experiential process – the medium once again is the message. As Guijt (2010) adds: "Learning entails not just pragmatic problem-solving but also reflection on the process by which this happens and the underlying perspective on knowledge. Seen like this, learning requires capacities for critical reflection, identifying assumptions, seeking evidence about what is going well or not, analysing multiple lines of evidence, relating evidence to expectations, and analysing and negotiating possible consequences. These processes all require connecting people and their perspectives." (p. 282).

While it is by no means certain or guaranteed, social learning *can* inherently progress sustainability. Additionally, the intra-disciplinary spin-offs from successful involvement in a learning culture – of trust, efficacy (both at an individual and collective, community level), social and ecological resilience (Tompkins and Adger 2004), community development, social cohesion and social capital (O'Brien 1997) are themselves profound. In fact, there is evidence that a generalized belief in one's self-efficacy may be central to psychological resilience (Lightsey 2006) and as "… an emergent cultural property in social systems analogous to ecological resilience in ecological systems". (Doubleday 2008, p. 242)

But the crucial nature of the learning process to the environmental resource crises that we now face is only now being understood as scholar-practitioners, who implicitly are further exploring the nature of Hardin's ecolacy, now grapple with the facilitation of transformational learning and the impact that this can have on environmental management (Pahl-Wostl 2006, 2009; Pahl-Wostl et al. 2007; Berkes 2009; Reed et al. 2010).

From a learning perspective, change can be driven by the need for practical improvements, strategic adjustments and changes, and as a result of rethinking core driving values. These different ways of learning are referred to as single-, double- and triple-loop learning, respectively. Single loop learning impacts on the behaviour or actions we choose to take to improve a situation – to 'do things right'. Double-loop and triple-loop learning, on the other hand, take a 'higher order' or 'meta-perspective': Double-loop has us reframing the problem – reflecting on our assumptions (mind sets or thinking patterns) to ask 'Are we doing the right things?', while triple-loop learning 'bumps it up another level' to ask 'Why do we have this problem?' or 'Are we dealing with the right problem?' which forces us to take a transformative perspective, reflecting on the beliefs and values that trigger our attitudes in the first place (see Guijt 2010; Dyball et al. 2007; Keen et al. 2005; Senge 1990; Keen and Mahanty 2006).

A well facilitated learning process can help people shift into a transformative, or meta perspective. As Fernandez-Gimenez et al. (2008) comments, communities can learn to anticipate and adapt to changing conditions, and better appreciate the complexity of linked social and ecological systems (Walker et al. 2002; Walker and Salt 2006). Within the reflective process, there can be doubt, hesitation, and perplexity – but rather than threatening self-efficacy learners are challenged to generate creative alternatives that preserve self-efficacy (Marquardt and Waddell 2006). Every shift can bring an 'a-ha' success experience, another 'small win' (Weick 1986). Shared understanding becomes then part of a collaborative learning process (Daniels and Walker 2001) – a soft infrastructure that facilitates deeper trust and commitment, nurtures collective efficacy and builds on social capital. It is this learning that is central to the governance processes that can develop when local citizens and government collaborate. The greater the success experienced the more the collective efficacy, or capacity to create change is nurtured and the greater the development of social and institutional capacity.

Our discussion so far has sought to establish the interrelationship between our first three variables of soft infrastructure – institutional capital, social capital and the learning process. We now examine two examples of soft infrastructure in order to further explore the nature of governance, our fourth variable, and the ways in which civil society actors and local government agencies might interact and cooperate for mutual benefit. We firstly examine the ways in which the governance process has unfolded in European cities as part of the process of developing policies to secure more sustainable development. We then move on to explore in more detail the processes of collaborative governance, drawing upon the example of Fiordland in New Zealand.

5.5 Governance Through Local Government and Civic Engagement

A recent study of local sustainable development in 40 European towns and cities concluded that sustainable development policy achievement was most likely when high levels of local government institutional capital coincided with high levels of local social capital, linked together with a process of governance (Evans et al. 2005). A key conclusion of this research was that effective *governing* for sustainability was achieved when 'good' *government* (i.e. high levels of institutional capital) promoted effective processes of *governing*, or governance. It should also be emphasised that this study concluded that the role of local government was central in the localities studied, as initiator and provider of support in the governance process. Moreover, as both Lowndes and Wilson (2002) and Khakee (2002) have argued, this research confirmed that the process of local government engagement with local civil society in turn helps to build and reinforce institutional capital.

This study concluded that, in the case of sustainable development policy, the *intensity* of tangible policy achievement is almost always linked to a high level of dialogue between local government and civil society (Evans et al. 2006). In this sense the process of *governance* is central to sustainable development policy achievement (see Fig. 5.1). It should be noted that governance as defined here, should be understood as an all-embracing concept encompassing all forms of government-civil society interaction including the more formally constituted collaborative style governance which we consider further below.

Utilising the Weberian conception of ideal types, this research concluded that it is possible to identify four 'ideal types' of possible outcome in terms of sustainable development policy achievement dependent upon the relative contributions of social and institutional capital. These are brought together through a process of governance (see Fig. 5.2), reinforcing the crucial importance of engaging citizens, interest organizations and other stakeholders in the sustainable development policy process. This implicitly supports the notion that the learning process fundamental to this engagement can nurture further capacity building and sustainability achievements.

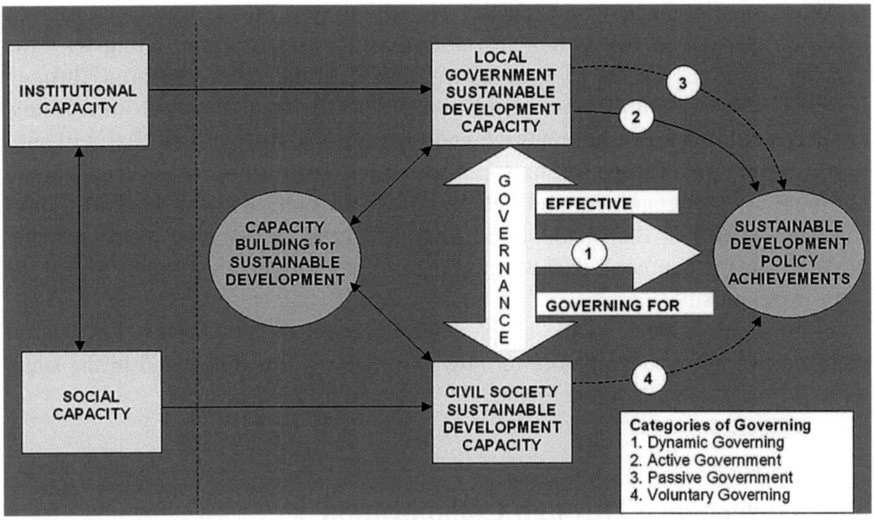

Fig. 5.1 Dynamic institutional capacity and sustainable development policy achievement (Source: Evans et al. 2005)

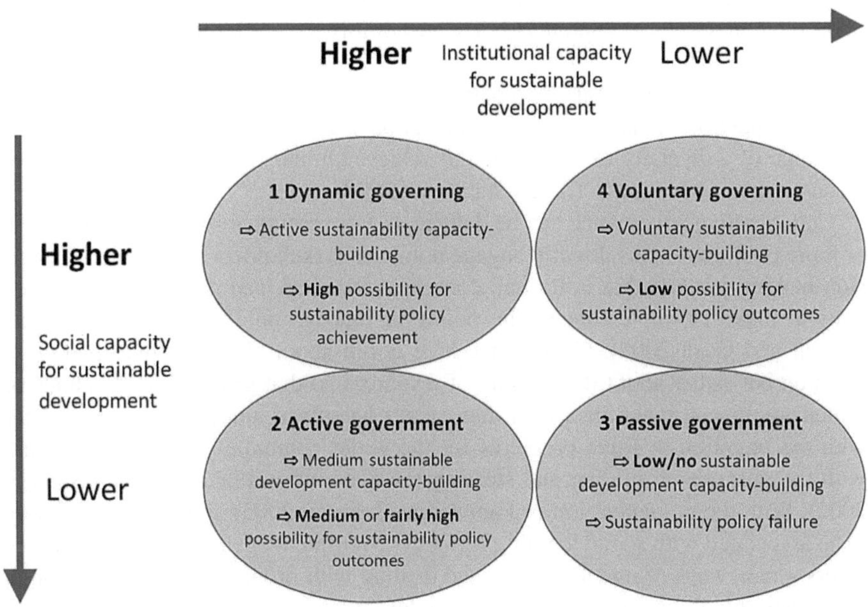

Fig. 5.2 The relationship between social and institutional capacity, capacity-building measures and sustainable development policy outcomes (Source: Evans et al. 2005)

Thus, to refer back to our earlier discussion, the processes of governance may be regarded as one crucial variable of soft infrastructure, along with the varying amounts of social and institutional capital, that nurtured through learning, determine its formative existence. The key issue becomes one of how to successfully achieve both a civic and institutional engagement that will support dynamic governing. But how can we create institutions of governance that facilitate civic engagement and better address the complex problems of today? The answer lies in the introduction of transformative learning processes into governance mechanisms and we propose collaborative governance to be one such 'mechanism'.

In the final sections of the chapter, we focus on the conditions that support collaborative governance before turning to the example of Fiordland in the south west of New Zealand's South Island.

5.6 Governance Through Collaboration

National debates on how we should deal with climate change, energy security and many other hazards (or resource risks) that threaten our ecosystems often overlook the newer forms of governance that have emerged over the last decades, in favour of traditional, hierarchical approaches. The newer forms we refer to place importance on a more collaborative approach – bringing "…public and private stakeholders together in collective forums with public agencies to engage in consensus oriented decision making … harnessing the innovation that can be achieved through the exchange of a diversity of viewpoints to achieve climate, energy security or other resource risk innovations." (Ansell and Gash 2008)

Collaborative governance can be defined as a governing arrangement where one or more public agencies directly engage non-state stakeholders in a collective decision making process that is formal, consensus oriented, and deliberative and that aims to make or implement public policy or manage public programs or assets (Ansell and Gash 2008). Governance here is not about one individual making a decision but rather about the pooling of resources so that groups of individuals or organizations can make decisions that cannot be solved alone – coming together with the intention to solve problems for the wider community (Zadek 2008) and with the purpose of guiding and steering the community (Takahashi and Smutny 2002). With the co-generation of knowledge (Johnsen 2005), this process is likely to be associated with social learning for both government agencies and civil society actors – new ways of understanding and dealing with problems generated through the harnessing of insights and local knowledges previously unavailable that work to develop both social and institutional capacity.

In fact Sørensen and Torfing (2007, pp. 3–4) assert that policy as an "…attempt to achieve a desired outcome, is a result of governing processes that are no longer fully controlled by government, but subject to negotiations between a wide range of public, semi-public and private actors, whose interactions give rise to a relatively stable pattern of policy making that constitutes a specific form of regulation, or mode of coordination." Innes and Booher (2010) herald this collaborative approach as governance for resilience.

Research on the success of this form of collaboration indicates that there are a number of commitments to be made by those involved in the process[2] – including a commitment to:

1. Diversity: The inclusion and full participation of all relevant stakeholders is required for coherent and innovative patterns of action to emerge. This means that the collaborative process includes not only those who have power because they are 'deal makers' or 'deal breakers' but also those who could be affected by outcomes of the process – so that representation is from across the community.
2. Interdependence: The parties involved in the process are committed to participate positively – with each participant having something that the others want. This ensures that participants maintain a level of interest and are energized to engage with one another and push for agreement and mutual gain. Mutual gain can involve a push towards a common vision or superordinate goal, ensuring that traditionally conflicting groups work together and towards a common end.
3. Authentic face-to-face dialogue: This requires that
   - Participants are involved in a deliberative process characterized by direct engagement so that the parties can test to be sure that claims are accurate, comprehensible and sincere.
   - Deliberations cannot be dominated by those with power outside the process.
   - Everyone involved must have equal access to all the relevant information and an equal ability to speak and be listened to, i.e. all kinds of people have a real voice and listening becomes as important as speaking!
   - Knowledge is co-produced through interaction and shared inquiry – it relies on what participants know from their everyday lives and not just on specialized, scientific expertise.
   - Participants have time to explore a range of views about the nature of an issue and have the opportunities to sort out what is going on around them – crucial to developing a sense of ownership of the issue at stake.
4. Facilitative leadership: This brings the parties together and assists them to overcome pre-existing conflicts, builds trust and creates a foundation for working relationships. This is crucial if participants are to examine publicly the attitudes – values, beliefs and assumptions – underlying their own views and the views of others.
5. Creation of an ongoing transformational learning culture: As participants are engaged in authentic dialogue they ask questions, listen, interact and learn about one another – assumptions are tested, thinking patterns are challenged and new ways to move forward are progressed.

---

[2] This section on commitments is informed by, and includes quotes from, the work of Innes and Booher (2010), McCoy and Scully (2002) and Baines and O'Brien (2012).

6. Belief in the collaborative process: Change starts with participants who have the capacity to think not only about alternative ways of working but also to imagine their successful implementation – participants must believe themselves capable of accomplishing significant social change (Bandura 1997). Both individual and collective efficacy expectations are vital here.

Possibly the most stringent requirement however, is that in the development of strategies to nurture resilience and progress sustainability, the collaboration outperforms feasible alternative arrangements in the creation of value for the community as a whole (Zadek 2008).

The collaborative process is not only about finding new ways forward for mutual gain – it is also about building community and institutional or agency capacity to be resilient in the face of inevitable new challenges. The anticipated outcomes of the collaborative process would include the development of future proofing strategies. We can anticipate this will mean:

- The development of long-lasting solutions with solid commitment to them from the participant stakeholders.
- Transaction cost savings as agreement is reached in advance between the interested parties reducing the need for litigation.
- The development of change generating solutions that cannot be easily ignored by government – the power of stakeholder consensus providing for a powerful lobby group.
- The identification of shared values, trust and even the expectation of long-term friendships among stakeholders traditionally in conflict.
- The ability, through the new found trust, to overcome future conflicts that enable parties to work together to find equitable solutions with mutual gain.
- Regional development achieved that will help prepare the community for change – for new ways of working – when ecological, social or economic conditions make the existing system untenable.

It is now of interest to consider a case where efforts to establish a collaborative governance process have been ongoing for almost two decades. While the developments have not been formally researched within the parameters we have discussed, it is possible from discussion with stakeholders and government agencies, and access to legislation and administrative records to determine progress.

5.7 Collaborative Governance in Fiordland, New Zealand

The Fiordland region can be found in the South-West corner of New Zealand. A spectacular landscape – it is filled with the dramatic peaks, sheer rock faces, thundering waterfalls and stunning reflections of tranquil fiord waters – matched underwater by the equally unique environment of outstanding natural values. The fiords have become national icons and annually attract large numbers of tourists - Milford alone attracts over half a million tourists per year while also supporting important commercial and recreational fisheries, notably rock lobster and blue cod (FMM Act 2005).

But this unique marine region is at risk primarily from human-mediated movement – recreational and charter boats, commercial fishing vessels and cruise ships – that have the capacity to transport invasive marine species to the region. It is not surprising then that people within, or associated with the region have been keen to work towards and ensure its ongoing protection. For not only is there value derived from the region from non-commercial recreational activities such as boating, diving, and aesthetic enjoyment of Fiordland's natural values, but also, indirectly through its existence value – by simply knowing that the fiords exist.

It is now over two decades since stakeholders in the region lobbied both local and central government to take actions to ensure the protection of the marine region. In 1995, they formed the Guardians of Fiordland – the formation being a direct response to the perceived risks of adverse impacts on the Fiordland area resulting from increasing human use, the devastation it could have on natural values, the urgent need for improved and integrated management of the area, and a strong conviction that the community be more involved in the risk management of Fiordland's marine environment. Membership included a diversity of stakeholders: commercial and recreational fishers, environmentalists, charter boat and tourism operators, scientists, and Maori – the Tangatawhenua, the people of the land – in this case from the Ngāi Tahu iwi.[3]

At their first meeting (September 1995), "…introductions were accompanied by suspicious glances from members who felt bound to protect their particular interests in Fiordland. But when each participant was asked to share their vision for Fiordland's fisheries for the next 10–20 years the suspicion turned to surprise as one by one the entire group expressed the same desire."[4] This was the first step in building an authentic and trusting engagement. Consensus was reached over shared values and the Guardians adopted a super-ordinate goal – a vision that was to be the basis of change generating solutions. This was: "That the quality of Fiordland's marine environment and fisheries, including the wider fishery experience, be maintained or improved for future generations to use and enjoy."–a vision that supported the development of an integrated management strategy for Fiordland's marine environment. Further, developing their sense of shared understanding was the single most important activity, and examples included:

- The map – a short, relationship building exercise that tapped into the combined 250 years of Guardian knowledge by identifying how each fished and otherwise used the fiords. This not only revealed patterns and trends but also made it very difficult for participants to hold back, particularly if 'their patch' was being discussed.

[3] The tribal group most prominent in the South Island.

[4] Personal communication (notes) from L. Teirney, past Fiordland Marine Guardian.

- A survey of the 'old codgers' whose families had been in the region since the 1800s provided a fascinating history and the context within which the development of current fisheries could be understood.
- A survey of 'current codgers' who were perceived to understand the 1990s conditions of individual fiords that then became the basis of the fisheries framework and management recommendations developed in the strategy – an important step that shifted the focus of participants away from blame-games to a focus on what still had to be achieved collectively (Teirney 2006, p. 3).

Initially, there was a fisheries focus, but with funding for the strategy development eventually coming from a sustainable management group, the emphasis changed to take a more holistic approach. This meant that in 2000, wider agency participation became important and officers from conservation, fisheries and local government, among others, found themselves not only working with ease across boundaries outside their usual legislative constraints, but also in the position of accessing vast amounts of information that the Guardians had accumulated in their first years working together. As one of the key facilitator's noted at the time: "My role has been to guide an enthusiastic group, endowed with first-hand knowledge, experience and a passion for the Fiordland marine environment through a process that would deliver a strategy acceptable to both central and local government. During the journey the Guardians have grown, acquired new skills, refined their debating and negotiating abilities. The end result is an effective unified group being recognised as a successful model for community initiatives."[5]

From our earlier discussion, we could say that this unique collaboration was an example of high social capacity for sustainable development. In this case, social capacity was facilitated by motivated individuals from within the community – the Guardians worked as transformational leaders whose sense of belief in change was high and whose attitude and values impacted on the general efficacy of stakeholder groups and organizations. A collective belief developed: change was needed and could also be created if the sustainability of the region was to be ensured. The vision guided the Guardians in developing a Fiordland Marine Conservation Strategy which was published in September 2003.

A major success for the Guardians was gaining wider stakeholder agreement to proposals for the integrated management of the Fiordland marine environment. Generally, this involved each stakeholder group relinquishing benefits in the interests of ensuring the quality and sustainable management of the Fiordland marine environment and fisheries. Self-interest became secondary to the requirements of the commons. The Guardians of Fiordland referred to this as the "gifts and gains" process. For instance, commercial fishers offered to withdraw from the inner fiords. Removing commercial fishing pressure was expected to not only benefit the fish stocks but also give customary, recreational and charter boat fishers, exclusive fishing access to the calm fiord waters. They in turn, in becoming solely responsible for sustainability of fish stocks within the fiords, proposed lower bag limits, removal of

[5] Personal communication (notes) from L. Teirney, past Fiordland Marine Guardian.

bulk harvesting methods and a cap on the accumulation of catch. And as a return gift to commercial fishers for leaving the fiords, non commercial fishers moved to prevent an increase in non commercial harvesting pressure in Fiordland. This gave commercial fishers greater security along the outer fiords and open coast. A balance of gifts and gains had been achieved that was supported by all the fishing interests.

Secondly, recreational and customary fishers agreed to withdraw from a number of areas of special significance identified inside the fiords -without extractive activities these areas qualified as representative of the fiords. Removing fishing opportunities was possible with the goodwill of the fishers. Along other parts of New Zealand's coastline the animosity over the creation of some marine reserves has persisted for many years. So in Fiordland, environmental interests were gifted whole fiords and arms of fiords without conflict. In return their gift to the fishers was some assurance that they would not continue advocating for the removal of further areas for reserve status in the medium term. (Teirney 2006, p. 4) In the final strategy eight representative areas were identified that have since become marine reserves and protection was agreed for over 20 small, discrete areas designated as 'china shops' or areas of particular biological interest or fragility, some of which became no anchoring zones. These results would indicate that as an approach to governance substantial progress in effective risk management and a reduction in risk exposure had been negotiated. However, the wider civic engagement in the process that occurred after the draft strategy was released for consultation (800 drafts were distributed and 430 submissions received) was not enough to achieve behavioral change in all quarters. Regulations were also required to support the work of the Guardians and after almost a decade (September 2004), the Government agreed to develop special legislation – Fiordland (Te Moana o Atawhenua) Marine Management Act 2005 – and give effect to many of the Strategy's recommendations. The Government also agreed "…to amend fisheries regulations for non-commercial harvesting, to exclude commercial fishing from large areas of the internal waters of Fiordland, and to implement a range of other non-legislative measures." (FMM Act 2005, pp. 2–3)

The Strategy provided a framework for the future management of the Fiordland marine environment. The Government's commitment to the Strategy and its acknowledgement of the value of community input into management was further recognised through the establishment by the Act of a new body – the Fiordland Marine Guardians (as distinct from the earlier Guardians of Fiordland) – a statutory role that provided an on-going involvement in the management of Fiordland. This move built on the collaboration and mutual understanding that had occurred prior to the Act. A key issue was the continued involvement of several of the original Guardians into the new statutory role that ensured that those that had been party to the development of the Strategy could share their understanding with the new people in the work of implementation.

To this extent the procedures by which the Guardians had worked, and still are working, mirrored in part, the process of collaborative governance (as in Fig. 5.3) that includes not only the focus on iterative 'learning-by-doing' and the integration of diverse knowledge systems, but also collaboration and power sharing across

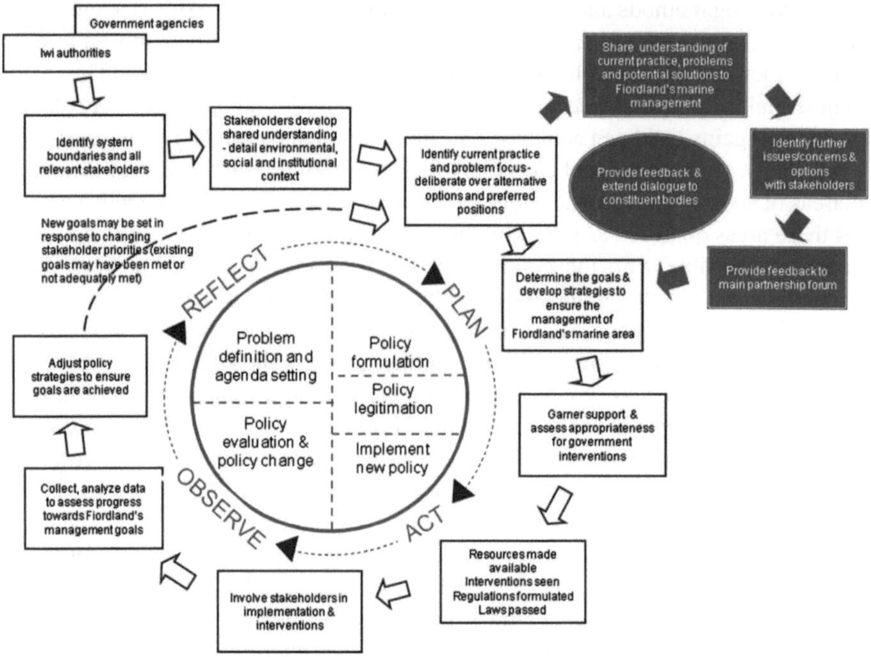

Fig. 5.3 A collaborative governance framework explains the work developed under the Fiordland (Te Moana o Atawhenua) Marine Management Act (Adapted from the work of Reed et al. 2006 and Baines and O'Brien 2012)

community, regional and national levels. This is an evolutionary approach that anticipates both social and institutional capacity building to achieve change.

From a risk management perspective, we can note that the collaborative process adopted by the Guardians, and then Government, is very much an inclusive participatory way of defining and articulating the objectives and goals of risk management. It was not a 'top-down' process, nor a 'bottom-up' process but rather the meeting of a diversity of people who came to talk, listen, understand one another's points of view, overcome their differences, search for common ground and set priorities for action. As a genuine face-to-face dialogue, where humour prevailed over formality, this occurred throughout the process of risk characterization, from problem framing through to detailed risk assessment and then on to the risk management and decision implementation (as anticipated in the work of Stern and Fineberg 1996). In the development of the Strategy, the definition of the risk problems, their assessment and their management were jointly developed. And this work has continued.

Nowadays, it is also clear that the advice and recommendations from the Guardians carry some authority. Guardians can not only take the initiative to advise Ministers, but also all persons (including management agencies) exercising powers or carrying out functions in the Fiordland region must take into account any advice or recommendations provided by the Guardians.

However, despite the innovative aspects of the Fiordland legislation there are limits to the Act's empowerment of the collaborative approach. The Guardians, entrusted with facilitating and promoting the integrated management of the FMA, have in the past experienced frustration in their efforts to achieve change. In essence, the Guardians are in the age-old situation of needing to protect the marine 'commons' from open access conditions, exacerbated, they have perceived, by the policies and/or lack of policies implemented at a national level. Central government is not always in the position to support regional initiatives at the expense of national priorities.

When the Fiordland legislation was initially drafted it was considered innovative and well before its time. It promoted a collaborative spirit and gave powers to the Guardians to advise and recommend to government agencies and Ministers regarding the welfare of the Fiordland Marine Area. But in the last 10 years, international understanding of these processes has developed. An adoption of some of the more innovative governance strategies to sustain the Fiordland social-ecological system could well provide a 'governance for resilience' model (Innes and Booher 2010), potentially an alternative to the relationship established by the present legislation that can allow concerns of regional stakeholders (Guardians and agencies) to be sidelined by central government policy decisions. A future review of the FMM Act might well consider whether a change in the legislation to enable an enhanced across scale collaborative and power sharing process would be timely.

5.8 Conclusions

This chapter has discussed the concept of 'soft infrastructure for sustainable development', referring to those institutions, networks, and social and psychological processes which foster civic engagement in public life which in turn helps nurture a resilient and sustainable society. The concept is a helpful heuristic device for securing greater understanding of how individuals, community organizations and local governments and associated agencies might move towards a more sustainable and resilient society.

Institutional processes are to institutional capital as engagement processes are to social capital. The processes on both counts are working as mechanisms that impact on the learning capacity of both agencies and communities. As people come together to address issues, soft infrastructure develops and networked individuals and groups can be immersed in an action-reflective process and civic engagement is a successful outcome. A learning culture is established that impacts on both institutional and social capacity via the development of efficacy expectations at both individual and collective/group levels. The more effective the governance process, the more effective the development of capacity and the more resilient and sustainable society.

At a deeper level success is dependent on developing levels of knowledge, trust and confidence to deal with difficult or 'wicked' issues that in turn create a sense of efficacy, a belief in the effectiveness to create change, either at an individual or collective level. It is therefore possible to encourage a capacity to change (whereby both social and institutional capacity are built) by involving people in an experiential process of civic engagement. The key issue is that for civic engagement to be successful there must be citizen involvement in an experiential learning process that governance can facilitate or reinforce. A culture of learning established within governance mechanisms can directly impact on the capacity building or learning potential of a community. This is a governance style that contributes to the soft infrastructure that nurtures the development of a resilient and sustainable society.

Given the challenges which are currently facing local and central governments, particularly in the heightened context of austerity politics, it will be even more important to recognise how the processes of soft infrastructure can contribute to addressing societal problems. In particular, it is increasingly clear that a combination of *dynamic governing*, and a process of *transformational learning*, as outlined above has the potential to secure substantial and effective collaborative responses to the challenges inherent in establishing resilient and sustainable societies.

References

Adger, N. 2000. Social and ecological resilience: Are they related? *Progress in Human Geography* 24(3): 347–364.

Ansell, C., and A. Gash. 2008. Collaborative governance in theory and practice. *Journal of Public Administration Research and Theory* 18(4): 543–571.

Arnstein, S. 1969. A ladder of citizen participation. *Journal of the American Institute of Planners* 35(4): 216–224.

Baines, J., and M. O'Brien. 2012. *Reflections on the collaborative governance process of the land and water forum*. Research report prepared for the Ministry for the Environment, PO Box 10362, Wellington 6143. New Zealand.

Bandura, A. 1977. Towards a unifying theory of behavioural change. *Psychological Review* 84: 191–215.

Bandura, A. 1993. Perceived self-efficacy in cognitive development and functioning. *Educational Psychologist* 28(2): 117–148.

Bandura, A. 1997. *Self-efficacy: The exercise of control*. New York: WH Freeman.

Berkes, F. 2009. Evolution of co-management: Role of knowledge generation, bridging organizations and social learning. *Journal of Environmental Management* 90: 1692–1702.

Bourdieu, P. 1986. The forms of social capital. In *Handbook of theory and research for the sociology of education*, ed. J.G. Richardson. New York: Greenwood.

Coleman, J.S. 1988. Social capital in the creation of human capital. *American Journal of Sociology* 94: S95–S120.

Collier, P. 1998. *Social capital and poverty*, Social capital initiative working paper 4. Washington, DC: The World Bank.

Dale, A., and L. Newman. 2006. Sustainable community development, networks and resilience. *Environments* 34(2): 17–27.

Daniels, S., and G.B. Walker. 2001. *Working through environmental conflict: The collaborative learning approach*. Westport: Praeger.

Dewey, J. 1938. *Experience and education*. Kappa Delta Pi. Republished 1963. New York: Collier Books.
Doubleday, N. 2008. Culturing adaptive co-management: Finding "keys" to resilience in asymmetries of power. In *Adaptive co-management*, ed. D. Armitage, F. Berkes, and N. Doubleday, 228–248. Vancouver: UBC Press.
Dyball, R., V.A. Brown, and M. Keen. 2007. Towards sustainability: Five strands of social learning. In *Social learning towards a sustainable world*, ed. A.E.J. Wals, 181–194. Wageningen: Wageningen Academic Publishers.
Evans, B., M. Joas, S. Sundback, and K. Theobald. 2005. *Governing sustainable cities*. London: Earthscan.
Evans, B., M. Joas, S. Sundback, and K. Theobald. 2006. Governing local sustainability. *Journal of Environmental Planning and Management* 49(6): 849–867.
Falk, I., and S. Kilpatrick. 2000. What is social capital? A study of interaction in a rural community. *Sociologia Ruralis* 40: 87–110.
Fernandez-Gimenez, M.E., H.L. Ballard, and V.E. Sturtevant. 2008. Adaptive management and social learning in collaborative and community-based monitoring: A study of five community-based forestry organizations in the western USA. *Ecology and Society* 13(2): 4. http://www.ecologyandsociety.org/vol13/iss2/art4/
FMM Act. 2005. Fiordland (Te Moana o Atawhenua) Marine Management Act 2005. Public Act 2005 No. 36. http://www.mfe.govt.nz-issues-biodiversity-fiordland-fiordland-marine-management-act.pdf
Freire, P. 1972. *Pedagogy of the oppressed*. New York: Continuum.
Guijt, I. 2010. Accountability and learning: Exploding the myth of incompatibility between accountability and learning. In *Capacity development in practice*, ed. J. Ubels, N. Acquaye-Baddoo, and A. Fowler, 277–291. London: Earthscan.
Hahn, K. 1949. *The loyalties by which we live*. Reprinted from an address to the conference on rural life at home and overseas. Aberdeen: Aberdeen University Press.
Hardin, G. 1978. An ecolate view of the human predicament. In *Global resources: Perspectives and alternatives: XIV nobel conference*, ed. C.N. McRostie, 49–71. Baltimore: University Park Press.
Hopkins, R. 2008. *The transition handbook: From oil dependency to local resilience*. White River Junction: Chelsea Green Publishing.
Innes, J.E., and D.E. Booher. 2010. *Planning with complexity: An introduction to collaborative rationality for public policy*. New York: Routledge.
Johnsen, H.C.J. 2005. Action research and knowledge co-generation: A not so dangerous liaison with conventional social research. *Artificial Intelligence & Society* 19: 513–551.
Keen, M., and S. Mahanty. 2006. Learning in sustainable natural resource management: Challenges and opportunities in the Pacific. *Society and Natural Resources* 19: 497–513.
Keen, M., V.A. Brown, and R. Dyball. 2005. *Social learning in environmental management: Towards a sustainable future*. London: Earthscan.
Khakee, A. 2002. Assessing institutional capital building in a Local Agenda 21 process in Gothenborg. *Planning Theory and Practice* 3(1): 53–68.
Kolb, D.A. 1984. *Experiential learning: Experience as the source of learning and development*. Englewood Cliffs: Prentice-Hall.
Lightsey, O.R. 2006. Resilience, meaning, and well-being. *The Counselling Psychologist* 34: 96–107.
Lowndes, V., and D. Wilson. 2002. Social capital and local governance: Exploring the institutional design variable. *Political Studies* 49(4): 629–647.
Marquardt, M., and D. Waddell. 2006. Optimizing the power of action learning. In *Action learning, leadership and organizational development in the public services*, ed. C. Riggand and S. Richards. New York: Routledge.
McCoy, M.L., and P.L. Scully. 2002. Deliberative dialogue to expand civic engagement: What kind of talk does democracy need? *National Civic Review* 91(2): 117–135.

O'Brien, M. 1990. *Northland wilderness experience: An experiential programme for the youth of Taitokerau*. Research report, Education Department, University of Auckland, Auckland.
O'Brien, M. 1995. Understanding community participation in conservation. In *The role of networks*, ed. D.A. Saunders, J.L. Craigand, and E.M. Mattiske, 209–212. Chipping Norton: Surrey Beatty & Sons.
O'Brien, M. 1997. Working on conservation issues: A perspective on social capital. In *Social capital and policy development*, ed. D. Robinson. Wellington: Institute of Policy Studies, Victoria University.
Orr, D. 1992. *Ecological literacy: Education and the transition to a post modern world*. Albany: State University of New York Press.
Pahl-Wostl, C. 2006. The importance of social learning in restoring the multifunctionality of rivers and floodplains. *Ecology and Society* 11(1): 10. http://www.ecologyandsociety.org/vol11/iss1/art10/
Pahl-Wostl, C. 2009. A conceptual framework for analysing adaptive capacity and multi-level learning processes in resource governance regimes. *Global Environmental Change* 19: 354–365.
Pahl-Wostl, C., M. Craps, A. Dewulf, E. Mostert, D. Tabara, and T. Taillieu. 2007. Social learning and water resources management. *Ecology and Society* 12(2): 5. http://www.ecologyandsociety.org/vol12/iss2/art5/
Putnam, R. 2000. Bowling Alone: The collapse and revival of American community. New York, NY: Simon and Schuster.
Putnam, R., R. Leonardi, and R. Nanetti. 1993. *Making democracy work: Civic traditions in modern Italy*. Princeton: Princeton University Press.
Reed, M.S., E.D.G. Fraser, and A.J. Dougill. 2006. An adaptive learning process for developing and applying sustainability indicators with local communities. *Ecological Economics* 59: 406–418.
Reed, M.S., A.C. Evely, G. Cundill, I. Fazey, J. Glass, A. Laing, J. Newig, B. Parrish, C. Prell, C. Raymond, and L.C. Stringer. 2010. What is social learning? *Ecology and Society* 15(4): r1. http://www.ecologyandsociety.org/vol15/iss4/resp1/
Rydin, Y., and M. Pennington. 2000. Public participation and local environmental planning: The collective action problem and the potential of social capital. *Local Environment* 5(2): 153–169.
Schneider, P.T., M. Marschall, M. Mintrom, and C. Roch. 1997. Institutional arrangements and the creation of social capital: The effects of public school choice. *The American Political Science Review* 91(1): 82–98.
Senge, P.M. 1990. *The fifth discipline: The art and practice of the learning organization*. New York: Doubleday.
Sørensen, E., and J. Torfing. 2007. *Theories of democratic network governance*. New York: Palgrave Macmillan.
Sterling, S. 2004. Vignette 4.1 The learning of ecology, or the ecology of learning? In *Key issues in sustainable development and learning: A critical review*, ed. W. Scott and S. Gough, 68–70. London: Routledge Falmer.
Sterling, S. 2007. Riding the storm: Towards a connective cultural consciousness. In *Social learning towards a sustainable world*, ed. A.E.J. Wals, 63–82. Wageningen: Wageningen Academic Publishers.
Stern, P.C., and H.C. Fineberg. 1996. *Understanding risk: Informing decisions in a democratic society*. Report to the US National Research Council, Washington, DC.
Takahashi, L., and G. Smutny. 2002. Collaborative windows and organizational governance: Exploring the formation and demise of social service partnership. *Nonprofit and Voluntary Sector Quarterly* 31: 165–185.
Teirney, L. 2006. The Guardians' Fiordland Marine initiative: A unique journey. Key note address at the *Coastal Dune vegetation network conference*, Invercargill.

Tompkins, E., and N. Adger. 2004. Does adaptive management of natural resources enhance resilience to climate change? *Ecology and Society* 9(2): 10.

Uphoff, N. 1999. Understanding social capital: Learning from the analysis and experiences of participation. In *Social capital: A multifaceted perspective*, ed. P. Dasgupta and I. Seregeldin. Washington, DC: World Bank.

Walker, B., and D. Salt. 2006. *Resilience thinking: Sustaining ecosystems and people in a changing world.* Washington, DC: Island Press.

Walker, B., S. Carpenter, J.M. Anderies, N. Abel, G. Cumming, M. Janssen, L. Lebel, J. Norberg, G.D. Peterson, and R. Pritchard. 2002. Resilience management in social–ecological systems: A working hypothesis for a participator approach. *Conservation Ecology* 6(1): 14. http://www.consecol.org/vol6/iss1/art14.

Weick, K.E. 1986. Small wins: Redefining the scale of social issues. In *Redefining social problems*, ed. E. Seidman and J. Rappaport, 29–48. New York: Plenum Press.

Zadek, S. 2008. Collaborative governance: The new multilateralism for the 21st century. In *Global development 2.0 can philanthropists, the public, and the poor make poverty history?* ed. L. Brainard and D. Chollet, 1–11. Washington, DC: Brookings Institute.

Bob Evans is Professor of Environmental Planning at the University of Waikato. He has worked as a town planner in local government, the private sector and for community organizations and was previously Professor and Director of the Sustainable Cities Research Institute, Northumbria University. He has published widely on local sustainable development, land use and environmental planning, with an emphasis upon issues of governance and environmental justice. He has undertaken numerous major research projects for the European Commission, and has worked on research projects in Japan, China and New Zealand. Recent research has included work on local governance for sustainability; civic engagement in renewable energy projects, and the development of soft infrastructure for urban sustainability.

Marg O'Brien works as a Social Ecologist in the Cawthron Institute, New Zealand's largest independent community-owned research centre. Marg has 30 years of experience working as a behavioural change expert -understanding people, their attitudes and their behaviours, the impact of social and institutional arrangements, and how change can be achieved. She is presently involved in government funded research on sustainable settlements and the development of soft urban infrastructure and the use of indigenous knowledge (Mātauranga Maori) in resource management. She has also more recently been involved in national research on collaborative governance in the management of freshwater in New Zealand.

Chapter 6
Understanding and Conceptualizing Risk in Large-Scale Social-Ecological Systems

Helen Elizabeth Allison

"One of the most enduring problems of the human race is that each person must continually take actions that require knowledge of the future. ....Human understanding of any event, even one witnessed directly, is obscured, warped, tainted by all sorts of unrecognized conceptual and perceptual biases. Hence the dilemma: one needs to foresee the future in order to make wise decisions, but one cannot even be very certain about the past or the present."

(Meadows and Robinson *1985)*

6.1 Introduction

In this chapter I take a cognitive approach to investigating how we perceive risk. The process of developing a new conceptualization begins with identifying and naming those aspects of the current dominant worldview which have become redundant and that do not serve our purposes going forward into the new century. We then have to create a new conceptual framework including a set of assumptions, worldviews and mental models about how we currently understand the world to operate. I discuss the development of theories of organization, systems and resilience, and Jungian theory of psychological type. These theories will help guide our thinking about how we know the complex world around us and will help us as individuals and societies, to deal with the turbulent systemic dynamics that we will face in the years to come. Small adaptive changes in how we understand the dynamics will be too little to effect the changes required, rather we need innovative and creative change. For change to occur I suggest an important step is to integrate the diversity of knowledge by creating a new type of systems framework, an 'Ontological-Driven Risk Governance System'. When created this will help teams to clarify and make explicit individuals' conceptions and understanding of risk, particularly when fundamental sets of underlying personal values are not articulated. There are moments

H.E. Allison (✉)
School of Veterinary and Life Sciences, Murdoch University,
90 South Street, Murdoch, WA 6150, Australia
e-mail: helenallison@ozemail.com.au

U. Fra.Paleo (ed.), *Risk Governance*, DOI 10.1007/978-94-017-9328-5_6

in time when people have a greater tendency to think about change and none more so than at the change of a century or a millennium as we have now. These thresholds magnify the perception of cultural, social and the symbolic significance of change (Slaughter 1996).

Humans have developed many conceptual ontological and epistemological models or frameworks to understand organization in its many forms, for example in the social sciences (Burrell and Morgan 1979; Quinn and Rohrbaugh 1983); information systems (Jayaratna 1994); ecology and social-ecological systems (Allison and Hobbs 2006). How we conceptualize what we know (ontology) and how we understand our construction of knowledge (epistemology) is moving towards more holistic or systemic worldviews (Holling 1973; Prigogine and Stengers 1997; Ackoff 1999; Matthies et al. 2001; Gunderson and Holling 2002; Allison and Hobbs 2006; Fischer-Kowalski and Rotmans 2009; Fazey 2010). Ontology and epistemology reflect how we think about the real world and has an important role to play in defining, analyzing and communicating about the subject under investigation and ultimately how we make decisions about levels of risk and appropriate actions to take. Communicating risk may be difficult because it has been observed that different people may create different conceptual models even when given the same information (Meadows and Robinson 1985; Soffer and Hadar 2007) which can lead to different representations of the real world. Not only that but Jayaratna (1994) identified that the translation of a model between the creator and the user of a model may be quite different depending on a range of factors in their mental constructs including perceptual processes, values and ethics, motives and prejudices, reasoning ability, experiences, skills and knowledge sets, structuring processes (including methodologies) and roles which together are used to construct their mental models about how they see the world and how they make their decisions.

The construct of personal epistemology (Hofer and Pintrich 2002), its psychological and intellectual development (Piaget 1970) and influencing factors including education, play an important role in perception, analysis and decision making concerning risk. More specifically, this includes beliefs about (a) the certainty of knowledge, (b) the simplicity of knowledge, (c) the source of knowledge, and (d) the justifications for knowing (Bendixen and Rule 2004). When we encounter new information the adequacy of our epistemological beliefs will determine what and how we make meaning of the information and therefore how we justify our decisions. As the amount of information and the interconnectedness increases, our personal epistemology may be challenged. An important question is: *will and how do these deep seated beliefs change?* Bendixen and Rule (2004) synthesized many of the theoretical models of personal epistemology and proposed a model for how one's beliefs might change. One proposal is that epistemic doubt is involved, as the ability to doubt one's beliefs is required for developmental change. However, epistemic doubt alone does not imply a progression through the mechanism of change. This idea is similar to the conceptual change literature indicating that even if individuals process information at a high level of engagement, conceptual change is still not a guarantee (Dole and Sinatra 1998). Epistemic doubt can be an impetus for epistemic change, but change also requires the willingness and capacity to change

(Gallopin 2002; Bendixen and Rule 2004). Our knowledge is always evolving and we can follow changes from the past to our current thinking and then propose an alternative for the future.

As a consequence, today we all face ontological insecurity (Giddens 1990) because we do not have the experience and therefore are not prepared for these new risks and our future is unclear. Often these risks occur due to our contemporary understanding of science and technology based on temporal and spatial scales that are too narrow (Holling et al. 2002a; Stoffle and Arnold 2008) identified as the paradox of 'The Trap of the Expert' (Holling et al. 2002b). This 'Trap', also known as the 'silo effect', is caused by the highly specialized knowledge of experts without reference to related disciplines, their interactions or to the broader context.

6.2 The Past

In the immediate past, the twentieth century was dominated by an ontology and epistemology of science that promoted perceptions of stability or at least more stable than our view of the future. Our form of thinking and worldview, for the most part, was narrowly reductive and analytical, and it has influenced the way we approach decision making processes involving risk. In times that are perceived to be stable, when the future will be similar to the past, traditional ways and methods can be used to find solutions to problems that have been successful in the past. However, they have embedded in their design the ideas and assumptions of the people who developed them, and they also have bounded rationality that places limitations and biases on these assumptions (Wells and Claxton 2002; Capper and Williams 2004).

Psychologists have documented a wide variety of errors and biases in people's observations of data, and the judgments and interpretations based on them. For example, when deciding what evidence is relevant to testing a hypothesis, people often ignore relevant information such as probabilities and information that does not confirm their hypothesis. When examining data, people often perceive correlations that are not there, fail to perceive strong correlations that are there, perceive patterns in data that are in fact random and see what they expect to see whether it is there or not. When drawing inferences from data, people sometimes conclude that the data supports their theory even when it strongly supports the exact opposite theory and are too willing to base firm conclusions on information that is incomplete or unrepresentative (Doyle 1997). Even the act of measurement introduces distortion, delays, biases, errors and other imperfections, some known, some unknown and some unknowable (Sterman 2000). No one knows the real rates of change in environmental or social variables. Instead they are all estimates of the real data based on sampled, averaged and importantly, delayed measurements. Many social variables are difficult to measure or we use surrogate measures. Above all, measurement is an act of selection and selection is subjective based on our worldview, what we value and what we can measure. Hence, social variables have often been ignored until recently and now attempts are being made to include them as important variables within the context of the system. Our senses and information systems select but a tiny fraction of possible information. Based on these experiences we select indicators and assess the factors that are highly valued by society, as opposed to what will help us to make better decisions in a complex world. In real world situations which are dynamically complex, policy resistance can be attributed to such factors as treating dynamic situations as an event-orientated situations; limited knowledge of the feedback processes, treating feedback processes as side-effects rather than endogenous feedback processes. Because of this and other pressures, policy makers dealing with complex systems will often rely on convenient, simple and linear indicators (Muneepeerakul and Qubbaj 2012) based on a mechanical metaphor.

The mechanical metaphor applied to biological systems is a myth, there is no balance of nature and humans are a component of the system and not set apart from it. These systems are controlled by multiple feedback processes with delays, especially in ecological systems, which can cause counterintuitive affects, often appearing far in time and space. This mechanical metaphor misconception was first incorrectly perpetrated for biological systems by Tansley (1935). However, in the mid 1970s a new cohort of ecologists identified the fatal flaw of assuming systems were in equilibrium and the stability of systems was questioned. The metaphor of a mechanistic world with order and control that lead us astray in our thinking was rejected and a new group of systems ecologists emerged (Botkin 1990).

6.3 Future

Increasingly the future is perceived to be more complex with greater connectivity, uncertainty and ambiguity than before. Consequently, it will require a very different way of knowing and understanding to make effective decisions to deal with problems in the future, identified as higher order thinking (Fazey 2010). What is emerging in the twenty-first century is the need for a change of frame, a better way of envisioning the future and a way to understand risk. Not only for the safety of societies but also in our increasingly litigious societies there is a need to understand how policy will be able to deal with uncertainty and risks.

There is a convergence of ideas from many disciplines that identifies that change must first occur in our minds in order to deal with complexity and the risks imposed by these changes. Will imminent global social disturbance be enough to overcome the cultural and cognitive hurdles (Jahoda et al. 1970) for society to adopt systemic approaches more appropriate to deal with complex wicked problems? This is not only about changing the way we think but initially there is a need to unlearn the habitual ways of thinking. Is it now safe to talk about ontology, epistemology in a systems framework in the sciences or will the dominant simplistic mechanical metaphor prevail and prevent the adoption of complex systems as the mainstream worldview?

6.3.1 Systemic Risks

In acknowledgement of the importance of systems thinking the International Risk Governance Council focused on the deficits of systemic risk, rather than simple risks (International Risk Governance Council 2009). This is because systemic risks – defined as those risks that affect the functionality of systems upon which society depends and that have impacts beyond their geographic and sector origins – provide a greater challenge for risk governance and thus greater scope for the occurrence of deficits (International Risk Governance Council 2009). A risk governance deficit is a failure in the identification, framing, assessment, management and communication of a risk issue or of how it is being addressed (International Risk Governance Council 2012). The International Risk Governance Council categorizes the deficits into two types, Cluster A – Assessing and understanding risks – and Cluster B – Managing risks (International Risk Governance Council 2009). In this chapter I am only concerned with Cluster A which describes 10 deficits that can arise when there is a deficiency of either scientific knowledge or knowledge about values, interests and perceptions by individuals and organizations (Table 6.1).

6.4 Social Systems

There can be no doubt about the increasing social connectivity in the global system and we are unable to foresee the consequences and the risks that this will produce. The internet illustrates the exponentially increasing rate of social connectivity, for example, there were more data transmitted over the Internet in 2010 than the entire history of the Internet from 1982 through to 2009. In 2011, there were four billion connected devices around the world and Intel expects this number to increase to 15 billion by 2015 and 50 billion by 2020 (Skaugen 2011). It was clear to the creators of the Internet that in order to be able to communicate across diverse types of knowledge, a mechanism was needed to structure the information so the Ontological-Driven Information System was developed (Guarino 1998). Indeed there are people in many areas who are recognizing the need to integrate their vocabularies around knowledge, e.g. genetics (Gene Ontology Consortium 2004). Similarly there is clearly a need to develop a way to integrate knowledge across the sciences and I suggest there is a need to develop an Ontological-Driven Risk Governance System, see below.

Despite change in the scope towards more systemic perspectives for dealing with complex problems, two theoretical perspectives of organization linked to culture have persisted: the subjective-objective dichotomy and the model of communication for risk. Consequently a cross-cultural analysis of risks is required because understanding risk, like other experiential phenomena, is informed by socially and culturally structured and historically conditioned conceptions and evaluations of the world (Stoffle and Arnold 2008; Wadsworth 2008; Boholm 2009). Both individuals

Table 6.1 The deficits of risk management

Assessing and understanding risk	
A1–A3	Address difficulties involving the gathering and interpreting of knowledge about risks and perceptions of risks
	A1. The failure to detect early warnings of risk because of erroneous signals, misinterpretation of information or simply not enough information being gathered
	A2. The lack of adequate factual knowledge for robust risk assessment because of existing gaps in scientific knowledge or failure to either source existing information or appreciate its associated uncertainty
	A3. The omission of knowledge related to stakeholder risk perceptions and concerns
A4–A6	Have to do with disputed or potentially biased or subjective knowledge and have the effect of making it difficult to judge whether a risk needs specific attention or action
	A4. The failure to consult the relevant stakeholders, as their involvement can improve the information input and the legitimacy of the risk management process (provided that interests and bias are carefully managed)
	A5. The failure to properly evaluate a risk as being acceptable or unacceptable to society
	A6. The misrepresentation of information about risk, whereby biased, selective or incomplete knowledge is used during, or communicated after, risk assessment, either intentionally or unintentionally
A7–A9	Focus on knowledge related to systems and their complexity
	A7. A failure to understand how the components of a system interact or how the system behaves as a whole, thus failure to assess the multiple dimensions of a risk and its potential consequences
	A8. A failure to recognise fast or fundamental changes to a system, which can cause new risks to emerge or old ones to change
	A9. The inappropriate use of formal models as a way to create and understand knowledge about complex systems (over- and under-reliance on models can be equally problematic
A10	Is about the incompleteness and inadequateness of knowledge

Cluster A – assessing and understanding risk
Source: International Risk Governance Council (2009)

and organizations respond to challenges posed by risk by reasserting their own habituated behavior (Cameron and Quinn 1999). Although an organizational culture is necessary and forms the characteristics that make the organization effective, the paradox is the need for stability but also the need to be able to respond to a turbulent environment and to change behaviour. Many of the models of organization identify the dichotomies of competing factors (Burrell and Morgan 1979; Quinn and Rohrbaugh 1983; Blann and Light 2000; Gunderson and Holling 2002). No one factor or phase can ensure success over time, rather a system has to cycle through multiple adaptive phases to be resilient in the long term.

Allison and Hobbs (2006) examined the literature that is relevant to some of the fundamental problems of organization, including the dynamics of organizational change, the management of uncertainty and the tensions that arise out of competing values and different perspectives in the social systems. Some of these frameworks are static spatial frameworks using axes describing two dichotomies which identify four categories of organizational preferences (Burrell and Morgan 1979; Quinn and Rohrbaugh 1983; Goldspink and Kay 2004) others have added a third dimension (Blann and Light 2000). The need to account for stability and change and identify effectiveness of an organization meant simple static indicators were chosen based on simple linear models. Criteria were found to be either for identifying the means or the ends for the effectiveness of organization. The selection being dependent on the mental constructs of the individual researcher. The Burrell and Morgan (1979) framework of theories of organization are still a useful starting point for discussion. The Burrell and Morgan framework is generated by two intersecting dichotomies, the philosophy of science along one axis (the subjective – objective dimension) and a theory of society along the second axis, (the sociology of radical change and the sociology of regulation). Similarly the competing values model of Quinn and Rohrbaugh (1983) uses the juxtaposition of three value dimensions, control-flexibility, internal-external focus and means versus the ends that underlie conceptualizations of organizational effectiveness. The adaptive cycle metaphor (Gunderson and Holling 2002) shows the sequential progression between apparently competing phases in a temporal cycle (Allison and Hobbs 2006) that are required to create resilience in social-ecological systems.

Increasing complexity means that the future is unknowable as rates of change increase and feedback processes make prediction and direction of change increase. The consequence is that the fundamental capacity for learning in twenty-first century will be to develop supple and nimble minds (Claxton 2004). It will be necessary for peoples' minds to become more flexible. Knowing 'what to do when you do not know what to do' and increased creativity will be the fundamental capacity for learning (Claxton 2004). Creative capacity is now identified as a key economic driver of the future (Department of Education 2004). More time may have to be spent in interpreting the situation and being allowed time for error-making to understand the complexities of any situation. This is called 'staying in the grey' and learning from the process rather than focusing on quick solutions and easy successes (McWilliam 2009).

6.5 Complexity

In contrast to conceptualizing systems as simple and linear, evolutionary cycles have been shown to be ubiquitous in nature and have been identified in systems created by human society including the economy (De Greene 1993; Carry 1996). Given the knowledge that systems are not in equilibrium but go through cycles of prosperity, recession, depression and recovery, the limitations of stationary analysis were well understood in the early 1900s by Kondratiev investigating industrial global cycles (Alexander 2002) and later taken up by the economist, Schumpeter in the 1950s (Schumpeter 1950). Subsequently the concept of the economy as self-exciting system or what is now known as a complex adaptive system (Waldrop

1992) was developed. Social-ecological systems may display resilience or alternatively cross a threshold and enter a maladaptive functional state, in which there is reduced ability to continue to produce the goods and services required by society.

Resilience is a term that is in common use as well as emerging in the ecological and social sciences and increasingly appearing in policy (Allison and Hobbs 2006). Interestingly its popularity emerged just as society lost confidence in the ability of social-ecological systems to persist in the face of new risks. Perhaps the two are related. As disasters increase in frequency and intensity, the issue of survival becomes increasingly salient. Resilience is applied in this analysis at the scale of the social-ecological system, not an individual or small group phenomenon or only in the ecological system. According to Holling (1973) and Berkes et al. (2003) resilience can be understood in terms of three characteristics:

- The amount of change the system can undergo and still retain the same controls over its function and structure;
- The degree to which the system is capable of self-organization; and
- The ability to build and increase the capacity for learning and co-adaptation.

Resilience is thus concerned with the magnitude, frequency, and kinds of disturbance that can be absorbed or buffered without the society and culture undergoing fundamental changes. In human terms the simple question is: Will we still be here, largely unchanged after the risk event?

Questions remain about the extent to which resilience thinking is useful as a generalized set of concepts that can predict and explain real-world processes (Scheffer et al. 2001; Walker et al. 2006; Allison 2011). Explanation and prediction are, however, only some of the many potential uses of a conceptual tool. Fazey (2010) suggests resilience is more of an awareness raising tool which can be used to identify key questions or perhaps asking better questions, facilitating design to reduce vulnerability, aid communication, motivate and engage interest in sustainability, develop new policies aid conceptual change, and particularly in tertiary education develop higher order thinking skills.

Much of the difficulty arising in complex systems is in the translation from the conceptualization to the implementation. In error, prevention of perturbations is still often a major goal of system management. This is unfortunate, not only because disturbance is a natural component of systems that promotes diversity and renewal processes, but also because it distracts attention from the underlying structural problem of resilience. The main implication of the insights presented here is that efforts to reduce the risk of unwanted perturbations and the potential to cross a threshold should address the gradual changes that affect resilience rather than merely control disturbance. The challenge is to sustain a large stability domain which allows fluctuations with a safe range, rather than to control and minimize all fluctuations. Stability domains typically depend on slowly changing variables such as land use, nutrient stocks, soil properties and biomass of long-lived organisms. These factors may be predicted, monitored and modified if the complexity worldview is adopted. Therefore, dynamics caused by endogenous factors, called manufactured risk by Giddens (1999) may be surprisingly unsurprising (Allison and Hobbs 2006) and which should have been a predictable surprise (Bazerman and

Watkins 2004) if a holistic perspective had been adopted. In contrast, stochastic events that trigger state shifts (such as hurricanes, droughts or disease outbreaks) are usually difficult to predict or control. Therefore, building and maintaining resilience of desired social-ecological system states is likely to be the most pragmatic and effective way to manage ecosystems in the face of increasing environmental change (Scheffer et al. 2001).

An important question is whether systems are becoming more complex with increasing numbers of connections between its parts or is it the fact that we now conceive them to be more complex. None-the-less it is argued that the easy work is finished as scholars are now confronted with questions that defy easy categorization or solution by traditional disciplinary frameworks. Many research problems facing today's society require coordinated efforts from multiple disciplines. Consequently the rubric in science is interdisciplinarity giving rise to scientific institutes that are multidisciplinary and a vast array of literature on how interdisciplinary science should be organized (Rhoten and Parker 2004). Interdisciplinarity, although difficult to define can be interpreted broadly as an active, multi-faceted learning process between researchers from different disciplines creating common ground for a special purpose. Interdisciplinary approaches are regarded as necessary in environmental research, especially in view of global environmental change (Bradshaw and Bekoff 2001). However, "the step from an appealing idea to an operational method is large indeed" (Karlqvist 1999) leading some to argue a lack of genuine interdisciplinarity in environmental research (Bruce et al. 2004; Fazey 2010).

One significant obstacle to the adoption of an integrative science approach is its lack of academic kudos. Surveys show concerns among researchers about perceptions of interdisciplinary science as second-rate. A sense of superiority within each discipline and the view that other disciplines are less rigorous or important also presents a barrier to full acceptance of interdisciplinarity. However, some scientists, known as polymaths, are recognized as having a tendency towards greater interdisciplinarity. The term 'polymath' usually refers to very learned scholars who are distinguished not only by genius in particular fields of interest, but also by their ability to traverse different fields of specialization and to recognize their interconnections. Polymaths can enhance the effectiveness of interdisciplinarity through their knowledge and understanding of different disciplinary languages, epistemologies and methodologies (Karlqvist 1999; Robinson 2005; Young and Marzano 2010). Whether an individual has an inherent broad way of thinking such as a polymath or a narrower and in depth perspective I suggest that each is inherent in an individual's psychological makeup and in part can be explained through Jungian theory of psychological type.

6.6 Psychological Type

The theory of psychological type developed by Carl Jung has been widely used in organizational development (Briggs-Myers 1999) to identify normal differences between individuals. Applying psychological type theory to organizational change helps people understand how people with different personality types influence and shape how organizations deal with change. Organizations that seek continuous innovation are learning what a powerful tool personality type theory is (Killen and Williams 2009). In the fields of memory, cognition, concept formation and problem solving two different types of cognitive processes have been recognized (Beyler and Schmeck 1992). The one traditionally emphasized is most commonly called analytic, rule-based, or schema-based, while the other is given a label such as holistic, episode-based, analogy-based, or simply 'nonanalytic' (Brooks 1978). A preference for one of these cognitive styles will have far reaching effects on how individuals learn (Beyler and Schmeck 1992), how individuals see and interpret the world around them and how they make their decisions. These cognitive processes map onto the Jungian theory of personality. Analytic processing involves a narrower focus of attention, retention of facts and details, noticing differences, more interest in parts than wholes, and preference for ordered (usually sequential) presentation of information (Beyler and Schmeck 1992; Allison and Hobbs 2006). In contrast, holistic processes seem to involve habitual preferences for a broad focus of attention, formation of impressions, noticing similarities, more interest in wholes than in component parts, and preferences for more random less orderly presentation of information.

When human actions cause changes in regional social-ecological systems, cause and effect may be distant in time and space. Consequently different ways of knowing are required, ways that encourage integration, reflection and learning, and multiple ways of knowing (Allan and Curtis 2005). However, research in social-ecological systems has neglected the potential role that individual and group differences play in decision making particularly with increasing levels of participatory (Allison and Hobbs 2010). In an increasingly complex and pluralistic world in which we will be required to increase our efforts to deal with social change, understanding individual personality will be a catalyst to realize human potential (Myers et al. 1998) in order to understand and manage those changes. Knowing individual preferences is particularly important in communication between people with different ways of seeing the world (Myers et al. 1998; Berens 2001; Crestani 2001; Dale 2002) and will be particularly important when change requires communication about paradigmatic change, transformation and innovation as opposed to small incremental and adaptive change (Garfield et al. 2001; Killen and Williams 2009).

There is interest in better understanding differences between groups and even different national psyche (van Rooyen 2007; Ball 2009; Allison and Hobbs 2010). Data archives of psychological type using the reported results on the well-known Myers-Briggs Type Indicator® (MBTI®) show the distribution of the 16 type profiles for adult populations. Not all of the 16 types are equally represented in the archive samples and it is acknowledged that the sample is not necessarily representative. In the Australian archive sample (Ball 2009), Table 6.2 (males) and Table 6.3 (females) it is apparent that the distributions of types are different. For males the 'Introverting, Sensing, Thinking, Judging' type (ISTJ) (21.3 %) is the most frequently occurring type, with a 15 times higher frequency than the lowest frequency type Extroverting, Sensing, Feeling Perceiving (ESFP). However, for females the

Table 6.2 Percentages of the personality types in the Australian archive data – males (n = 12,645)

ISTJ	ISFJ	INFJ	INTJ
21.3	3.9	2.6	8.7
ISTP	ISFP	INFP	INTP
5.6	1.9	3.2	6.4
ESTP	ESFP	ENFP	ENTP
4.7	1.4	4.1	6.8
ESTJ	ESFJ	ENFJ	ENTJ
15.8	3.1	2.1	8.4

Source: Ball (2009)

Table 6.3 Percentages of the personality types in the Australian archive data – females (n = 9,513)

ISTJ	ISFJ	INFJ	INTJ
9.7	12.8	6.0	5.3
ISTP	ISFP	INFP	INTP
2.7	4.9	8.4	4.1
ESTP	ESFP	ENFP	ENTP
2.5	3.8	9.1	5.4
ESTJ	ESFJ	ENFJ	ENTJ
7.7	8.2	4.8	4.6

Source: Ball (2009)

highest frequency is for Introverting, Sensing, Feeling, Judging types (ISFJ) (12.8 %). The distribution of types in males and females show a different pattern, there is an indication that Australian males are predisposed toward preferences for Sensing, Thinking, and Judging with the two highest frequencies ISTJ and ESTJ making up 37.1 % of the male population, whereas the female population shows a more even distribution than the male population. In the female population three out of the four of the top four frequencies favor a Feeling preference in contrast to the male population in which the Thinking preference occurred in the top eight. The profiles for Australian males and females are relatively similar to those of the USA population (Ball 2009); both populations are dominantly Sensing, Thinking and Judging types. Type distributions of large groups of managers and leaders, the decision makers in organizations, from a wide variety of cultures demonstrate an over-representation of Thinking and Judging preferences compared to the general population (Myers et al. 1998). In general the group differences of type distribution in organizations will influence the dominant way people take in information and how they prioritize information to make decision in relation to risk. Group differences were also found across organizational scales in decision makers in natural resource management in Australia (Allison and Hobbs 2010). Decision makers at the local organization scale were over represented by Sensing and Judging preferences whilst at the regional and national scales, mostly government policy officers, members were overrepresented by iNtuitive and Thinking preferences.

The significance of the differences in the representation of types is that each type has a general set of characteristics that influence how they perceive the world and how they relate to and communicate with each other. Consequently risk will be perceived very differently by different psychological types and difficulties may arise in the communication between different types.

6.7 Ontological-Driven Risk Governance System

Complexity theory and the ancillary theories to understand order at the edge of chaos (Waldrop 1992) have progressed considerably since the 1990s. The genealogy of systems science (Allison and Hobbs 2006) identifies the linkages from the beginnings of cybernetics and control engineering to general systems theory, complexity theory and resilience in social-ecological systems. Regardless of such theoretical progress many case studies and attempts to manage complex systems from health to natural resource systems have proved unsuccessful. As the literature points out most of these problems are due to the difficulties of capturing and knowing the structure of organization and understanding the relationships between the ecological and social systems across time and space including delays in the system. The causes underlying such problems are diverse and difficult to identify. None-the-less it is plausible to assume that at the heart of these problems is the difficulty of conceptualizing an organizational system and its real-world domain.

Just as new ways of thinking about complex systems was brought about by people from different disciplines looking for commonality (Waldrop 1992), the complexity of the worldwide web with its diversity of knowledge (ontology) has been instrumental in new ways of integrating and understanding information across the internet. It was realized that in information systems research, involving interdisciplinary teams, there was a need to communicate differences in initial conceptualizations. Ontologies are rapidly becoming mainstream in information systems researched by the Semantic Web and Artificial Intelligence communities (Guarino 1998) and gene systems (Gene Ontology Consortium 2004) as a means to create conceptual models of the real world that are formalized and semantically accurate, and that are linguistically accurate in the sense of meaning. It is considered that a research focus on Ontological-Driven Information Systems has helped the information systems field develop rapidly (Guarino 1998).

It is important to understand language in different contexts as these are likely to affect the meaning. Semantics includes how meaning is constructed, interpreted, clarified, obscured, illustrated, simplified, negotiated, contradicted and paraphrased. Consequently it is necessary to understand epistemology, as it underlies semantics in a fundamental way. Research on ontology and epistemology and semantics has not been invoked in research in the risk governance community. Ontologies have the potential to contribute to and better inform the diverse information required to understand risk governance at the nexus of a systemic approach to decision making processes to natural and technological risk in social-ecological systems. Mapping of information systems onto the real world system is the key to accurately modeling the system of interest because all we have is the map of the organization (de Cesare et al. 2003). Risk governance is a systemic approach to decision making processes associated with natural and technological risks. It is based on the principles of cooperation, participation, mitigation, sustainability and adopted to achieve more effective risk management that is convergent with other public and private policies. It seeks to reduce risk exposure and vulnerability by filling gaps in risk policy, in order to avoid or reduce human and economic costs caused by disasters (Paleo 2009).

The message from industry is that increasing numbers of university graduates will be working in digitally enhanced environments where there are few transportable blueprints for project design and management. If higher education is to play a key role in capacity building for graduates' professional workforce futures, then a learning environment of induction into knowledge needs radical reworking into a learning environment in which teachers and students work as co-creators and co-assemblers (and dissemblers) of transdisciplinary knowledge applications for digital work futures (McWilliam 2007). Consequently there is a need to integrate multiple types of knowledge.

In the field of risk governance in which there is integration across diverse fields of knowledge I wonder if it is now safe to again talk about systems thinking and ontology and epistemology in order to integrate and create successful interdisciplinary teams (Wadsworth 2008). I propose that we have a requirement for an Ontological-Driven Risk Governance System to integrate across different types of information.

6.8 Conclusion

The frameworks and metaphors described in this chapter are the heuristic models that represent our knowledge, and form the shared reality about how we understand the world around us. It is clear that our past understanding of how we conceptualize the context and structure of the world, define problems, identify solutions and implement actions has failed to deal with the increasingly complex social-ecological systems. The paradox of the competing values of stability and flexibility, the means versus the ends and the factors that may confer effectiveness and resilience will be found in new conceptualizations. Systems and resilience thinking are now more than ever finding traction in science and policy. The new models based on systems thinking and resilience stress the importance of assuming change and explaining stability instead of assuming stability and explaining change. The emphasis on different ways on knowing reality is a social process. Consequently the role of the individual, the unit that makes up society and collective decision making must be investigated to identify the distinctive qualities that this introduces. I suggest that the habitual ways of understanding the world are linked to the dominant ontology and epistemology which in turn is linked to culture and the distribution of psychological type. Individuals have a predisposition to view the world in a certain way and science has been instrumental in promoting only one paradigm, the rubric of disciplinarity and stability. More systemic and resilience approaches have challenged this dominant view but they remain subordinate. In order to progress to a systemic and dynamic worldview in which systems go through a natural series of growth, recession, depression and recovery, changes must take place in our educational systems to inform our future decision makers. Education must change to prepare future decision makers to be self-reflective and to understand their innate preferences, to think critically and have a greater capacity integrate information from diverse sources, to learn and make risk management decisions in complex and dynamic systems.

References

Ackoff, R.L. 1999. *Ackoff's best: His classic writings on management*. New York: Wiley.

Alexander, M.A. 2002. *The Kondratiev cycle. A generational interpretation*. Lincoln: Writers Club Press.

Allan, C., and A. Curtis. 2005. Nipped in the bud: Why regional scale adaptive management is not blooming. *Environmental Management* 363: 414–425.

Allison, H. 2011. *Mixed ontologies and epistemologies in resilience. Addressing some of the criticisms about resilience*. Paper presented at the Resilience 2011 Conference. Resilience, Adaptation and Transformation in Turbulent Times. Arizona State University, Arizona, USA.

Allison, H., and R. Hobbs. 2006. *Science and policy in natural resource management: Understanding system complexity*. Cambridge: Cambridge University Press.

Allison, H., and R. Hobbs. 2010. Natural resource management at four social scales: Psychological type matters. *Environmental Management* 45: 590–602.

Ball, I.L. 2009. The sum of us: Australian data on the distribution of psychological types. *Australian Psychological Type Review* 11(1): 23–25.

Bazerman, M.H., and M.D. Watkins. 2004. *Predictable surprises: The disasters you should have seen coming, and how to prevent them*. Boston: Harvard Business School Press.

Bendixen, L.D., and D.C. Rule. 2004. An integrative approach to personal epistemology: A guiding model. *Educational Psychologist* 391: 69–80.

Berens, L.V. 2001. *Understanding yourself and others. An introduction to interaction styles*. Huntington Beach: Telos Publications.

Berkes, F., J. Colding, and C. Folke (eds.). 2003. *Navigating social-ecological systems. Building resilience for complexity and change*. Cambridge: Cambridge University Press.

Beyler, J., and R.R. Schmeck. 1992. Assessment of individual differences in preferences for holistic-analytic strategies: Evaluation of some commonly available instruments. *Educational and Psychological Measurement* 523: 709–719.

Blann, K., and S. Light. 2000. *The path of last resort. Adaptive environmental assessment and management AEAM Nine heuristics of highly adaptive managers*. Retrieved from: http://www.iatp.org/AEAM/primer.htm.

Boholm, A. 2009. Speaking of risks: A matter of context. *Environmental Communication: A Journal of Nature and Culture* 33: 335–354.

Botkin, D.B. 1990. *Discordant harmonies: A new ecology for the twenty-first century*. New York: Oxford University Press.

Bradshaw, G.A., and M. Bekoff. 2001. Ecology and social responsibility: Re-embodiment of science. *Trends in Ecology and Evolution* 16(8): 460–465.

Briggs-Myers, I. 1999. *Introduction to type. A guide to understanding your results on the Myers-Briggs Type Indicator®*, 6th ed., by Linda K. Kirby and Katherine D. Myers. Palo Alto: Consulting Psychologists Press, Inc./Australian Council for Educational Research.

Brooks, L. 1978. Nonanalytic concept formation and memory for instances. In *Cognition and categorization*, ed. E. Rosch and B.B. Lloyd, 169–211. Hillsdale: Erlbaum.

Bruce, A., C. Lyall, J. Tait, and R. Williams. 2004. Interdisciplinary integration in Europe: The case of the fifth framework programme. *Futures* 36: 457–470.

Burrell, G., and G. Morgan. 1979. *Sociological paradigms and organizational analysis: Elements of the sociology of corporate life*. London: Heinemann.

Cameron, K.S., and R.E. Quinn. 1999. *Diagnosing and changing organizational culture: Based on the competing values framework*. San Fransisco: Jossey-Bass, a Wiley imprint.

Capper, P., and B. Williams. 2004. Enhancing evaluation using systems concepts. *American Evaluation Association*. http://users.actrix.co.nz/bobwill/activity.doc. Accessed 3 Aug 2014.

Carry, A. 1996. Uncertainty in the writings of Kondratiev. In *Uncertainty in economic thought*, ed. C. Schmidt, 126–145. Cheltenham: Edward Elgar.

Claxton, G. 2004. Learning is learnable (and we ought to teach it). In *Ten years on*, ed. Sir J. Cassell. National Commission for education report, UK.

Crestani, I. 2001. *Scientists talk to the military: Impact of communication preferences on client relationships*. Dissertation, Charles Sturt University.

Dale, V.H. 2002. Science and decision making. In *Understanding and solving environmental problems in the 21st century. Toward a new integrated hard problem science*, ed. R. Costanza and S.E. Jorgensen, 139–152. Amsterdam: Elsevier.

de Cesare, S., M. Lycett, and R.J. Paul. 2003. Actor perception in business use case modeling. *Communications of the Association of Information Systems* 12: 223–241.

De Greene, K.B. 1993. *Policy challenges in a world of nonlinearity and structural change. A systems-based approach to policy making*. Boston: Kluwer.

Department of Education. 2004. *Employability skills for the future*. Canberra: Australian Government Publishing Service.

Dole, J.A., and G.M. Sinatra. 1998. Reconceptualizing change in the cognitive construction of knowledge. *Educational Psychologist* 332(3): 109–128.

Doyle, J.K. 1997. The cognitive psychology of systems thinking. *System Dynamics Review* 133: 253–265.

Fazey, I. 2010. Resilience and higher order thinking. *Ecology and Society* 153: 9. http://www.ecologyandsociety.org/vol15/iss3/art9/.

Fischer-Kowalski, M., and J. Rotmans. 2009. Conceptualizing, observing, and influencing social–ecological transitions. *Ecology and Society* 142: 3. http://www.ecologyandsociety.org/vol14/iss2/art3/.

Gallopin, G.C. 2002. Planning for resilience: Scenarios, surprises, and branch points. In *Panarchy: Understanding transformations in human and natural systems*, ed. L.H. Gunderson and C.S. Holling, 361–392. Washington, DC: Island Press.

Garfield, M.J., N.J. Taylor, A.R. Dennis, and J.W. Satzinger. 2001. Modifying paradigms – Individual differences, creativity techniques, and exposure to ideas in group idea generation. *Information Systems Research* 123: 322–333.

Gene Ontology Consortium. 2004. The gene ontology GO database and informatics resource. *Nucleic Acids Research* 32(Suppl 1): D258–D261.

Giddens, A. 1990. *The consequences of modernity*. Stanford: Stanford University Press.

Giddens, A. 1999. *Runaway world: How globalization is reshaping our lives*. New York: Routledge.

Goldspink, C., and R. Kay. 2004. Bridging the micro–macro divide: a new basis for social science. *Human Relations* 575: 597–615.

Guarino, N. 1998. *Formal ontology and information systems. Formal ontology in information systems*. Paper presented at the FOIS'98, Trento.

Gunderson, L.H., and C.S. Holling (eds.). 2002. *Panarchy: Understanding transformations in human and natural systems*. Washington, DC: Island Press.

Hofer, B.K., and P.R. Pintrich (eds.). 2002. *Personal epistemology: The psychology of beliefs about knowledge and knowing*. Mahwah: Lawrence Erlbaum.

Holling, C.S. 1973. Resilience and stability of ecological systems. *Annual Review of Ecology and Systematics* 4: 1–24.

Holling, C.S., L.H. Gunderson, and G.D. Peterson. 2002a. Sustainability and panarchies. In *Panarchy: Understanding transformations in human and natural systems*, ed. L.H. Gunderson and C.S. Holling, 63–102. Washington, DC: Island Press.

Holling, C.S., L.H. Gunderson, and D. Ludwig. 2002b. In quest of a theory of adaptive change. In *Panarchy: Understanding transformations in human and natural systems*, ed. L.H. Gunderson and C.S. Holling, 3–22. Washington, DC: Island Press.

International Risk Governance Council. 2009. *Risk governance deficits. An analysis and illustration of the most common deficits in risk governance*. Geneva.

International Risk Governance Council. 2012. *Risk governance deficits – Project overview*. Retrieved from: http://www.irgc.org/-Risk-Governance-Deficits-.html.

Jahoda, M., K.L.R. Pavitt, H.S.D. Cole, and C. Freeman (eds.). 1970. *Models of doom: A critique of the limits to growth*. New York: Universal Publishing.

Jayaratna, N. 1994. *Understanding and evaluating methodologies. NIMSAD: A systemic framework*. London: McGraw-Hill.

Karlqvist, A. 1999. Going beyond disciplines: The meanings of interdisciplinarity. *Policy Sciences* 324: 379–383.

Killen, D., and G. Williams. 2009. *Introduction to type and innovation*. Mountain View: CPP.

Matthies, M., H. Malchow, and J. Kriz (eds.). 2001. *Integrative systems approaches to natural and social dynamics*. Heidelberg: Springer.

McWilliam, E. 2007. *Unlearning how to teach. Creativity or conformity? Building cultures of creativity in higher education*. Paper presented at the conference organized by the University of Wales Institute, Cardiff in collaboration with the Higher Education Academy, Cardiff.

McWilliam, E. 2009. *The creative workforce: How to launch young people into high-flying futures*. Sydney: UNSW Press.

Meadows, D.H., and J.M. Robinson. 1985. *The electronic oracle: Computer models and social decisions*. Chichester: Wiley.

Muneepeerakul, R., and M.R. Qubbaj. 2012. The effect of scaling and connection on the sustainability of resource-population system. *Ecological Economics* 77: 123–128.

Myers, I.B., M. McCaulley, N.L. Quenk, and A.L. Hammer. 1998. *MBTI manual: A guide to the development and use of the Myers-Briggs type indicator*. Camberwell: Consulting Psychologists Press Inc./Australian Council for Educational Research.

Paleo, U.F. (ed.). 2009. *Building safer communities. Risk governance, spatial planning and responses to natural hazards*. Amsterdam: Ios Press.

Piaget, J. 1970. *Genetic epistemology*. New York: Columbia U.P.

Prigogine, I., and I. Stengers. 1997. *The end of certainty: Time, chaos, and the new laws of nature*. New York: Free Press.

Quinn, R., and J. Rohrbaugh. 1983. A spatial model of effectiveness criteria: Towards a competing values approach to organizational analysis. *Management Science* 29: 363–377.

Rhoten, D., and A. Parker. 2004. Risk and rewards of an interdisciplinary research path. *Science* 306: 2046.

Robinson, A. 2005. A polymath's dilemma. *Nature* 438: 291.

Scheffer, M., S. Carpenter, J.A. Foley, C. Folke, and B. Walker. 2001. Catastrophic shifts in ecosystems. *Nature* 413: 591–596.

Schumpeter, J.A. 1950. *Capitalism, socialism and democracy*. New York: Harper and Row.

Skaugen, K. 2011. *High order bit*. Paper presented at the Web 2.0 Summit 2011, San Francisco.

Slaughter, R.A. (ed.). 1996. *New thinking for a new millennium*. London: Routledge.

Soffer, P., and I. Hadar. 2007. Applying ontology-based rules to conceptual modeling: A reflection on modeling decision making. *European Journal of Information Systems* 16: 599–611.

Sterman, J.D. 2000. *Business dynamics. Systems thinking and modeling for a complex world*. Boston: Irwin McGraw-Hill.

Stoffle, R.W., and R. Arnold. 2008. *Facing the unimaginable: The limits of resilience and the risk society*. Paper presented at the TG04 sociology of risk and uncertainty at the first international sociology association forum on sociology and public debat, Barcelona.

Tansley, A.G. 1935. The use and abuse of vegetational concepts and terms. *Ecology* 16: 284–307.

van Rooyen, J. 2007. *South African MBTI form M data supplement*. Johannesburg: Jopie van Rooyen & Partners SA Pty Ltd.

Wadsworth, Y. 2008. Is it safe to talk about systems again yet? Self organizing processes for complex living systems and the dynamics of human inquiry. *Systemic Practice and Action Research* 213: 153–170.

Waldrop, M.M. 1992. *Complexity: The emerging science at the edge of order and chaos*. New York: Simon & Schuster.

Walker, B.H., J.M. Anderies, A.P. Kinzig, and P. Ryan. 2006. Exploring resilience in social-ecological systems through comparative studies and theory development: Introduction to the special issue. *Ecology and Society* 11(1): 12. http://www.ecologyandsociety.org/vol11/iss1/art11.

Wells, G., and G. Claxton (eds.). 2002. *Introduction: Sociocultural perspectives on the future of education*. Oxford: Blackwell.

Young, J., and M. Marzano. 2010. Embodied interdisciplinarity: What is the role of polymaths in environmental research? *Environmental Conservation* 37(4): 373–375.

Helen Elizabeth Allison received her doctorate on the resilience of regional social-ecological systems from Murdoch University, Australia, her Masters in marine science from James Cook University, Australia, and Bachelor's degree with Honours in Zoology from Aberdeen University, Scotland. Helen currently works with groups of interdisciplinary scientists in dynamic and interlinked social and environmental contexts to support real-world decision making. She is the author of the book *Science and Policy in Natural Resource Management: Understanding System Complexity* and papers on systems thinking and personality type. Helen is Honorary Senior Research Scientist at Murdoch University.

Chapter 7
Interest, Interest, Whose Interest Is at Risk? Risk Governance, Issues Management, and the Fully Functioning Society

Robert L. Heath and Katherine McComas

7.1 Introduction

Risk management, including risk governance, is an ancient topic that has recently become a highly focused and robust discipline. As it changes and evolves, many points of analysis come into play, and paradigms push and shove one another. As Renn (2009, p. 80) set the scope and purpose of this discipline, he conceptualized it in the following way:

> The ultimate goal of risk communication is to assist stakeholders and the public at large in understanding the rationale for a risk-based decision, and to arrive at a balanced judgment that reflects the factual evidence about the matter in hand in relation to the interests and values of those making this judgment.

By the logic of Renn's challenge, principles and strategies of risk management and risk communication both provide a rationale for societal risk governance as well as derive their shape and purpose from the dynamics of risk governance as it incorporates science, culture, and aligned interests. Conceptualized in that way, discussions of risk governance raise the question: By what authority and in whose interests does risk governance operate?

Such positioning and reconceptualization result from the notion that risk management may feature the role of risk governance as individual organizations adjusting their affairs to manage various risks, some of which they create. A focus on the individual organization's role and ability to create meaning can presume an atomistic and agentic organization approach to risk governance. It can reason that how risks

R.L. Heath (✉)
School of Communication, University of Houston, Houston, TX 77204-3002, USA
e-mail: rheath@uh.edu

K. McComas
Department of Communication, Cornell University, 339 Kennedy Hall, Ithaca, NY, USA
e-mail: kam19@cornell.edu

U. Fra.Paleo (ed.), *Risk Governance*, DOI 10.1007/978-94-017-9328-5_7

play out and are 'managed' (which includes analysis, governance, and communication) is the responsibility of individual organizations exhibited various standards of corporate social responsibility. These can, as the logic goes, either be organizations that create risks (such as nuclear power generation) or have the mission and vision to manage risks on behalf of others (such as environmental regulation, public health and emergency management).

The narrow organization-as-agency paradigm pushes responsibility for risk governance onto the shoulders of such organizations, which thereby are expected to know and wisely manage risks. But it can, and often does, fail to ask in whose interests control is brought to bear or even possible regarding certain risks related to their management and decisions about who benefits and loses in the context of each risk. The same logic brings spotlight focus onto organizations that are presumed to or explicitly charged with managing risks such as global sustainability, environmental change, and infectious disease. Such mandates can be as narrow as government agencies that work to communicate with and motivate individuals to make wise personal health choices, which collectively tend to have public health consequences. Tobacco use is one example, as is vaccination and other means of dealing with communicable disease – especially that which can reach pandemic stages.

Issues of the sort highlighted in this introduction suggest many permutations on risk, responsibility, and locus of control that are fundamental to the dynamics of risk governance. On such matters, the history of risk management is as ancient as human society. However, a case can be made that the nineteenth century brought new challenges, advanced by the twentieth century, which have set the stage for a necessary discussion of risk governance in the twenty-first century. Central to that topic is the paradox that risks and their governance tends to require insights into the compatibility and collisions of interests within the decision-making capabilities of societal infrastructures and the meaning they create – as well as the meaning that empowers or disempowers various voices within the scope of the management of specific risks. Such challenges center attention on the willingness and ability individually and collectively of institutions and individuals to exert control of risks in the public interest.

Risk governance has emerged as a concept begging continuing insights, definitions, and practical application. It centers attention on how risks are perceived, scientifically assessed, culturally interpreted, and discussed in infrastructures. Such discussion necessarily pits some interests against others. It presses for insights into how societies create infrastructures and craft meanings that make collective decisions. Relevant to such perplexities, the literature of deliberative democracy postulates that risk governance is best when it is collaborate, integrative, and focused to achieve a fully functional society. That perspective does not deny the importance of the agentic organization working to manage and communicate about its risks or risks for which it is responsible, but it reasons that discourse is best in an agentic society where the abilities and challenges, both private or public, are brought to bear for the collective good.

With that thought in mind, this chapter offers insights into the challenges and gains from effective risk management based on the logic provided by leading paradigms in risk management and communication as well as the study and practice of issues management. Framed in this way, risk governance requires the infrastructural and interpretive means to analyze conflicting interests and foster the general need for collective self-efficacy and legitimacy as a foundation for achieving a fully functioning society.

This analysis begins with a brief historical review starting in the post-Civil War era in the United States. In the last decades of that century, the USA witnessed a steady move to become an industrialized mass production mass consumption society. As corporations grew in size and enjoyed virtually unlimited risk self-governance, critics of such processes and practices began a battle to define what levels of various risks are safe and how fair safe is. That battle continues today. It is the essential dialectic of the self-interest of individual organizations against the aggregate risk management of society: The rationale of society is the collective management of risk.

This chapter features that dialectic as history and context for self-governance. In addition, it argues that risk management is both a technical and cultural challenge whereby forces of power discourse (institutionalized structures and co-created meanings) are engaged to abate and correct the imbalance of risk benefits and risk bearing. Because of the changing dynamics of the field, the 'new paradigm' of risk management is on solid footing but in need of continued development as it asks: Interest, interest, whose interest is at risk? By that emphasis, risk is not only approached as a matter of sound science but also as an alignment or collision of interests.

A collective approach to risk governance postulates that knowledge and balanced interests are fundamental to the appropriate control of risks in the public interest. As such science has the responsibility of identifying and calculating risk, but the final risk management decisions are best when embedded in a risk communication paradigm that is dialogic rather than monologic. As such, dialogues of risk decision making are no better or worse that the quality of infrastructures in which risk decisions are made in discursively and culturally sensitive ways.

The elaboration of those themes into a theory of risk governance begins with a discussion of risk management.

7.2 Risk Management: Foundations of Analysis, Communication, and Governance

The industrial and global political might of the USA, starting at the end of the nineteenth century, not only parallels the modern era of risk governance in other countries but also set new challenges into place. For instance, that era gave the world mechanically rolled, mass produced cigarettes, which facilitated mass consumption of a product which eventually became one of the nation's and world's greatest public health risks. In the latter part of the nineteenth century and early twentieth century industrial efforts worked to shrug off risks onto other interests, including workers, consumers, and the general public.

As a counterbalance to the growing might of industry and the creation of myriad risks, social movements such as Progressivism worked for legislation and regulation to bring substantially greater control over worker safety, community health, and public safety in areas such as transportation safety and safe food and medicine. Such trends suggest the role of scientific investigation of risks coupled with cultural assessments of those risks. And, in that dynamic, cultures as well as science tend to evolve and become intertwined in complex infrastructures which provide the power resources of risk governance.

Taking on the battle of risks in the last half of the twentieth century, critics voiced concern about nuclear weaponry and war of total annihilation. WWII brought about a new level of industrialization and global reach which set new risk standards and challenges. From that era, the current risk analysis discipline started, essentially tied to the rise of corporativization, which perhaps was nowhere more powerful than in the United States, where endless analysis is brought to bear regarding the identification of risk, management of risk, communication about risk, and weighing of the costs and benefits associated with risk.

Each year focuses risk analysts' attention on an unending array of risks – whether to foresee them or to respond once they manifest themselves by producing varying degrees of harm (micro-biology is one of the new disciplines in this regard). Some risks are timeless, such as diseases like malaria and influenza. Others are new or emerging. Technologies are created and used to reduce risk, such as bisphenol-a (BPA), but these technologies can also create additional risks. One of the uses of BPA is in bottles used to market water and other beverages. BPA is also an ingredient in the plastic that lines cans to reduce the damage from acidic foods such as tomatoes. Such plastics can add value to the quality of plastics but have been indicted as producing long-term health risks especially in children. (As a note, it is worth considering how a retailer such as Wal-Mart decided to move away from products, including those containing BPA, as a move to maintain its reputation as a responsive and responsible vendor of products associated with scientific controversy related to health risks, especially involving children.) Nuclear generating plants designed to one level of safety fail when both that level is exceeded and the maintenance and operation of the plant has not kept pace with the needs for community safety. Mining safety may increase, but miners still perish at work especially when management policies and practices override safety. Bankers and designers of mortgage instruments and marketing plans can create a financial system that leads to global financial calamity – strangely enough wrapped in the mantra of system risk reduction. This list goes on and on.

So risks confront humans and all for which they have dominion. Disciplines and professions have developed since the dawn of humanity to address and deal with such risks: Science, religion, politics, and even conventional wisdom often called 'old wives tales.' The point is that humans recognize risks, believe something can be done to foretell, mitigate, and even avoid them. This recognition also motivates people to either look for leaders who are assigned the power and responsibility to analyze risks and take or prescribe corrective action. As important is the realization that 'ordinary' people want to be part of such decision making (or leave such heavy lifting to risk arbiters who are well positioned to mitigate risk as part of the risk governance process). Those choices are among the many options that result in battles and risk decision paradigms relevant to risk governance.

To that end, and because of several now iconic risk manifestations, people looked to science for guidance, insight, and solutions. Starting in the 1980s, 'hard' science and social science of risk analysis, management, and communication began in earnest to become well established and woven into the fabric of public and private life. In several ways, those approaches to risk are not new. In fact, it can be said that the rationale for society is the collective management of risk because of the timelessness and universal experiences of death and disease, accident and battle and even childbirth. Infrastructural changes occur in such matters whereby government agencies are developed to manage risks, as are private sector NGO (non-governmental organization) risk arbiters, such as the Union of Concerned Scientists.

By that logic, a powerful risk governance proposition is developing and being tested. It argues that societies that do best in collectively managing risks advance better than those that do not. In fact, one can argue that one of the causes of societies to end or be absorbed is explained by their failure to manage vital risks in competition with the superior risk management ability of other societies.

In the 1980s, risk governance led to refinements regarding the concept of risk and the assumption regarding its governance, often focused on either government or industry. The assumption was that if profit corrupts industry to be indifferent to risk management, public policy solutions must force industry to exert greater levels of control over industrially created risks.

The logic of that approach often presumed that the responsibility of either industry or government, depending on circumstances, was related to the obligation to inform. That principle of risk communication grew out of the era of 'right-to-know' and 'failure to inform' logics made public, for instance, by the asbestos controversy. That paradigm, as powerful as it was, could then allow an industry group to proclaim: We have informed the public so if they continue to engage in risky behavior we have satisfied our legal and ethical responsibility. Such a management and communication model can feature the role of information without acknowledging the reality that facts become meaningful only through interpretation. Such interpretation can have a scientific bias, or the bias of some scientists. And, it can have a cultural bias that may ignore, downplay, and augment the perceived seriousness of some risk. Thus, identification and assessment of risk is a vital aspect of risk governance.

As more and more academics and practitioners examined the topic, it became much more than a matter of information and even informed consent to what came by the late 1980s to be called, risk democracy. One reason for moving beyond a risk information paradigm was this: "Risk messages necessarily compress technical information, which can lead to misunderstanding, confusion, and distrust" (National Research Council 1989, p. 3).

That limitation, and many more, challenged those who examined the problem to eventually believe that rather than placing the locus of responsibility on one entity, the principle supporting such responsibility should be shared risk governance. Thus, the new paradigm of risk analysis, management, and communication became captured in the concept of risk democracy (National Research Council 1989).

That move had liabilities as well as advantages. One of the liabilities was a more cumbersome process. As the National Research Council (1989, p. 5) observed, "Communicating with citizens about risks can increase their desire to participate in or otherwise influence decisions about the control of those risks". Even though citizens may not know and may be incapable of appreciating the science of risk, they are presumed to deserve to have a role and voice in deciding what risks are safe, safe enough (Fischhoff et al. 1978), and how fair is safe enough (Rayner and Cantor 1987).

Addressing questions of that kind has led to an even higher sense of democracy, what has been termed deliberative democracy (Palazzo and Scherer 2006; Scherer and Palazzo 2007). By that logic, it is not only an ethical responsibility for organizations of all kinds to engage with citizens but also the essence of their legitimacy and the foundation of the authority by which they operate. They become legitimate based on how willing and able they are to engage as well as the quality of the processes and outcomes of that engagement. Collectively, this leads to a greater ability to control risk and prevent or mitigate its manifestation. That sort of principle is fundamental to any reasonable definition of risk governance.

Where risk communication and governance was in terms of academic and practitioner thinking in the 1980s and where it is today has substantial relevance for the continuing and evolving discussion of risk governance. As implied above, in the 1980s, it tended to be source-based and therefore organizationally agentic. That means that the responsibility for risk management and communication was largely considered to be the responsibility of key organizations. In part that logic arose from the asbestos cases which reasoned in court decisions that the industry had failed to warn people of the health risks. A similar logic arose from iconic cases such as the release of MIC (methyl isocyanate) in Bhopal, India. Key cases that led to the modern approach to risk management and communication tended to focus the responsibility on one or more organizations – with some vague justification based on the public interest.

The decades surveyed in this opening section featured a growing and shifting interest in defining and imposing higher standards of control, often under the umbrella of corporate responsibility and responsiveness. To that end, infrastructures have been created and destroyed in the tug of war over risk governance, Similarly, the vocabulary of risk, with its implications for the creation and shifts in power, has been a battleground, one never independent of collisions and new as well as shifting alignments of interests. In such infrastructures, voices of industry, government, and activists (NGOs) contested levels and locus of responsibility for identifying, mitigating, and assigning the burden and benefits of risks. Such efforts led to the development of various paradigms of risk understanding, perception, and control, which is the topic of the next section. These paradigms can narrowly be seen as different approaches to risk management and communication but more broadly analyzed as the pillars for effective risk government.

7.3 Paradigmatic Foundations of Risk Governance

Early efforts to increase the public capacity or efficacy to engage in risk governance led to the development of the Mental Models Approach (MMA) to risk communication (Morgan et al. 2002). That approach assigns the responsibility to scientists and key organizations, often businesses but also government agencies, to determine levels of risk (how safe is safe) and communicate them to key publics. According to MMA, message design strategies focus on the lack of knowledge, understanding, or agreement on key technical matters between experts (on behalf of organizations) and some part of the public (targeted audiences and/or concerned publics, especially risk bearers). The fundamental reasoning is that risk communication must effectively narrow the gap between what experts know and believe compared to what key publics know and believe.

MMA proponents reason that experts can and must accurately assess each risk's probability of occurrence, its impact, the entities affected, and means for its mitigation. They then should assess the gap in understanding and agreement between the scientists and other segments of any population. Although many contexts and various decision heuristics are routine in risk analysis, one is to calculate the probability of risk occurrence, the likely bearers of the risk manifestation, and assessment of whether the magnitude of harm outweighs the benefits associated with the risk. Finally, they should use communication strategies and tactics needed to narrow the gap so that targeted audiences/concerned publics come to accept the assessment and scientific conclusions preferred by experts.

As solid as that approach is, it is flawed for many reasons. One of which is the tendency for experts to disagree. Another is due to the recurring changes through scientific research in regard to standards used to assess risks, risk occurrence, risk causes, and risk mitigation techniques. A third is that lay audiences may not trust what the experts say and therefore seek alternative sources of information. Such changes suggest that risk management and communication (essential to effective risk governance, as well as the result of effective risk government) is a work in progress, a dialogic rather than monologic process.

Tensions between dialogic engagements versus monologues pose not only ethical challenges but are also at the heart of the risk governance challenge The question is who decides what risk is safe and safe enough. Is that decision best left to experts, often serving corporate or government agencies? Or, should the public, especially risk bearers, be the ones to engage in debate and regulation of safety and public health? Can they raise questions and demand higher standards of risk assessment and management? Can they call for postponing industrial decisions that are not well supported by science? These are the sorts of challenges scientists encounter which may lead them to argue that lay audiences lack the knowledge and analytical skills needed to make sound science assessments and that such populations' risk perceptions are so distorted that sound science cannot prevail.

Substantial amounts of research over the years have addressed various publics' perception of risks and cognitive heuristics that influence how individuals understand, accept, or reject risks -and their scientific assessment. Several key themes arose from such studies, one of which was the issue of gender. Studies found that women tend to be less tolerant of risks to hearth and home than are men, for instance. Risks can be less tolerated if the bearers are children; studies suggest as well that the level of exposure to some hazard is less for older individuals than is the case for those who have young and tender tissue.

Sensitive to questions regarding lay perceptions and acceptance/rejection of risks, scholars such as Covello (1992) and Slovic (1987, 1992, 2000; Fischhoff et al. 1978) initiated a massive research agenda to understand psychometric assessments of risk. Such work concludes that perceptions are not uniform across risk conditions and demographically distinguishable groups. Risk perception is complex, as is risk (in)tolerance. Thus, risk governance can assume that surveys, focus groups, and other tools are vital to the MMA approach or persons who are affected by risks should be included in decisions regarding their assessment, tolerance, and acceptance.

Scientific and social scientific assessments of risks and various groups' perception of them added value to the discipline but tended to privilege science, even social science, and those interests which funded the research and analysis. That kind of bias fosters an organizational agency as opposed to a community empowerment. Governance, in that way, tended to empower science, behavioral science, and the organizations which used such tools in their risk management and communication. But the dynamics of risk governance could, and did, not include the voice of those who were and could be affected by the risks. They could not affect those entities which created, interpreted, and assessed the risks and assigned the role of risk bearers.

Set against those disciplines, the paradigm of cultural interpretation and management of risk developed. It grew out of anthropology and reasoned that societies develop to collectively manage risk. By that logic, insights could be gained to reveal how roles (such as professions) developed specific to risk management. Thus, societies developed warriors which eventually translated into police, military, fire, emergency first responders, and such. Over time cultural sensitivity informed the investigative insight into risk, its management and governance.

That conceptualization, role dependent, played out in permutation. Various roles developed: religious leaders, as well as politicians, teachers, accountants, medical professions, engineers, and technicians came to be professional categories that played roles and performed functions to manage risks. All of these elements of society, and one can also include artists, were created to manage and communicate about risks. For that reason, discussions of risk society sought to understand how the quality of societies depends on how well each recognizes and develops functions and structures (as well as social constructions). Skilled risk assessment and management determines societies' success – or failure. That could and has been used to explain how some societies thrived and others even became extinct.

Drawing on the work of Edward Evans Pritchard and Emile Durkheim, Mary Douglas (1986, 1992, 1997; Douglas and Wildavsky 1983) was one of the intellectual pioneers in the contemporary development of the cultural interpretation of risk. Such works were based on the description of society, and, as critical theory developed, provided analytical tools to assess power and ethics. All of this analysis has substantial relevance for risk governance because it asks: whose science should be allowed to assess risks and whose management decisions should guide the actions of society? Such thinking provokes interest in a wide array of voices speaking to the issue of how safe is a risk… and how fair.

Douglas' work was joined by others such as Ulrich Beck's (1992, 1995, 1996). One of Beck's arguments was that modern risks, such as radiation release, are no longer merely a challenge to and a political matter of a community or country. Risks often cross borders, and peoples of various cultures and countries experience similar risks and the challenges of their management. Such reasoning brought him to analyze the risk society with a particular interest in the topic being transboundary. The nature of the impact changed the conception of the public sphere where discourse and decision occur. Those advances forced reconsideration of the limits and challenges of risk governance. The scope and voices, as well as structures and functions at operation in each public sphere become important to discussions of risk governance. So too are the meanings that define and empower the infrastructures as well as emerge from the dialogue.

The cultural tradition in risk management has deepened and broadened. As Tansey and Rayner (2009, p. 53) observed, "By focusing on the inherently political character of risk controversies, it offers an approach to the interpretation of risk issues that contrasts starkly with atomistic economic, engineering and psychometric approaches". This profound observation sets a solid foundation for the discussion of the complexities and perplexities of risk governance.

One of the principles that flow from this analysis is driven by the recognition of and assumptions relevant to free will. The issue of risk governance is quite different if believed to flow from some immutable force and a decision heuristic that doubts free will. Thus, one can find advocates of risks as 'determined by a god.' They can even be the result of 'intelligent design.' One key challenge, however, is to see the paradox of risk management as not only the problematic of individuals' free will but also the complexity of society's decision making ability that can either recognize free will, or seek to deny it.

In that way, the cultural theory of risk advanced the analysis and gave grounding for an institutional and infrastructural approach. In this regard, work by Ortwin Renn and others is particularly relevant. As we pursue the dynamics of risk governance, then, we see the substance of science, and the discussion of perceptions, as either being incorporated into or challenged by the meaning structures and roles that are identified as discussed as risk culture.

One of the proponents of an infrastructural approach to risk, Renn (2001, 2008, see also 1992) has long worked to draw attention to the means by which each society identifies, interprets, manages, and communicates about risks. That means that risks are analyzed, managed, and communicated about within the infrastructures of society. Such societies require scientific analyses and communication within the spirit of MMA. They also are contextualized by the psychometric analysis of risk perception. They are culturally defined, including the normative ideal of empowering all interests to voice their judgments of risks, their analysis, management, and communication. In this way, scientific investigation and conclusion must survive public scrutiny, no matter how unwise that analysis might be. And the paradigm of communication shifts from monologue to dialogue. Thus, we have the rationale for risk governance as risk democracy.

Renn (2009, p. 80) concluded that "the ultimate goal of risk communication is to assist stakeholders and the public at large in understanding the rationale of a risk-based decision, and to arrive at a balanced judgment that reflects the factual evidence about the matter at hand in relation to their own interests and values." This is a powerful principle, fraught with tensions and perils. It suggests that risk governance depends both on the nature and quality of institutions and the meanings that shape them as well as result from them.

Risk infrastructures are power structures; risk governance is a matter of whose interests are championed and whose interests can or do suffer disproportionally as risk bearers. Risk governance is a process by which science, however sound, becomes the focus of consideration and interpretation. Rights, interests, and values become captured in the discourse and are shaped by the meaning structures at play and resolve themselves into new vocabularies.

One dimension of such structures is the systemically idiosyncratic structures and functions by which individuals perform various roles of risk management. Power by this line of analysis is in part a matter of the inclusion or exclusion as well as the hierarchical order and role of the decision makers. Thus, the institutionalization of risk deliberation, assessment, and management enables risk governance in various ways and to different levels of satisfaction by ritualistically assigning and responding to authorized decision makers.

A second aspect of the risk management and communication power of each society results from and leads to the meaning generated to interpret, mitigate, assign responsibility, and determine tolerance levels. Such meanings can allow or confound minds' ability to perceive, interpret, and assess risk. Similarly, meanings can focus on what various risks mean for individual and collective identities and relationships; individuals' identities can be shaped in terms of risk dynamics. And finally, meaning not only defines the risk tolerance accepted or rejected by society but also has substantial implications for the resilience of society in the face of risk. Seen then as structure and meaning, power explains how organizations and individuals bend reality to their interests and their interests to reality (Heath et al. 2010).

Rather than a monologic approach to risk communication as the outreach aspect of management and sound scientific analysis designed to 'convince,' the outcome power discourse is to help communities to make sound choices and wise decisions. Thus, Renn continued, "It is rather the purpose of risk communication to provide people with all the insights they need in order to make decisions or judgments that reflect the best available knowledge and their own preferences" (p. 80). Such analysis acknowledges that people bring to such discussions and decisions their own insights, however sound or flawed.

Such dialogue has many dimensions, but three of them are (1) science-based factual evidence and probabilities, (2) institutional performance, expertise, and experience, and (3) conflicts about worldviews and value systems (Renn 2009,

p. 81). All three are essential, but for purposes of decisional (governance) efficiency the third item in the list can in practice be slighted and even ignored. However that occurs or does not is vital to the character of each instance of risk governance.

The challenge of risk governance shouts for a kind of discourse that brings the voice of all interests to the public arena in meaningful and empowering ways. That is the only way in which perceptions and evaluations of risk can be brought to bear on the identification, evaluation, and policy formulation that meets the legitimacy standard of deliberative discourse. Such a shift in approach toward deliberative democracy leads to analysis of questions and the search for answers regarding fairness and competency in citizen participation (Renn et al. 1995). Policy, legitimacy, and communicative publics (including NGOs) become the problematic to be solved (Jones 2002).

Deliberative democracy presumes that political haggling, economic bargaining, and ethical discourse come together in varying degrees of harmony and tension. Such value laden discourse is framed in the larger sense of ethics which defines and sets benchmarks for the legitimacy of organizations which may create but regardless are expected to manage risks.

Such deliberation is likely to assess the legitimacy of organizations, roles, individuals, and meanings through comparisons of alternative grand narratives, those that lead to and away from risk manifestation. Relevant to such narratives is the willingness and values of stakeholders agreeing with narratives of risk management about which they deserve to engage (Scherer and Palazzo 2007). As such, risk management is not only a matter of positivism but also, probably even more importantly, a matter of engagement through collisions and alignments of interests and interpretations. Central to such challenges is not merely the spirit of discourse, as risk governance, but also the structures in which such discourse can meaningfully occur.

Risk democracy as a foundation for risk governance rests on many premises. Featuring an infrastructural approach to risk management and communication, Heath et al. (2009) offered what can be seen as a truism essential to the motive for community building through risk communication structures:

> There would be no discipline called risk communication if all of the people of any relevant society perceived the same risks, perceived them in the same way, and reacted to them as of one mind. Instead, we are confronted at the basis of analysis with the reality that differences of many kinds and for many reasons account for the discipline and all of what makes it interesting. (pp. 472–473)

Scherer and Cho (2003) offered additional rationale for advancing the analysis of risk management, communication, and participative governance beyond the conception of risk as a matter of individual cognitive mechanisms, even the purview of ostensibly sound science. They found that "community networks may function in ways similar to organizational networks." For that reason "our understanding of risk perceptions may also be improved by further exploration of social networks" (p. 267). In that regard, the important conclusion is not merely to see networks as a means for the efficient flow of information or even the structured and institutionalized approach to risk management, but to view networks as the interdependent loci of participative decision making, thus risk governance. Efficiency could even be a liability if the volume of the flow was essentially a 'data' dump and because information does not come with pre-packaged and universally accepted interpretation and decisional heuristics (Hadden 1989).

Relevant to such thinking, McComas and colleagues (McComas et al. 2009; Trumbo and McComas 2003) have worked to conceptualize and link public participation and decision making through risk communication. McComas et al. (2009) dug deeply into the polarity of analysis regarding less public deliberation/discussion-oriented studies as compared to more public deliberation/discussion-oriented studies. Such analysis offered insights into this tension, "how much control over risk management decisions should agencies or authorities cede to members of the public" (p. 367). Similarly, such tensions center on the types of knowledge at play in regard to various risks. And, the discussion of risk governance must consider the types of risks as an essential element for the conceptualization and solution of the challenge.

Advancing the discussion of the conditions for participation, McComas et al. (2009) postulated several considerations, among which were incentives to motivate participants: Rational incentives, socio-economic incentives, and relational incentives. A corollary to recognizing these incentives is acknowledging the disincentives to participate, which include individual and infrastructural constraints or tensions surrounding the discursive process. These seem to be pillars for participation, risk governance, to build community efficacy and ensure representative engagement, substantive input, and legitimate outcomes.

That theme will be continued in the next section which explores the problematic of legitimacy (as well as companion concepts such as credibility and trust) as a fundamental battle over control and the challenge of risk governance.

7.4 Legitimacy and Discursive Processes of Interests

Exploring the issue of corporate legitimacy as deliberation, Palazzo and Scherer (2006) proposed "a fundamental shift to moral legitimacy, from an output and power oriented approach to an input related and discursive concept of legitimacy." According to their reasoning, that approach "creates a new basis of legitimacy and involves organizations in processes of active justification vis-à-vis society rather than simply responding to the demands of powerful groups" (p. 71). Another way of unlocking the challenge, a traditional approach to risk governance, conceptualizes organizations as engaging in business planning relevant to risks which then require monologic efforts to achieve concurrence with the plans as being legitimate since they manage risks in ways so that benefits outweigh harms.

Rather than an agentic, science based understanding of such matters, the new challenge is to link legitimacy with deliberative democracy. All voices count dialogically, and the presumption for the start of governance is that one organization or even several actually are not politically legitimate without public vetting. "This shift also signifies the necessary transition from a cognitive and pragmatic approach to a moral approach in more and more legitimacy challenges of corporate decision-making" (Palazzo and Scherer 2006, p. 82).

Organizational decision making is the central and motivating concept in the development of issues management. This discipline grew by name from its origins in the risk controversies in post-WWII USA. The discipline developed as a reactive posture to widespread criticism of private and public sector business models relevant to nuclear war/nuclear energy, environmental impact, identity and fairness issues relevant to myriad demographic profiles, war and weaponry, corporate colonialism, and sustainability. This is not the full list, but suffice it to say, such investigation of legitimacy and the targeting of a legitimacy gap (Sethi 1977) was the essential rationale for social movement activism and thus the rationale for issues management (Heath and Palenchar 2009). One of the central questions underpinning the legitimacy challenges addressed through issues management is how organizations, especially large ones whether private or public sector, accurately perceived and wisely and ethically managed risks relevant to their mission and vision.

Risk governance presumes the requirement of legitimacy (and thus the concept of gap). By that logic, organizations derive their right to operate (authority) from the will of the community where they operate. These artificial citizens are therefore socially constructed entities; the rationale for their existence is that they add value and do no unacceptable harm to the community. That last sentence highlights three concepts relevant to risk governance: A right to operate, added value, and production of no unacceptable harm (and in fact management of risks in the community and individual interests). Corporativism can lead to risks or to role specific expectations that organizations are expected to understand, manage, and communicate effectively in the face of risk. In either case, corporativization presumes the tendency to interpret, manage, and allot risk bearing to the interest of organizations rather than to ways that make society more fully functioning. Instead of the corporate entity serving society, by that logic, society needs to bend to the service of the organization.

Insights into various kinds of legitimacy (Palazzo and Scherer 2006) open the risk governance analysis to the sorts of adjustment organizations and community members need to align their interests and feel individual and collective efficacy (Bandura 1997) in the face of risks and the counterbalance and allotment of their harms and benefits. Palazzo and Scherer identified several kinds of legitimacy:

- Pragmatic: calculations of risk harm-benefit ratios.
- Cognitive: structures, functions, and leadership behaviors that are relevant to sociocultural conditions and contexts.
- Moral: ethical justification of the organization as part of a community.

These become the tensions over legitimacy regarding the organization's willingness and ability to manage risk in ways that lead to and reflect a fully functioning society.

Institutional theory, as a modern corporate and political theory, presumes that organizational legitimacy derives from the structure and functions, as well as rituals, organizations need to produce the outcomes that justify their legitimacy in society. One view of that can feature the virtue of the agentic organization. A postmodern view, built upon reflective management, presumes that any organization can be sufficiently agentic to the extent that its commitment to and justification for existence is focused on the agency of society, the fully functioning society (Heath 2006). That theme is further developed in the next section.

7.5 Risk Governance: Politicization of Legitimacy, Control, Support/Opposition, and Efficacy

Relevant to risk and risk governance, the requirements for an organization's service to achieve a fully functioning society are inseparable from the authority by which the organization works to achieve its mission and vision. Risk governance presumes the politicization of organizations and risks. It elevates the discussion beyond fact and information which are never trivial to risk decisions but are not sufficient alone to support the demands of risk governance. Issues of efficiency and modern management are likely to be resolved in favor of community interests rather that individual interests. Deliberative democracy presumes a constructive dialogue that leads to support/opposition of risk analysis and risk tolerance. And risks are brought within the best kind and amount of control available to the collective judgment of society – its risk governance.

Sound science must accept the reality that it is politicized and must therefore sustain the analysis of fact and conclusion in a sociocultural context. That context has as its foundational rationale the control measures necessary to maximize risk/benefit ratios that depend on psychometric evaluations and cultural judgements. The infrastructure in which this discourse transpires must depend on a process that achieves collective learning, understanding, and fairness. Those are the principles of legitimacy, intertwined with standards of corporate social responsibility (CSR), as well as the guidelines for the collective management of risk.

Risk governance depends on the quality of discourse, the infrastructures in which it occurs, and the ability for the society to become fully functioning with a joined spirit that risk management is not an individual but collective endeavor. So too the CSR standards need to lead to appropriate understanding of the conditions and requirement of control in the face of various degrees of uncertainty. That reality recognizes that risk management and risk governance combine because they are essentially the collective management of uncertainty. Timeless considerations have focused on the ability to recognize, know, and bring reasonable amounts of control to the uncertainty of risk and the magnitude of risk manifestation of various risk bearers.

The essential challenge of risk governance is to distribute risk benefits and harms in ways that achieve community efficacy. In this way, standards of legitimacy presume that efficacy has at least three components: Expert efficacy (sound science, wise and responsible management, and even effective emergency management), self-efficacy (identity and protocols needed by individuals to respond favorably in the face of risk manifestation), and community efficacy (the collective response based on expert and individual self-efficacy) as a collective response of isolates and interdependencies.

7.6 Conclusions

Risk management is the essential rationale for society. For that reason, the discussion of risk governance combines the other pieces of the risk puzzle so that they become a sociopolitical whole. As such, the dialogue over risk asks not only whose interests are at play but how the system and meaning of risk governance serves interests that benefit when aligned. Support for that conclusion results from the reality that sound science has a constructive but not imperial role to play in risk assessment and management. It is a foundational element for discourse, because it is essential to understanding how safe is safe, but even that analysis lacks the sociocultural status of deciding safety as a community construct rather than a probabilistic estimate.

Perceptions, sensitive to psychometrics, and judgments, sensitive to cultural interpretations, are vital to the evaluation of the fairness of risk: How fair is safe? This discussion does not occur in a vacuum or in isolation. It is a collective way of thinking, because ultimately science, evaluation, and judgment are community sensitive topics. If the role of society is the collective management of risk, then a risk governance perspective is needed for the other aspects of risk management and communication to have system and make sense.

The outcomes of the risk governance process are not only understanding and judgment but also support/opposition and community efficacy. It is a political question, one that examines the legitimacy of the roles individuals and organizations play, their efficacy in that regard, and the control that is collectively achievable on a risk-by-risk basis. In that way, risk governance is best when it aspires to result in fully functioning societies.

References

Bandura, A. 1997. *Self-efficacy: The exercise of control*. New York: Freeman.

Beck, U. 1992. *Risk society: Towards a new modernity*. London: Sage.

Beck, U. 1995. Freedom from technology. *Dissent* 42(4): 503–507.

Beck, U. 1996. Risk society and the provident state. In *Risk, environment and modernity: Towards a new ecology*, ed. S. Lash, B. Szerszynski, and B. Wynne, 27–43. London: Sage.

Covello, V.T. 1992. Risk communication: An emerging area of health communication research. In *Communication yearbook 15*, ed. S.A. Deetz, 359–373. Thousand Oaks: Sage.

Douglas, M. 1986. *How institutions think*. London: Routledge and Kegan Paul.

Douglas, M. 1992. *Risk and blame*. London: Routledge.

Douglas, M. 1997. The depoliticisation of risk. In *Culture matters: Essays in honour of Aaron Wildavsky*, ed. R.J. Ellis and M. Thompson, 121–132. Boulder: Westview Press.

Douglas, M., and A. Wildavsky. 1983. *Risk and culture: An essay on the selection of technological and environmental dangers*. Berkeley: University of California Press.

Fischhoff, B., P. Slovic, S. Lichtenstein, S. Read, and B. Combs. 1978. How safe is safe enough? A psychometric study of attitudes towards technological risks and benefits. *Policy Sciences* 8: 127–152.

Hadden, S.G. 1989. Institutional barriers to risk communication. *Risk Analysis* 9: 301–308.

Heath, R.L. 2006. Onward into more fog: Thoughts on public relations research directions. *Journal of Public Relations Research* 18: 93–114.

Heath, R.L., and M.J. Palenchar. 2009. *Strategic issues management*, 2nd ed. Thousand Oaks: Sage.

Heath, R.L., M.J. Palenchar, and H.D. O'Hair. 2009. Community building through risk communication infrastructures. In *Handbook of risk and crisis communication*, ed. R.L. Heath and H.D. O'Hair, 471–487. New York: Routledge.

Heath, R.L., J. Motion, and S. Leitch. 2010. Power and public relations: Paradoxes and programmatic thoughts. In *Sage handbook of public relations*, ed. R.L. Heath, 191–204. Thousand Oaks: Sage.

Jones, R. 2002. Challenges to the notion of publics in public relations: Implications of the risk society for the discipline. *Public Relations Review* 28: 49–62.

McComas, K.A., J. Arvai, and J.C. Beasley. 2009. Linking public participation and decision making through risk communication. In *Handbook of risk and crisis communication*, ed. R.L. Heath and H.D. O'Hair, 364–385. New York: Routledge.

Morgan, M.G., B. Fischhoff, A. Bostrom, and C.J. Atman. 2002. *Risk communication: A mental models approach*. Cambridge: Cambridge University Press.

National Research Council. 1989. *Improving risk communication*. Washington, DC: National Academy Press.

Palazzo, G., and A.G. Scherer. 2006. Corporate legitimacy as deliberation: A communicative framework. *Journal of Business Ethics* 66: 71–88.

Rayner, S., and R. Cantor. 1987. How fair is safe enough? The cultural approach to technology choice. *Risk Analysis* 7(1): 3–9.

Renn, O. 1992. Risk communication: Toward a rational discourse with the public. *Journal of Hazardous Materials* 29: 465–519.

Renn, O. 2001. The role of risk communication and public dialogue for improving risk management. In *Environmental risk planning and management*, ed. S. Gerrard, R.K. Turner, and I.J. Bateman, 312–337. Cheltenham: Edward Elgar.

Renn, O. 2008. *Risk governance: Coping with uncertainty in a complex world*. London: Earthscan.

Renn, O. 2009. Risk communication: Insights and requirements for designing successful communication programs on health and environmental hazards. In *Handbook of risk and crisis communication*, ed. R.L. Heath and H.D. O'Hair, 80–98. New York: Routledge.

Renn, O., T. Webler, and P. Wiedemann (eds.). 1995. *Fairness and competence in citizen participation: Evaluating models for environmental discourse*. Dordrecht: Kluwer Academic.

Scherer, A.G., and H. Cho. 2003. A social network contagion theory of risk perception. *Risk Analysis* 23: 261–267.

Scherer, A.G., and G. Palazzo. 2007. Toward a political conceptualization of corporate responsibility: Business and society seen from a Habermasian perspective. *Academy of Management Review* 31: 1096–1120.

Sethi, S.P. 1977. *Advocacy advertising and large corporations: Social conflict, big business image, the news media, and public policy*. Lexington: Lexington Books.

Slovic, P. 1987. Perception of risk. *Science* 230: 280–285.

Slovic, P. 1992. Perception of risk: Reflections on the psychometric paradigm. In *Social theories of risk*, ed. S. Krimsky and D. Golding, 117–152. Westport: Praeger.

Slovic, P. (ed.). 2000. *Perception of risk*. London: Earthscan.

Tansey, J., and S. Rayner. 2009. Cultural theory and risk. In *Handbook of risk and crisis communication*, ed. R.L. Heath and H.D. O'Hair, 53–79. New York: Routledge.
Trumbo, C.W., and K.A. McComas. 2003. The function of credibility in information process for risk perception. *Risk Analysis* 23: 343–353.

Robert L. Heath is Professor Emeritus at the School of Communication University of Houston. He has engaged in risk communication studies since the early 1990s. Heath is one of the academic pioneers in examining the history and theoretical foundations of strategic issues management. He is the author and editor of 18 books as well as hundreds of papers, articles, and chapters on issues management, crisis communication and management, risk communication and management, public relations, and corporate discourse practices and processes. He co-edited the *Handbook of risk and crisis communication* (2010) with H. Dan O'Hair and *Responding to crisis: A rhetorical approach to crisis communication* with Dan P. Millar.

Katherine McComas is Associate Professor at the Department of Communication, Cornell University. She is the author of over five dozen peer reviewed articles and book chapters on risk communication, environmental communication, trust and credibility, science communication, and public engagement. She is an editorial board member for the journals *Science Communication and Environmental Communication* and an Associate Area Editor for risk perception/communication for the journal *Risk Analysis*. Her research interests lie in how people communicate about health, science, and environmental risks, the incentives and barriers people face in the context of risk communication and how it influences people's attitudes and behaviors.

Chapter 8
The Emergence of Civil Society: Networks in Disasters, Mitigation, and Recovery

Daniel P. Aldrich

8.1 Introduction

On 11 March 2011, a tremendous earthquake occurred off Japan's northeastern shore, registering 8.9 on the Richter scale and causing office buildings in Tokyo – more than 200 miles away – to sway like trees in a wind. The quake itself did little damage in terms of casualties – fewer than 5 % of the 19,000 or so victims died from collapsed homes, buildings, or structures. Instead, the earthquake triggered a towering tsunami which, within 40 min, had decimated villages in the coastal prefectures of Iwate, Miyagi, and Fukushima. The quake and the 40-ft tall tsunami disrupted the backup cooling systems of the Fukushima Dai-ichi nuclear reactors operated by the Tokyo Electric Power Company (TEPCO). While the seismic activity automatically triggered a shutdown of the three operating plants (out of the six on site), residual heat within them spiked upwards of 2,000° F despite attempts to keep them cool. The zircalloy fuel rods holding the nuclear fuel (uranium oxide pellets) melted, dropping the fuel onto the floor of the pressure chambers; engineers believe that there were complete meltdowns in the three reactors and these may have created cracks in thick steel plate floors.

Despite initial hesitation about the proper course of action, TEPCO and the government (along with members of the US military) worked to pump salt water into the complex to reduce the temperature (in a procedure labeled by engineers as a 'feed and bleed' process), vented the reactors to the atmosphere to reduce the pressure, and sought to evacuate residents who were at risk of exposure to radiation. At one point, more than a quarter of a million people evacuated from their homes, either due to the destruction of the tsunami (which traveled as far inland as 2 miles)

D.P. Aldrich (✉)
Department of Political Science, Purdue University,
100 N. University Street, West Lafayette, IN 47907, USA
e-mail: daniel.aldrich@gmail.com

U. Fra.Paleo (ed.), *Risk Governance*, DOI 10.1007/978-94-017-9328-5_8

or due to concerns about radiation exposure. More than 85,000 people living in and around Fukushima prefecture may not be able to return to their homes for years, if ever.

Japan's 3/11 catastrophe typifies what experts call a 'compound disaster' – an event which interacts with technological, political, social, and geographic conditions and magnifies the overall negative externalities. Several other recent high profile disasters also qualified as compound disasters; Hurricane Katrina, where the bulk of the damage to the city of New Orleans came not from the storm itself, but from the structural failure of the levees built by the Army Corps of Engineers, is another. Due to high engineering standards and mitigation plans, the Tōhoku quake itself caused little damage to either people or facilities across Japan, but in combination with the lax regulation on coastal construction, the placement of generators and batteries at sea level, shortage of alternative transportation infrastructure, and the location of nuclear power plants in highly vulnerable areas within meters of the ocean, the earthquake triggered a crisis which is still ongoing. The destruction of thousands of homes and lives along the Tōhoku area coast has raised questions about Japan's long term plans for coastal development, accelerated demographic changes in the nation's periphery, and raised questions about the economic viability of many fishing communities. The ongoing nuclear crisis has resulted in the banning of a number of agricultural products from the area (including beef, tea, and rice), anxiety among parents and residents about radiation exposure, and an undermining of confidence in Japan's nuclear energy program.

The disaster has had an impact far beyond Japan's shores, prompting other advanced industrial democracies such as Germany, Italy, and Switzerland to move away from atomic energy despite large scale investments in the industry. In Japan, that nation's long term energy plans are shifting away from a focus on nuclear power to one involving safer alternative and renewable sources, including solar and geothermal power. Much of the critical discussion about Japan's response to the quake and ongoing nuclear risk has focused on the work of the central government, large scale nongovernmental organizations (NGOs) such as the Red Cross, and private firms, such as TEPCO and the other regional power monopolies. The central government, for example, works as a unitary, not federal government, holding the purse strings for local governments, which cannot carry out much of the recovery work without Tokyo's support. Critics have singled out these actors for tremendous criticism, ranging from the fact that it took the Red Cross almost 4 months to start distributing 169 billion yen in donations to victims, to the slowness and inflexibility of the central government's handling requests from local communities for aid and logistics.

However, the ham-handed responses from the government and the slow release of data and crucial radiation information from TEPCO have brought about a silver lining: the visible emergence of the effectiveness of civil society in post-disaster settings. The critical role of citizens and residents – and not formal state or private sector institutions – can be seen in disasters around the world, ranging from their role as first responders to their long term involvement in city planning. Furthermore, disasters may serve as catalysts for less active social networks, pushing them to become more active in situations where the government and other providers have failed to perform well. In Japan, for example, the government's slow and incomplete response to the Tōhoku disaster has brought about new forms of transparent 'citizen science' along with a more informed and active citizenry which is willing to directly confront the government. This chapter will underscore the role played by networks of residents, volunteers, and even the mafia in disaster response along with post-crisis recovery and mitigation.

8.2 Social Resources in Disaster

A number of recent academic works have underscored the role of norms, trustworthiness, group solidarity, and mobilization after crises and disasters in different time periods and cultures. Rieko Kage began her 2010 book *Civic engagement in postwar Japan* by showing that levels of recovery after World War II varied tremendously among the 47 prefectures in Japan. Using quantitative data on the number of jobs, hospitals, elementary schools, and library books in each prefecture, she demonstrated how connections among citizens – measured through proxies such as the volume of mail per capita – interacted with other variables to help drive the process of rebuilding. State assistance by itself would not have been able to effectively guide recovery, but by interacting with local citizens, voluntary groups, and other horizontal associations, planners were better able to accelerate the growth process. Local civil society was able to better guide state resources where they were most necessary and avoid duplication and unnecessary investments. Kage's book also used case studies of the Japanese YMCA and judo clubs to show how civic engagement and volunteerism flourished in Japan's new democratic environment and further contributed to its post-war growth. In precisely the environment where we might expect social capital to fail – a society where most urban centers had been reduced to rubble, where millions of citizens had been stationed abroad as soldiers or colonists, and where the government had suppressed democratic thought and activity – it provided a core component of Japan's postwar 'economic miracle.'

Emily Chamlee-Wright's 2010 book *The cultural and political economy of recovery: Social learning in a post-disaster environment* emphasized local narratives, norms, and expectations in New Orleans following the 2005 Hurricane Katrina. Through approximately 300 or so interviews with survivors in locations such as the Village de L'est, the Lower Ninth Ward and Lakeview, Chamlee-Wright found that different neighborhoods had very different stories of their own evacuation, survival, and renewal. Residents in these communities drew on different cultural toolkits for handling the trauma of destruction. Some neighbors had strong self-help ethics which drew neighbors and even strangers together into collaborative efforts to tear down moldy dry wall, work collectively to restart electricity and other utilities, and push local institutions to restart. Other areas of New Orleans seemed to be waiting for aid from the authorities, and displayed less cooperative behavior and a slower pace of regrowth over time. Her focus on "the structure of socially embedded resources" illuminates how, more than aid from the government or damage from the disaster, local norms and social resources strongly determined the trajectory of recovery in the Big Easy (Chamlee-Wright 2010, p. 16).

Rick Weil and his research team have interviewed more than 7,000 residents across New Orleans and gathered data about civic engagement, community work, participation, crime, and recovery, among other factors. His research, like that of Chamlee-Wright, demonstrated the deep connections between positive recovery outcomes and a whole battery of factors related to volunteerism, mobilization, and other collective resources. At the census tract, the indicators of associational involvement, civic engagement, service performance, holding social trust, and attendance at public meetings positively correlated with higher levels of repopulation (Weil 2011). Areas whose citizens did not vote, get engaged in group activities, or work for a common cause are those where blight, crime, and slow recovery remain most evident after the storm. His research hammers home the message: recovery is not primarily a function of government aid, or damage, or wealth before disasters. Rather, social relationships are the drivers of recovery.

8.3 Three Categories of Social Capital

These and other scholars have sought to categorize the role played by the ties, connections and networks that bind us together – resources known by social science as social capital – in the process of disaster recovery and mitigation. Initially named by L.J. Hanifan in the early twentieth century, expanded upon by Pierre Bourdieu (1986) and James Coleman (1988), the field of social capital exploded after Robert Putnam popularized it in his study of the civic roots of Italian economic development (1993) and his follow up research on declining levels of engagement in the United States (2000). Researchers have categorized social capital into three types, depending on the types of connections in the relationship.

The first and most common form of network is bonding social capital. Sociologists have long argued that we connect most regularly and easily with individuals who are similar to us; this condition is called homophily (c.f. Rueff et al. 2003), and predicts, for example, that middle age Japanese housewives are likely to have class friends of similar age and gender. Bonding social capital connects these kinds of kin, family, and friends and can be seen as a "horizontal" connection between individuals of the same power levels. It is strongest in homogeneous communities and often results in closed networks which are relatively stable over time because they are composed of like-minded, similar-value holding residents. Residents in the primarily African-American neighborhood of the Lower 9th Ward in New Orleans, for example, are more likely than counterparts in the Lakeview neighborhood to know their neighbors' names, engage in regular communal group activities, and have higher levels of trust in nearby residents (Elliott et al. 2010). Granovetter's work envisioned these types of connections as 'strong ties,' in contrast to weak ones (1973); in such networks, daily or weekly direct contact is the norm. Studies have demonstrated that bonding social capital does not necessarily generate strong economic growth or motivate entrepreneurial behavior (Gittell and Vidal 1998). Instead, bonding social capital has helped poorer communities 'get by,' but not necessarily 'get ahead' (Narayan 1999; Woolcock 2000). Bonding social ties reinforce tendencies toward homophily but provide few resources beyond often geographically circumscribed boundaries.

The second type of social capital is labeled bridging social capital, and it connects between often isolated or enclosed groups. Parent Teacher Association (PTA) meetings, which bring together individuals who may normally not have encounters in their regular social circles, can foster this type of connection. Other formal institutions and associations such as hobby and sport clubs, employment groups, unions, and schools may create bridging social networks. Bridging social capital has the capacity to cross cultural, religious, and racial lines and reduce conflict among ethnic groups.

Ashutosh Varshney has argued that communities with more bridging organizations experienced fewer (often deadly) Hindu-Muslim riots over the late twentieth century (2002). Through direct contact on a regular basis through cross-cultural voluntary associations, both Hindus and Muslims could reduce ethnic tension, smooth over differences, and solve potentially explosive conflicts before they reach the streets. Granovetter (1973) has argued that 'weak ties' between people who may have little or no direct regular contact, but instead provide extra local, extra-network resources, are most critical in processes such as job searches. That is, many people find jobs not through their personal friends, but rather through friends of friends and colleagues. As a result of research on bridging social capital, institutions such as the Concord Project have sought to "bring together people with fundamentally opposing views or identities for the purpose of promoting civil society while recognizing group differences" (Nelson et al. 2003, p. 1). The Concord Project was founded on the belief that bridging social capital acted as a resource which would create a more harmonious society.

The third and final type of social capital comes when individuals and nodes are connected not through standard horizontal ties – when the connections stretch between people of similar status and power – but rather through vertical ties (Szreter and Woolcock 2004). Scholars have titled this connection as linking social capital. Residents of the Village de L'est neighborhood in northwestern New Orleans rarely interacted with city government officials, NGO leaders, or decision makers before Hurricane Katrina's advent in 2005. Once the storm decimated the city and flooded their neighborhood, though, they quickly worked to find allies and politicians within the power structure who could assist them in their drive to re-establish their community. Their search for power brokers in the administration contrasted with members of the Lakeview community, many of whom already had direct personal ties with city government officials well before the storm.

Similarly, local residents in poverty-stricken villages in the Indian state of Tamil Nadu rarely met their local 'collector;' that is, the government representative who travels throughout sub-regions to connect citizens with government services. The government has published public lists with email and phone contact information for these officials (see http://www.tn.gov.in/telephone/collectors.html) to better connect residents with assistance. Linking social capital provides access to translocal resources, information, and services that are often embedded with decision makers who operate above the day-to-day lives of most residents. Given these different types of social networks, how do social resources focus after crisis?

8.4 Social Capital Mechanisms

My work in disaster recoveries in India, Japan, and the United States has demonstrated that social networks provide assistance before, during, and after disaster through at least three overlapping mechanisms: exit vs. voice, the provision of informal insurance and mutual assistance, and the overcoming of barriers to collective action (Aldrich 2012a). The first decision each survivor must make is to either return to a damaged home or to uproot stakes and relocate to a new community. Of course, in some cases the government will not allow survivors back in; this has been the situation for Japanese survivors of the tsunami whose homes were within 20 km of the Fukushima Dai-ichi nuclear power plants. For those citizens, while they can break the police cordon around the area and return illegally, the radiation levels in these 'hot spots' have created an environment unsafe for long term habitation.

But for the vast majority of disaster survivors, they must select either 'exit' or 'voice.' These terms are borrowed from the work of Albert Hirschman (1970); he discussed the decision making heuristics of disappointed clients and customers who can either stop patronizing a business (exit) or complain to improve levels of service (voice). Exit in a post-disaster setting means relocating to a new community, possibly quite far from one's initial home. Internally displaced people from New Orleans settled as far away as Arkansas and Massachusetts in the diaspora after Hurricane Katrina; many others settled in Houston, Texas. Many of these displaced New Orleanians remain in their new communities some 6 years after the storm.

The decision to stay and rebuild whatever the costs, or to start over in a new location, is heavily influenced by social networks. For individuals who are only loosely connected to a location, or for those who have fewer friends, family, and kin in the area, relocating may be less painful and hence more likely. If one's job has been lost, and one has few friends in the area, restarting life a new location may be a refreshing start. On the other hand, for residents who feel tightly bound to a sense of belonging to local networks, or have friends or family who are staying in the damaged area, relocation may not be an option even if the rebuilding will be an expensive and slow process. Their love for their neighborhood and their connections to their neighbors have them eschew exit and use their voice; they will join with other concerned local citizens to have their needs met by the government. Research on Tokyo residents after the 1923 earthquake which leveled roughly half of the capital city has shown that neighborhoods in which people worked together on common causes – through political demonstrations, riots, and voting, for example – were the ones which regrew population levels most effectively (Aldrich 2012b).

The next mechanism by which social capital works following a crisis is through the provision of informal insurance or mutual assistance. Many people around the world receive services, such as information, medical assistance, child care, food, and water from a combination of public and private organizations. Government welfare offices, free or private medical clinics, and public or private childcare providers are in high demand around the world. Following a severe crisis, these institutions may not be open or able to provide such services to their constituents. After hurricane Katrina in New Orleans, for example, there were almost no grocery stores, hospitals, gas stations, or kindergartens in the city itself. However, members of social networks can serve as substitute – even if temporary – providers of these services when standard providers cannot.

When local stores are shut down and residents need power tools to remove moldy dry wall, they ask neighbors if they can borrow them. If a day care is shut down, local mothers may form a round-robin daycare group until a more permanent solution can be found. Importantly, disaster survivors in need of information about logistics such as restarting utilities, registering a home as damaged, or signing up children in a new school system, will seek out this information from their networks. Similarly, survivors who need damaged homes repaired by contractors will ask their friends and neighbors for the names of those who can be trusted. Those survivors with fewer contacts – that is, lower levels of bonding, bridging, and linking social capital – are at risk of being outside critical information loops and missing the boat (or deadlines, as the case may be) in important areas.

The third mechanism by which social capital alters the recovery trajectory is through lowering the barriers to collective action. During and after disasters communities may have a number of commonly held goals and targets – deterring looting, removing debris, and pressuring local and national authorities to dedicate more resources to their area. Successful accomplishment of these goals requires more than just individual action and will, however. In order to deter looting, local citizens must systematically work to keep out potential thieves and coordinate their watches over the area. The creation of a clean, debris-free neighborhood requires all local homeowners and renters to move their garbage out of the area and keep others from dumping in their neighborhood.

Mancur Olson (1965) and other social scientists established decades ago that there are high barriers to collective action – individuals would often prefer to free ride on the accomplishments of others. Given that highly motivated residents may act independently and take on broader responsibilities, the motivation to shirk for less outgoing individuals is quite high. Why would I use up valuable time and resources to join a community watch patrol if I believe that my eager-to-serve neighbor will fill in for my absence? Why bother walking my garbage an additional hundred yards to a special dumpster if I can toss it over into a ravine or onto the street? These problems require collective obligation, an expectation of surveillance, and a long term perspective.

Neighborhoods with higher levels of bonding social capital are ones where successful collective action is most likely. The aforementioned of Village de L'est in New Orleans returned to their damaged homes far earlier than other communities and found themselves without power and other utilities. They contacted the local utility in New Orleans, Entergy, and were told that they must submit a petition with several hundred names in order to show that they had sufficient numbers to justify these actions. They easily collected several hundred signatures within days of the request. In areas where trust is lower and neighbors have less contact, collective action becomes almost impossible. Following the 1995 Kobe earthquake, the local government offered condominium owners free debris removal if they could provide signatures of all owners within a set period of time. Unfortunately, very few condos took advantage of this offer because they did not have the information about the location of their neighbors.

8.5 Filling the Gap: Weak and Failed States

Some may imagine that social networks and social capital serve as important parts of recovery and mitigation only in advanced industrial democracies, such as Japan, Germany, or the United States. Observers may argue that social networks cannot play such strong roles in weak or failed states. But the crucial role of communities, mutual assistance, and norms of trustworthiness has emerged even in disaster responses in developing nations with weak governance structures, such as Haiti. On 12 January 2010 a 7.0 earthquake struck some 20 miles away from Haiti's capital Port-au-Prince, collapsing buildings, homes, and roads. Estimates of casualties go as high as 300,000, with more than a million residents made homeless by the event (Reuters News, 12 January 2011). Immediately following the quake, neighbors, friends, and family began to dig in the rubble to try to free those trapped underneath. As we have seen in past disasters, the first responders were not professional fire fighters, police officers, or uniformed military personnel, but instead those who knew where to find the bodies. People who were able to use shovels, hands, and other improvised implements tried to save those survivors buried in the ruins. Beyond the emergence of members of civil society as first responders, post-earthquake Haiti provides another example of the power held by communities and norms after crisis.

A number of Haitian residents whose homes survived the quake opened them to internally displaced people, providing shelter and food despite the lack of a mandate from the state or supplies from the private sector. People set up pots of food to share with passersby, and sought to provide what creature comforts that they could. In this environment of communal sharing, individuals seen as violating collectively held norms of cooperation and mutual aid were punished severely. Several reporters captured the gruesome scene as a crowd distributed their form of justice to a suspected thief, who was beaten to death and then dragged through the streets (17 January 2010, The Guardian). This unfortunate outcome arose out of the recognition by locals of the need to work together and protect each other from outsiders who might upset the delicate balance of demand and supply.

Civil society and social networks will often move to fill in a vacuum left by a weak, absent, or poorly performing state. These networks may not be the 'good' social capital that many citizens hope to see active in their backyards. Instead, well organized criminal groups, especially highly disciplined and well-resourced mafia associations, regularly emerge after disaster to provide supplies, deter looting, and assist victims. The motivation for these groups to undertake philanthropic activities is unclear. After Japan's triple-disaster in March 2011, when asked why gangsters worked so hard to deliver supplies to the area, a Japanese mafia boss was quoted as saying "It takes too long for the arm of the government to reach out here so it's important to do it now" (Reuters, 25 March 2011). His words implied that these organizations acted out of a desire to improve the public good. Critics – especially in law enforcement agencies – say that these are attempts to rehabilitate tarnished public images; one observer argued that the yakuza simply enjoy publicity (National Post, 23 July 2003). Whatever the reason, the involvement of these non-governmental, 'uncivil' civil society organizations in disaster recovery is well documented.

In Japan, the mafia group known as the yakuza is well organized and operates visibly, especially in cities such as Kobe and Osaka. In these urban metropolises, local yakuza groups may even hang out their 'shingle' for all to see. On the 17 January 1995, at around 5:45 am, an earthquake registering 6.8 on the Richter scale struck near the densely populated city of Kobe. Close to 6,500 people lost their lives as buildings, homes, and roads across the area collapsed. Fires broke out, and while volunteer fire brigades organized to combat the blazes, many trucks could not reach victims or their homes in time due to narrow streets which were blocked with rubble. The government moved slowly to call out the Japanese Self Defense Forces (SDF) for assistance. In the meantime, as citizens dug with their hands and other simple tools in the rubble, the local yakuza clans organized the delivery of supplies (Fukushima 1995). Some have claimed that the Yamaguchi-gumi criminal syndicate – one of the largest in Japan – was "one of the most responsive forces on the ground" (Adelstein 2011). While these groups operate beyond the law and have been implicated in crimes including stock market manipulation, extortion, and prostitution, they effectively provided post-disaster assistance. Beyond the work of these peripheral social groups post-disaster, recent events in Japan have triggered a renaissance in civil society.

8.6 The Surge in Civic Activities in Post-3/11 Japan

Residents who believe that their governments and dedicated aid organizations are performing poorly may take action when such citizens in the past were reluctant to do so, and this has been the case in post 3/11 Japan. Outsiders have regularly seen Japanese civil society as weak, with civil society organizations (CSOs) only coming into their own since the 1980s (Hirata 2002). Observers have categorized civil society in Japan as "underdeveloped in the arenas that promote democratic agenda" and "underprofessionalized" as it lacks a large cadre of trained professionals in the field (Lowry 2008).

Past disasters in Japan have created ‘silver linings’ of volunteerism. After the 1995 Kobe earthquake, for example, more than one million volunteers flooded into the city to assist, and this outpouring of volunteers triggered a radical shift in Japan’s laws towards nonprofit organizations (NPOs) (Haddad 2007). In 1998 the government passed the so-called NPO Law (Law for the Promotion of Specified NonProfit Activities) and followed up with the 2001 tax reform which provided tax privileges to the NPOs (Pekkanen 2006). These regulations made it far easier for groups across Japan to receive administrative and financial benefits if they registered with the government and were officially recognized. In the late 1990s, there were fewer than 1,000 recognized NPOs in Japan, but by 2005 more than 20,000 had signed up, and by 2011 more than 41,000 were in the system. Despite this tremendous growth in the number of registered NPOs, in international perspective, these numbers (especially as a percentage of the population) are quite small, and NPOs remain constrained by very small budgets, staffs, and membership levels.

The Tōhoku disaster and the resulting Fukushima nuclear crisis have altered the civil society landscape for many Japanese residents, however. For the first few days after earthquake and tsunami struck, many people living nearby the Fukushima Dai-ichi received contradictory information. The Fukushima government initially ordered those within 2 km to leave at 7 p.m. on the day of the tsunami, and then the next day the government expanded this evacuation zone to 20 km. On March 16 the United States government advised US citizens within 80 km of the plant to seek shelter elsewhere, while the Japanese government told citizens within 20–30 km not to leave their homes (BBC News, 16 March 2011). Specifically, then-Prime Minister Naoto Kan told residents, “Please do not go outside. Please stay indoors. Please close windows and make your homes airtight.” The US government, at the same time, assisted US military and diplomatic personnel with evacuation from the country. Government authorities argued that they relied on information provided to them by the Tokyo Electric Power Company (TEPCO), which was slow in revealing radiation levels at and around the plant. Later TEPCO argued that failed sensors caused it to under report radiation exposure both to workers at the plant and nearby residents.

Many Fukushima prefecture residents argued that they felt betrayed by years of reassurances that accidents at Japanese nuclear power plants were not possible. One older evacuee told reporters, “We knew it was close by, but they told us over and over again that it was safe, safe, safe, safe. I can’t believe them now. Not at all” (AP News, 15 March 2011). Mothers of children have been especially vocal about their distrust of the government and the private sector’s statements. One mother, who planned on moving some 20 miles further from the plant, told her interviewer that “When the explosion happened, they didn’t say anything about it being dangerous. We don’t trust the media either, since the nuclear plant operator sponsors many newspapers and television stations” (Lim 2011).

Parents of school aged children have argued that the government did little to deal with the pressing issue of decontamination of playgrounds and topsoil near schools, where children regularly play. As a result, many school administrators and parents have kept children indoors during the spring and summer because of fears about radiation exposure. To avoid having tremendous amounts of land declared hazardous, the government has relaxed standards from pre-Fukushima levels in another move which had many Japanese citizens wondering about 'safe' levels of exposure (Watts 2011). The central government recently released additional funding to help remove topsoil from schools in and around the area, but for many parents, it was too little, too late. In Fukushima, each resident can now receive a full body radiation scan along with an invitation to participate in a longitudinal health study on the effects of radiation exposure on health.

In this blizzard of anxiety and confusing and often contradictory claims, many citizens have stepped forward and begun to work to produce clear information through transparent citizen science. Combining 'street science' with crowdsourcing social media tools, the new NGO Safecast project enables volunteers to post radiation level readings to a public website. The maps on Safecast's website utilize more than 600,000 data points collected by volunteers across Japan wielding Geiger counters; by working collectively, they have created detailed, up to date color-coded radiation level maps of Japan. Where the bureaucratic response to a nuclear accident has been to evacuate citizens in concentric circles, these bottom-up maps have shown instead that wind patterns, topology, and geography create hot (and cold) spots in nonlinear, non-predictable ways. Beyond a group of core volunteers who use their cars to take continuous readings, the project enables anyone within Japan to submit their own Geiger counter readings. Sociologists and anthropologists studying Japan often stress the norms of obedience and submission to authority (Nakane 1967; Miyamoto 1995); the creation and maintenance of the Safecast project has been one way that an independent and engaged civil society has emerged in the post 3/11 environment.

Another interesting site of a more active civil society has been the interaction between Fukushima citizens and government decision makers. Past research on conflict in Japan has stressed that Japanese parties usually seek to privatize conflict, removing it from the public realm and thus 'saving face' for participants (Pharr 1990). Hence many studies have shown that Japan has fewer marches and demonstrations than Western nations, and that surveys underscore many Japanese citizens are far less likely to participate in such public protest than counterparts abroad. Therefore recent videos of open strife between local residents and government bureaucrats over the issue of radiation exposure and contamination show how much has changed. Audience members at public meetings asked questions such as, "As other people do, Fukushima residents have a right to avoid radiation exposure and live healthy lives, right?" to which the attending bureaucrats could only respond, "The government has tried to reduce radiation exposure as much as it can." Amazingly, audience members then raised their voices to express their displeasure, shouting out that the civil servants had not answered the question. This and other uncomfortable confrontations in post 3/11 Japan – such as the TEPCO president's visits to temporary shelters where evacuees challenged his handling of the situation – illuminate a society in which citizens may no longer accept the government's or private sector's answers as gospel. A final sign of a more active citizenry have been the large scale marches against nuclear power in Tokyo and other major urban centers. In mid-September 2011, some 50,000 people marched in the Sayonara Nuclear Power rallies, undertaking some of the largest demonstrations in that nation in decades (The Economist, 24 September 2011). While the 3/11 disaster may have put many lives at risk, it may also be opening up a new era in civil society-state relations where local residents, the social networks that bind them, and civil society organizations work together to enact policies in their best interest.

8.7 Conclusions

This chapter has pushed for the recognition of social resources as critical ones in response to disasters. Where standard government and NGO policies on disaster recovery have pushed for physical infrastructure repair and maintenance, such as the repair of bridges, homes, and ports, this approach to crisis has instead seen social infrastructure as the most important factor. Recent research in a number of disaster sites, including post World War II Japan and post Katrina New Orleans has found qualitative and quantitative evidence that social networks matter. The different types of social capital – bonding, which links like-minded people, bridging, which brings together different races, ethnicities, and religions, and linking, which connects people at different power levels – work to assist survivors during and after crises. Social capital works through reducing the use of 'exit' and strengthening 'voice,' the provision of mutual assistance, the overcoming of barriers to collective action.

In developing nations like Haiti and advanced industrial democracies like Japan, social networks provide critical resources, advice, and motivation to survivors and guide them in their post-disaster behaviors. In Japan, after years of relative inactivity, the 1995 Kobe earthquake and the 2011 Tōhoku earthquake have created new forms of participation and altered the legislation relating to civil society. Where the Japanese government and private sector have been slow to provide information, citizen scientists have used social media tools to create transparent databases on radiation exposure levels across the country. Where citizens previously attended government-run meetings which were run as 'rituals of assent' (Gusterson 2000), they now openly challenge decision makers and push them to answer questions directly, along with gathering by the tens of thousands to show their displeasure with central government policies.

Further, much work on civil society in Japan has emphasized the tight connections between civil society organizations and the central government (c.f. Schwartz and Pharr 2003). Many observers would argue that trust in government regularly correlates strongly with trust in other citizens. The new resurgence in civic activism following 3/11 has challenged these conventional approaches to the field, and shown that many citizens now act autonomously in direct opposition to government policies. Where before the state's intervention in civil society may have gone unchallenged, many activists have raised their voices against the 'nuclear village' (*genshiryoku mura*) constructed with deliberate intent by central government decision makers. Citizens who before had faith that the government was acting in their best interests have been pained by revelations that decision makers deliberately withheld critical data about radioactive contamination.

Decision makers in NGOs and governments alike must understand that social networks will continue to play important roles in both mitigation and recovery from disaster. Whereas some disaster responses, such as the random placement of survivors in temporary shelters, actually damage existing social networks, future responses should incorporate social capital into their implementation. In an era when the number and costs of disasters will increase because of global climate change, social networks around the world will continue to serve as the engines of recovery.

Acknowledgments Much of the research underpinning this chapter was carried out on an Abe Fellowship (2007–2008) from the Center for Global Partnership, the Japan Foundation, and the Social Science Research Council. Erik Cleven, Paul Danyi, Urbano Fra and anonymous reviewers provided helpful criticism.

References

Adelstein, J. 2011. Yakuza to the rescue. *The Daily Beast*. March 18.

Aldrich, D.P. 2012a. *Building resilience: Social capital in post-disaster recovery*. Chicago: University of Chicago Press.

Aldrich, D.P. 2012b. Social, not physical, infrastructure: The critical role of civil society in disaster recovery. *Disasters: The Journal of Disaster Studies, Policy and Management* 36(3): 398–419.

Bourdieu, P. 1986. The forms of capital. In *Handbook of theory and research for the sociology of education*, ed. John G. Richardson, 241–258. New York: Greenwood.

Chamlee-Wright, E. 2010. *The cultural and political economy of recovery: Social learning in a post-disaster environment*. London: Routledge.

Coleman, J. 1988. Social capital in the creation of human capital. *The American Journal of Sociology* 94(Supplement: Organizations and institutions: Sociological and economic approaches to the analysis of social structure): S95–S120.

Elliott, J.R., T.J. Haney, and P. Sams-Abiodun. 2010. Limits to social capital: Comparing network assistance in two New Orleans neighborhoods devastated by Hurricane Katrina. *The Sociological Quarterly* 51(4): 624–648.

Fukushima, G.S. 1995. *The Great Hanshin earthquake*, JPRI occasional paper no. 2 (March). Oakland: Japan Policy Research Institute.

Gittell, R., and A. Vidal. 1998. *Community organizing: Building social capital as a development strategy*. Thousand Oaks: Sage.

Granovetter, M. 1973. The strength of weak ties. *American Journal of Sociology* 78(6): 1360–1380.

Gusterson, H. 2000. How not to construct a radioactive waste incinerator. *Science, Technology & Human Values* 25(3): 332–351.

Haddad, M.A. 2007. Transformation of Japan's civil society landscape. *Journal of East Asian Studies* 7: 413–443.

Hirata, K. 2002. *Civil society in Japan: The growing role of NGOs in Tokyo's aid and development policy*. New York: St. Martin's Press.

Hirschman, A. 1970. *Exit, voice and loyalty: Responses to decline in firms, organizations and states*. Cambridge: Harvard University Press.

Kage, R. 2010. *Civic engagement in post war Japan: The revival of a defeated society, 1945–55*. New York: Cambridge University Press.

Lim, L. 2011. After crises, Japanese lose faith in government. *National Public Radio*. June 1.
Lowry, C. 2008. *Civil society engagement in Asia: Six country profiles*. Honolulu: East West Center, Asia Pacific Governance and Democracy Initiative.
Miyamoto, M. 1995. *Straitjacket society: An insider's irreverent view of bureaucratic Japan*. New York: Kodansha America.
Nakane, C. 1967. *Tateshakai no ningen kankei* [Human relations in a vertical society]. Tokyo: Kodansha Publishers.
Narayan, D. 1999. *Bonds and bridges: Social capital and poverty*, Policy research working paper series, no. 2167. Washington, DC: World Bank.
Nelson, B.J., L. Kaboolian, and K.A. Carver. 2003. *The Concord handbook: How to build social capital across communities*. Los Angeles: UCLA School of Public Policy and Social Research.
Olson, M. 1965. *The logic of collective action: Public goods and the theory of groups*. Cambridge: Harvard University Press.
Pekkanen, R. 2006. *Japan's dual civil society: Members without advocates*. Palo Alto: Stanford University Press.
Pharr, S.J. 1990. *Losing face: Status politics in Japan*. Berkeley: University of California Press.
Putnam, R. 1993. *Making democracy work: Civic traditions in modern Italy*. Princeton: Princeton University Press.
Putnam, R. 2000. *Bowling alone: America's declining social capital*. New York: Simon & Schuster.
Rueff, M., H.E. Aldrich, and N.M. Carter. 2003. The structure of founding teams: Homophily, strong ties, and isolation among U.S. entrepreneurs. *American Sociological Review* 68(2): 195–222.
Schwartz, F.K., and S.K. Pharr. 2003. *The state of civil society in Japan*. Cambridge: Cambridge University Press.
Szreter, S., and M. Woolcock. 2004. Health by association? Social capital, social theory, and the political economy of public health. *International Journal of Epidemiology* 33: 650–667.
Varshney, A. 2002. *Ethnic conflict and civic life: Hindus and Muslims in India*. New Haven: Yale University Press.
Watts, J. 2011. Fukushima disaster: It's not over yet. *The Guardian*. September 9.
Weil, R. 2011. Rise of community organizations, citizen engagement, and new institutions. In *Resilience and opportunity: Lessons from the U.S. Gulf Coast after Katrina and Rita*, ed. A. Liu, R.V. Anglin, R. Mizelle, and A. Plyer, 201–219. Washington, DC: Brookings Institution Press.
Woolcock, M. 2000. *Friends in high places? An overview of social capital*, Development research insights 34. Brighton: Institute of Development Studies.

Daniel P. Aldrich is an Associate Professor of political science at Purdue University's Department of political science, a member of the Mansfield US-Japan Network for the Future, and, during the 2011–2012 academic year, an American Association for the Advancement of Science (AAAS) fellow at USAID. He is the author of *Site Fights: Divisive Facilities and Civil Society in Japan and the West* (2010) and *Building Resilience: Social Capital in Post-Disaster Recovery* (2012) along with more than 40 peer reviewed articles, book chapters, reviews, and general interest pieces. His research interests include post-disaster recovery, the siting of controversial facilities, the interaction between civil society and the state, and the socialization of women and men through experience.

Chapter 9
Risk Governance and the Integration of Different Types of Knowledge

Bruna De Marchi

9.1 Defining and Contextualizing Risk Governance

The term "governance" which was once used in restricted disciplinary or professional circles to denote areas of common theoretical or practical interest (e.g. corporate governance) was extended gradually in its scope, becoming of common use but, I suspect, not of common understanding. A trendy utterance doesn't necessarily denote a new concept generating innovative practices. Some of the new users adopt the word as an unnecessary and inappropriate substitute for government, others as a declaration of intents, a statement of principles. For example, in its White Paper, the European Commission (CEC 2001) refers to "European governance" as the rules, processes and behavior that affect the way in which powers are exercised at European level, particularly as regards openness, participation, accountability, effectiveness and coherence. It also adds that these five "principles of good governance" reinforce those of subsidiarity and proportionality. In this case the term acquires a positive connotation (good governance) indicating a path away from obsolete or unsatisfactory practices, but its explanation seems to me rather contorted and somewhat circular.

Here I will adopt the 1995 definition of the Commission on Global Governance; its main merit is of being stated in comprehensible and unambiguous terms, thus facilitating either accord or disagreement with its content. "Governance is the sum of the many ways individuals and institutions, public and private, manage their common affairs. It is a continuing process through which conflicting or diverse interests may be accommodated and cooperative action may be taken. It includes formal institutions and regimes empowered to enforce compliance, as well as informal arrangements that people and institutions either have agreed to or perceive to be in their interest." (Commission on Global Governance 1995, p. 2).

B. De Marchi (✉)
Centre for the Study of the Sciences and the Humanities (SVT),
University of Bergen, Allegaten 34, P.O. Box 7805, N-5020 Bergen, Norway
e-mail: brunademarchi@gmail.com

U. Fra.Paleo (ed.), *Risk Governance*, DOI 10.1007/978-94-017-9328-5_9

This is a very comprehensive characterization, which includes multiple social actors and diverse social practices and applies to different levels of aggregation, ranging from small interest groups to the international arena. It also acknowledges the presence of both conflict and cooperation in human endeavours and incorporates the idea of ongoing change in the design and implementation of strategies for managing "common affairs" in a democratic setting. The definition has also the advantage of being perfectly applicable, with virtually no change, to a wide range of fields and topics, including that of risk. Thus "risk governance" can be described as the various ways in which all interested subjects manage their common "risk affairs", more specifically, within the purpose of this book, those related to the environment.

9.2 From Expert Calculations to Integrated Approaches

Risk is conceived technically as something that can be calculated and expressed quantitatively, most commonly in probabilistic terms. Risk assessment styles differ according to the issues at hand and the disciplinary fields involved, but they are all based on calculations which should produce scientifically sound results possibly applicable to policy, management and communication. The distinction between risk assessment and risk management was traditionally based on the pretended exclusively scientific nature of the former vs. the politically and value constrained character of the latter. Risk communication, the last phase of a linear process, was customarily devoted to correct the distorted perceptions of lay people, unable or unwilling to accept the verdict of the experts.

It has been a long time now, since a number of scientists engaged in risk assessment have recognized the uncertainties involved in the endeavour, thus entering a debate about the relation between facts and values that in the previous decades had been restricted to philosophers of science (e.g. Rudner 1953). This new awareness is well represented by the nuclear physicist Alvin Weinberg who coined the term trans-science (Weinberg 1972) to describe problems which can be expressed scientifically but cannot be solved scientifically. In the same year, Harvey Brooks (1972) then dean at Harvard school of Engineering and Applied Sciences, argued that trans-science was not restricted to normal scientific practices (laboratory and field experiments) but also to the then new techniques of computer simulation models. Some years later, the definition of trans-science was made widely popular by William Ruckelshaus (1984) who had served as EPA (Environmental Protection Agency) administrator when it was created in 1970 and then again in the early 1980s. He claimed that most of the problems he had to face during his tenure in office shared these characteristics.

Although positions remained distant and often incompatible (Jasanoff 1987), an open discussion on the role of scientific inputs in policy decisions progressively became to be perceived as both legitimate and urgent. Moreover it was not limited to risk issues but moved across disciplinary fields and policy issues to embrace the overall relation between science and society. By now a very rich literature exists on themes such as the past and present co-evolution of science and the Modern State; the rational-actor paradigm in decision-making and its limitations, and the use of science and expertise in the legislative, judicial and administrative contexts (Jasanoff 2004; Tallacchini 2005; Cranor 2006; Wynne et al. 2007). Much of it points to the problematic nature of expert advice, and highlights the differences between curiosity-generated research and mandated (Salter 1988) or issue-driven science (Funtowicz and Ravetz 2008) where the criterion of quality shifts from truth or Popperian falsification to robustness (Nowotny et al. 2001).

The relatively new expression "risk governance" seems to reflect this novel state of things where experts must justify their theoretical assumptions and technical procedures when they present risk assessment results and illustrate their implications for policy. Moreover, they must defend their advice or testimony in public, facing questions about both the results and the methods of their work. In other words, no longer does "science speaks truth to power" (Wildavsky 1979), rather it is one among many legitimate perspectives and inputs in policy processes. This of course doesn't mean that anyone can come to the forum advancing all kinds of ideas. It simply means that expert contributions, usually expressed quantitatively, are not easily accepted as objective facts but they are carefully scrutinized, starting from the criteria which were considered in the very framing of the problem (health, economy, ecology, ethics, etc.) and the weights which were assigned to them.

Appraisal, analysis, management, communication, and education remain separate activities, but none of them can be performed in isolation by some expert group without been exposed to public scrutiny and deliberation. Moreover, no one of them can be performed based only on objective facts, which are now seen as inevitably conditioned by value judgments. As Funtowicz and Ravetz put it in their characterization of Post Normal Science, "In the sorts of issue-driven science relating to the protection of health and the environment, typically facts are uncertain, values in dispute, stakes high, and decisions urgent. The traditional distinction between 'hard', objective scientific facts and 'soft', subjective value-judgments is now inverted. All too often, we must make hard policy decisions where our only scientific inputs are irremediably soft." (Funtowicz and Ravetz 2008).

If this claim is correct, as I will try to illustrate, the idea of risk governance cannot be reduced to calculable quantitative risk but must be interpreted broadly, as referring to situations characterized by uncertainty, even ignorance, and complexity, implying a plurality of irreducible perspectives.

9.3 Complex Systems and Risk Surprises

A quarter century ago, the German sociologist Ulrich Beck coined the term *Risikogesellschaft* in the book by the same title, which become very popular when its English translation, *The risk society*, was published (Beck 1986/1992). Since then, ideas that were previously aired only in restricted circles became broadly discussed: first and foremost those concerning the unforeseen negative effects of technologies intended to improve our quality of life and/or to respond to environmental challenges.

The existence of risks generated by technologies which were not purposely designed to be aggressive (like in the military field) had been already experimented through a long list of accidents; Seveso in 1976 and Bhopal in 1984 being just two of those which captured wide public attention and which also influenced the regulation of "major-accident hazards" in Europe, starting from the early 1980s onwards. The debate of "manufactured risks" (Giddens 1999) become progressively more reflexive, questioning not only the experts' capacity of control over supposedly well known technological systems, but the very possibility of understanding their overall functioning, including in their interactions with the natural world.

Coming from organizational studies, the sociologist Charles Perrow contributed greatly to promote the idea that complex systems cannot be addressed with the analytical and practical tools used for simple ones. In his influential book *Normal accidents* (Perrow 1984/1999) he claimed that with "high risk technologies", i.e. systems characterized by high complexity and tight coupling, accidents are inevitable, though not necessarily frequent. Discrete failures can interact in unexpected and unrecognized ways and move from one part of the system to another, possibly leading to its breakdown before those in charge are even able to detect the origin of the problem and the ways in which it escalated. Redundancy, which is included in well designed systems so that a single fault doesn't prevent their functioning, also contributes to increasing their complexity and consequently their vulnerability. To illustrate his thesis Perrow discusses a number of accidents in many sectors, including chemical, petrochemical and nuclear plants, air and marine traffic, dams, mines, etc.

In the early 1990s the expression "natech" was coined to signify natural events which trigger technological emergencies s and has since entered the vocabulary of analysts and practitioners (Showalter and Fran Myers 1992; Steinberg et al. 2008; Menoni and Margottini 2011). The list of "natechs" is endless, as virtually any severe natural event which impacts on a human system has the potential to disrupt its technological devices, including domestic appliances, industrial equipments, lifelines, etc. And the more technologically advanced is the system, the greater is the damage potential. The failure of the nuclear plant at Fukushima Dai-ichi following the earthquake and tsunami that struck Japan on 11 March 2011 is one of the last dramatic examples of what can happen in a highly industrialized and technologically advanced society. Interestingly enough, in a book published 4 years before such catastrophe, Perrow provided a very detailed description of a similar possible failure, also based on chronicles of poor maintenance, lack of foresight and culpable negligence in the US, where accidents were sometimes avoided just out of pure luck (Perrow 2007).

Another very instructive example dates from May 2010, when the eruption of Iceland volcano Eyjafjallajökull disrupted aviation traffic for many weeks. Besides confirming the disquieting power of nature, that episode showed how tightly interconnected is the world in which we live and how vulnerable we all are, independent on where the crisis starts. The technological transport system that daily moves millions of people and enormous quantities of goods from one part to another of the globe did not operate because of something happening in an island located at remote latitude on the atlas and possibly absent from the mental maps of a large majority of people. The obvious fact of finding everyday consumption products on supermarkets' shelves was no longer so obvious with, for example, consumers in Europe starting to realize that coffee is not produced at their latitudes and Ecuador cultivators of long stem roses favored by Russian purchasers worrying about the impossibility to ship them to destination. Also, those travelling for work or leisure suddenly found it hopeless to calculate distance in terms of hours of flight. Reverting to a metrics of miles meant experiencing one's destination as not only more remote, but also not assured.

Taking inspiration from Perrow (1984) and reflecting of episodes such as the ones just mentioned, it seems advisable to start conceptualizing the world as a complex, tightly coupled system, where unexpected or unforeseen interactions between apparently separate units may generate high risk surprises. Natural systems in themselves are extremely complex and our capacity of understanding them (to say nothing of controlling them) is limited rather than increased by our more and more pervasive interferences (accidental or carefully designed) with their functioning. Indeed such interventions cannot but increase system complexity and consequently amplify our ignorance.

9.4 Recognizing Uncertainty and Ignorance

Climate change is a paradigmatic example of the pitfalls involved both in problem framing and the collection of undisputable evidence. Originally framed in terms of atmospheric chemistry, the issue was gradually redefined as a multidimensional one, with anthropic pressure as a key feature. It followed that the contributions of a number of disciplines and a multiplicity of expertise were required for its understanding and effective management. In a process, perceived as urgent, controversy between parties became harsher and harsher, with a polarization of positions which has been simplistically described (sometimes by the same contenders) as a confrontation between two opposing factions: believers vs. skeptics. Many of the stakeholders have qualified the issue as one of Post-Normal Science as there is no way to solve the problem technically and controversy revolves mainly around conflicting values and the treatment of the "uncertainty monster" (Van der Sluijs 2005), i.e. on how to deal with "the confusion and ambiguity associated with knowledge versus ignorance, objectivity versus subjectivity, facts versus values, prediction versus speculation, and science versus policy." (Curry and Webster 2011).

Recently, the IPCC addressed the issue of extreme events and disasters in relation to climate change. In the summary for policy makers (IPCC 2011) two metrics are used for communicating the degree of certainty in key findings: the former expressed in qualitative terms, the latter in quantitative ones. Literally: (a) "Confidence in the validity of a finding based on the type, amount, quality, and consistency of evidence (e.g., mechanistic understanding, theory, data, models, expert judgment) and the degree of agreement." and (b) "Quantified measures of uncertainty in a finding expressed probabilistically (based on statistical analysis of observations or model results, or expert judgment)" (IPCC 2011, p. 16). The degrees of confidence are illustrated by different shades of grey in a double entry table with evidence on the *x* axis and agreement on the *y* axis. The levels of certainty instead are summarized in a template where seven verbal expressions are coded into seven probability intervals. Thus, for example, the top extreme, "virtually certain" indicates that a certain outcome has a 99–100 % probability; the lowest one, "exceptionally unlikely" that it has a 0–1 % probability. In the middle of the continuum, the expression "about as likely as not" signals a 33–66 % probability. These equivalences bring back uncertainty in the domain of risk, ignoring Knight's distinction between the two, based precisely on the possibility of quantifying the latter but not the former (Knight 1921). Moreover, as Risbey and O'Kane remarked already, ignorance is left out of the picture. "Ignorance – they write – is an inevitable component of climate change research, and yet it has not been specifically catered for in standard uncertainty guidance documents for climate assessments" (Risbey and O'Kane 2011, p 755).

The quantification of uncertainty in terms of probability remains a quite arbitrary operation, whereas the main contribution of the report resides in its attempt to integrate "perspectives from several historically distinct research communities studying climate science, climate impacts, adaptation to climate change, and disaster risk management." (IPCC 2011, p. 1). For example many contributions from the social sciences are derived from individual case studies conducted with qualitative techniques of investigation, which neither produce data amenable to statistical treatment nor allow for extensive generalizations. Nonetheless such contributions often provide descriptions and insights as important as those derived from data amenable to numerical treatment. Precisely because "Each community brings different viewpoints, vocabularies, approaches, and goals" (Ibidem), collapsing their findings into the language of probabilities conceals the irreducible nature of uncertainty, which is intrinsic to the phenomena under consideration, dynamic and subject to constant change.

More generally and from a pragmatic point of view, if action in risk matters is justifiable only in terms of predictions of the future, "legitimate doubt about the predictions may remain until they have been empirically verified, that is, when it is too late" (Funtowicz and Strand 2011, p. 2). The authors thus call for a different principle of legitimating public action, "decoupling the concept of responsibility from the aspirations of control over Nature and the future" (Funtowicz and Strand 2011, p. 1).

Since the early days of their involvement in disaster research, social scientists addressed the reciprocal influences between humans and their environment. For most pioneers, the main concern was not the precise quantification of such interactions but the detection of clear signs of their existence, in a time when research on natural hazards was the almost exclusive domain of the physical sciences.

As early as 1934 the geographer Gilbert White, who later funded the Hazard Center at the University of Colorado, wrote in his Ph.D. dissertation that "Floods are acts of God, but flood losses are largely acts of man". Progress in theoretical speculation and empirical research prompted the idea that humans are to blame not only for the consequences, but sometimes also for the causes of disasters previously conceived as "natural". Within the tradition of sociology of disasters, scholars always looked at natural and human phenomena as connected and conceived of them as a combination of hazard and vulnerability, both needing consideration for promoting effective preparation and response (Quarantelli 1998). Nowadays, as geographer James Mitchell has effectively described, "Urban hazards and disasters are becoming an interactive mix of natural, technological, and social events" (Mitchell 1999, p. 484). This definition can be extended beyond the urban setting the author specifically refers to, especially if one takes into account that the geographical and temporal space in which the perverse consequences of the combination of such events appear is not necessarily the same as the one in which the events of any such type have occurred.

Many examples can be added to those already mentioned, an instructive one being the Katrina disaster. Over several decades and possibly centuries the territory of New Orleans and its surroundings was misused and abused so that vulnerability to floods and hurricanes was largely increased (Colten 2005). Over the years the engineering system devised to protect the city became more and more complex, due to subsequent and often uncoordinated changes and additions and also more and more obsolete, due to lack of maintenance and scarce resources. A "normal accident" was inevitable in such a poorly known and poorly managed highly complex and tightly coupled system. Failures in the levees were the sudden manifestation of events long waiting to happen, so that the Katrina disaster was a "highly anticipated surprise" (Colten and De Marchi 2009).

Although the scientific uncertainty on a phenomena such as hurricanes has been considerably reduced and we have achieved good knowledge and an increased capacity of forecast and monitoring, still a great deal of ignorance remains on the vulnerability of the system under threat, of the myriad of its physical and human components and their multiple interactions. In spite of massive resources put in their prevention, disasters continue to cause extended economic losses, disruption of ecological systems and huge suffering for individuals, communities and societies. We are still largely unprepared to face events that we have seen occurring repeatedly for decades and in some cases for centuries such as hurricanes, earthquakes, volcanic eruptions, oil spills, chemical and nuclear accidents, etc. And yet we nourish the hope (or the illusion) that we can anticipate all possible occurrences by the use of sophisticated models and precise calculations, that we can prevent and restore damage by more and more advanced technologies.

9.5 Different Types of Knowledge

The considerations exposed so far are not to be taken as statements against scientific research and technological development. On the contrary, they are an invitation to reconsider the context in which they presently take place, when the most urgent problems seem to be of remediation, facing the perverse effects, the unwanted consequences of technological progress and economic development. As stated before, the present state of affairs seems incompatible with a framing of hazardous contingencies only in terms of calculable risks. If progress in knowledge has shed light on many previously obscure phenomena, it has also continually revealed new areas of ignorance, which are likely to be increased rather than reduced by our growing power of manipulation of nature. The question is then how to act in the face of the irreducible uncertainty embedded in many present day threats; uncertainty, that is, which cannot be reduced by progress in research. If taming the "uncertainty monster" proves impossible, we must learn to live with it, and recognize that it must be given an explicit place in tackling present risks (Van der Sluijs 2005). This change requires a deep revision, though definitely not the abandonment, of the beliefs that have accompanied the success of Western civilization for over three centuries. Whilst we can still subscribe to Descartes' plea for acquiring knowledge which is useful for life (*parvenir à des connaissances qui soient fort utiles à la vie*) the time has come to revisit his idea of the purpose of such knowledge and at the same time ask ourselves which kind of knowledge is presently useful for life. Can we still be sure, in the twenty-first century, that it is possible and desirable for us to become "masters and possessors of nature" (*maîtres et possesseurs de la nature*) as the French philosopher asserted in his 1637 *Discours de la méthode?* (Descartes 1637). Or should we reflect with humility on both our successes and failures and extract lessons for moving towards a desirable and sustainable future?

I maintain that some lessons might be gained also from a re-consideration of traditional, local knowledge, which is instead ignored and discredited when progress is equated to scientific discovery and technology driven control. It would be a mistake to assume that local knowledge is necessarily contrary or alternative to scientific knowledge or, put it the other way around, that the latter is contrary or alternative to the former. Although they are achieved by different means and may be grounded on different types of evidence, they both provide clues to be taken into consideration in decisions about risk issues. Not necessarily can they be reconciled, but the a-priori dismissal of popular beliefs on the assumption that they have no scientific grounding is definitely to be avoided.

The very names of certain localities, for instance, evoke that they have been either dangerous or safe places at the occasion of past events, such as floods or landslides (De Marchi and Scolobig 2012). A telling and tragic example is that of the "Monte Toc" in the Italian Alps, the site of "the most deadly landslide in Europe in recorded history" (Petley 2008). In the local dialect the name of the mountain hinted to a loose terrain and embodied a knowledge gained through centuries of experience and oral transmission. Such knowledge was not dissimilar to the one derived from observations and calculations performed by the (few) geologists who discouraged the construction of a huge dam in that location. Both went unheard and on 9th October 1963 an enormous mass of material from the mountain slope slid into the Vajont reservoir generating a wave of water of about 30 million cubic metres, which destroyed several villages causing 2,500 deaths, immense economic and environmental damage, and everlasting grief. A mixture of technological hubris and selfish interests thus were at the origin of what at first sight might appear as a "natech" disaster, but on further reflection is definitely to be considered of a "techna" type. It was indeed because of the dam construction works that a landslide of gigantic proportions occurred, destroying everything … except the dam, which is still standing, idle and useless, in a moonlike landscape.

There are very many examples of the tendency of experts to ignore so called lay knowledge (which indeed is a different kind of specialized knowledge) and I will mention just a few. In his by now classical article about sheep farming in Cumbria after the Chernobyl nuclear accident, Brian Wynne (1996) shows how the experts sent by the government to assess the radioactive fallout failed to consider the shepherds' knowledge about the composition of the local terrains and the grazing habits of the cattle. This resulted in inaccurate evaluations with negative repercussions on the local farming economy.

I learned of a similar case of lack of humility from the part of the experts when I was invited to a workshop at Værøy, in the Lofoten islands, a few years ago. After the sessions, I used to walk with fellow participants along a deserted runway to a trendy bar for a beer. The bar had been the former check-in hall of the now dismissed local airport, whose story was summarised on a text hanging on a wall. It read approximately as follows: "The Værøy airport was built at Nordland despite strong warnings from local residents about the area being exposed to stormy weather, in particular strong gusts of wind along the mountainside. It was inaugurated in 1986, with great publicity. After 4 years, in 1990, a plane crashed, five people died and the Aviation Authority recommended the airport be closed. And it has remained close since. Already during an Episcopal visitation in 1750, the bishop had noticed that braces had been placed on Værøy church's northern wall in order to support it against the gusts from a terribly high mountain". I saw the same piece of writing in other locations in the island, and the memory of both the accident and the warnings seemed to be lively and shared.

Another example of the importance of local knowledge, in this case for preventing damage, was brought to public attention after the tsunami that hit Japan in March 2011 by a journalist who signaled the existence of a number of stone tablets on the hillside by the coast. "Carved on their face – he writes - are stark warnings such as: 'Do not build your homes below this point', or 'seek higher ground after a strong earthquake'. All such tablets are over a century old and most were erected after a tsunami that killed 22,000 people in 1896" (Fackler 2011).

Present lifestyles, including increased mobility, tend to make traditional knowledge less and less important, not so much with regard to its content (as the examples above show) but because of the ways it is usually transmitted, i.e. orally from one generation to the next or through written documents of limited and informal circulation. Yet, even when the original witnesses of past occurrences are no longer there and their heirs have moved away, there is room for the social sciences and the humanities to use their tools of investigation not only for exploring present attitudes, perceptions and behaviors, but also for digging into the past, interrogating ancient chronicles and testimonies.

9.6 Implications for Risk Communication

Attention to different types of knowledge and inclusion of multiple perspectives in the management of "common risk affairs" will confirm the multi-faceted nature of the issues at hand and reveal the impossibility to do away with uncertainty and ignorance. It is very unlikely that different needs, interests and understandings are easily reconciled, and by no means that would occur on the basis of quantitative assessments and numerical calculations. We have to live with the awareness that there are no simple solutions to complex problems. Actually, we have to become suspicious of simple solutions, as they might be the right answers to the wrong questions.

A change of attitude towards the "uncertainty monster" will have relevant implications for risk governance, including risk communication. As to the latter, approaches in terms of calculable risks have implied the progressive contraposition between a superior and an inferior form of knowledge (or perhaps between knowledge and superstition) as well as between competent experts and ignorant lay citizens. As stated by Mary Douglas, in risk analysis a great effort was put in "trying to turn uncertainty into probabilities" (Douglas 1985, p. 42). And the idea of calculable risks goes hand in hand with that of expert systems (Giddens 1990) which can regulate and control them.

Ian Hacking has asked about the circumstances that made it possible for probability to be "discovered", studied, and partially formalized centuries before, to become so largely applied in the nineteenth century. The systematic collection of statistical data – he answered – originated the possibility of finding regularities in a world where the deterministic vision had been progressively eroding, opening up the frightening possibility of a lawlessness world dominated by chance. Statistics and probabilities were applied extensively to both natural and human phenomena which thus were brought "under the control of natural or social law" (Hacking 1990, p. 10).

Frightening as it might have been, the thought of having to deal with chance solicited observation and attention to environmental signals and promoted collaboration for alerting and protecting those exposed to it. Until not so long ago, even at our latitudes different tones of the church bells informed people of impending dangers and the necessity to take previously arranged actions. Alert systems are nowadays much more sophisticated, often to the point of being transformed into expert systems not directly accessible to the general public. Consequently their usefulness is subject to the capacity of specialized disaster management organizations of instantly translating digital warnings into practical information and advice and of transmitting them to those at risk. Whereas successful transmission depends largely on technical factors, effective translation requires the sharing of linguistic and cultural codes, which frame reciprocal expectations about personal engagement, agency, trust, and responsibility.

Recent research has shown that people's personal involvement in risk prevention activities is related to perception of expert systems, broadly defined. Among residents exposed to flood risk, a strong tendency was observed to overlook personal protection measures and to delegate responsibility for safety to appointed emergency services (Kuhlicke et al. 2011; De Marchi and Scolobig 2012). Thus, paradoxically, the more efficient are professional organizations, the more people conceive of risk prevention has someone else's business, consequently becoming more vulnerable and dependent on external aid. Similarly, the existence of structural devices (dams, levees, barriers, etc.) tends to discourage the respect of non structural measures, such as regulations and restrictions to land use. By their very presence engineering works convey the message that a technological fix is possible, that danger can be eliminated once translated into calculable and manageable risk. Once this passage is interiorised, it is difficult to accept, or even to understand the existence of what the experts call "residual risk", an expression which prompts associations with concepts and experiences that people were encouraged to forget such as danger, fear, uncertainty and lack of control. Engineering artifacts not only perform a material function (e.g. containing a river) but also a communicative one, symbolizing the control of humans over nature.

Of course the point is not to do away with dedicated agencies of engineering solutions, but to combine them with citizens' awareness and preparation. In this endeavor, traditional knowledge about local dangers and connected caution would prove precious, had they not been largely dispelled by the language or risk, as well as changes in demography, mobility and life styles.

When notions of danger, uncertainty, and ignorance are exorcised as it has long been the case, the collapse of levees and dams, the accidental releases of dangerous substances from laboratories, chemical or nuclear sites not only cause physical harm, but also affect confidence in expert systems and diminish trust in designers, regulators and managers. As Nobel laureate Joseph Stiglitz recently put it, "Experts in both the nuclear and finance industries assured us that new technology had all but eliminated the risk of catastrophe. Events proved them wrong: not only did the risks exist, but their consequences were so enormous that they easily erased all the supposed benefits of the systems that industry leaders promoted." (Stiglitz 2011).

9.7 Which Tasks? Whose Responsibility?

Recent major disasters, such as the Deepwater Horizon oil spill in 2010 and the Fukushima Dai-ichi nuclear accident in 2011 clearly showed that high technology systems can fail due to inappropriate design coupled with "lack of imagination" in foreseeing likely future contingencies. They also dramatically revealed that when something goes wrong nobody knows what to do, except reassuring people that actually things are not that bad after all, as denounced, among others, by journalists Tom Diemer (2010) and Geoffrey Lean (2011).

On August 3rd, 2010, over 3 months from the April 20 rig explosion that generated the still uncontained Gulf of Mexico oil spill, an irritated President Obama denounced on the *Today* show both the attempts to minimize its consequences from the part of the BP and the incompetence of the experts. As to the latter, he used a colorful expression which reveals his conception of their role and responsibilities:

"We talk to these folks [experts] because they potentially have the best answers, so I know whose ass to kick" (Diemer 2010). President Obama's expectation was that those who (should) know better are capable if not of avoiding accidents, at least of understanding their dynamics and implementing effective containment measures.

In situations like these, the relation between scientific expertise and policy seems to shift from advice to blame, highlighting the key importance of the science-society co-evolution (Jasanoff 2004), as hinted in Sect. 9.2 of this chapter.

The contested characterization of such relation popped up for public scrutiny even more dramatically in connection with a 2009 earthquake in Italy. The story is worth telling for its many and different implications. On April 6th, a 6.3 moment magnitude Mw earthquake devastated the city of L'Aquila, capital of the Abruzzo region and some neighboring municipalities. More than 300 people were killed, about 1,600 injured and tens of thousands left homeless, with an estimated damage of some 10 billion Euros.

The main shock had been preceded by a large number of sporadic low magnitude tremors (technically a seismic swarm) over the previous months, which had understandably alarmed the residents. In such tense atmosphere, an unofficial warning by a technician formerly working in a laboratory of the National Research Council captured large media attention. On the basis of measurements of radon he had performed, he insisted that a major earthquake was soon going to occur. This outraged the then head of the Civil Protection Department, Guido Bertolaso who reported him to the authorities for diffusing alarming news and subsequently convened a meeting of the Commissione Grandi Rischi (Major Risks Commission), a consultative organ of the National Service for Civil Protection composed of experts in seismic, volcanic, hydrological and other risks.[1] The meeting took place on March 31st in L'Aquila and was followed by a press conference where no specific measures of protection were suggested to the citizens while it was reaffirmed that no scientifically sound method exists to predict earthquakes. Repeated reassurance was provided to journalists and residents that the seismic situation in L'Aquila was normal and actually favorable because of the continuous discharge of energy due to the seismic swarm (Hall 2011).

Following the major earthquake which occurred just a week after, some relatives of the victims brought five members of the Commission and two government officers to court, on the accusation of multiple manslaughter (*omicidio colposo plurimo*) and injuries for failing to provide complete and precise information which might have saved many people's lives. In particular, the allegation was of not having

[1] In the official website of the Italian Civil Protection Agency, the Commissione Grandi Rischi (Major Risks Commission), short for Commissione Nazionale per la Previsione e Prevenzione dei Grandi Rischi (National Commission for Forecast and Prevention of Major Risks) is defined as the formal liaison structure (struttura di collegamento) between the National Service of Civil Protection and the scientific community. Its activities are of a techno-scientific and advisory type and include providing guidance in connection with the forecast and prevention of the different risk situations (*attività consultiva, tecnico-scientifica e propositiva in materia di previsione e prevenzione delle varie situazioni di rischio*).

http://www.protezionecivile.gov.it/jcms/it/commissione_grandi_rischi.wp

taken into account (or duly communicated) the elements of risk derived, for example, from the state of vulnerable buildings, including public ones, which could and should have been monitored from close and possibly evacuated. The trial began in late September 2011 and about a year later the seven defendants were sentenced to six years in prison and to pay huge compensations to the victims. The verdict has not been applied, pending the outcome of the appeal presented by the defendants.

This affair has at least two points of major relevance in relation to the topics discussed in this paper, one very specific and one more general. The first one is which was the question posed (explicitly or implicitly) to the Commissione Grandi Rischi. Certainly it was not about the possibility of a short term prediction, as the negative answer was already known and it would not have been necessary to consult the Commission in that respect. Then, was it about the measures to be taken to minimize the potential damage to the local residents? Or was it, as some circumstances seem to indicate, about ways of reducing their psychological stress, after months of minor tremors? In other words, was the problem framed in terms of public safety or of public control? Some relatives of the victims now claim that because they believed the official information (science speaking truth to power), their dear ones and themselves failed to apply the protective measures that had been transmitted from one generation to the next as part of the local culture of a seismic area.

The question comes to mind whether for saving lives fear in an impending, unpredictable danger may be more effective than belief in scientific probabilistic assessments. And indeed fear is a mechanism that keeps animals (including humans) alert in case of danger, as opposed to panic, which triggers life threatening behaviours. Unfortunately, avoiding panic is such a major concern for public authorities that, as sociologists discovered long ago, they often fail to make the appropriate distinctions, addressing their attention in the wrong direction (Quarantelli and Dynes 1972).

The second point worth exploring, concerns the tasks and the responsibilities of scientists when appointed as advisers in matters of risk and safety. Their expertise is requested not simply to report information on the present state of knowledge in their disciplinary field, which moreover is often incomplete or controversial as it was with seismic swarms in the case just discussed (Grandori and Guagenti 2009). Often, they are expected to provide an informed opinion taking into account not only the evidence derived from their specific discipline, but also the contextual factors which favour or constrain different courses of action. This is not a one person endeavor and permanent communication needs to be established between various disciplines, forms of expertise and types of knowledge, including local one. Certainly the ultimate responsibility on what is to be done doesn't rest with the scientists alone, but the quality of a decision cannot be evaluated separately from the process which led to it, which requires transparency and accountability.

In the case of the Deepwater Horizon mentioned above, we saw how Obama shifted register, from advice to blame. It is not surprising then that Willy Aspinall, professor in natural hazards and risk science at Bristol University, commenting the L'Aquila case recommends seismologists (and other scientists as well) to check their legal position before providing advice (Aspinall 2011).

Nowadays scientific inputs are key in virtually any area of policy and scientists themselves vindicate not only that they must be heard, but also that their research must be publicly funded (be it in seismology, genetics or anything else). Their requests are not advanced on the ground that research is interesting for investigators themselves, but that it is useful for society at large. This implies that the demarcation between value-free scientific assessments and politically constrained decisions is often very fuzzy and moreover suggests that "the scientist should consider a broadened professional role in which he or she is obliged to take on board the wider uncertainties in the professional's decision-making role" (Faulkner et al. 2007, p. 696).

9.8 Conclusion

The governance of present day risks requires the recognition of their complex nature and the awareness that they cannot be fully understood nor managed with traditional risk assessment tools. Similar to past civilizations, also our advanced technological one must come to grips with uncertainty and ignorance. The claim that "Integration of local knowledge with additional scientific and technical knowledge can improve disaster risk reduction and climate change adaptation" (IPCC 2011, p. 14) is now part of official documents and policy orientations.

The awareness of living in a dangerous world calls for personal and collective memory of past experiences and transfer mechanisms from one generation to the next, through oral transmission, written chronicles, symbols and signs enshrined in languages and dialects.

While revisiting our past to learn useful lessons, we must also realize that previous events and actions constrain our choices and ways forward. As a human species, we have colonized the planet to its most unfriendly areas and have grown exponentially in numbers, also thanks to our ability to manipulate our environment. The use of the territory has changed significantly worldwide (e.g. with the progressive abandonment of agriculture) redesigning the maps of vulnerability and inequality and exposing more and more people, especially the poor, to old and new risks.

As stated before, with increased social and geographical mobility, local knowledge often becomes obsolete and is no longer transmitted from one generation to the next. However new "knowledge communities" have been emerging of people who, despite physical distance, share similar concerns about risk issues. Any new disaster offers inputs for reflection on both past and future trends, like in the words pronounced by Iceland President Ólafur Grímsson after the Eyjafjallajökull volcano eruption. In an interview to the Daily Telegraph he said: "In modern societies like Britain and Europe, there has been disengagement between people and nature. There has been a belief that the forces of nature can't impact the functioning of technologically advanced societies … But, in Iceland, we learn from childhood that forces of nature are stronger than ourselves, and they remind us who are the masters of the universe." (Sherwell and Sawer 2010).

Some three and a half centuries from Descartes' death, we need to reconsider with some humility his (and our) conception of humans as the alleged "masters and possessors of nature".

References

Aspinall, W. 2011. Check your legal position before advising others. *Nature* 477: 251. doi:10.1038/477251a. (Published online 14 September).

Beck, U. 1986. *Risikogesellschaft.* Frankfurt: Suhrkamp. (*Risk society, towards a new modernity.* London: Sage, 1992).

Brooks, H. 1972. Letter to the editor (Science and trans-science). *Minerva* 10: 484–486.

CEC (Commission of the European Communities). 2001. *European governance. A white paper.* COM (2001) 428 final. Brussels. 25 July 2001. http://eurlex.europa.eu/LexUriServ/site/en/com/2001/com2001_0428en01.pdf

Colten, C. 2005. *An unnatural metropolis: Wresting New Orleans from nature.* Baton Rouge: LSU Press.

Colten, C.C., and B. De Marchi. 2009. Hurricane Katrina: The highly anticipated surprise. In *Città salute e sicurezza. Strumenti di governo e casi di studio*, ed. M.C. Treu, 638–667. Sant'Arcangelo di Romagna: Politecnica Maggioli.

Commission on Global Governance. 1995. *Our global neighbourhood.* Oxford: Oxford University Press.

Cranor, C. 2006. *Toxic torts: Science, law and the possibility of justice.* Cambridge, MA: Cambridge University Press.

Curry, J.A., and P. Webster. 2011. Climate science and the uncertainty monster. *Bulletin of the American Meteorological Society* 92(12): 1667–1682. doi:10.1175/2011BAMS3139.1.

De Marchi, B., and A. Scolobig. 2012. Experts' and residents' views on social vulnerability to flash floods in an Alpine region. *Disasters* 36(2): 316–337. doi:10.1111/j.1467-7717.2011.01252.x.

Descartes, R. 1637. *Discours de la méthode.* Sixième partie. http://www.webphilo.com/textes/voir.php?numero=453061240.

Diemer, T. 2010. Obama sought to know 'whose ass to kick' on oil spill. *Politics Daily,* August 6. http://www.politicsdaily.com/2010/06/08/obama-sought-to-know-whose-ass-to-kick-on-oil-spill/.

Douglas, M. 1985. *Risk acceptability according to the social sciences.* London: Routledge and Kegan Paul.

Fackler, M. 2011. Tsunami warnings, written in stone. *New York Times* online, April 20.

Faulkner, H., D. Parker, C. Green, and K. Beven. 2007. Developing a translational discourse to communicate uncertainty in flood risk between science and the practitioner. *Ambio* 36(7): 692–703.

Funtowicz, S., and J. Ravetz. 2008. Post-normal science. *The encyclopedia of earth.* http://www.eoearth.org/article/Post-Normal_Science.

Funtowicz, S., and R. Strand. 2011. Change and commitment: Beyond risk and responsibility. *Journal of Risk Research* 14(8): 995–1003. doi:10.1080/13669877.2011.571784.

Giddens, A. 1999. Risk and responsibility. *Modern Law Review* 62(1): 1–10.

Grandori, G., and E. Guagenti. 2009. Prevedere i terremoti: la lezione dell'Abruzzo. *Ingegneria Sismica* XXVI(3): 56–61. (With an extended English abstract, Earthquake Prediction. Lesson from l'Aquila Earthquake).

Hacking, I. 1990. *The taming of chance.* Cambridge: Cambridge University Press.

Hall, S.S. 2011. Scientists on trial: At fault? *Nature* 477: 264–269. doi:10.1038/477264a. (Published online 14 September).

IPCC. 2011. *Summary for policymakers of the special report on managing the risks of extreme events and disasters to advance climate change adaptation* (SREX). http://ipcc-wg2.gov/SREX and www.ipcc.ch

Jasanoff, S. 1987. Contested boundaries in policy-relevant science. *Social Studies of Science* 17(2): 195–230.

Jasanoff, S. (ed.). 2004. *States of knowledge. The co-production of science and the social order*. London/New York: Routledge.

Kuhlicke, C., A. Scolobig, S. Tapsell, A. Steinführer, and B. De Marchi. 2011. Contextualizing social vulnerability: Findings from case studies across Europe. *Natural Hazards* 58: 789–810.

Lean, G. 2011. The nuclear industry must understand that the unexpected can happen – Even in Britain. *The Telegraph*, April 5.

Menoni, S., and C. Margottini (eds.). 2011. *Inside risk strategies for sustainable risk mitigation*. Milan: Springer.

Mitchell, J.K. (ed.). 1999. *Crucibles of hazard: Megacities and disasters in transition*. Tokyo: United Nations University Press.

Nowotny, H., P. Scott, and M. Gibbons. 2001. *Re-thinking science. Knowledge and the public in an age of uncertainty*. Cambridge: Polity.

Perrow, Ch. 1984/1999. *Normal accidents: Living with high-risk technologies*. New York: Basic Books. Republished in 1999 with a new afterword and a postscript on the Y2K problem. Princeton: Princeton University Press.

Perrow, Ch. 2007. *The next catastrophe: Reducing our vulnerabilities to natural, industrial, and terrorist disasters*. Princeton: Princeton University Press.

Petley, D. 2008. *The Vaiont (Vajont) landslide of 1963*. The Landslide blog. www.landslideblog.org/2008/12/vaiont-vajont-landslide-of-1963.html. Last accessed July 2012.

Quarantelli, E.L. (ed.). 1998. *What is a disaster? A dozen perspectives on the question*. New York: Routledge.

Quarantelli, E.L., and R.R. Dynes. 1972. When disaster strikes (It isn't much like what you've heard or read about). *Psychology Today*, 67–70.

Risbey, J.S., and T.J. O'Kane. 2011. Sources of knowledge and ignorance in climate research. *Climatic Change* 108: 755–773. doi:10.1007/s10584-011-0186-6.

Ruckelshaus, W. 1984. Risk in a free society. *Risk Analysis* 4: 157–162.

Rudner, R. 1953. The scientist qua scientist makes value judgments. *Philosophy of Science* 20(1): 1–6.

Salter, L. 1988. *Mandated science. Science and scientists in the making of science*. Dordrecht: Kluwer Academic.

Sherwell, P., and P. Sawer. 2010. Volcano crisis: Why did they land us in a mess like this? Interview to Iceland President Ólafur Grímsson, *The Sunday Telegraph*, published online: 6:38PM BST, April 24.

Showalter, P., and M. Fran Myers. 1992. *Natural disasters as the cause of technological emergencies: A review of the decade 1980–1989*. Boulder: Natural Hazard Research and Applications Center, University of Colorado.

Steinberg, L.J., H. Sengul, and A.M. Cruz. 2008. Natech risk and management: An assessment of the state of the art. *Natural Hazards* 46(2): 143–152.

Stiglitz, J.E. 2011. *Gambling with the planet*. http://www.project-syndicate.org/commentary/stiglitz137/English. 2011-04-06.

Tallacchini, M. 2005. Legalising science. *Health Care Analysis* 10(3): 329–337.

Van der Sluijs, J. 2005. Uncertainty as a monster in the science-policy interface: Four coping strategies. *Water Science and Technology* 52: 87–92.

Weinberg, A. 1972. Science and trans-science. *Minerva* 10: 209–222.

Wildavsky, A. 1979. *Speaking truth to power: The art and craft of policy analysis*. Boston: Little Brown & Co.

Wynne, B. 1996. May the sheep safely graze? A reflexive view of the expert-lay knowledge divide. In *Risk, environment and modernity: Towards a new ecology*, ed. S. Lash, B. Szerszynski, and B. Wynne, 44–83. London/Beverly-Hills: Sage.

Wynne, B., U. Felt, M. Callon, M.E. Gonçalves, S. Jasanoff, M. Jepsen, P.-B. Joly, Z. Konopasek, S. May, C. Neubauer, A. Rip, K. Siune, A. Stirling, and M. Tallacchini. 2007. *Science and governance. Taking European knowledge society seriously*. European Commission Directorate-General for Research, EUR 22700, Office for Official Publications of the European Communities.

Bruna De Marchi sociologist, is presently a Guest Researcher at the Center for the Study of the Sciences and Humanities at the University of Bergen, Norway. She also works as a private consultant after being in charge of the Mass Emergencies Programme at the Institute of International Sociology of Gorizia, Italy, for many years. De Marchi has been a consultant to governmental and international agencies and private companies, including the WHO, the EC, the Italian Department of Civil Protection and the Italian National Seismic Service. She has broad research interests, the main ones in the social dimensions of environmental change, human system response to disasters, risk perception, communication and governance, public participation in health, environment and disaster policies.

Chapter 10
Educational Governance in Disaster Risk Reduction

Rajib Shaw

10.1 Introduction

The goal of developing 'disaster-resilient communities' is widely understood to depend heavily on the success of disaster education and that the integration of both formal and non-formal/informal education through school is one way in ensuring that these messages reach into every family and community so that the learning can be sustained into the future generations (Petal 2008). There still remain several challenges on educational policy, since there is not a single policy which covers on different sector of educational issues. Recognizing the significance of education in disaster risk reduction, suitable policy is necessary to integrate disaster risk reduction into education.

The importance of education in disaster risk reduction has been emphasized in several international governance and policy agenda, frameworks, conferences as well as UN programs. Chapter 36 of Agenda 21, on Education, Awareness and Training stated "Education, including formal education, public awareness and training, should be recognized as a process by which human beings and societies can reach their fullest potential" (UNEP 1992). Furthermore, the UN/ISDR System Thematic Cluster/Platform on Knowledge and Education argued that "Education for disaster risk reduction is an interactive process of mutual learning among people and institutions. It encompasses far more than formal education at schools and universities, and involves the recognition and use of traditional wisdom and local knowledge for protection from natural hazards" (UN/ISDR 2005). In the 2006 Review of the Role of Education and Knowledge in Disaster Risk Reduction pointed out that "Education, knowledge and awareness are critical to building the ability to reduce losses from natural hazards, as well as the capacity to respond to and recover

---

R. Shaw (✉)
Graduate School of Global Environmental Studies, Kyoto University,
Yoshida Honmachi, Sakyo-ku, Kyoto 606-8501, Japan
e-mail: shaw.rajib.5u@kyoto-u.ac.jp

U. Fra.Paleo (ed.), *Risk Governance*, DOI 10.1007/978-94-017-9328-5_10

effectively from extreme natural events when they do, inevitably, occur" (Wisner 2006). The Second Asian Ministerial Conference on Disaster Risk Reduction (2007, India) urged governments to make school safety and the integration of disaster risk reduction into school curricula a priority on the national agenda (UN/ISDR 2007a). The Third Asian Ministerial Conference on Disaster Risk Reduction (2008, Malaysia) recognized education as an essential contribution to effective implementation of disaster risk reduction and concrete impact tin terms of shifts in behaviors art the local level, where communities are most vulnerable to disasters (UN/ISDR 2008). Last but not least, the UNESCO Education for Sustainable Development (ESD) programme emphasized that 'Education is the primary agent of transformation towards sustainable development, increasing people's capacities to transform their visions for society into reality' (UNESCO 2005a, b).

Citing several examples from Asian countries, this chapter highlights the key challenges and draws a framework of educational governance in disaster risk reduction.

10.2 Hyogo Framework for Action and Implications to Disaster Education

The Hyogo Framework for Action (HFA) 2005–2015 prepared by the UN/ISDR emphasizes the role of formal and non-formal education and awareness raising as an important component for disaster risk reduction. Following the adoption of HFA in 2005, various educational materials in the form of booklets, handbooks, textbooks, posters, activities, games and practices were developed (UN/ISDR 2006). 'Disaster Reduction, Education and Youth' was the UN World Disaster Reduction Campaign in 2000 and in 2006–2007, there was the UN/ISDR campaign on 'Disaster risk reduction begins at school', both campaigns addressing the emphasis of integrating disaster risk reduction into education. The 2006–2007 'Disaster Risk Reduction Begins at School' campaign reaffirms the priority for action three of the HFA on Use knowledge, innovation and education to build a culture of safety and resilience at all levels (Box 10.1). UN/ISDR highlighted not only the importance of integrating disaster risk reduction into formal education, at the same time, emphasized the importance of community participation in order to achieve sustainability within the community (UN/ISDR 2006). In addition to providing education, school buildings could also serve as temporary shelter for the community following disasters, thus, the safety of the buildings is important to ensure the safety of students as well as continuation of education following disasters and the campaign also promotes safe construction and retrofitting of school buildings to withstand natural hazards (UN/ISDR 2006).

Box 10.1: Key Activities of HFA Priority 3

(i) **Information management and exchange**

(a) Provide easily understandable information on disaster risks and protection options, especially to citizens in high-risk areas, to encourage and enable people to take action to reduce risks and build resilience. The information should incorporate relevant traditional and indigenous knowledge and culture heritage and be tailored to different target audiences, considering cultural and social factors.
(b) Strengthen networks among disaster experts, managers and planners across sectors and between regions, and create or strengthen procedures for using available expertise when agencies and other important actors develop local risk reduction plans.
(c) Promote and improve dialogue and cooperation among scientific communities and practitioners working on DRR (Disaster Risk Reduction), and encourage partnerships among stakeholders, including those working on the socioeconomic dimensions of DRR.
(d) Promote the use, application and affordability of recent information, communication and space-based technologies and related services, as well as earth observations, to support DRR, particularly for training and for the sharing and dissemination of information among different categories of users.
(e) In the medium term, develop local, national, regional and international user friendly directories, inventories and national information-sharing systems and services for the exchange of information on good practices, cost-effective and easy-to-use DRR technologies, and lessons learned on policies, plans and measures for DRR.
(f) Institutions dealing with urban development should provide information to the public on disaster reduction options prior to constructions, land purchase or land sale.
(g) Update and widely disseminate international standard terminology related to DRR, at least in all official United Nations languages, for use in program and institutional development, operations, research, training curricula and public information programmes.

(ii) **Education and training**

(h) Promote the inclusion of DRR knowledge in relevant sections of school curricula at all levels and the use of other formal and informal channels to reach youth and children with information; promote the integration of DRR as an intrinsic element of the UN Decade of Education for Sustainable Development (2005–2015).

(continued)

Box 10.1 (continued)

(i) Promote the implementation of local risk assessment and disaster preparedness programmes in schools and institutions of higher education.

(j) Promote the implementation of programmes and activities in schools for learning how to minimize the effects of hazards.

(k) Develop training and learning programmes in DRR targeted at specific sectors (development planners, emergency managers, local government officials, etc.).

(l) Promote community-based training initiatives, considering the role of volunteers, as appropriate, to enhance local capacities to mitigate and cope with disasters.

(m) Ensure equal access to appropriate training and educational opportunities for women and vulnerable constituencies; promote gender and cultural sensitivity training as integral components of education and training for DRR.

(iii) **Research**

(n) Develop improved methods for predictive multi-risk assessments and socioeconomic cost–benefit analysis of risk reduction actions at all levels; incorporate these methods into decision-making processes at regional, national and local levels.

(o) Strengthen the technical and scientific capacity to develop and apply methodologies, studies and models to assess vulnerabilities to and the impact of geological, weather, water and climate-related hazards, including the improvement of regional monitoring capacities and assessments.

(iv) **Public awareness**

(p) Promote the engagement of the media in order to stimulate a culture of disaster resilience and strong community involvement in sustained public education campaigns and public consultations at all levels of society.

Source: UN/ISDR (2007b)

Lessons learnt from the experiences include (i) education is a process for effective disaster reduction, (ii) knowledge, perception, comprehension and actions are the four important steps, (iii) schools and formal education play an important role in knowledge development, (iv) family, community and self-education are important for comprehension of knowledge and implementing risk reduction actions and (v) holistic education includes actions at local level, as well as its policy integration (Shiwaku et al. 2007).

10.3 Progress of Disaster Education Over Last Several Years

Responding to the call from the UN/ISDR 2006–2007 campaign, various international and/or regional conferences and workshops on school safety and school education were held, and countries have developed national action agenda addressing the issue on integrating disaster risk reduction into the education curriculum as well as ensuring safety of school buildings. The International Conference on School Safety, held from 18–20th January 2007 in Ahmedabad, Gujarat, India, recognizes the importance of every child to receive education and living in safe and sustainable environment and adopted the Ahmedabad Action Agenda for School Safety which aims to achieve 'Zero Mortality of Children in School from Preventable Disasters by the year 2015'. To achieve its goal, the action agenda are outlined under immediate priority and long term accomplishments (i.e. by 2015) which four priority areas namely (i) disaster education in school, (ii) disaster resistant school infrastructure, (iii) safe school and community environments and (iv) advocacy and government policy on school safety (Ahmedabad Action Agenda 2007). In the same year in October, the Asia-Pacific Regional Workshop on School Education and Disaster Risk Reduction was held in Bangkok, Thailand, whereby 304 participants from 24 countries in the Asia and Pacific region came together to discuss on ways to improve resilience of school communities struck by disasters or in hazard prone areas. The output of the workshop was the adoption of Bangkok Action Agenda which focuses on four priority areas namely (i) integrating disaster risk reduction into school education, (ii) strengthening disaster education for community resilience, (iii) making schools safer and (iv) empowering children for disaster risk reduction (UN/ISDR 2007c, d). Both agendas, the main focus was on integrating disaster risk reduction into school education and ensuring the safety of school buildings, effectively complement the core indicators of the HFA, particular with the specific guidance on making schools disaster resilient and promoting participation of communities and children in the disaster risk reduction initiatives.

The Kashimir earthquake in 2005 affected over 3.5 million people, killed 87,350 and injured approximately 70,000 people in Pakistan. Over 17,000 school-age children died in the collapsed schools and over 20,000 injured. Eight thousand schools out of 9,000 in the earthquake affected regions in northern Pakistan were damaged beyond repair (HRMP 2009). In 2008, the International Conference on School Safety held in Islamabad issued the Islamabad Declaration on School Safety which specified that 'policies, guidelines, implementing and monitoring mechanisms are needed. This translates into actions that address identifying resilient school needs, retrofitting existing structures, creating evacuation plans and safe havens, improving community and student awareness through outreach and simulations. Selection of safe sites, design and construction technologies and materials also apply to the larger built environment' (Islamabad Declaration on School Safety 2008). The Declaration emphasizes the importance of safe school buildings to achieve disaster resilience in the education sector. The Declaration emphasized on the aspect of safe school buildings and seeks to achieve its aim through establishment of policies,

Table 10.1 Summary of the three national initiatives – Ahmedabad Action Agenda, Bangkok Action Agenda and Islamabad Declaration on School Safety

National initiative	Priority areas
Ahmedabad Action Agenda for School Safety	Disaster education in school
	Disaster resistant school infrastructure
	Safe school and community environments
	Advocacy and government policy on school safety
Bangkok Action Agenda	Integrate disaster risk into school education
	Strengthen disaster education for community resilience
	Make schools safer
	Empower children for disaster risk reduction
Islamabad Declaration on School Safety	Identify school structural and non-structural vulnerabilities
	Retrofit existing structures
	Create evacuation plans and safe havens
	Improve community and student awareness through outreach

partnerships between national government and local entities and community, reaffirming the call from the HFA on enhancing disaster resilience at schools. Action plans address school structural and non-structural vulnerabilities and strongly encourages community participation. Community involvement is necessary since they are the first responders to disaster situations and partnership will allow transfer of knowledge and practices, ensuring its continuation among the individuals as well as the community. Table 10.1 summarizes the three national initiatives described.

In 2009, the UN launched the 'One Million Safe Schools and Hospital' campaign to address and advocate the need to ensure that buildings, such as schools, are built in compliance to the safety standards to enhance disaster resilience. "People in unsafe schools, hospitals and health facilities are at the greatest risk of losing their lives," UNISDR said and continued that "Children in schools and the sick in hospitals and health facilities are the most vulnerable people in times of disaster" (UNC 2010a, b).

As discussed in the previous section, the importance of Education in Emergencies has also been advocated in recent years. The Inter-Agency Network for Education in Emergencies (INEE), a online network of educational stakeholders of non-governmental organizations, UN agencies, donors, practitioners, researchers and individuals, has also developed a similar document entitled 'A Handbook of Minimum Standards for Education in Emergencies, Chronic Crises and Early Reconstruction' as referred to as the 'INEE Minimum Standards'. Launched in 2004 at INEE's Second Global Inter-Agency Consultation on Education in Emergencies and Early Recovery in South Africa, the handbook is designed to provide governments and humanitarian workers the tools necessary to address the Education for All and UN MDG (INEE 2002; UN 2010). The INEE Minimum Standards consists of five categories namely (i) 'Access and Learning Environment' focuses on partnerships to promote access to learning opportunities as well as inter-sectoral linkages, (ii) 'Teaching and Learning' focuses curriculum, training, instruction and assessments, as the essential factors to promote effective teaching and learning, (iii) 'Teachers and other Education Personnel' focuses on the administration and management of human resources in the field of education, which includes recruitment and selection, conditions of service, as well as supervision and support, (iv) 'Education Policy and Coordination' focuses on policy formulation, planning, implementation and coordination and (v) 'Community participation' and monitoring and evaluation which are applicable to all the categories (INEE 2002).

In the effort to promote the Education in Emergencies, the UNESCO International Institute for Education Planning (IIEP) developed the "Guidebook for Planning Education in Emergencies and Reconstruction", consisting of five sections namely (i) general overview [introduction, Introduction, Prevention of conflict and preparedness for disaster, Challenges in emergencies and reconstruction, Capacity building Education for all in emergencies and reconstruction] (ii) access and inclusion [Rural populations, Gender, Ethnicity/political affiliation/religion, Children with disabilities, Former child soldiers, Learning spaces and school, facilities, Open and distance learning, Technology, Non-formal education, Early childhood development, Post-primary education], (iii) teachers and learners [Identification, selection, recruitment of teachers, education workers, Teacher motivation, compensation and working conditions, Measuring and monitoring teachers' impact, Teacher training: teaching and learning methods, Psychosocial support to learners], (iv) curriculum and learning [Curriculum content and review processes, Health and hygiene education, HIV/AIDS preventive education, Environmental education, Landmine awareness, Education for life skills peace, human rights and citizenship, Vocational education and training, Textbooks, educational materials and teaching aids], and (v) management capacity [Assessment of needs and resources, Planning processes, Project management, Legal frameworks, Community participation, Structure of the education system, Data collection and education management information systems, Budget and financial management, Human resources: ministry officials, Donor relations and funding mechanisms, Coordination and communication] (UNESCO 2010a).

The Guidebook is primarily targeted at the staff of ministries of education, national, provincial and district level planners and managers, in disaster affected areas. The following are the three core aspects in education in emergencies emphasized by IIEP namely: (i) Preparing the planning for emergency – an education sector diagnosis is necessary to examine the risk and vulnerabilities of the education system to both human and manmade disaster, which in turn help in developing budgeting for contingency planning, capacity strengthening within the education system, (ii) Responding to an emergency – including rapid assessment of education needs, coordination of education actors, capacity gap analysis for reconstruction and (iii) Continuing education during and after an emergency – depending on the context, but strategies in ensuring continued provision of education such as temporary learning spaces as well as advocacy to donors, agencies, private sectors on funding education (UNESCO 2010b).

The Global Assessment Report [GAR 2011] has made a bi-annual assessment of the progress of the HFA priority areas (Fig. 10.1), which shows that the least progresses are confided to HFA 3 and 4. Compared to HFA 4, the HFA 3 progress is better in last 2 years [2009–2011], which points to increasing awareness in the

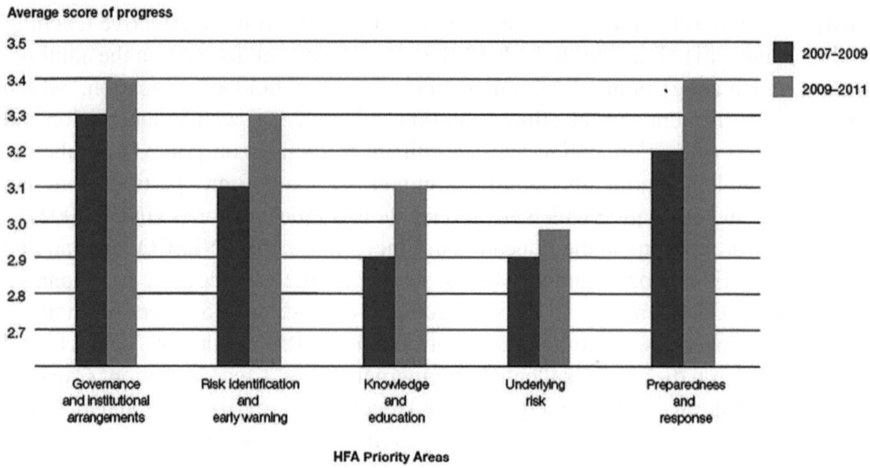

Fig. 10.1 Global progress of HFA priority areas (Source: GAR 2011)

need of implementation of disaster education programs. As discussed in this section, nations have recognized the importance of integrating disaster risk reduction in education and safe buildings and have adopted national action agendas and declaration to promote school safety. However, is educational curriculum and safe building structures sufficient in achieving disaster resilience in schools? Possibly not, it needs different aspects of disaster education decision making, related to educational governance. This is argued in the next part of the chapter.

10.4 Disaster Education Policies and Programs

10.4.1 OECD Draft Policy

OECD is an international organization that conducts monitoring, analyzing and forecasting thereby providing public policy guidelines to help governments foster prosperity and fight poverty through economic growth and financial stability, and consists of 33 member countries, as well as international organizations (OECD 2001). The 'Draft Policy Handbook on Natural Hazard Awareness and Disaster Risk Reduction Education' in 2009 suggests nine fundamental principles for disaster education as well as a strategy consisting seven focus areas (OECD 2009). Among them some of the highlight areas are:

- Natural hazard awareness and disaster education efforts should aim at encouraging voluntary risk reduction activities
- Appropriate risk communication techniques should be adopted to reach the targeted audiences and induce the desired changes in behavior and perception. Specific and realistic for local conditions messages are strongly recommended.

- Continuous monitoring and periodic evaluations of awareness and education efforts should be conducted to assure accountability and transparency to increase public confidence outcomes.

In addition, it emphasized three crucial areas that require specific attention, which includes: gain scientific knowledge of hazards and risk hazard mapping, prioritization], identify the desired behavior and perception changes [awareness, public education], and identify roles, methods tools [stakeholder involvement and international cooperation].

10.4.2 Pakistan National Education Policy of 2009

Pakistan, in particular the northern highlands, falls in the seismic active zone, is prone to earthquake as well another natural hazards such as floods, snow and ice avalanches, landslides and river erosions (HRMP 2009). As discussed in the earlier section, the Islamabad Declaration on School Safety was adopted at the International Conference on School Safety in Islamabad in 2008. In the following year, the Pakistan National Education Policy was issued (Ovington 2010). The policy includes various aspects from strengthening education curricula, training to infrastructure and preventive measure. Curriculum, especially of social studies, geography, languages, and literacy shall include themes on emergencies, natural disasters and trauma management based on latest international best practices and shall also include information about response in an emergency or disaster. A repository of all emergency related materials, manuals, guidelines, minimum standards and research pertaining to education shall be maintained at the teachers training institutions, schools, college and universities. Teacher education and training curricula shall include provisions to enable the teachers to address education in emergencies. Furthermore, Education in Emergencies (action 7) and rehabilitation measures (action 8) were mentioned as well.

10.4.3 SEEDS School Safety Initiative in India

SEEDS India recognizes the role of schools in the community as being very important. In fact, it would be befitting to call schools cradles of the society. Children are a dynamic and powerful force of change and are supporters in creating awareness in the community. They can contribute in a unique manner with energy and vision to find local solutions. School children should be encouraged to take up tasks, which make them realize their importance as necessary stakeholders in the process of change. SEEDS School Safety Initiative (SSSI) endeavors to create a safe environment for children. Recognizing the immense potential of children as potent agents of change, the initiative is trying to tap this powerful resource to bring about a larger change. Since its inception in 2005, SSSI has spread its culture of safety across the

Indian states of Gujarat, Himachal Pradesh, Jammu and Kashmir, Rajasthan, Tamil Nadu and Orissa, Delhi and the highly vulnerable Andaman and Nicobar Islands and the neighboring countries of Maldives, and Afghanistan. The school safety campaign also helps other NGOs and partners by training them on the tools and techniques to further the cause of school safety. As clearly mentioned in the Ahmedabad Action Agenda the SSSI is steadfast to ensuring "Zero Mortality of Children in Schools from Preventable Disasters by the year 2015." SSSI can be described as a good practice because it is a pioneering initiative that seeks to promote a culture of disaster safety in schools– that are the most suitable areas for sowing enduring good habits and practices. To this end, for instance, SSSI has a four-pronged approach in earthquake-prone areas: (1) structural retrofitting of school buildings to prevent their collapse in future earthquakes; (2) implementing non-structural mitigation measures to avoid injuries from falling hazards in schools; (3) education on safe infrastructure for school management staff and construction workers; and (4) preparing school disaster management plans and training school communities in immediate response, evacuation and first aid (SEEDS 2010).

10.4.4 Earthquake Safety Education Program of Iran

Another example of disaster education programme is the Earthquake safety education programme in Iran. The programme includes integration of disaster education into the formal education and performs non-formal education activities, at the same time, involving students at the local context as well as engaging participation from the community. Iran, being located near the faults, has high risks of seismic hazards and has suffered several earthquakes resulting in severe economic losses in the past. As such, the government has initiated the 'School Earthquake Safety' initiative, 'School Safety Act' as well as the 'Earthquake Safety Education in School' (IIEES 2006). The Earthquake Safety Education programme is led by the Iranian Ministry of Education (MOE) in collaboration with IIEES as well as other public and civic organizations. Disaster lessons are integrated within science, geography, literature and other curricula with preparedness and practice books designed for different grades and handbooks for teachers (Petal 2008). Safety drills are one of the most commonly performed activities. Every year, the IIEES, Iranian MOE, National Committee for Natural Disaster Reduction), Iranian Red Crescent Society, Iran National Television and Radio, and other related national organizations organizes these drills to prepare students and staff for appropriate and rapid responses during earthquakes. There is also a broadcast of an 'Earthquake safety alarm' is on national and local radio. Furthermore, a "School Earthquake Safety Councils" involving parents and teachers voluntarily serves as a form of disaster risk reduction and preparedness efforts at the individual school level (Petal 2008). In this case study, it shows the integration of disaster education into the formal education, at the same time, non-formal activities (i.e. safety drills) being conducted to increase preparedness and response among students. In addition, it programme is not only limited within school boundaries but also reaches out to parents who voluntarily involves themselves in the 'school earthquake safety councils' with the teachers in their effort to achieve disaster risk reduction and preparedness.

10.4.5 Myanmar Education Recovery Program (MERP)

In 2010, the author and colleagues worked with the Ministry of Education, Union of Myanmar together with UNESCO to develop the MERP initiative which seeks to enhance resilience in the education sector in Myanmar by addressing Disaster Risk Reduction and Emergency Preparedness as an integral part of education, and integrates them in all the HFA's five priorities for action, through a community-based, participatory and multi-sector approach. The initiative consists of a comprehensive training package Disaster Reduction in Education for township education officers, school principals, teachers and students. The content of the training material addresses the five priorities for actions, as seen through an education lens, to help reduce risk, mitigate and prevent the impacts in the education sector, namely: (i) developing institutional base for disaster risk reduction in education, (ii) identifying, assessing and monitoring disaster risks in the education sector, (iii) building a culture of safety through disaster risk reduction in education, (iv) reducing the underlying risk factors in the education sector and (v) preparing for effective emergency response and recovery in education (UNESCO 2010c).

10.5 E-HFA: Future Perspective of Governance in Disaster Education

An integrated approach is necessary to ensure disaster risk reduction is incorporated into not only the schools but into the education sector as a whole. The approach should not only consider education curricula and safe school buildings but should also address legislative measures (i.e. having formal guidelines for implementation and funding), proper early warning systems and risk assessments, training of qualified professionals, promoting community involvement as well as measures taken to prepare community in responding to disasters. The MERP initiative of Myanmar developed by the author, addresses HFA in the education sector, and provides the basis for the integrated approach for disaster education. The document 'Words into Action: A Guide for Implementing the Hyogo Framework', by the UN/ISDR, provides advice on useful strategies for implementing the HFA. The Guide describes 22 suggested tasks that are organized to help address and guide the implementation of the HFA's five Priorities for Action. Gwee (2011) did initial research on the applicability of the 22 tasks in the education sector to enhance the educational governance. She found that out of the 22 tasks, 16 tasks are related to the education sector, and it is considered as E-HFA

(Education in Hyogo Framework for action) (Box 10.2). The 16 tasks identified should be performed at all levels (i.e. national, local, community/school) to achieve sustainable implementation. Gwee et al. (2011a) studying the educational system in Taiwan and bringing some examples from different other Asian countries, has summarized the roles of national and local governments in the educational governance in the disaster risk reduction, especially implementing the E-HFA.

Box 10.2: Proposed 16 Tasks Relevant to the Education Sector

Priority 1: Developing institutional base for disaster risk reduction in education

1. Engage in multi-stakeholder dialogue to establish the foundation for diasster education.
2. Create or strengthen mechanism for systematic coordination for diasster education.
3. Assess and develop the institutional basis for disaster education.
4. Prioritize disaster risk reduction and allocate appropriate resources for disaster education.

Priority 2: Identifying, assessing and monitoring disaster risks in the education sector

5. Establish risk assessments for the education sector
6. Strengthen early warning in the education sector through effective communication and dissemination mechanism.

Priority 3: Building a culture of safety through disaster education

7. Develop public programme to raise awareness of disaster risk reduction
8. Include disaster risk reduction in the education system
9. Develop disaster risk reduction training and learning at community level
10. Enhance dissemination of disaster risk reduction information

Priority 4: Reducing the underlying risk factors in the education sector

11. Environment: Understand sustainable ecosystem, environmental and natural resources management
12. Establish measures to incorporate disaster risk reduction in urban and land-use planning
13. Structures: Strengthen mechanisms for improved building safety and protection of critical facilities in the education sector
14. Disaster recovery: Develop a recovery planning process that incorporates disaster risk reduction

Priority 5: Preparing for effective emergency response and recovery in education

15. Build on disaster preparedness capacities and mechanisms in the education sector
16. Assess disaster response preparedness capacities and mechanisms through strengthened planning

In HFA 1, the four key focuses are namely: (i) multi-stakeholder dialogue, (ii) systematic coordination, (iii) institutional basis and (iv) allocation of resources. With regards to HFA 1 focus area (i) and (ii), establishment of a multi-stakeholder group at all levels is strongly recommended so as to develop a pool of disaster risk reduction specialists. Furthermore, regular meetings to discuss on disaster risk reduction issues in the education provides a platform for the members to be updated with the current progress and share experiences with each other. In terms of institutional basis, development of appropriate policy, Acts and/or guidelines by the MOE [Ministry of Education] to promote disaster risk reduction into the education curricula provides a standardized framework for implementation. Local education department plays the role in ensuring proper implantation and compliance to the policies, Acts and/or guidelines. A feedback mechanism between the MOE and local education department would be helpful for long-term monitoring of the effectiveness and efficiency of the implementation progress. In addition, the MOE may decide on a national disaster month and coordinate with the related departments in organizing various activities to promote awareness raising. The education department may also implement and organize activities in accordance to the local context to promote disaster risk reduction in the community during the 'disaster month'. Last but not least, it would be recommended that specific funds are allocated for disaster education purpose to ensure that disaster risk reduction funds are available.

HFA 2 focuses on risk assessments and early warning through effective communication. Risk assessments standards and evaluation system should be established by the MOE. Risk assessment should not only be limited to seismic capacity but also all hazards. The local education department and schools plays the role in ensuring proper implementation and strict compliance to the standards. Furthermore, community participation in risk assessment should be encouraged. In terms of early warning systems, coordination between MOE and other related national departments as well as between the national and local governments is necessary. Coordination with the media (e.g. TV and radio) would be helpful. Within the school, PA system is necessary for emergency announcement. Practice of disaster calendar may help cultivate disaster preparedness among students.

In terms of HFA 3, which focuses on disaster education itself, the key areas are (i) public awareness, (ii) incorporating disaster risk reduction into the education curricula, (iii) training and (iv) dissemination of disaster risk reduction information. To achieve HFA 3, it is essential that MOE develops a set of disaster educational materials that can be integrated into the curricula. Integration may be through (i) curricula integration (i.e. specially developed units, modules or chapter concentrating on disaster risk reduction), (ii) extracurricular integration (i.e. national and/or local campaigns), (iii) curriculum infusion (i.e. complete set of units focusing on disaster risk reduction) and (iv) stand alone course (i.e. specialized course curricula focused on disaster risk reduction, which could be on a one-time basis) (Petal 2008). Local education department and schools ensures proper integration and/infusion based on the local context. At the local level, community mapping may be encouraged which also serves as an opportunity for the community people to interact and exchange information. At the schools, activities such as town-watching, hazard mapping and field-trips may help reinforce and/or expand the knowledge acquired from the classroom curriculum. Parental participation in school activities (e.g. evacuation drills, first-aid training) may also help enhance awareness among the community. Students may be asked to perform simple tasks at home (e.g. ensuring home furniture are well-anchored, preparing home emergency kit etc.) which could serve as an opportunity for the students to interact with their parents and family members, at the same time, spread the preparedness practice and disaster awareness within the family. In order that teachers are equipped with the knowledge and skills to teach, it is important for MOE to develop suitable disaster risk reduction training programmes for the teachers. In addition, governmental officers (i.e. MOE and/or local education department), academicians and/or practitioners (i.e. working on disaster risk reduction) may also attend such training courses.

HFA 4 focuses on land-use planning, building safety and protection of facilities. In addition, the priority for action also stresses on the need for disaster recovery planning (Patel n.d.; WBDG 2010; WMO 2007). Similar to risk assessment in HFA 2, land-use planning should not be limited to seismic capacity but also consider risks of other hazards such as typhoons, floods and land-subsidence. Suggestions regarding considerations on land-use planning, building safety as well as structural and non-structural safety (i.e. facilities, furniture etc.) are summarized in Table 10.1. The standards for land-use planning, building safety as well as structural and non-structural safety would be determined by MOE and enforced at local levels through the local education department and schools. Regular monitoring and evaluation is necessary and would be conducted by the local education department. Following disasters, class disruption poses a severe problem to students. Proper post-disaster recovery planning, which includes measures in ensuring class continuity, disaster recovery actions are essential and it is necessary for MOE to provide proper guidelines for these purposes. Local education department and schools enforces these guidelines and may modify them according to the local context.

Last but not least, preparing for effective emergency response and recovery in education mentioned in HFA 5. The emphasis in this priority for action would be: (i) enhancing disaster preparedness capacities and (ii) assessing disaster response preparedness capacities. To achieve this, proper pre-disaster recovery planning, assessing the preparedness capacities is essential, which in turn helps prepare appropriate response capacities. Areas to be considered would include establishing standards on evacuation shelters so as to assess suitability of schools being used as an evacuation shelter as well as establishing guidelines on setting up specific task forces and their respective responsibilities as well as feedback mechanisms during emergencies. Guidelines and standards should be developed by the MOE and enforced by the local education department and schools. Within the community and schools, creating a local evacuation map may serve to enhance preparedness and response among the community people and students.

When considering disaster education, it should not only be limited to the education curricula, but should include the relating issues such as structural and non-structural safety, legislative measures supporting the integration, implementation as well as sufficient funding, proper early warning systems and risk assessments, training of qualified professionals etc. Therefore, an integrated approach is necessary and the suggestions made seek to address the various issues so as achieve the aim of enhancing disaster resilience in the education sector.

10.6 Application of E-HFA in Local Governance

An analysis of made in 123 out of 184 schools in the Yunlin province of Taiwan, which is prone of different types of disasters like earthquake, typhoon, flood, landslide, and land subsidence (Gwee et al. 2011b). Schools were classified into four regions; coastal, mountainous, rural, plain land and urban to observe if DRR education differs between the locations. Figure 10.2 shows the overall findings from the questionnaire analysis, details of which can be found in Gwee et al. (2011b). Upper and lower column implies that suggested indicative activity was and was not performed respectively. Number of columns varies depending on the questions that were asked on the indicative activity. Each column represents a question asked on the activity for the task. Similar observations were seen in the four areas with HFA 5 being weak in general and the following areas were found to be weak: (i) funds allocation for DRR activities (HFA 1 task 4), (ii) use of disaster calendar as a form of early warning practice (HFA 2 task 6), (iii) dissemination of DRR information (HFA 3 task 10), (iv) disaster recovery (HFA 4 task 14) and (v) preparedness for emergency response (HFA 5 task 15 and 16).

This is the first attempt to analyze the different aspects of DRR activities in educational sector in a collective way, which leads to governance decision. Thus, the results re-establish the argument, which was posed in the beginning of the chapter

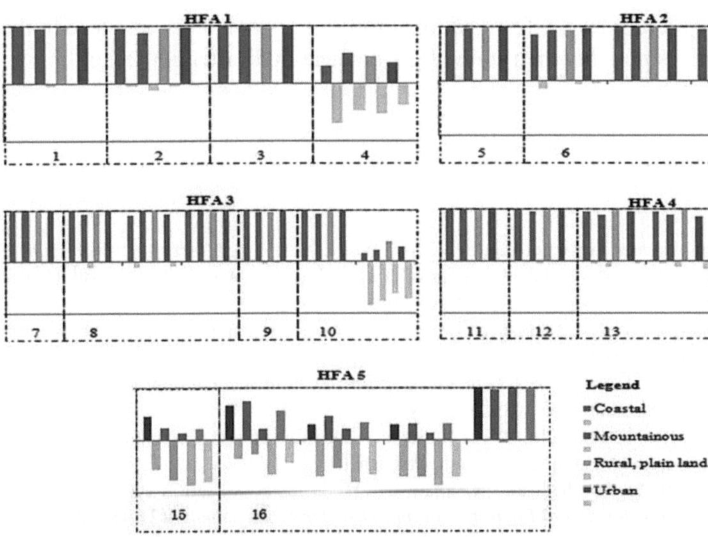

Fig. 10.2 Analysis of school governance with E-HFA in Yunlin province

that only making the school safer would not achieve the resilience of the education sector. It need a holistic approach of different sectors, and needs to implement different aspects of HFA tasks.

10.7 Way Ahead

The ideas and approach posed in this chapter is rather new, and need to be tested in different countries in different context. The chapter, from the review of evolution of disaster education has pointed out that there has been good progress made by the countries globally and locally to raise the awareness of people and children, and to take specific actions on strengthening the school buildings after different disasters. The recent East Japan earthquake and tsunami of 2011 also posed different challenges in the education sectors (Shaw 2011), which includes: location, structural and functional issues related to school buildings, management issues of teacher [especially school principal] during and after the disasters, relation of school of local communities, decisions of local educational board to restart the education, and to enhance the resilience of the education sectors. Several of these issues are related to the educational governance, which is the key point of this chapter.

The MERP program of UNESCO developed by the author and colleagues was the first attempt of using the HFA as a specific framework to prepare training modules for the educational administrators. In Yunlin province, the first attempt was made to use E-HFA for the provincial level to understand the governance issues of education sector in the disaster risk reduction. Educational governance needs emphasis on different aspects of disaster risk reduction issues, starting from institutional to risk assessment, to teacher training, to emergency education. This needs to be practiced at different levels, from national to local governments, and also at the school level. This is possibly the future of governance in disaster education.

References

Ahmedabad Action Agenda for School Safety. 2007. *International conference on school safety*, 18–20 January, 2007. Ahmedabad, Gujarat, India. http://www.ineesite.org/toolkit/docs/doc_1_87_Agenda_for_School_Safety.pdf. Accessed 7 May 2010.

GAR. 2011. Chapter 4: Progress against the Hyogo framework for action. In *Global assessment report*, 74–96. Geneva, Switzerland.

Gwee, Q. 2011. *Integrated disaster risk reduction approaches in the educational sector in Yunlin county, Taiwan*. Master thesis, Kyoto University.

Gwee, Q., R. Shaw, and Y. Takeuchi. 2011a. Disaster education policy: Current and future. In *Disaster education*, 23–44. Bingley: Emerald Publisher.

Gwee, Q., Y. Takeuchi, W. Jet-chau, and R. Shaw. 2011b. Disaster education system in Yunlin county, Taiwan. *Asian Journal of Environment and Disaster Management* 3(2): 189–204.

Habitat Risk Management Programme (HRMP). 2009. *A programme of the Aga Khan Planning and Building Service, Pakistan*. http://www.preventionweb.net/files/13409_13409HRMPBrochure1.pdf. Accessed 29 Nov 2010.

Inter-Agency Network for Education in Emergencies (INEE). 2002. *INEE minimum standards*. http://www.ineesite.org/. Assessed 15 Feb 2011.

International Institute of Earthquake Engineering and Seismology (IIEES). 2006. *Case study: IRAN. School disaster risk reduction: think globally, act locally*. http://www.iiees.ac.ir/iiees/English/Publicedu/school_safety_iran_case_study_davos.pdf. Accessed 20 Nov 2010.

Islamabad Declaration on School Safety. 2008. *International conference on school safety Islamabad, Pakistan*, 14–16 May 2008.

Organization for Economic Cooperation and Development (OECD). 2001. http://www.oecd.org/. Accessed 9 Dec 2010.

Organization for Economic Cooperation and Development (OECD). 2009. *Draft policy handbook on natural hazard awareness and disaster risk reduction education*. http://www.oecd.orgdatao-ecd15043730809.pdf. Accessed 30 Nov 2010.

Ovington, G. 2010. *DRR in education in DRR high level experts' group meeting*, 10 Nov 2010, Islamabad, Pakistan. UNICEF, UNESCAP. http://www.unescap.org/idd/events/2010_Pakistan_Floods_II/Unicef_Education.pdf.

Patel, T. n.d. *Risk and vulnerability assessment of health facilities: Focus on structural and non-structural elements*. http://www.preventionweb.net/files/11214_Riskandvulnerability-assessmentofhea.pdf. Accessed 20 Dec 2010.

Petal, M. 2008. *Concept note: Formal and informal education for disaster risk reduction*. http://www.riskred.org/activities/ddredislamabad.pdf. Accessed 20 Nov 2010.

SEEDS. 2010. *Let us make school safer*. SEEDS report, 70 pages. New Delhi, India.

Shaw, R. 2011. Linkages between school and community based disaster education: Experiences from 2011 East Japan earthquake and tsunami. In *Proceedings of the international conference, UN habitat*, Fukuoka, October 2011.

Shiwaku, K., R. Shaw, R.C. Kandel, S. Shrestha, and A. Dixit. 2007. Future perspective of school disaster education in Nepal. *Disaster Prevention and Management* 16(4): 576–587.

UN. 2010. *Millennium development goals*. http://www.un.org/millenniumgoals/. Accessed 19 Nov 2010.

UN News Centre (UNC). 2010a. *UN seeks to improve safety of schools and hospitals in natural disasters*. http://www.un.org/apps/news/story.asp?NewsID=34321&Cr=quake&Cr1=. Accessed 25 Nov 2010.

UN News Centre (UNC). 2010b. *Indonesia commits to boost safety of schools and hospitals in UN-backed scheme*. http://www.un.org/apps/news/story.asp?NewsID=35469&Cr=disaster&Cr1=&Kw1=sumatra&Kw2=&Kw3=. Accessed 25 Nov 2010.

UN/ISDR. 2005. *Thematic cluster/platform on knowledge and education*. Available at: http://www.unisdr.org/eng/task%20force/working%20groups/knowledge-education/knowledge-education.htm. Accessed 10 Nov 2009.

UN/ISDR. 2006. World disaster reduction campaign. *Disaster risk reduction begins at school*. http://www.unisdr.org/eng/public_aware/world_camp/2006-2007/pdf/WDRC-2006-2007-English-fullversion.pdf. Accessed 28 Jan 2010.

UN/ISDR. 2007a. Regional analysis on disaster risk reduction education in the Asia-Pacific region in the context of Hyogo framework for action priority 3 implementation. In *Proceedings of the 2nd Asian ministerial conference on disaster risk reduction*, 7–8 November 2007, New Delhi, India. Available at: http://nidm.gov.in/amcdrr/acdrrbro.pdf. Accessed 29 Nov 2010.

UN/ISDR. 2007b. *Hyogo framework for action 2005–2015: Building the resilience of nations and communities to disasters*. http://www.unisdr.org/eng/hfa/docs/Hyogo-framework-for-action-english.pdf. Accessed 28 Jan 2010.

UN/ISDR. 2007c. *Bangkok Action Agenda*. Outcome of the Asia-Pacific regional workshop on school education and disaster risk reduction, 8–10 October 2007. http://www.unisdr.org/asia-pacific/ap-whatnew/Regional-thematic-report-education-agenda.pdf. Accessed 25 Jan 2010.

UN/ISDR. 2007d. *Words into action: A guide for implementing the Hyogo framework*. Geneva: UN/ISDR.

UN/ISDR. 2008. Regional analysis on disaster risk reduction education in the Asia-Pacific region in the context of Hyogo framework for action priority 3 implementation. In *Proceedings of the 3rd Asian ministerial conference on disaster risk reduction*, 2–4 December 2008, Malaysia.

http://www.preventionweb.net/…/3089_3089FirstAnnouncementAsianMinisterialConference.pdf. Accessed 29 Nov 2010.

UNEP. 1992. *Agenda 21/Chapter 36: Promoting education, public awareness and training*. Available at: http://habitat.igc.org/agenda21/a21-36.htm. Accessed 12 Jan 2010.

UNESCO. 2005a. Education for sustainable development. *Strategy*. http://www.unesco.org/en/esd/strategy/. Accessed 9 Dec 2010.

UNESCO. 2005b. *UN decade of education for sustainable development*. http://www.unesco.org/en/education-for-sustainable-development/decade-of-esd/. Accessed 10 Nov 2010.

UNESCO. 2010a. *International institute for education planning, guidebook for planning education in emergencies and reconstruction*. http://www.iiep.unesco.org/fileadmin/user_upload/Cap_Dev_Technical_Assistance/pdf/Guidebook/Guideboook.pdf. Accessed 20 Jan 2011.

UNESCO. 2010b. *International institute for education planning, education in emergencies*. Available at: http://www.iiep.unesco.org/en/focus-on-education-in-emergencies/education-in-emergencies-an-introduction.html. Accessed 20 Jan 2011.

UNESCO. 2010c. Myanmar education recovery programme (MERP). *Building the resilience in the education sector to disasters*. http://www.unescobkk.org/en/education/education-units/esd/projects/post-conflict-and-post-disaster-responses/unesco-asia-pacific-initiatives/myanmar-education-recovery-project/. Accessed 1 May 2010.

Whole Building Design Guide (WBDG). 2010. *Resist natural hazards*. http://www.wbdg.org/design/resist_hazards.php. Accessed 20 Dec 2010.

Wisner, B. 2006. *Let our children teach us! A review of the role of education and knowledge in disaster risk reduction. Inter-agency task force cluster group on education and knowledge*. http://scholar.google.co.jp/scholar?q=2006+Review+of+the+Role+of+Education+and+Knowledge+in+Disaster+Risk+Reduction&hl=ja&as_sdt=0&as_vis=1&oi=scholart. Accessed 30 Nov 2010.

World Meteorological Organization (WMO). 2007. *Formulating a basin flood management plan: A tool for integrated flood management*. http://www.apfm.info/pdf/ifm_tools/Tools_Basin_Flood_Management_Plan.pdf. Accessed 20 Dec 2010.

Rajib Shaw is a professor in the Graduate School of Global Environmental Studies of Kyoto University, Japan. He works closely with local communities, NGOs, governments and international organization, including United Nations, especially in the Asian countries. He is currently the Chair of the United Nations Asia Regional Task Force for Urban Risk Reduction. His research interests are in community based disaster risk management, climate change adaptation, urban risk management, and disaster and environmental education. He has published several books in the field of disaster and environmental management. Shaw is also the Chief Editor of the *Asian Journal of Environment and Disaster Management*.

Chapter 11
The Disaster Epidemic: Research, Diagnosis, and Prescriptions

Thea Dickinson and Ian Burton

11.1 Why an Epidemiology of Disasters?

So called 'natural' disasters have been viewed as single and infrequent or isolated events in particular places. By contrast disease epidemics are understood to have common underlying causes. This does not refer to the bacteria or the viruses which are almost always present. It refers to the factors which allow or promote the spread of the diseases. Similarly, it is not the earthquake that causes the disaster. It is the conditions in which the earthquake occurs that is truly responsible.

It has not been the practice to search for or even think about the common causes among different kinds of disasters (earthquakes, tsunamis, tropical cyclones) in widely scattered places (Haiti, Fukushima, New Orleans). This paper challenges the deficiency in disaster research, and uses a comparison with epidemics to help illustrate the case.

Take for example the Zimbabwean cholera epidemic of 2008–2009. There were almost 100,000 reported cases over a 12 month period and over 4,000 deaths (WHO 2009). It was not the bacterium that caused so many cases and fatalities, but a number of factors in the economy, society and environment of Zimbabwe. The epidemic in Zimbabwe started well before the first infection and can easily be dated back to the global recession (2007), if not earlier, when the financial crisis made maintaining shelves with vital medicine impossible and

T. Dickinson (✉)
Department of Physical and Environmental Sciences, University of Toronto,
1265 Military Trail, Toronto, ON M1C 1A4, Canada
e-mail: thea.dickinson@rogers.com

I. Burton
School of the Environment, University of Toronto, 26 St Annes Rd,
Toronto, ON M6J 2C1, Canada
e-mail: burtoni.ian@gmail.com

U. Fra.Paleo (ed.), *Risk Governance*, DOI 10.1007/978-94-017-9328-5_11

increasing human capital flight, where doctors (who should have been first responders to the crisis) left Zimbabwe for other more promising nations. These economic factors peaked at the start of the rainy season and heightened the potential for a cholera outbreak (which thrives in a wet environment) and led to an additional shortage – this time of necessary water purification chemicals that could have hindered an expansion of the outbreak. Soon the sanitation system collapsed, leading families, many whom were already stricken with immune suppressing viruses, to try and find financial means for fuel to boil water since the government was no longer able to guarantee the water they were providing was safe. Once this core political-economic system was broken, the number of cholera cases started to increase exponentially in December 2008; a time of year when many families come together for holiday gatherings – increasing the transmission of infection. The public health care system already had a shortage of both doctors and medicine and was suddenly inundated with thousands of cholera cases. This increased pressure led to a national emergency being declared in December, 2008. It would not be until later in 2009 before the cholera epidemic would become under control (further reading: Chambers 2009; Mason 2009; Nelson et al. 2009; Musemwa 2010; Ahmed et al. 2011)

A similar set of causes is always present in epidemics, and the profession of epidemiology has grown up to address them. There are two broad categories of epidemiologists – clinical epidemiologists and research epidemiologists. The clinicians are in fact the practitioners or the managers, and the researchers supply the knowledge and understanding on which the practice can be based. A discipline like epidemiology, directed at common and underlying causes, is needed for so-called 'natural disasters'. Just as it is not the virus that causes the epidemic, it is not the natural hazard that causes the disaster but the way in which the hazard or risk is managed or mismanaged. The professionals engaged as practitioners and researchers in the disaster field do not yet see themselves as partners in a common enterprise. Movement in this direction begins with research. When sufficient forensic research on disasters is developed that demonstrates the common, root, and underlying causes then the practice of disaster risk management will be radically transformed.

11.2 What Is a 'Natural' Disaster?

In this paper we are concerned with what are commonly referred to as 'natural' disasters. A starting point for this analysis is the recognition that the term 'natural' disaster is misleading. It suggests that the magnitude and frequency of disasters are driven by natural processes. What are commonly referred to as 'natural' disasters are to a very large extent the result of human behaviour, choices, and decisions. What is natural – or mostly natural – is the triggering event. An earthquake can be safely regarded, for most purposes, as a natural event. So too can a tropical cyclone. Hence it is the event that is natural and not the consequences. A more appropriate term therefore is 'natural hazard'. It is important to add the caveat that even natural hazards are often not completely natural. Floods can be exacerbated by deforestation, and increasingly extreme weather events are being influenced by anthropogenic climate change.

It has been understood for a long time that human activities can increase the frequency and magnitude of some natural hazards. Deforestation, urbanization, and other changes in land use and land cover can increase run-off and hence the nature of hydrological events including floods. The recent Intergovernmental Panel on Climate Change Special Report, *Managing the Risks of Extreme Events and Disasters to Advance Climate Change Adaptation (SREX)* (IPCC 2012) defines disasters as, "severe alterations in the normal functioning of a community or a society due to hazardous physical events interacting with vulnerable social conditions, leading to widespread adverse human, material, economic, or environmental effects that require immediate emergency response to satisfy critical human needs and that may require external support for recovery" (IPCC 2012). Hazards, on the other hand, are simply the danger. Or more definitively, a phenomenon, event or occurrence that has the potential for causing injury to life or damage to property or the environment. Hazards include earthquakes, floods, tornadoes, volcanic eruptions, landslides and many others. Conversely, disasters are largely the consequence of inappropriately managed risks and a product of hazards, exposure, and vulnerability (IDNDR 1992).

Natural hazards events are therefore not the principal cause of disasters; rather, people, communities and nations live and build in exposed areas often populated by the poorer and more vulnerable social groups. What might remain simply a natural extreme or event is transformed into a largely man-made disaster. While hazard events themselves (such as volcanic eruptions, hurricanes or earthquakes) are not preventable, the associated damages and disasters can be mitigated or even prevented by human development choices. The term 'natural disaster', used almost universally in media reportage should thus be stricken from the lexicon of crisis terminology and replaced with more precise language which reflects the human role in the generation of disasters.

This distinction between natural hazards and man-made disasters is important because unless its usage is corrected in daily professional life, within civil society, and in media there will continue to be a common and persistent misunderstanding that disasters are caused by nature or are *Acts of God. Acts of God* is the term commonly applied in legal processes, especially by insurance companies, to 'natural' disasters events – those viewed exterior to human control. The threat of climate change is altering the landscape for the use of this term. There are very real legal implications and ramifications for insurance and reinsurance industries if *Acts of God* turn is corrected to *Acts of Man*; responsibly shifts, bringing the need for changes to behaviour, policy and governance. Thus the magnitude of the Hurricane Katrina disaster, which devastated New Orleans and adjacent areas in 2005 killing more than 1,800 people, and displacing many more, cannot simply be attributed the hurricane. The flooding of 80 % of the city and economic damages estimated at one $100 billion were consequences of human choices and decisions (Glantz 2008; Grossi and Muir-Wood 2006; Kates et al. 2006).

Similarly, the catastrophic consequences of the Fukushima disaster in north east Japan in March 2011 were triggered but not caused by the offshore earthquake and consequent tsunami. Funabashi (2012) refers to Fukushima as a "man-made calamity [caused by] technological failures…derived from the failure of multiple social safeguards". Again, the earthquake was the triggering event, but it was human choice driven by many contributory factors (that likely include economic benefits, level of safely standards, technological and social safeguards, quality and design of infrastructure) that created a level of exposure and vulnerability subsequently in the wisdom of hind sight have come to be seen as unacceptable.

Of course it can be argued that without such triggering events there would have been no disasters. It is also true that many such triggering events are not followed by disasters due to the lack of exposure and vulnerability. Lack of exposure and vulnerability is not accidental. It results to a large extent from human choices and decisions. In this paper when we speak of disasters we are referring to naturally triggered disasters. As Gilbert White wrote in 1945, "Floods are acts of God, but floods are largely acts of man" (White 1945).

11.3 The Disaster Epidemic

Both the frequency and magnitude of naturally triggered disasters is increasing worldwide. The evidence for this assertion is based upon scientific publications and data collected by EM-DAT and Munich Re. Figure 11.1 illustrates how in the past 30 years geophysical events have remained fairly constant, while meteorological (storm), hydrological (flood and mass movement) and climatological events (extreme temperatures, drought and forest fires) have all increased, shifting from under 400 events per year in 1980 to over 1,000 events per year in 2007. Statistically speaking, 4,130 disasters were documented between the years 2002 and 2011, with over 1 million lives lost, costing a recorded $1.2 trillion. In 2011, the globe witnessed over 300 hundred disasters impacting the lives of 200 million citizens (UNISDR 2012).

So why are disasters on the increase in both magnitude and frequency? Until recently it could be argued that there was no significant change in the frequency or magnitude of the triggering events themselves. However, since the discovery and recognition of anthropogenic climate change it has become common to question whether a changing climate could account for part of the increase (IPCC 2012).

Climate change compounds man-made vulnerability and exposure, where anthropogenic (man-made) climate change means acknowledging that not only are disaster losses the result of human action but hazardous events themselves are now partially due to human actions. Climate change is and will continue to alter the intensity, frequency, duration, spatial extent, and timing of extreme weather and climate events (IPCC 2012); shifting the probability distribution of an events occurrence – moving the mean, increasing the variability and changing the symmetry of present and future climates. Irrefutable evidence is available dating from 1950

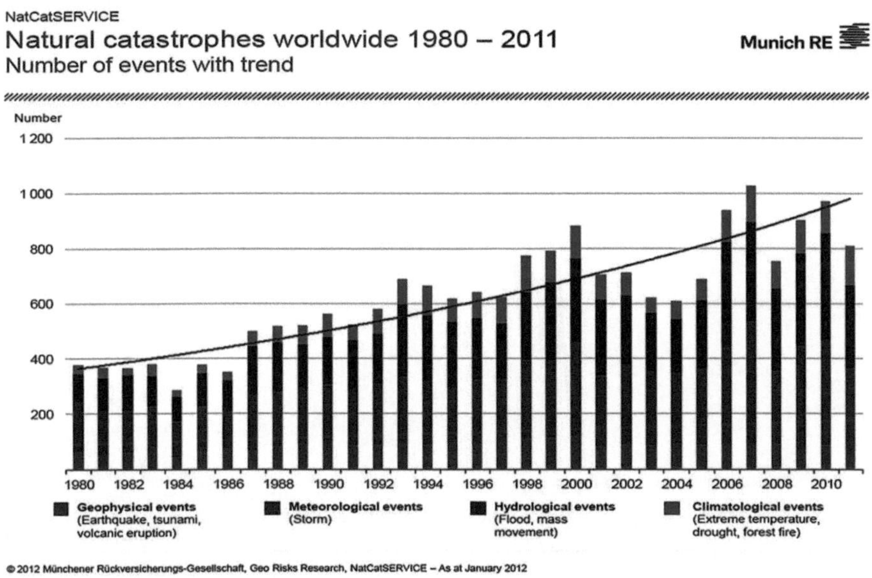

Fig. 11.1 Natural catastrophes worldwide 1980–2011 (MunichRe)

demonstrating that extreme events are changing, however the confidence of these changes varies by hazard type and depends on the quality and quantity of data available. For example, floods are the most frequently reported hazard, with earthquakes being the easiest to detect (and possibility overly-measured and reported) and droughts being the hardest to measure and report. Data and accuracy in reporting also fluctuates by global region.

Figure 11.2 shows a variation of Fig. 11.1, where all disasters (drought, extreme temperatures, famine, flood, insect infestation, slides, volcanic eruption, wave and surge, wild fires and windstorms) are added and compared against the number of earthquakes per year. The combined disasters are seemingly epidemic in proportion to the earthquakes (where the increase could be justified by increasing technological capacity to detect disasters since the 1960s). Figure 11.2 also shows that when earthquakes are compared to climatic disasters (floods and cyclones) earthquakes remain relatively static while climatic disasters have an increasing trend in the past 30 years.

A second observation in disaster trends is an increase, over the past 30 years, in overall and insured losses from natural catastrophes (Fig. 11.3).

"A sequence of devastating earthquakes and a large number of weather-related catastrophes made 2011 the costliest year ever in terms of natural catastrophe losses. At about US$ 380bn, global economic losses were nearly two-thirds higher than in 2005, the previous record year with losses of US$ 220bn. The earthquakes in Japan in March and New Zealand in February alone caused almost two-thirds of these losses. Insured losses of US$ 105bn also exceeded the 2005 record (US$ 101bn)" (MunichRe 2012).

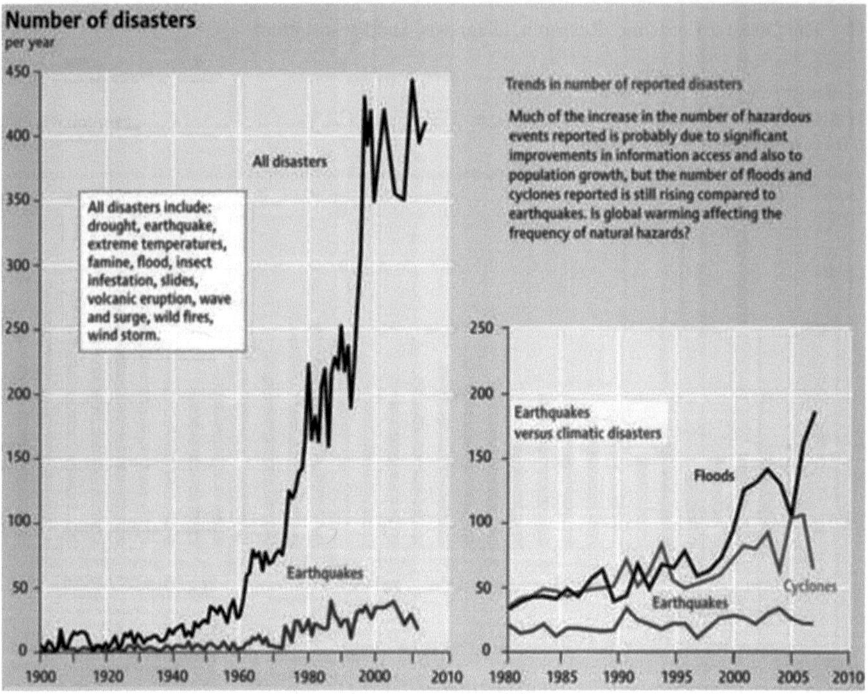

Fig. 11.2 Number of disasters per year, 1900–2010 (CRED) (Source: CRED Annual Disaster Statistical Review 2006, 2007)

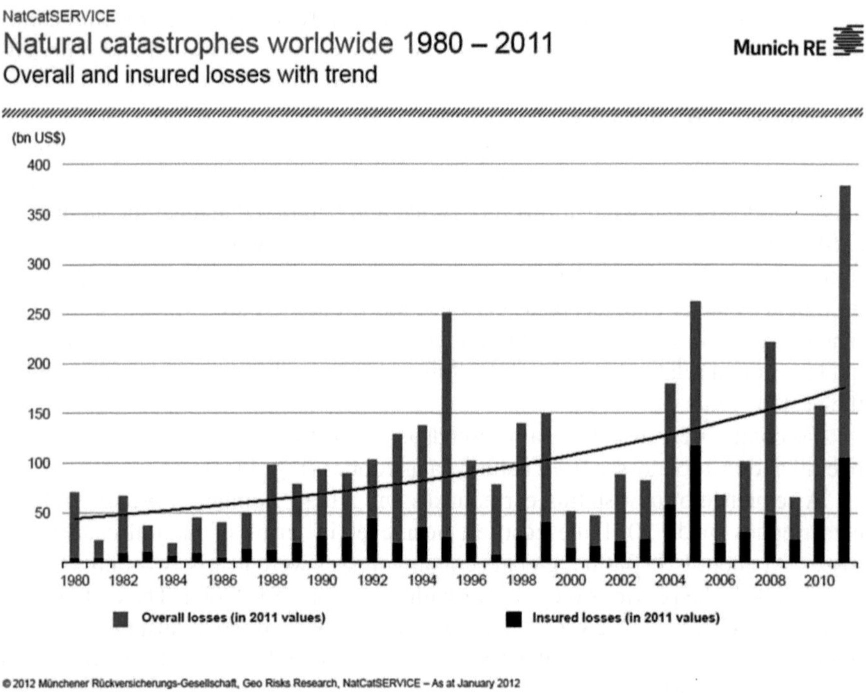

Fig. 11.3 Overall and insured losses with trend 1980–2011 (MunichRe)

And while greater loss may be explained by larger populations, and greater wealth including tangible assets, these observations do raise many questions, including what role does technological and scientific understanding play in expanding the ability to measure and report disasters? And what role does population growth, changes in land use planning and infrastructure development, and anthropogenic climate change play in the frequency of natural hazards and disaster losses? Irrespective of the answers, Fig. 11.1 reveals an almost exponential increase in disasters which cannot be easily justified or overlooked.

11.4 Types of Disasters

It has been the practice for some time to make a distinction between 'natural' disasters and man-made or technological disasters. This distinction is breaking down because as we have argued the magnitude of so called 'natural' disasters is not primarily the result of natural processes. Further, with the advent of climate change and the detection of anthropogenic climate change as a small but growing factor in the frequency of triggering events, even these events are falling into the man-made category.

There is also a growing interaction between the initial triggering events and ensuing technological, social, economic, and environmental consequences. Large disasters almost always have effects well beyond the site of initial occurrence.

Cascading disasters occur when hazards affect interconnected and interdependent systems. These typically include critical infrastructure (power, communications, transportation, water supply systems) and lifesaving services (emergency, government and health services). These systems are mutually dependent on one another. When one system fails or is impacted it cascades and impacts one or more other systems which can results in a complete failure in functioning of an entire community, state or nation and threaten all forms of life and normalcy (see Fig. 11.4).

In March 2011, an earthquake with a magnitude of 9.0 Mw struck off the Pacific coast of Tōhoku; the quake subsequently triggered a powerful tsunami with 40.5 m waves (133 ft) which flooded the shores of Japan leading to the meltdown of the Fukushima Daiichi Nuclear plant and consequently released radioactive isotopes. Almost all forms of critical infrastructure were lost – emergency services were hampered by radioactivity concerns, flooding, lack of transportation and viable infrastructure, and frigid weather conditions. An exclusion zone was immediately put in place around the Plant requiring the relocation and migration of thousands of residents whose health was acutely impacted by the hazardous events and whose long term quality of life was being threatened by carcinogenic radioactive material seeping into the water and atmosphere. The disaster continued to cascade with economic impacts both immediate and long term; businesses across Japan promptly halted, with major automotive plants overseas going under stocked over radioactive concerns and lack of critical infrastructure to manufacture and transport goods. Long term anxieties over seafood and ecosystem services continue to diminish economic gain and endanger human health.

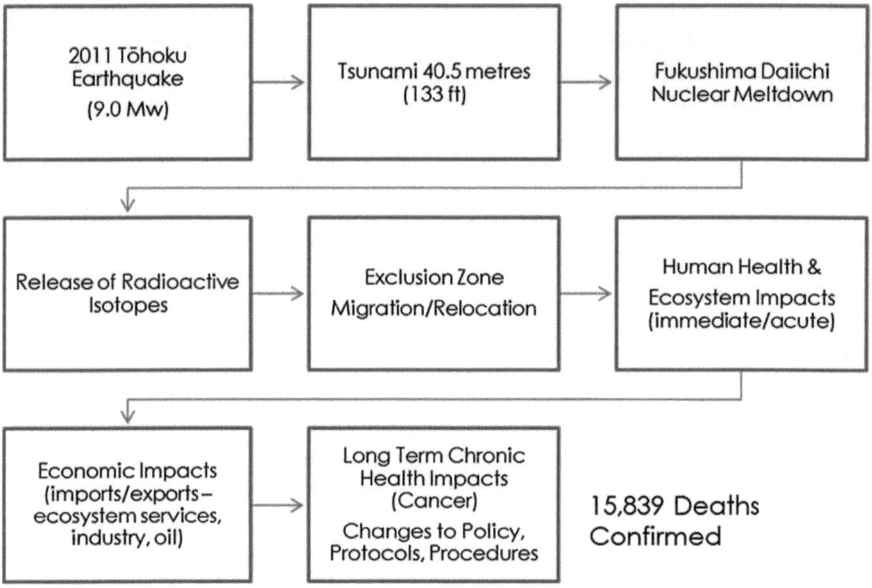

Fig. 11.4 Interconnected and cascading effects from combined natural and technological disasters (simplistic schematic for illustration purposes)

And the cascading effects continue as policy surrounding nuclear power is being revisited and altered. An energy white paper approved by the Japanese Cabinet in October 2011, says "public confidence in safety of nuclear power was greatly damaged" by the Fukushima disaster – the last working nuclear reactor in Japan was shut down in May 2012. Newly formed protocols and procedures now require each reactor to withstand earthquakes and tsunamis before they can be restarted, in the meantime, this disaster has resulted in a large trade deficit requiring the importation of greater amounts of oil and reliance on other forms of energy.

Our increasing reliance on interconnected and interdependent systems in a globalized society has created a greater vulnerability that once was not so prevalent. We have shifted our connections from local and regional to national and international. Globalization has a profound effect on the spatial impact of previously localized disasters. The earthquake in Japan was not an isolated event, its consequences spread across the globe from migrating/relocating needs, to a 1 % shrink in economic growth in Japan resonating to China, Malaysia, India, Singapore, Philippines and beyond.

The 2011 floods in Thailand were estimated to have caused a 2.5 % decrease in global industrial production. Growing dependence on information technology and telecommunications has also increased our mutual reliance – increased interdependencies – however, it has benefited society by allowing for increased awareness (for those with technological access), spreading word of disaster and aiding in response efforts. These cascading events and interlinkages added an increase complexity in the disaster risk management and risk reduction community. Plans need to address not mutually exclusive hazards and events, but the multiple dimensions of disasters and the widespread and compounding impacts they can have globally.

11.5 Disaster Risk Reduction: History and Response

Over the last two decades there has been growing efforts to move the disaster research and humanitarian response communities in the direction of disaster risk reduction.

11.5.1 History

In 1989, the United Nations General Assembly adopted a resolution that designated the 1990s as the International Decade for Natural Disaster Reduction (IDNDR), demonstrating a commitment to disaster reduction. This was the first major collective international attempt to reduce disaster impact, particularly within hazard-prone developing countries. Each country was encouraged to establish National Committees for the Decade and about 120 were established.

In 1994, at the Yokohama Conference on Risk Reduction called the Yokohama Strategy and Plan of Action, was developed and adopted to provide policy guidance with technical and scientific assistance and cooperation. In 2000, at the end of the Decade the IDNDR was succeeded by a new institutional arrangement called the United Nations International Strategy for Disaster Reduction (UNISDR), which broadened the scope to include increased public commitment and linkages to sustainable development. The approach of the ISDR system is to advocate the use and scaling up of tools and measures to reduce disaster risk and to encourage collaboration between disaster reduction and climate change managers, for instance by undertaking joint disaster risk reduction and adaptation planning and programming. Partners in the system engage in capacity-building of climate change and disaster risk actors, awareness-raising at community and national levels, advocacy with climate change delegates to promote the integration of the disaster risk reduction approach in international climate policy, and the production and dissemination of risk assessment and management tools. The ISDR secretariat provides information and guidance on disaster risk reduction as a tool to manage climate risks and adapt to climate change, both to inform international policy deliberations and to assist governments and other parties to reduce climate-related vulnerabilities and risk. It undertakes global reviews of disaster risk and progress on risk reduction and facilitates the compilation, exchange, analysis and dissemination of good practices and lessons learned in disaster risk reduction.

In January 2005, 168 governments adopted the Hyogo Framework for Action (HFA) 2005–2015: Building the Resilience of Nations and Communities to Disasters at the United Nations World Conference on Disaster Reduction in Kobe, Japan

(WCDR). The Framework was unanimously endorsed by the UN General Assembly. The HFA is the key instrument for implementing disaster risk reduction, adopted by the Member States of the United Nations. The HFA's Strategic Goals include the integration of disaster risk reduction into sustainable development policies and planning; development and strengthening of institutions, mechanisms and capacities to build resilience to hazards; and the systematic incorporation of risk reduction approaches into the design and implementation of emergency preparedness, response and recovery programs.

The Framework also provides five areas of Priorities for Action:

1. Ensure that disaster risk reduction (DRR) is a national and local priority, with a strong institutional basis for implementation
2. Identify, assess and monitor disaster risks and enhance early warning
3. Use knowledge, innovation and education to build a culture of safety and resilience at all levels
4. Reduce the underlying risk factors
5. Strengthen disaster preparedness for effective response at all levels (UNISDR 2011)

The fulfillment of the HFA requires the commitment and involvement of a broad array of stakeholders including all levels of government, regional and international organizations, civil society, NGOs and the private sector, and the scientific community.

According to the Framework states are responsible for developing national coordination mechanisms; conducting baseline assessments on the status of disaster risk reduction; publishing and updating summaries of national programmes; reviewing national progress towards achieving the objectives and priorities of the Hyogo Framework; working to implement relevant international legal instruments; and integrating disaster risk reduction with climate change strategies. Regional organizations are expected to promote programmes for disaster risk reduction; undertake and publish regional and sub-regional baseline assessments; coordinate reviews on progress toward implementing the Hyogo Framework within the region; establish regional collaborative centres; and support the development of regional early warning mechanisms. International organizations are depended to encourage the integration of disaster risk reduction into humanitarian and sustainable development programmes and frameworks; strengthen the capacity of the United Nations system to assist disaster-prone developing countries with disaster risk reduction initiatives; support data collection and forecasting, information exchange, and early warning systems; supporting States' own efforts with coordinated international assistance; and, strengthen disaster management training and capacity building. The overarching ISDR System is responsible for developing a matrix of roles and initiatives related to the Hyogo Framework; facilitating the coordination of actions at the international and regional levels; developing indicators of progress to assist States in tracking their progress towards implementation of the Hyogo Framework; supporting national platforms and coordination mechanisms; stimulating the exchange of best practices and lessons learned; and, preparing reviews on progress toward achieving the Hyogo Framework objectives.

As a result of the adoption of the HFA, global efforts to address DRR have become more systematic. In 2009, the first biennial Global Assessment Report on Disaster Risk Reduction (GAR) was released. The report found that since the adoption of the HFA, progress towards decreasing disaster risk is varied. Countries have been making progress towards increasing capacity, developing institutional systems and legislation to combat DRR; and early warning systems have been implemented in many areas. However, progress is still required in mainstreaming DRR into planning and development. The findings continued that current DRR governance arrangements do not allow for the integration of risk reduction into development. Further, at national and international levels, policy and institutional frameworks for climate change adaptation and poverty reduction are faintly connected to those for DRR. Underlying risk factors including poverty, ecosystem decline, poor governance systems and vulnerable livelihoods are difficult but possible for countries to address using mechanisms (e.g., micro insurance) to increase resilience.

11.5.2 Present

Until climate change there was little recognition of the possibility of common causes for diverse disaster events. Disasters have been widely been perceived as unique and local events or sometimes regional as in the case of droughts. This social construction of disasters has had the adverse consequence of focusing attention on two dimensions of disasters and the neglect of a third. First overwhelming attention has been given to the natural science of disaster triggering events and process. Disaster research has meant physical or natural science research, often linked to improved forecasting and warning systems. Secondly attention has been drawn to the aftermath of disasters as in the process of emergency relief and disaster relief, recovery, and rehabilitation. The neglected area has been the opportunity to search for understandings of potential common causes, and to develop policies and practices of disaster risk reduction. Far more money is spent on natural science research, forecasting and warning, and on emergency response, humanitarian relief, rehabilitation and reconstruction than on searching for and addressing the deeper and underlying causes of the creation of exposure and vulnerability.

When disaster risk management and vulnerability reduction measures are lacking or insufficient humanitarian responses are necessary. Hazards often pose greater challenges to impoverished communities or countries whose economies are dependent on a limited number of vulnerable sectors (e.g., drought/flood prone agriculture). Without greater emphasis placed on disaster risk reduction and management, humanitarian efforts in the form of emergency aid and rebuilding infrastructure are essential. In a 2012 survey of international (humanitarian) aid groups (NGOs), greater than half of the agencies stated more needs to be done in the area of disaster risk reduction and what is deeply needed is putting emphasis on reducing exposure to disasters as opposed to funding grand humanitarian efforts after a disaster strikes. The International Federation of Red Cross and Red Crescent Societies (IFRC) stated in its *2009 World Disasters Report* that, "$1 spent on prevention saves $4 on emergency response".

Currently, global funding for disaster risk reduction and management continues to receive suppressed amounts when compared with humanitarian aid, and it is a fraction of total development aid. To illustrate, for the years 2000–2009 only US$3.7 billion out of US$363 billion of development aid was for disaster risk reduction – translating into 1 % of total funds (GHA Report 2012). Funding for disaster risk reduction is also not spread evenly among vulnerable countries or provided in a way proportional to vulnerability. Funding continues to be inconsistent and does not increase year to year – fluctuating from $121 million in 2000, US$809 million in 2007 and US$338 million in 2009.

In the past decade there have been only two instances when disaster risk reduction funding achieved 2 % of official development aid. And with disaster damages in the most vulnerable countries totalling US$74 billion with only US$3.7B in risk reduction funding (a mere 5 %) there is great potential to reduce the damages and losses from disasters especially reflecting on the statistic that every $1 spent saves $4 on emergency response and recovery. It should also be mentioned that there exists a partially untapped opportunity to merge international funding for climate change adaptation and disaster risk reduction.

11.6 The FORIN Project

Partly in the light of the type of analysis and diagnosis presented here, the International Council for Science (ICSU) has joined with the International Social Sciences Council (ISSC) and the ISDR to develop a new research programme on disasters, under the title of Integrated Research on Disaster Risk (IRDR). Projects being developed include an assessment of the state of knowledge and knowledge gaps in integrated research on disaster risk; data needs; the behavioural and psychological dimensions of disaster risk; and of particular relevance to the epidemiology of disasters the FORIN Project (the Forensic Investigation of Disasters) (Burton 2010).

The FORIN Project is developing with collaborating institutions a series of case studies designed to examine the roots and underlying causes in the growth of disaster risk. A common framework and template for the forensic investigations has been prepared (Integrated Research on Disaster Risk 2011) titled *Forensic Investigations of Disasters: The FORIN Project.*

11.7 Improved Governance of Disaster Risk

In our view these recent initiatives (outlined in the section: Disaster Epidemiology History and Response) are on the right track, but are deficient and have much further to go in at least two specific directions. First, there is a need to strengthen disaster risk governance at all levels. Second, we need to expand the knowledge base to incorporate a much wider understanding of the root causes of disasters in both spatial and temporal dimensions. These final sections discuss the need for improved governance and risk management decisions.

The UNDP states, the "appropriate governance for disaster risk management is a fundamental requirement if risk considerations are to be factored into development planning and if existing risks are to be successfully mitigated" (UNISDR 2004). Good governance structures are an essential key to addressing disaster risk and informing responses as they help shape efficiency, effectiveness, equity, and legitimacy (UNISDR 2011). Governance in disaster risk management includes the process to design, implement and evaluate strategies, policies and measures to first improve the knowledge and understanding of risks associated with disaster, develop and advance disaster risk reduction and promote technology and information transfer and cultivate an environment where there is ongoing progress to increase disaster preparation, response and recovery. The intention is to improve human security, well-being, quality of life, resilience and sustainable growth and development. To measure governance effectiveness, the World Bank developed and defined six governance indicators:

- Voice and accountability;
- Political stability and absence of violence;
- Government effectiveness;
- Regulatory quality;
- Rule of law, and
- Control of corruption

And while these may not be completely transferable as disaster risk reduction indicators, a country, state or community with a high level of political stability and accountability, has a higher probability that disaster risk will be reduced and preparedness will be increased. When governance improves so too does managing disaster risk.

A suitable disaster risk governance system at the international level should be made up of a few core elements including a foundational framework for risk reduction (e.g., Hyogo), a bevy of international organizations (e.g., UNISDR) and funding mechanisms to provide financial assistance for the undertaking of reducing exposure to natural and man-made hazards. And to ensure the effectiveness of these elements entails the following (suggested) core requirements: promote awareness and understanding of disaster risk and reduction measures, create processes and methods to increase adaptive capacity, facilitate the development, acquisition and transfer of technology to reduce vulnerability (often from developed to developing country, and if necessary incentivise the transfer), and provide technical and development assistance to countries and regions needing support. Currently, however, each of the elements (framework, organization and funding) is neglecting at least one of its core requirements.

There are two types of inclusiveness essential for successful governance in risk reduction: multi-hazard and multi-input. A multi-hazard governance approach ensures the inclusion of all hazard types (natural and man-made). Figure 11.4 showed the interconnection and cascading effects from natural and technological disasters demonstrating the need for such comprehensiveness. If policies or measures are made without the consideration of more than one hazard type or outcome or are not exhaustive, the consequences -as in the case in Japan- can be catastrophic. Ensuring a multi-hazard platform allows an initial opportunity to decrease the likelihood of cascading effects from natural and man-made hazards. Successful governance systems also require multi-inclusion. No group is an island to themselves. Naturally triggered disasters, while in some instances may be local, are by no means isolated. Input is needed from all vested stakeholders including research institutions, commercial enterprises, local representatives (the public), non-profit and community based organizations and all levels of government. These comprehensive groups of representatives, whose differing yet complimentary roles, merge to reduce and manage disaster risk more effectively by improving design, implementation and evaluation of disaster risk reduction.

Increasing knowledge is another way to strengthen governance designed to manage disaster risk. Past experience, both local and international, is needed to grasp the complexities of disasters and to manage systems more realistically. Currently, data required to make proper decisions and strategies for disasters and disaster risk reduction is deficient at the local level (IPCC 2012). This limits the amount to which governance structures can improve –good governance requires good and accurate data. National data and projection centres are a primary institution in supporting the challenges in predicting future trends in extreme weather events. As data improves, so does the foundation for understanding risks associated with disasters; and addressing knowledge gaps with greater observation and research capacity will translating into more effective risk reduction strategies.

For good governance to develop and advance disaster risk reduction, it obliges the integration of climate change adaptation and disaster risk management. Integration, however, is required across scales for an international risk governance system to result in tangible change and significant reduction in vulnerability (IPCC 2012).

11.8 Post 2015

We have argued that the increase that is now being observed in naturally triggered disasters is akin to an epidemic. Public health epidemics (and pandemics) prompt a vigorous search for a common cause or causes. It is not assumed that disease outbreaks occurring hundreds or even thousands of kilometres apart are independent events. By convention and practice the same approach has yet to be applied in the field of naturally triggered disasters. The Hyogo Framework is designed to extend to 2015 and will then be updated, redesigned or redeveloped. Also IPCC 5th Assessment Report is due to be in 2014. There will now be an opportunity to develop a more integrated and comprehensive programme that aligns disaster risk reduction and climate change adaptation and increases the focus of international cooperation on disaster risk reduction. The current mandate of the IRDR can be revised to take into account the preliminary results of forensic disaster investigations, and the recognition of the common and underlying causes of disasters as conveyed in the concept of disaster epidemiology.

References

Ahmed, S., P.K. Bardhan, A. Iqbal, R.N. Mazumder, A.I. Khan, M.S. Islam, and A. Cravioto. 2011. The 2008 cholera epidemic in Zimbabwe: Experience of the icddr,b team in the field. *Journal of Health, Population and Nutrition* 29(5): 541–546.

Burton, I. 2010. Forensic disaster investigations in depth: A new case study model. *Environment* 52(5): 36–41.

Chambers, K. 2009. Zimbabwe's battle against cholera. *Lancet* 373(9668): 993–994.

Funabashi, H. 2012. Why the Fukushima nuclear disaster is a man-made calamity. *International Journal of Japanese Sociology* 21(1): 65–75.

GHA. 2012. *Global humanitarian assistance report*. http://www.globalhumanitarianassistance.org/wp-content/uploads/2012/07/GHA_Report_2012-Websingle.pdf.

Glantz, M.H. 2008. Hurricane Katrina as a 'teachable moment'. *Advances in Geosciences* 14: 287–294. Accessible at: http://www.adv-geosci.net/14/287/2008/.

Grossi, P., and R. Muir-Wood. 2006. *Flood risk in New Orleans: Implications for future management and insurability*. London: Risk Management Solutions.

IDNDR. 1992. *Internationally agreed glossary of basic terms related to disaster management. United Nations*. Geneva: DHA/IDNDR.

Integrated Research on Disaster Risk. 2011. *Forensic investigations of disasters: The FORIN project*, IRDR FORIN publication no. 1. Beijing: Integrated Research on Disaster Risk.

IPCC. 2012. *Managing the risks of extreme events and disasters to advance climate change adaptation. A special report of working groups I and II of the Intergovernmental Panel on Climate Change*, ed. C.B. Field, V. Barros, T.F. Stocker, D. Qin, D.J. Dokken, K.L. Ebi, M.D. Mastrandrea, K.J. Mach, G.-K. Plattner, S.K. Allen, M. Tignor, and P.M. Midgley. Cambridge: Cambridge University Press.

Kates, R.W., C.E. Colten, S. Laska, and S.P. Leatherman. 2006. Reconstruction of New Orleans after Hurricane Katrina: A research perspective. *Proceedings of the National Academy of Sciences* 103(40): 14653–14660.

Mason, P.R. 2009. Zimbabwe experiences the worst epidemic of cholera in Africa. *Journal of Infection in Developing Countries* 3(2): 148–151.

MunichRe. 2012. *Review of natural catastrophes in 2011: Earthquakes result in record loss year*. Reinsurance Press release, 4 January 2012. http://www.munichre.com/en/media_relations/press_releases/2012/2012_01_04_press_release.aspx.

Musemwa, M. 2010. From 'sunshine city' to a landscape of disaster: The politics of water, sanitation and disease in Harare, Zimbabwe, 1980–2009. *Journal of Developing Societies* 26(2): 165–206.

Nelson, E.J., J.B. Harris, J.G. Morris Jr., S.B. Calderwood, and A. Camilli. 2009. Cholera transmission: The host, pathogen and bacteriophage dynamic. *Nature Reviews Microbiology* 7(10): 693–702.

UNISDR. 2004. *Guidelines for mainstreaming disaster risk assessment in development, disaster risk reduction for sustainable development in Africa*. Geneva: United Nations International Strategy for Disaster Reduction.

UNISDR. 2011. *Global assessment report on disaster risk reduction*. Geneva: United Nations International Strategy for Disaster Reduction.

UNISDR. 2012. *Towards a post-2015 framework for disaster risk reduction.* http://www.unisdr.org/files/25129_towardsapost2015frameworkfordisaste.pdf.
White, G. 1945. *Human adjustment to floods*, Department of Geography research paper no. 29. Chicago: The University of Chicago.
WHO. 2009. *Cholera in Zimbabwe – Update 4. Global Alert and Response (GAR).* http://www.who.int/csr/don/2009_06_09/en/.

Thea Dickinson is a partner in Burton Dickinson Consulting and a Ph.D. Candidate at the University of Toronto. She has over 10 years experience in environmental and biomedical research and has authored *The compendium of adaptation models for climate change* (2007) and co-authored several chapters in publications including *Climate change adaptation in Canada: A multi-level mosaic, Climate change and human health in cities.* Thea has worked with and consulted for the UNDP and UNEP, Red Cross/Red Crescent, Danida, Environment Canada and many other organizations.

Ian Burton is an Emeritus Professor in the Department of Geography and the School of the Environment, University of Toronto. Burton is a pioneer in the field of adaptation to climate change and extreme events and disasters during his work with the Meteorological Service of Canada. He has been working with colleagues to improve the integration of climate change adaptation with disaster risk reduction, and has been a Lead Author for three of the *IPCC Assessments and the Special Report on Extremes* (SREX). He is currently co-chair of the Working Group on Forensic Disaster Investigations within the programme Integrated Research on Disaster Risk. With Robert Kates and Gilbert F. White in 1978 he authored the reference book *The environment as hazard.*

Chapter 12
An Evaluation of Risk Management and Emergency Management. Relying on the Concept of Comprehensive Vulnerability Management for an Integrated Perspective

David A. McEntire

12.1 Introduction

According to Kathleen Tierney, "the [disaster] field has yet to develop a coherent theoretical perspective" (Tierney 1999). This problem still plagues researchers and professionals today. For instance, in the United States, concepts such as civil defense, disaster resistance, sustainable development, disaster resilience and homeland security have guided professionals over time (McEntire 2005). Surprisingly, there have been insufficient efforts to critically evaluate these concepts and integrate them in a more holistic manner. A similar challenge is also noticeable when risk management and emergency management are compared and contrasted.

With this introduction in mind, the following chapter will discuss the concepts, processes and weaknesses of risk management and emergency management. It will explain that a more holistic approach to disasters requires an integration of these concepts, and concludes with an exploration into what this policy guide may look like. Specifically, the paper introduces the notion of 'comprehensive vulnerability management (CVM).' It attempts to illustrate how this concept incorporates vital aspects of both risk management and emergency management while possibly overcoming the weaknesses of each individual perspective.

12.2 Understanding Risk Management

Risk management, like many other concepts, has been defined in a variety of ways and there is consequently no universal consensus on what it actually means. For instance, Gavrilescu suggests that risk management "can be defined

D.A. McEntire (✉)
Emergency Administration and Planning, Department of Public Administration,
University of North Texas, 1155 Union Circle, #310617, Denton, TX 76203-5017, USA
e-mail: mcentire@unt.edu

U. Fra.Paleo (ed.), *Risk Governance*, DOI 10.1007/978-94-017-9328-5_12

as an organized, systematic decision-making process that efficiently identifies risks, assesses or analyzes risks, and effectively reduces or eliminates risks … [in order to achieve] program goals" (2002). Others equate risk management to the estimation of likelihood as well as the consequences of unpleasant surprises or the management of opportunities and adverse effects (Holt 2004; Kunreuther 2002; Hodges 2000). Perhaps giving up due to the diversity of opinions on the matter, Kloman simply states that "risk management is the management of risk" (Kloman 2000).

In spite of the wide variety of viewpoints, there is, however, a general agreement that risk management is related to making decisions about probability and impacts while also implementing choices to maximize utility and minimize costs. In terms of disasters, risk management can thus be considered as a method of determining how to reduce both the probability and impacts of these deadly, destructive and disruptive events.

If the definition of risk management is contested, so, too, is the specific process of managing risks. As an example, Gavrilescu regards risk management a simple "decision-making process through which choices can be made between a range of options, which achieve the required outcome" (Gavrilescu 2002). He also gives a great deal of attention to risk perception, or the interpretation of potential occurrence as well as likely outcomes (Gavrilescu 2002). Kunreuther (2002) likewise asserts that the 'psychological and emotional factors' have to be included in risk management in recent years. In contrast, Reynolds and Seeger (2005) give priority to risk communication, or the sharing of knowledge about what might happen and its probability. Biocca also agrees about the value of risk communication. "Great attention should also be paid to improve risk communication. Communication between the stakeholders (experts, decision makers, political and social leaders, media, groups of interest and people involved) is possibly the best condition to be successful in … risk management" (Biocca 2005).

Other scholars accept a multi-faceted approach that includes a number of components. Plattner (2005) discusses three phases: risk analysis, risk evaluation and risk management. Tchanakova (2002) also accepts three aspects of risk management, but labels these as risk identification, risk analysis and risk control. O'Brien et al. (2010) incorporate four features in their model of risk management: identifying, prioritizing, warning and informing. Kunreuther (2002) mentions the processes of risk assessment, risk perception, choice under conditions of uncertainty, and risk management. Several of the explanations are even more detailed and complicated:

- Hodges (2000) includes in his model establishing the context, identifying risks, analyzing risks, evaluating risks, and treating risks.
- Kloman (2000) asserts that risk management is "the systematic application of management policies, procedures, and practices to the tasks of identifying, analyzing, assessing, treating and monitoring risk."
- Gavrilescu (2002) identifies many components such as defining the entity to be managed, hazard identification, consequence and likelihood analysis, risk estimation, risk assessment, risk treatment.

Due to his desire for simplicity, Kloman (2000) suggests that the theoretical processes of risk management are often too lengthy. And, he may be justified in making this argument. The majority of proposals deal with common elements such as risk identification, risk analysis, and risk treatment. First, risk identification is a listing of what type of hazards may occur and how bad they could be. Second, risk analysis is a determination of which hazards should be given the highest priority (in terms of attention and resources). Third, risk treatment is the implementation of measures to deal with the hazard in the most effective way possible. This final step is fairly complicated however. Risk treatment may include avoiding risk, reducing the likelihood, reducing consequences, or transferring risk (Hodges 2000). It may also consist of accepting the risk, eliminating the risk, or reducing the risk, among other things (Gavrilescu 2002). Freeman and Kunreuther (1994) likewise give attention to the transfer of risk through legal liability and the use of insurance to spread risk, reduce variance, segregate risks, encourage loss reduction measures, and monitor and control risk. In other words, doing nothing, moving to a safer location, building in a more resistant manner, closing a dangerous factory, relying on insurance or planning on how to respond are all potential preferences for risk treatment.

As can be seen, risk management is a very proactive approach and many regard it to be the solution to the problem of more frequent and intense disasters. Although this statement is technically accurate, it would be difficult to assume that risk management is without limitation or controversy. Theoretical expositions and empirical observation reveals significant challenges regarding such approaches (McEntire 2009a). There are a few reasons why risk management is at times called into question. For example, people sometimes believe disasters will not happen to them so they are not willing to undertake protective measures (Kunreuther 2002, 2006). On other occasions, "a 'zero-risk-society' is something the public demands … [but] in the real world … it is extremely difficult to present 'zero risk' as a viable option" (Nakayachi 2000). Risk is also considered to be less important when it is perceived to be voluntary (Frewer 1999), and risk management is difficult because it deals with low probabilities and requires conflicting choices about multiple hazards (Kunreuther 2002; Nakayachi 2000). In addition, "disasters are … [sometimes] unpredictable" and risk management does not always recognize this fact (Somers and Svara 2009). Plattner (2005) and Kunreuther (2002) also declare that risk management is technical, and may also ignore other social, political and economic concerns. Similarly, Tierney believes "risk estimates are social constructs" which are consequently subject to all sorts of oversight, mistakes and manipulations (Tierney 1999). Furthermore, decision makers may "have little interest in undertaking costly protective measures" (Kunreuther 2002, 2006) and citizens tire of continuous payments to reduce risk (Nakayachi 2000). Therefore, risk management might incorrectly assume that disasters can be prevented and that humans will be able to eliminate their negative consequences. Risk management may consequently overlook, ignore, or downplay and be distinct from crisis management activities (see Reynolds and Seeger 2005). If this is an accurate assumption, it would be illogical to base disaster policy on risk management alone.

12.3 Comprehending Emergency Management

Much like risk management, emergency management also has various challenges relating to definitions, processes and limitations. The conceptual clarity of emergency management has often been called into question (McEntire 2006). For instance, emergency management is often regarded to be a misnomer because it really deals with disasters and not necessarily smaller emergencies. In other ways, emergency management is an oxymoron since it seemingly suggests that it is possible to manage all disasters in an effective manner. Emergency management thus shares conceptual challenges as does risk management.

Interestingly, emergency management has been defined by some "as the discipline dealing with risk and risk avoidance" (Haddow and Bullock 2006; Waugh 2000). In addition, concentrating on risk is regarded to be a principle of emergency management (Blanchard et al. 2007). Although emergency management is certainly related to elements of risk management, there may be reason to view these two fields as being somewhat separate and distinct. For example, there are numerous definitions that consider emergency management in a completely different light. Hoetmer asserts that emergency management is "the discipline and profession of applying science, technology, planning and management to deal with extreme events that can injure or kill large numbers of people, do extensive damage to property, and disrupt community life" (1991). The National Fire Protection Association declares that emergency management is "an ongoing process to prevent, mitigate, prepare for, respond to, and recover from an incident that threatens life, property, operations, or the environment" (2007). The Principles of Emergency Management Working Group also regard emergency management as "the managerial function charged with creating the framework within which communities reduce vulnerability to hazards and cope with disasters" (Blanchard et al. 2007). Therefore, many definitions typically focus on efforts to reduce potential disasters, anticipate their consequences, and react to them in an effective manner. In this sense, emergency management could be construed as a very broad concept.

Nevertheless, the process of emergency management has not always been complete or consistent over time, and it has even become more controversial in recent years. For instance, emergency management traditionally placed a great deal of emphasis on preparedness and response activities (Rubin 2007). In countries such as the United States, planning was a high priority during the civil defense era and emergency managers seemed to assert that their efforts were constrained to response when disasters occurred. In this sense, emergency management was a very reactive profession.

Later on, the Federal Emergency Management Agency introduced the 'comprehensive emergency management' (CEM) concept in 1979 which retained preparedness and response initiatives but added emphasis on mitigation and recovery phases. The goal of CEM was thus to encourage emergency managers to do more to prevent disasters and minimize their impacts in addition to helping communities rebound when negative consequences occurred. Nevertheless, this 'comprehensive' objective was never fully achieved (McEntire 2005). Emergency managers were either overwhelmed with planning activities or did not and/or would not fully engage in mitigation efforts. Even when Project Impact -a program to promote the Disaster Resistant Community- was introduced, local and state emergency managers misunderstood mitigation or did not give equal attention to this proactive phase of emergency management. While recovery did become part of the lexicon in emergency management, it certainly did not achieve the same importance as preparedness and response did.

After the 9/11 terrorist attacks, the phases of emergency management became even more muddled. The Department of Homeland Security introduced two additional terms to the four phases of emergency management (McEntire 2009b). The notion of 'prevention' suggested the imperative of eliminating or preempting terrorist attacks through intelligence operations, counter-terrorism operations, border control, law enforcement and/or prosecution. The idea of 'protection' implied a determination to guard critical infrastructure to discourage attacks or minimize damage through target hardening. The prevention and protection phases in homeland security are obviously aggressive measures, but they may at times remain outside of or even conflict with the emergency management realm (e.g., closed/top secret military communications or a law enforcement approach to attacks and disasters). The rise of homeland security also underscored the need to react to potential weapons of mass destruction, and negatively impacted the Federal Emergency Management Agency due to budget decisions and the development of new plans (Tierney 2006; Ekici and McEntire 2007). These types of issues have made it difficult to determine the appropriate range of emergency management responsibilities and if emergency managers can see beyond preparedness and response functions.

In light of these conceptual and practical challenges relating to the identity and boundaries of emergency management, the field is beset with numerous weakness and problems. For instance, because risk assessment and mitigation are not always regarded to be an individual or integral phase of emergency management, disaster professionals have yet to really develop skills in this area (Pine 2009). Also, emergency managers have generally retained priority over preparedness and response phases. Therefore, emergency management is sometimes regarded by others to be a reactive and incomplete profession (Sikich 1996). It is true that comprehensive emergency management does promote a broader perspective for disasters and is therefore more inclusive of partners than risk management. However, the intentional or unintentional neglect of mitigation and recovery makes emergency management less inclusive than it could or should be (McEntire 2005). Finally, because of the overwhelming attention given to terrorism and the impressive resources devoted to homeland security, it is unclear if emergency management will address all types of disasters (McEntire 2009a, b). Emergency management is thus problematic like the risk management concept, but for different and perhaps even opposite reasons.

12.4 The Need for an Integrated Perspective

In light of the distinctions between risk management and emergency management (see Somers and Svara 2009), it looks as if there is a need for a new perspective that incorporates the strengths of each conceptual viewpoint. Those that study risk management favor this recommendation. Gavrilescu asserts that "emergency/contingency planning … should be fully integrated into … environmental risk management" (2002, p. 15). Hodges also recommends integrating risk management into emergency management. He asserts that "emergency management can be promoted more effectively through risk management; and emergency management should dovetail into the broader risk management process" (Hodges 2000, p. 13). Disaster scholars also see the value of integrating risk management into emergency management practices. Pine (2009) believes disaster decision making can be improved through risk management. O'Brien et al. (2010) also says the process of risk management builds capacity in emergency management.

If it is true that risk management and emergency management need to be combined, what should such a perspective be called and what should it look like? What are its respective strengths and weaknesses? As has been argued elsewhere, a holistic concept and policy guide is needed for disaster scholars and those involved in the disaster profession, and 'comprehensive vulnerability management' has been introduced as one such possibility (McEntire et al. 2010). Comprehensive vulnerability management is the careful design of decisions and implementation of policies that attempt to reduce both the probability and consequences of disasters. Under this concept, the likelihood of disasters can be reduced through efforts to eliminate 'liabilities' that are produced by human activity in the physical and social environments. Such activities may include, but are certainly not limited to, improved land-use planning, environmental protection, enforcement of safety regulations, a reduction in poverty, promotion of cultural understanding and tolerance to reduce terrorism, and the challenging of apathetic attitudes that are often witnessed towards future disasters.

In addition, the notion of comprehensive vulnerability management suggests that the impact of disasters must also be lessened by enhancing 'capabilities' that are required to be more effective. Examples of the essential actions for this to occur are disaster planning, training of first responders and other relevant parties, the holding of disaster exercises, networking with stakeholders, community education and engagement, cautious decision making during crises, and learning that influences recovery plans among other things.

Due to its emphasis on liabilities and capabilities, comprehensive vulnerability management is perhaps similar to risk management in that it is a proactive form of dealing with disasters. In addition, comprehensive vulnerability management takes into account all types of potential hazards, and it intentionally sets about to address vulnerability -the only thing humans really have control over in the disaster equation. Comprehensive vulnerability management is likewise related to emergency management in that it recognizes that disasters can never be fully eliminated. This proposed concept accordingly incorporates reactive measures that will always be required in our imperfect and uncertain world.

Although comprehensive vulnerability management is based on principles from both risk management and emergency management, it is not necessarily a difficult policy to implement. Comprehensive vulnerability management is a three-step process of: (1) assessing liabilities and capabilities, (2) reducing risk and susceptibility, and (3) raising resistance and resilience.

- First, it is imperative that public officials and others determine what factors may trigger and aggravate disasters as well as decipher their strategic and operational ability to deal with these potential vulnerabilities. This may include the identification of a population in floodplain, structures along a fault line not built to code, unawareness of safety regulations in an industrial plant, or insufficient detection of weapons of mass destruction. This stage may also evaluate laws and options for relocation, willingness for code revisions, knowledge of training protocols, and abilities to conduct epidemiological surveillance at local hospitals.
- A second step is to undertake activities to reduce physical risk and social susceptibilities. Reducing physical risk may require elevation of flood prone structures, increased insurance premiums to discourage development in earthquake prone areas, careful location of oil refineries, and tight control over materials that could be accessed for terrorist attacks. Minimizing social susceptibilities may necessitate public awareness of flood potential, education of public officials about the need for building retrofit, new laws regarding the production of chemicals, and the encouragement of tolerance for different opinions and religions.
- The final step to implement comprehensive vulnerability management is to promote resistance and resilience. Resistance can be improved by maintaining dams and levees, following modern earthquake construction practices, relying on redundant valves and safety mechanisms at industrial facilities, and installing window laminates to minimize shattered glass in bombing attacks. Resilience can be facilitated through the development of flood warning systems, the training of search and rescue teams, improved situational awareness and decision making, and better understanding of recovery programs and procedures, among other things. Thus, comprehensive vulnerability management may be considered as "holistic and integrated activities directed toward the reduction of emergencies and disasters by diminishing risk and susceptibilities [while also] building resistance and resilience" (McEntire et al. 2002).

As has been argued elsewhere, comprehensive vulnerability management has several strengths (McEntire et al. 2002). For instance, because the concept is based explicitly on vulnerability, it is related to all types of hazards -whether they are natural, industrial or anthropogenic in nature (see McEntire et al. 2010, for an example of how the concept is applied to different disasters). Even though vulnerability results from human decisions and activities, it has bearing on the impact of hazards such as hurricanes, famines, industrial fires, chemical spills, plane crashes and terrorist bombings. Also, since comprehensive vulnerability management accepts both proactive and reactive approaches to disasters, it incorporates the traditional phases of mitigation, preparedness, response and recovery as well as the Department of Homeland Security's phases of prevention and protection. In other words, comprehensive vulnerability management is proactive through risk and susceptibility reduction and the raising of resistance. However, this concept is also reactive in the sense that the resilience concept acknowledges that disasters will occur and require effective response and recovery operations. Furthermore, because everyone has a role in creating or reducing vulnerability, comprehensive vulnerability management includes the general populace as well as those in the public, private and non-profit sectors. CVM is accordingly based on the assumption that all individuals, groups, organizations, governments and nations must play a role in managing our vulnerability if disasters are to be reduced in frequency and intensity. Another advantage of the concept is that it accepts complexity. Because there are many variables that produce and interact with risk, susceptibility, resistance and resilience, comprehensive vulnerability management reflects the reality of disasters we are confronted with in our modern era. Ergo, CVM is distinct from simple concepts that only offer piecemeal solutions for the challenging problem of disasters. Finally, and perhaps most importantly for the purposes of this chapter, comprehensive vulnerability management is an integration of the risk management and emergency management concepts. In many ways, the concentration on liabilities and capabilities is built on, but expands from, FEMA's policy of integrated emergency management which advocated the assessment of risks, the assessment of capabilities, and the closing of the gap between them. If this is accurate, then comprehensive vulnerability management may truly combine aspects of risk management and emergency management.

The above advantages should not be taken to imply that comprehensive vulnerability management is without weaknesses. For example, the concept is relative new and is not known in many academic and professional circles. Also, because comprehensive vulnerability management is a novel idea, it may experience skepticism and even opposition by those who favor the other concepts of risk management and emergency management. Moreover, even though recommendations have been given on how to implement comprehensive vulnerability management (McEntire 2004), more research needs to be conducted on what this might look like and how it can be promoted.

12.5 Conclusion

As has been illustrated, neither risk management nor emergency management is a complete disaster perspective and they each fail to fully embrace logical assumptions. For its part, risk management is highly proactive but it seems to incorrectly assume that all disasters can be eliminated if only our assessments are sufficiently detailed and accurate. In addition, risk management appears to neglect many response activities that must be fulfilled during and after disasters. These include measures such as warning, evacuation, emergency medical care, and disaster assistance. In contrast, emergency management is largely reactive and almost presumes that there is virtually nothing that can be done to reduce the probability and consequences of disasters. Emergency management is often relegated to preparedness and response phases, thereby overlooking the value of mitigation before disasters or during the process of recovery.

Because risk management and emergency management may not be holistic disaster concepts, the notion of comprehensive vulnerability management has been discussed in this chapter. Comprehensive vulnerability management can be defined as policies and their implementation in such a way as to overcome vulnerabilities in the physical and social environments. It concentrates on the reduction of liabilities (such as risk and susceptibility) as well as the building of capabilities (such as resistance and resilience). Because this broad perspective takes into account so many variables, hazards, phases and stakeholders, it might prove to be useful for both scholars and practitioners alike. However, there is clearly a need for further conceptual development along with additional research about its feasibility for implementation. For this reason, the author invites additional studies on how to reduce to probability and consequences of disasters.

References

Biocca, M. 2005. Risk communication and the precautionary principle. *Human and Ecological Risk Assessment* 11: 261–266.

Blanchard, B.W., L.G. Canton, C.L. Cwiak, K.C. Goss, D.A. McEntire, L. Newsome, M.D. Selves, E.A. Sorchik, K. Stenson, J.E. Turner III, W.L. Waugh Jr., and D. West. 2007. *Principles of emergency management*. Retrieved October 25, 2011, from FEMA, Higher Education Program. http://training.fema.gov/EMIWeb/edu/emprinciples.asp.

Ekici, S., and D.A. McEntire. 2007. Reassessing FEMA and its role in preparing the nation: Recommendations to respond effectively to disasters. In *Understanding terrorism*, ed. H. Durmaz, B. Sevinc, A.S. Yayla, and S. Ekici, 344–351. Fairfax: IOS Press.

Freeman, P.K., and H.K. Kunreuther. 1994. *Environmental risk through insurance*. Norwell: Kluwer.

Frewer, L. 1999. Risk perception, social trust, and public participation in strategic decision making: Implications for emerging technologies. *Ambio* 28(6): 569–574.

Gavrilescu, M. 2002. Risk assessment and management. Tools for sustainable development. *Environmental Engineering and Management Journal* 1(1): 3–20.

Haddow, G.D., and J.A. Bullock. 2006. *Introduction to emergency management*, 2nd ed. New York: Butterworth Heinemann.

Hodges, A. 2000. Emergency risk management. *Risk Management* 2(4): 7–18.

Hoetmer, G.J. 1991. Introduction. In *Emergency management: Principles and practice for local government*, ed. G.J. Hoetmer and T.E. Drabek, xvii–xxxiv. Washington, DC: International City/County Management Association.

Holt, R. 2004. Risk management: The talking cure. *Organization* 11(2): 251–270.

Kloman, H.F. 2000. *Risk management standards*. Risk management reports. www. riskreports. com/standards.html

Kunreuther, H. 2002. Risk analysis and risk management in an uncertain world. *Risk Analysis* 22(4): 655–664.

Kunreuther, H. 2006. Risk and reaction: Dealing with interdependencies. *Harvard International Review* 28(3): 38–42.

McEntire, D.A. 2004. Implementing a new approach to emergency management: Managing our vulnerability to reduce the occurrence and impact of disasters. *ASPEP Journal* 11: 1–7.
McEntire, D.A. 2005. Why vulnerability matters: Exploring the merit of an inclusive disaster reduction concept. *Disaster Prevention and Management* 14(2): 206–222.
McEntire, D.A. 2006. Local emergency management organizations. In *Handbook of disaster research*, ed. H. Rodriguez, E.L. Quarantelli, and R.R. Dynes, 168–182. New York: Springer.
McEntire, D.A. 2009a. Revolutionary and evolutionary change in emergency management. *Journal of Business Continuity and Emergency Planning* 4(1): 69–85.
McEntire, D.A. 2009b. *Introduction to homeland security: Understanding terrorism with an emergency management perspective*. New York: Wiley.
McEntire, D.A., C. Fuller, C.W. Johnson, and R. Weber. 2002. A comparison of disaster paradigms: The search for a holistic policy guide. *Public Administration Review* 62(3): 267–281.
McEntire, D.A., G. Crocker, C. Peters, and E. Peters. 2010. Addressing vulnerability through an integrated approach. *International Journal of Disaster Resilience in the Built Environment* 1(1): 50–64.
Nakayachi, K. 2000. Do people actually pursue risk elimination in environmental risk management? *Risk Analysis* 20(5): 705–711.
National Fire Protection Association. 2007. *NFPA 1600: Standards on emergency manage-ment/business continuity programs*. Retrieved on October 25, 2011, from http://www.nfpa.org/assets/files/pdf/nfpa1600.pdf.
O'Brien, G., P. O'Kefe, Z. Gadema, and J. Swords. 2010. Approaching disaster management through social learning. *Disaster Prevention and Management* 19(4): 498–508.
Pine, J.C. 2009. *Natural hazards analysis: Reducing the impact of disaster*. Boca Raton: CRC Press.
Plattner, Th. 2005. Modeling public risk evaluation of natural hazards: A conceptual approach. *Natural Hazards and Earth System Sciences* 5: 357–366.
Reynolds, B., and M.W. Seeger. 2005. Crisis and emergency risk communication as an integrating model. *Journal of Health Communication* 10: 43–55.
Rubin, C.B. 2007. *Emergency management: The American experience 1900–2005*. Fairfax: PERI.
Sikich, G.W. 1996. *Emergency management planning handbook*. New York: McGraw-Hill.
Somers, S., and J.H. Svara. 2009. Assessing and managing environmental risk: Connecting local government management with emergency management. *Public Administration Review* 69(2): 181–193.
Tchankova, L. 2002. Risk identification. Basic stage in risk management. *Environmental Management and Health* 13(3): 290–297.
Tierney, K.J. 1999. Toward a critical sociology of risk. *Sociological Forum* 14(2): 215–242.
Tierney, K.J. 2006. Recent developments in U.S. Homeland Security policies. In *Handbook of disaster research*, ed. H. Rodriguez, E.L. Quarantelli, and R.R. Dynes, 405–412. New York: Springer.
Waugh Jr., W.L. 2000. *Living with hazards: Dealing with disasters: An introduction to emergency management*. Armonk: ME Sharpe.

David A. McEntire is a Professor in the Emergency Administration and Planning Program in the Department of Public Administration, University of North Texas. He teaches emergency management and homeland security courses, and has expertise in emergency management theory, vulnerability reduction, community preparedness, and response coordination. Among others, McEntire has published the books *Disaster response and recovery* (2006) and *Introduction to homeland security: Understanding terrorism with an emergency management perspective* (2008). He is currently studying the relationship between emergency planning and improvisation. He is the 2012 recipient of the Dr. B. Wayne Blanchard Award for Academic Excellence in Emergency Management Higher Education.

Chapter 13
Leadership and Collaborative Governance in Managing Emergencies and Crises

Naim Kapucu

13.1 Introduction

Collaborative governance (Ansell and Gash 2008; Freeman 1997; Huxham 2000) and policy networks (Streck 2002; Atkinson and Coleman 1992) are replacing traditional modes of governing and formulating public policies. Governance represents the altering form and role of the traditional public sector (Bevir et al. 2003) and reflects the limitations of the government by creating "the conditions for ordered rule and collective action" (Stoker 1998, p. 17). The key element of new governance structures is that networks now direct and control public policy and management compared to traditional governance structures where once formal policy making institutions in government dominated and controlled policies (Elderlein et al. 2010; Goldsmith and Kettl 2009; Kapucu et al. 2010; Osborne 2010; Peters and Pierre 2000). Applications of network or collaborative governance and implementation of public policy tools require significant leadership support, especially when the subject matter and situations relate to the stressful conditions of disasters and crises (Kapucu 2012). Recent crises such as the Japanese Earthquake and Tsunami, the Haiti Earthquake, the Pakistan Floods, and Hurricane Katrina demonstrate that among many functions, managing crises require timely implementation of plans, coordination of local, national, and international resources, and leadership support.

Collaborative governance and leadership perspectives will be applied to challenging cases such as natural disasters in this research. This paper will specifically analyze disaster and crises management in Haiti (2010 earthquake) and Pakistan (2010 flood). These two cases stand out as huge catastrophic events in 2010 and are referred to as mega-disasters as more than 95 % of international funding alone went to these two events (Ferris and Petz 2011).

N. Kapucu (✉)
School of Public Administration, University of Central Florida,
4364 Scorpius St., P.O. Box 161395, Orlando, FL 32816-1395, USA
e-mail: kapucu@ucf.edu

U. Fra.Paleo (ed.), *Risk Governance*, DOI 10.1007/978-94-017-9328-5_13

The key research questions for this study are: What role does collaborative governance play in formulating public policies, implementing policies and as tools in the context of disaster response? What role does leadership play in collaborative and multi-level governance in the context of catastrophic events? This research is valuable since natural disasters are not isolated as regional and local events. Managing disasters requires multi-level and collaborative governance. Leadership support from different levels including presidential and political leadership, international leadership, and community leadership in managing crises and developing and implementing policies in disasters is integral. Just as governance and collaboration is multi-level and layered in disasters, leadership is also multi-level and layered in the form of internal leadership (political, presidential), and external leadership in the form of US led or UN led leadership.

The introduction will be followed by a literature review and background section on effective leadership, collaborative governance and policy making. The literature review will be followed by a brief overview of the two cases: Haiti Earthquake of 2010 and Pakistan Floods of 2010. Both cases will be compared and later analyzed in the discussion section and final findings and conclusions will be summarized in the last part of the paper.

13.2 Literature Review and Background

Collaborative governance, networks and policy making and decisions within collaborative settings are popular areas of study. In this section a literature review on collaborative governance, public policy tools and leadership in the context of collaborative settings is provided. The literature review also discusses these concepts in the context of managing disasters and crises.

13.2.1 Collaborative and Multi-level Governance

Governance is a networked form of government which is complex and involves networks of public, nonprofit, and even private entities with diverse functions, stakes, skills, resources, and information (Kapucu 2011; Moynihan 2005). Collaborative governance is concerned with information sharing, resource sharing, and ultimately collective and consensus-oriented decision-making. Ansell and Gash (2008) define collaborative governance as "a governing arrangement where one or more public agencies directly engage non-state stakeholders in a collective decision-making process that is formal, consensus-oriented, and deliberative and that aims to make or implement public policy or manage public programs or assets" (p. 2). Ansell and Gash (2008) highlight that their definition of collaborative governance emphasizes collective and consensus-oriented decision-making that relates to public policies and public affairs, and thus places collaborative governance on a different plain, as compared to governance that can be equally applied to private dispute resolution and consensus-building for decision-making.

Stoker's (1998) five propositions pertaining to governance help to describe the main elements of governance structures. The five propositions are: governance is comprised of institutions and players from both the government and outside of the government; it reflects boundary spanning for addressing public issues; it reflects power dependencies between different actors and institutions that are involved in collaborative action; governance embodies self-governing networks made up of institutions and actors; and, governance structures don't solely rely on the capacity, authority and power of the government but understand that the government is there to be utilized as a more effective tool to guide successful governance.

Lynn et al. (2000) suggest that the study of governance is complicated due to the political, configurational, and loosely coupled nature of governance. The configurational nature of governance is reflected through various interrelated elements such as policy mandates, laws, statutes, and administrative and institutional rules, which either constrain or enable provision of goods and services for the public interest. The political interests that influence action need to be balanced amongst many conflicting stakeholders who are involved in a single policy decision or implementation. Moreover, the study of governance also needs to address the fact that both formal (through statutes, judicial decrees and laws, etc.,) and informal authority structures exist in the governance process. Thus, implying that the "links between formal authority and government operations may be loose and unreliable, especially if policy makers and administrators disagree over the means and ends of governance" (p. 236).

Global public policy networks are loose alliances of agencies and institutions belonging to different sectors, different levels of government, international players, and civil and nonprofit organizations (Reinicke 1999). According to Streck (2002), "recent trends in international governance indicate that the focus has shifted from intergovernmental activities to multisectoral initiatives – from governance at the international level to governance across different levels, and from a largely formal, legalistic process to a less formal, more participatory and integrated approach" (p. 3). Global public policy networks reflect global governance and have burgeoned to embrace the interdependent nature of tackling global problems. National economies rely on financial flows, knowledge exchanges, and movement of people and resources through tapping technological changes and economic integration to address complex problems. These global policy networks are flexible, dynamic, and transparent structures and are formed around common global issues such as public health, international crime and terrorism, environmental and climate change and other broad social issues (Streck 2002).

Streck (2002) argues that global public policy networks are opportunistic and surface when national policymakers don't have enough capacity, knowledge, and resources to tackle issues, or when global consensus building has to take place to address a certain problem. The 'strength of weak ties' as Reinicke (1999) suggests lies in mobilizing knowledge and information from geographically and culturally diverse and distant players that bring newer, more innovative ways of addressing issues. Moreover, in such global networks, leadership roles for effective deliberation are usually played by international organizations that act as hubs and conveners to ensure commitment and input of different stakeholders (Streck 2002). In the context of disasters, international humanitarian organizations such as the United Nations and World Bank play central roles and act as hubs and conveners.

Collaborative governance and networks are required to manage disasters. During disasters, public, nonprofit, and private players pool resources and collaborate extensively to ensure a crisis is managed well. However, as Kapucu (2008, p. 256) states, "organizing a cooperative effort, though, is almost as difficult as the problems that the initiative is created to address." Lack of coordination between different players in emergency management networks is the most obvious failure in disaster management. A key strategy to overcome this failure is to create interoperable systems and standardize communication protocols. Many initiatives in disaster management have followed a top-down mandated approach, such as the creation of U.S. Department of Homeland Security (DHS) to address communication and coordination failures (Schafer et al. 2008). Similarly, the 'cluster approach' is also a top-down, UN centered initiative that aims to offer timely and effective response and improve coordination between various responding agencies and actors in huge disasters (Kapucu 2011; Thomas and Rendon 2010). However, leveraging various resources effectively from different levels of government, different countries, and different sectors requires more than a top-down initiative. It requires intra- and inter-agency collaboration (Schafer et al. 2008). Collaborative emergency management is effective when there are pre-existing and trusting relationships (development of social capital) between public, non-profit and private organizations, strong relationships with the media, well-trained and equipped responding staff with strong communication and leadership skills and the political support and will of elected officials to make collaboration and effective management of disasters possible (Kapucu 2005, 2008).

A key issue of interest in collaborative governance research is that: "How can disperse governance regimes (across states, across municipalities within a state, across local offices or networks) be induced to converge on the achievement of particular policy objectives?" (Lynn et al. 2000, p. 235) In other words, how does governance succeed when stakeholders involved have conflicting and diverse ideas and views about policy objectives? This leads us to explore the literature on good governance practices.

13.2.2 Public Policy Tools in Collaborative Governance

A government's legitimacy, operational transparency, public accountability, and administrative competency reflecting a corruption-free, less bureaucratic process all point towards a country's good governance practices (Siddiquee and Mohamed 2007). Kapucu (2010), in his book on comparative governance reforms, presents a framework of four pillars of good governance which cover accountability, transparency, rule of law and citizen participation. Kapucu (2010) also adds that although a strong civil society is not among the four pillars of good governance, it is an integral part of the governance structure on which the four pillars rest. Thus, civil community and citizen engagement in governance are integral for good governance as new mechanisms of governance do not place citizens as clients and customers but as equal partners in collaborative endeavors (Vigoda 2002). In the dynamic environments of disasters, these pillars of governance are gravely important to ensure effective disaster relief and response efforts.

Collaborative arrangements involving a myriad of international players require transparent and accountable institutions and structures. Furthermore, delivery of the aid coming from international actors requires a strong civil society for better disaster relief. In most of the cases catastrophic disasters overwhelm the capacity of governments at all levels. Also, issues such as corruption, lack of transparency, and insufficient leadership undermines the capacity of governments to handle disasters. Consequently, the need for involvement of nongovernmental actors (i.e. private and nonprofit organizations) becomes apparent. In these cases, international actors prefer working with civic organizations beside the local governments and civic organizations serve as means of reaching disaster victim more effectively.

Although collaborative governance is promoted recently, it is difficult to generalize the factors that support good governance. What may work in one contextual setting in a country, may not necessarily work in another. Jung et al. (2009) study collaborative governance cases in United States and Korea, pertaining to negotiated policymaking and service delivery, and conclude that collaborative governance is more effective when it is applied to negotiation, consensus-building and the policy planning phase when its focus is to bring conflicting stakeholders to the same table in an informally structured way. However, during service delivery and implementation, collaborative governance is more effective when a formalized structure exists to support these policies and programs.

Policies concerned with complex and grave consequences cannot be compared with mundane administrative policy decisions. Kowert (2001) analyzes Ronald Reagan's decision-making in the Iran-Contra Affair and concludes that if decision-making power is concentrated in few people there is little dissent and conflict in formulating and implementing decisions as compared to a situation where many people are involved with an equal distribution of power, and different points of view. The more conflicts and diverse views within huge decision-making groups, the more challenging it is to arrive at effective policy decisions. Kowert (2001) also suggests that in larger decision-making groups, directive leadership gains importance.

Additionally, Kim (2010) suggests that what might work in developed countries might not necessarily work in developing countries. For example public policies and reforms in developed countries might not be directly imported and applied to developing countries, since the governance structures are typically different in terms of accountability, transparency, rule of law, technological advancements, civil society engagement, and levels of corruption. For instance, new governance infrastructures with advanced technologies in developed nations such as the United

States support citizen engagement opportunities and other innovations in policy options. Johnston (2010) claims that policy customization for diverse population needs in the US is not a far-fetched idea anymore. Technological advancements provide tools to tailor policy options to diverse populations and their unique requirements. Johnston (2010) gives an example of how in the near future it would be "possible to have an election in which some people could opt into a tax increase to pay for education and would be held accountable for their choice at tax time" (p. S124). Thus, e-governance and policy customization would work and succeed in developed countries, but not in countries with very little internet access and undemocratic and non-participative set-ups.

Thus, good governance has to be geared towards growth and development with the constraints and limitations existing in different settings. Another imperative question to ask is "How is good governance developed when the sufficient conditions of civil society engagement, transparency, and accountability are not met?" Kim (2010) proposes that trust building is essential in developing countries and this requires a more ethical and trustworthy leadership that is not merely involved in shortsighted governance reforms for political gains but is committed to develop policies and programs that are suitable for the growth and development of their particular country.

With technological changes and global interdependence, collaborative governance and global networked governance are here to stay. In order to understand these mechanisms of governing, many studies have been dedicated to exploring the various factors that influence governance in positive ways. Ansell and Gash (2008) carry out a meta-analysis of existing literature on collaborative governance to identify critical factors that influence new forms of governance. They conclude that one of the indispensable factors that provide support and facilitation for collaborative governance is leadership. According to them the mediation and facilitation required in the governance process is not possible without an honest broker and a facilitative leader. Vangen and Huxham (2003) also have a similar stance and emphasize that collaborative governance leaders might have to step in the governance process to steer a policy or issue agenda so that the process of governance continues to be smooth and achieves the necessary goals.

Ansell and Gash (2008) make the claim that effective collaborative governance will be highly constrained if effective leadership does not exist. Thus, leadership is an important factor that is closely related to new forms of governance.

13.2.3 Leadership and Governance

The success of collaborative governance "depends on the ability of leaders to organize structures, resources, and interactions" (Moynihan 2005, p. 7). Getha-Taylor et al. (2011) suggest that although leadership has been studied as a broad topic in many disciplines, there is a need to focus on the study of public leadership. Public leadership is different from general leadership research and theories as it aims to create public value. In the context of new governance structures, public leadership is a "boundary-crossing process with those we consider leaders often leading in the absence of formal authority" while "effectiveness rests on leadership within as well as across hierarchies" (Getha-Taylor et al. 2011, p. 84). According to their definition public leadership is not limited to political leaders. New structures of governance "require that individuals lead without hierarchical or designated authority" (p. 84). Thus, leadership today is boundary-spanning and more complicated and dynamic than ever. Public service leaders are also referred to as 'change agents' when they are proactively involved and engaged in bringing about positive change in the form of reforms and policies (Wallace et al. 2011).

Just as governance and accountability are closely linked, Wallis and Gregory (2009) suggest that the concepts of accountability and leadership are also very closely linked when concerned with public issues and policies. Their study investigates how leadership may be explored in response to demands for accountability in public administration. They believe this is an important area of study "since both leadership and accountability impinge on the concerns policy actors have with the question of who is to be given credit and assigned blame for public policy outcomes" (p. 252). Citizens expect that if a public policy fails to deliver on its promises or goals, someone should provide them an honest and transparent account of what went wrong and that sanctions were imposed in the most appropriate way.

Wallis and Gregory (2009) suggest that in governance and politics "success has a thousand fathers while failure is an orphan." Thus, in reforms such as New Public Management a lot of focus was laid on accountability and trust-building. Wallis and Gregory (2009) suggest that trust and accountability are necessary conditions but not sufficient to ensure responsible governance. Public value, honor and responsibility are also important elements that need to be placed within the realms of networked and new forms of governance. Similarly, Getha-Taylor et al. (2011) also propose that public leadership needs to reflect the strong commitment for creating and sustaining public interest and public values.

According to Wallis and Gregory (2009) public value-seeking leadership emphasizes responsibility more than accountability in networked governance structures. Public managers that are exercising public value-seeking leadership have to go beyond their formal roles as managers and officials to engage stakeholders in deliberative ways. Also new forms of governance require the input, support and commitment of various players involved in public value creating endeavors. Thus, leaders need to take the responsibility to utilize both informal and formal sources of authority to encourage sustainable commitment of important players.

Moreover, Wallis and Gregory (2009) also emphasize that public value-seeking leadership is inherent in networked governance where public managers and leaders take responsibility to mobilize networks and collaborative governance structures which create public value. These networks can have both horizontal and vertical elements that engage actors and have deliberative and democratic ways.

Many recent studies have also focused on delineating what role public or political leadership plays in public policy and policy change. Goldfinch and 't Hart (2003) analyze the economic liberalization movement in Australia in the 1980s and develop a theory of reformist political leadership which emphasizes the "crisis recognition and crisis management as a crucial vehicle for effecting reforms" (p. 235). According to Goldfinch and 't Hart (2003) much of the literature has focused on explaining policy reforms and public policies in the structural sense, without focusing on the role of policy makers, political leaders and decision-makers, creating a 'dehumanized account of governance' (p. 236). They propose that to understand reformist political leadership, four factors are very important. They call these four factors the four critical C's: Crisis, Communication, Commitment, and Coalitions. Reformist leaders create a sense of urgency and portray their proposed reforms and policy changes as addressing crisis situations which require urgency. These leaders communicate this sense of urgency well and reflect their political commitment to address the dire issues. They also create strong reform coalitions and cohesive teams of political actors and policy supporters that support policy changes and increase the chances of successful implementation of the policy.

13.2.4 Leadership and Collaborative Governance in Managing Disasters

Crisis leadership can take the shape of sympathizing and empathizing with victims, facilitating and leading response and recovery, coordinating between different response entities, providing correct and reliable information in disasters, and timely communication with citizens (Kapucu and Ozerdem 2013). The concepts of crisis and leadership create an important relationship that requires a deeper study.

The common way to study leadership in the context of disaster situations has focused on discussing and exploring the presidential and political leadership of individuals or emergency managers. During disasters, the public generally looks for leaders to make responsible and intelligent decisions to mitigate risks and threats. Uncertain and confused feelings during crises encourage the public to look towards a strong, transformational leader and alter their leadership expectations (Bligh et al. 2004). Boin and 't Hart (2003) studied President Bush's and Mayor Rudolph Giuliani's post 9/11 leadership. Their study shows that Bush and Giuliani's approval ratings and personal reputations improved tremendously after their response and proposed plans to the disaster. According to Boin and 't Hart (2003), "successful performance in times of collective stress turns leaders into statesmen. But when the crisis fails to dissipate and "normality" does not return, leaders are obvious scapegoats" (p. 544). Thus, poor disaster management perpetuates a move to find leaders to blame and, as critics suggest, advocating reform by leaders post-crisis is interpreted as a common tool or strategy to circumvent public criticism and blame.

The role of the media, in framing the nature of leadership, influences the perceptions of citizens and the international community about the effectiveness or ineffectiveness of leadership in the disaster-inflicted country. Littlefield and Quenette (2007) discuss that crisis leadership theory can be applied to see how the positive or negative decisions and response efforts of authorities and leaders can impact the disaster management perceptions of the public at large.

Public rhetoric can help to exhibit how Presidential addresses to the nation were altered to reflect more charismatic and transformational images of President Bush in the media following 9/11. Thus, the media at large also reiterated the portrayal of charismatic leadership. As a result the public approval ratings and opinions of the Presidential leader became very positive (Bligh et al. 2004). According to Bligh et al. (2004), 'leadership represents an intersection of forces and factors associated with the leader, the followers, and the context in which they are embedded' (p. 227). Similarly, these factors and forces are associated with charismatic leadership as well. They also suggest "in situations in which the social distance between the leader and the follower is great, it is the symbolic words and images that a leader is able to evoke that are largely responsible for subsequent charismatic or noncharismatic attributions, regardless of who actually crafted them" (p. 229).

Waugh and Streib (2006) also suggest that when we usually refer to leadership, we more narrowly relate to top-down traditional leadership, such as presidential leadership, or federal level leadership such as FEMA and DHS leadership (Waugh and Streib 2006). For instance, the response to Hurricane Katrina was strongly criticized and centered on the ineffective leadership at the federal level such as the FEMA and DHS leadership (Waugh and Streib 2006). Thus, along with government officials (Schneider 2005) organizations were also blamed for leadership failures in Hurricane Katrina. Although, criticism is usually aimed at federal players, we need to remember that disaster management and response is no longer a top-down operation and requires collaborative response and functions. The leadership failure in Hurricane Katrina started at the local and state level and became a nationwide issue as it became a catastrophe. For example, the mayor of New Orleans remained reluctant to order a mandatory evacuation. Moreover, local officials failed to coordinate necessary arrangements for those who did not have private transportation (Derthick 2007). Cases like Hurricane Katrina tell that a network approach that facilitates cross-sector, intergovernmental and multi-organizational coordination and cooperation is the new model (Waugh and Streib 2006).

Littlefield and Quenette (2007) carry out a study to explore how crisis leadership was portrayed through the media during Hurricane Katrina. They study the nature of crisis leadership as depicted through the media to reflect the legitimate sources of authority and leadership roles in managing crisis. Their research shows that five levels of leadership and authority were seen, which were: the military, the Department of Homeland Security (DHS)/FEMA, the President of United States, the federal government and the local government. Their analysis reflects that leadership in crisis is not limited to individuals but may be treated as a broader concept in the form of a leading agency, or even a level of government or sector. In fact leadership in disasters can be referred to as a multi-dimensional or multi-layered phenomenon where the leadership of politicians, federal agencies, and international agencies are all being scrutinized.

The literature shows that new forms of governance call for new types of reformist, facilitative and boundary-spanning leadership that lead to more deliberative ways to formulate and implement public policies. In the context of managing crises, collaborative governance structures require leadership support from a number of internal and external high-profile actors and organizations so that better decisions are made and effective disaster management plans and policies are formulated and implemented.

13.3 Case Studies

13.3.1 The 2010 Haiti Earthquake

The January 12, 2010 Haiti Earthquake is the largest earthquake to ever hit Haiti. Many areas of Haiti were severely damaged including Port-au-Prince, the capital of the country, where 50 % of the buildings were destroyed, out of which many were government buildings (Aon Benfield 2010; Alexander 2010; Margesson and Taft-Morales 2010). A third of Haiti's population of 9 million (Margesson and Taft-Morales 2010) was affected by the earthquake and around 316,000 people were confirmed to have died during the earthquake (Archibold 2011, January 13).

Infrastructure destruction and communication disruption led to great delays in response and relief efforts. For 2–3 days first responders, government officials and the general public faced challenges since main cell towers and power lines were damaged (Aon Benfield 2010). Poor building construction was the main reason for the devastation and damage caused by the earthquake. Although the government had received ample warnings by seismologists to improve the methods of construction utilized in the Caribbean plate boundary, the government failed to give importance to their warnings (Bilham 2010; Alexander 2010). Two days after the tragic incident took place, the Haitian government, the United Nations and other donor and relief organizations came together to discuss and coordinate their response and relief efforts (Margesson and Taft-Morales 2010). Many different players and actors were involved in responding to the Earthquake. In the following section we explore key actors and their leadership roles in managing the Haiti Earthquake disaster.

13.3.1.1 Political and Presidential Leadership

The role of a country's President is crucial in the immediate aftermath of a disaster as a sense of urgency for immediate help and relief is created by a formal presidential disaster declaration. In Haiti's case the situation was exasperated since the Presidential Palace collapsed and the President's private residence was also destroyed. Immediately after the earthquake the President was not able to meet with his Cabinet and had to operate from a room in the police headquarters. However, despite these grave conditions, President Preval did act promptly by appealing for help and assistance to the international community (Margesson and Taft-Morales 2010). President Preval also sought help for rescue and relief missions, for building an offshore vessel for medical aid and for generating electricity so that communication and coordination operations could be quickly restored.

The quick decision-making and initial presidential response translates into public perceptions of how political and presidential leaders are responding and catering to grave needs. General public expectations following a disaster are usually high. In the case of Haiti, initially some civilians were complaining that the President hadn't done enough for the country and felt that the US had to work in the place of the Haitian government to lead better response and relief activities (Margesson and Taft-Morales 2010).

13.3.1.2 US Involvement

Haiti's development and survival depends a great deal on external forces since its internal governance structures and institutions are fairly weak and reflect a weak disaster response capacity (Rencoret et al. 2010). The humanitarian relief effort was primarily led by the United States through an inter-agency taskforce developed under the Agency for International Development (USAID) (Margesson and Taft-Morales 2010). Since the US has led international relief coordination in Haiti, it is believed by many that the US should develop a well integrated relief and recovery strategy by incorporating three core principles that should guide its policy and strategy: "Strengthen Haitian government capacity at each stage of the recovery process; Ensure a coordinated U.S. approach so that our policies in other areas do not undercut our efforts to reduce poverty and promote sustainable economic growth; Make long-term development the primary objective" (Millner 2010, p. 4). Thus, Haiti and its fragile governance structures and institutions require solid leadership, which many believe can be provided by the United States of America.

Recovery and reconstruction in Haiti will be overseen by the Interim Haiti Recovery Commission (IHRC) which is headed by former US President Clinton and the Prime Minister of Haiti, Jean-Max Bellerive. The external setting of the commission reflects the limited capacity of the government and their weak governance institutions. However, it has been established that after 18 months of functioning, the IHRC will run completely under the supervision of the Haitian government (Millner 2010).

Ambassadors of Haiti are also important in terms of encouraging donations and seeking international help. The US Ambassador to Haiti, Mr. Kenneth Merten, quickly declared a disaster and helped to authorize $50,000 through the Office of Foreign Disaster Assistance (Margesson and Taft-Morales 2010). Moreover, important high-profile US personalities and leaders also played an integral role in such high profile devastating events. Former President Clinton and George Bush are working through the Clinton Bush Haiti Fund and raising awareness throughout the world about Haiti's needs. Clinton played a very important role as the co-chair of the Interim Haiti Recovery Commission (IHRC) and as a special UN Envoy for Haiti. His high-publicity visits to Haiti helped to attract media attention and encouraged individuals to donate and mobilize funds (Charbonneau 2010; Rencoret et al. 2010).

The general public in the US has also helped a great deal in providing relief funds and assistance. Almost 50 % of Americans donated to the Haiti relief funds and congress passed a debt cancellation legislation as well (Millner 2010). Being a US ally helps a country recover from a disaster. The US government provided tremendous help by sending response teams from USAID to assist in search and rescue operations. The US military also played a forefront role in Haiti during the relief efforts (Margesson and Taft-Morales 2010). Other nations such as Spain, France, Iceland, United Kingdom, etc. also helped in the relief efforts by sending their search and rescue teams.

13.3.1.3 International Humanitarian Organizations and Their Leadership Role

Leadership from international humanitarian organizations is also crucial for managing emergencies, especially in developing countries. The United Nations (UN) and its leadership have been critical to Haiti. Even in pre-disaster Haiti, the UN has had a strong presence, however, it suffered major losses from the earthquake as its headquarters in Haiti were destroyed and around 150 of its personnel were missing. Hedi Annabi, the head of the UN Stabilization mission in Haiti (MINUSTAH) was also amongst those who were killed in the earthquake. Thus, leadership deficits and delays at different stages of disaster response led to interruptions and holdups in immediate response (Margesson and Taft-Morales 2010). However, despite these initial challenges, the relief efforts were largely UN-centric. The UN established the Office for the Coordination of Humanitarian Affairs (OCHA) teams which helped in coordinating search and rescue operations. OCHA teams have worked closely with the Haitian government and have played a leading role in efforts to coordinate with the military and other actors providing donor support (Margesson and Taft-Morales 2010).

During humanitarian relief efforts, a cluster approach is usually applied to coordinate relief services amongst a number of responding organizations. In Haiti, initially 12 clusters were developed to guide relief in different sectors (Margesson and Taft-Morales 2010). These were: Emergency Shelter and Non-food Items (efforts led by International Organization for Migration), Food Assistance (under World Food Program), Health (under World Health Organization), Logistics (World Food Program), and Water/Sanitation (UN Children's Fund – UNICEF). However, the cluster approach has not been very successful. Although it provided clear leadership in the form of lead agencies, its performance and accountability issues remain of concern (Rencoret et al. 2010).

Bolton (2011) explores the political system of global governance in Haiti and highlights some shortcomings of the cluster approach in the context of making policies relating to relief and recovery efforts in post-disaster Haiti. According to Bolton (2011) the clusters work as 'key governance structures' as they help coordinate funds and form policies and decisions crucial for recovery and relief stages. Bolton (2011) goes as far as stating that "in the absence of clear governmental leadership from the Haitian parliament, they have become para-parliaments where policy is debated and discussed" (pp. 12–13). But the major issue of concern is that in national parliaments there is some degree of representativeness of the general public, however, in the cluster approach foreigners dominate decision-making and this whole policy making and decision making process lacks participatory and democratic elements. Moreover, meetings in the clusters are held in either French or English and only the educated elite of the country can understand these languages as around 20 % of the Haitians speak French while even less speak the English language. Thus, this globalized governance system of relief depicted via the cluster approach reflects "a form of trusteeship, rather than democratic representation" (p. 14).

Prominent nonprofit and non-governmental organizations like the Red Cross also play an important leadership role in disaster relief and response. The International Federation of the Red Cross and Red Crescent Societies (IFRC) worked with the Haitian Red Cross Society (HRCS) and American Red Cross in providing relief goods and services (Margesson and Taft-Morales 2010).

13.3.2 2010 Pakistan Floods

In July, 2010 Pakistan experienced the worst flooding since its creation. What began as monsoon rains ended up in disastrous floods impacting over 78 districts in the country (ADB 2010). In recent years the 'timber mafia' in Pakistan has contributed to massive deforestation in the country. This massive deforestation wiped out natural resources and trees which could have protected from this massive flooding by obstructing fast flood currents (Cedar 2011).

Around 20 million Pakistanis, which is more than one in ten Pakistanis, have been impacted by the disaster. At one point during the flooding, 1/4th of the country was covered by water, which is equivalent to the size of England (Solberg 2010). Farmers in rural areas have suffered the most as their plowing fields and seasonal crop supplies have been destroyed. Moreover, agricultural land destruction and dead livestock has led to enormous increases in food prices (Kronstadt et al. 2010). The UN Secretary General declared that this disaster is larger than the combined impact of the 2004 Asian Tsunami, the 2005 Kashmir Earthquake, and the 2008 Nargis Cyclone and the 2010 Haiti Earthquake (Solberg 2010).

13.3.2.1 Collaborative Governance and Leadership

Relief and response operations were collaboratively carried out by the government of Pakistan, the military, the United Nations, the United States Agency for International Development (USAID), the US government, etc. Along with these formal actors, groups of civilian volunteers, teams of physicians, local civil based organizations and NGOs lead the relief efforts in Pakistan. The National Disaster

Management Agency (NDMA), under the leadership of the Prime Minister of Pakistan, provides overall leadership and works with different branches of the government, international humanitarian organizations, NGOs and the military during disaster events (Kronstadt et al. 2010). The NDRMF provides a guide and framework for delineating roles and responsibilities for responding agencies. However, assimilation and elimination of roles and responsibilities of organizations needs to be carried out (ADB 2010).

13.3.2.2 Presidential, Political, and Military Leadership

Political and Presidential leadership is extremely weak in Pakistan's case. The President of Pakistan, Asif Ali Zardari, is known for being involved in corruption scandals and his team also lacks trust in the eyes of the public. Immediately after the flood, the Pakistan Prime Minister Gillani established a Relief Fund which only managed to collect $200,000 in the first 2 weeks. Pakistani citizens also preferred to donate to nonprofits and other well-reputed honest organizations, such as the Edhi Foundation, compared to the Prime Minister's fund for relief (Burki 2010). President Zardari was strongly criticized for his immediate response, or rather lack of response, since he was touring Europe when the disaster in Pakistan unfolded. His sincerity towards his people and country was questioned by the public when he continued to tour Europe despite the massive suffering and destruction at home.

In difficult times, the people of Pakistan do not trust their political leaders and turn to the military for support and help. During the floods, the military proved that they were in a much better situation and had the capacity to perform search and rescue operations while the civilian government seemed largely incapacitated and helpless to tackle a disaster of such a large scope.

13.3.2.3 Military Involvement

Pakistanis do not hesitate to look up to military forces when the civilian government fails to deliver. During disaster events, the military is seen as a trusted party by the general public. As discussed earlier, during the floods, the civilian government proved to be weak suffering from a lack of capacity to respond effectively and military leadership took over providing evacuation and search and rescue operations (Kronstadt et al. 2010).

13.3.2.4 US Involvement

The US government has been the largest donor in relief efforts. The United States Agency for International Development (USAID) has led the US response for the floods. Relief goods were provided via 30 US military helicopters and C-130 planes and 26 US mobile medical teams provided health care services since 450 health centers in Pakistan had been wiped out in the floods (Kronstadt et al. 2010).

The issue of how external, powerful, political, and humanitarian forces exert pressure in the form of branding while delivering relief goods to flood victims is an interesting dimension to explore. During relief operations in Pakistan last fall, the United States Agency for International Development (USAID) required that all foreign aid in the form of relief goods funded through the US government had to be explicitly labeled with US and USAID logos. This branding strategy was used to improve the general perceptions of the Pakistani public about the US government. However, in some regions where anti-western and anti-US sentiments are high, this branding strategy seemed to be in fact counterproductive and dangerous. Due to this branding strategy, some local NGOs felt pressured to follow and comply with these branding policies and some due to security concerns asked for waivers (Burns 2010).

Thus, leadership from USAID and the US government may be very strong, but may not be popular and well-respected amongst the Pakistani masses. This branding requirement and strategy was in fact a policy to try to develop better perceptions of the US government.

13.3.2.5 UN Involvement

Early on the international community did not pay too much attention to the seriousness and scope of the disaster. Aid was slow to arrive due to the corrupt reputation of officials and institutions in Pakistan. International organizations rarely trust the Pakistani government to provide aid and relief to deserving victims in a transparent manner. There is also the grave concern of donating to a country known for terrorist activities. Other reasons for slow aid can be attributed to donor fatigue, global recession, and the underestimation of the scope and scale of the disaster (Burki 2010; Warraich et al. 2011; IEG 2010).

It is only after the UN Secretary General's visit to Pakistan, that he urged the international community to help and provide as much assistance and funds as possible (Burki 2010). The United Nations Secretary General, Mr. Ban Ki-moon toured Pakistan on August 15 with President Zardari and encouraged and urged global support and assistance from developed nations.

The UN agencies and their leadership in midst of disaster relief and response has helped to save many lives in Pakistan. The UN developed an initial response plan that followed their popular cluster approach. Initially 12 sectors were developed through this approach which were: water and sanitation, health, shelter, agriculture, food, community restoration, protection, education, nutrition, logistics, coordination, and camp management. Each cluster is led by a lead agency responsible for coordinating operations and functions specific to that particular sector and cluster (Kronstadt et al. 2010). Although the cluster approach delineates the lead agencies and establishes their roles, the practical implications of this approach are marred with some weaknesses. In the case of the Pakistan floods, the cluster coordinators were not well-trained and experienced in the coordination of massive operations, the communication between clusters was weak and the duplication of functions continued to be a major problem within the clusters (Oxfam 2011).

Leadership in the country should understand that aid will provide short-term relief but long-term sustainability can be brought about by economic and tax reform in the country. Tax reform focusing on the increased taxation of rich, landowning elite should be emphasized (Warraich et al. 2011). Moreover, with the help of the international community and international organizations, Pakistan needs to invest in flood protection and adopt and develop flood prevention techniques (Cedar 2011).

A comparison table by Ferris and Petz (2011) is provided below that compares basic indicators of the two disasters. The table provides comprehensive background information about the economic indicators, response needs and functions and the underlying vulnerabilities and risks in both disasters (Table 13.1).

Table 13.1 Comparing two mega-disasters: The Haiti Earthquake and Pakistan Floods

	Haiti earthquake	Pakistan floods
Date of disaster	12 January 2010	Late July 2010
National population 2009	10 million	169.7 million
Deaths	316,000	1,985
Injured	Over 300,000	2,946
Displaced	Est. 1.8 million	Est. 6 million in need of shelter
Total affected/as percentage of total national population	3 million (30 %)	20.1 million (11.84 %)
Houses destroyed/damaged	188,383	913,307/694,878
Schools destroyed/damaged	1,300	10,044
Hospitals/health facilities destroyed/damaged	50	588
Original UN flash appeal launched	15 January: $575 million	11 August: $460 million
Donation per affected person received after 2 weeks of flash appeal	$157.16	$15.24
International pledges 2 weeks after flash appeal as percent of total appeal	82 %	57 %
Flash appeal funded 100 %	16 February (35 days)	24 September (44 days)
Revised Humanitarian Appeal	$1.4 billion for 1 year – Launched 18 February (includes the $575 million of the flash appeal)	$2.0 billion for 1 year – Launched 17 September (includes the $460 million of the flash appeal)
Funding of revised appeal 2 months after flash appeal/total % funded	$703 million (49 %)	$688.6 million (34 %)

(continued)

Table 13.1 (continued)

	Haiti earthquake	Pakistan floods
Funding of revised appeal 5 months after flash appeal/total % funded	$843 million (57 %)	$1,037 million (53 %)
Appeal by International Federation of the Red Cross/Crescent Society	$103 million	$74 million
Number of tents/plastic sheets distributed 3 weeks after	10,545/11,390 (February 3)	109,500/72,200 (August 23)
% of displaced receiving tents/tarpaulins (after 3 weeks)	0.017 tents/tarpaulins per targeted displaced person (1.3 million)	0.03 tents/tarpaulins per person in need of shelter (6 million)
Number of tents/plastic sheets distributed 2 months after	35,000/259,266 (March 15)	261,498/439,745 (October 14)
% of displaced receiving tents/tarpaulins (2 months after)	0.23 tents/tarpaulins per targeted displaced person (1.3 million); 63 % of targeted population (UN OCHA)	0.11 tents/tarpaulins per person in need of shelter (6 million); 25 % of targeted population (UN OCHA)
Number of tents/plastic sheets distributed 4 months after	45,000/376,000 (April 12)	352,000/690,000 (December 8)
Number of emergency shelters constructed	350 of 130,000 T-shelters (0.27 % completed approx. 5 months after quake); 39,219 of 110,440 T-shelters completed (35 % completed 12 months after quake	2,042 of 79,581 one-room shelters 7,977 of 53,711 transitional shelters completed (2.57 and 14.75 % completed approx. 5 months after floods)
IDP camps/sites	1, 555 (July); 1,199 (November)	4,788 (October); 188 (January)
IDPs in camps	1,536,447 (July)	1,800,000 (September)
	1,058,853 (November)	124,164 (January 2011)
Role of US military	Deployed 22,268 troops (at peak in February 2010), more than 300 aircraft and helicopters; 23 ships including the hospital ship USNS Comfort; initially controlled airport; rehabilitated the harbor; distributed aid;	26 helicopters in Pakistan and more offshore; as of October 14, 2010 the U.S. military had delivered 16 million pounds of relief supplies and food, and helicopters had rescued or transported about 22,000 people:
Health concerns	Traumatic injuries, including crushing injuries; high needs for surgery; infections; cholera epidemic;	Water-borne illnesses (diarrhea, cholera); skin-disease; acute respiratory disease; malaria; dengue fever;
Protection concerns	Trafficking of children; gender based violence in camps; general insecurity; forced evictions of IDPs;	Early reports of separated families; discrimination against lower castes, women-headed households;

(continued)

Table 13.1 (continued)

	Haiti earthquake	Pakistan floods
Shelter concerns	Land tenure issues; slow debris clearance; lack of reconstruction master plan;	Land markers washed away by floods; mud removal;
Political concerns	Election schedule and election crisis; limited government capacity (weak overall capacity before the quake, augmented by death of many government officials);	Potential strengthening of fundamentalist groups; destabilization and delegitimization of government; interference of conflict with disaster response;
Economic concerns	70 % of Haiti's GDP is generated in the Port-au-Prince area which was most heavily impacted by the disaster; massive destruction of housing and infrastructure;	Massive destruction of infrastructure; agriculture most affected sector with estimated $5 billion in losses, especially in crops but also widespread loss of livestock; high inflation;
Logistics	Destroyed airport, harbor, roads; generally poor infrastructure even before earthquake: roads, water, sewage and garbage disposal; 20 million cubic meters of debris	Destroyed roads, bridges; some areas only accessible by helicopter; 20 % of the country flooded, with some areas flooded for months;
Total GDP 2009	$6.5 billion	$166.5 billion
GDP per capita 2009 nominal	$733	$1,017
Estimated damage	$7.8 billion	Est. $8.74–10.85 billion
Estimated damage as percentage of GDP	119 %	5.2–6.5 %
Reconstruction pledges	March 31 – Donors pledge $9.9 billion of which $5.3 billion is pledged over 2 years (requested $3.9 billion)	August 22 – World Bank $0.9 billion; Asia Development Bank $2.0 billion (loans)
Corruption Perception Index 2009 (out of 180)	160	139
HDI 2009 (out of 182)	149	141
International Media stories 10 days after the disaster	More than 3,000 stories in both print and broadcast media respectively by day 10 and by day 20	320 broadcast news stories and 730 print news stories

13.4 Discussions

Many of the differences in response activities and types of resources mobilized can be explained by the different nature of the two events. Response to disasters depends a great deal on the onset of the disaster. For instance, the Haiti Earthquake was a sudden onset event, while the floods in Pakistan gradually built up in the shape of a national disaster (Ferris 2010b). Moreover, the resources and activities required in an earthquake are fairly different than those needed in floods. Search and rescue operations and rubble removal are urgent needs in an earthquake, while a timely evacuation is the primary requirement in floods. Moreover, debris removal requires heavy equipment and search and rescue teams while floods require boats and helicopters along with rescue teams (Ferris and Petz 2011). Thus, the diversity in the types of resources deployed, the number of deaths, displacement populations, and timing of international appeals is linked to the different nature of the two disasters.

However, the comparison table shows some interesting statistics. It shows that there were drastic differences in the international funding per affected person. In Haiti, every disaster-affected person received an international funding of \$948.37, while in Pakistan they received \$121.67 (Ferris and Petz 2011). Major reasons for this are a lack of effective media coverage, the amount of causalities, long-term displacements, the geographical proximity to US, the perceived level of corruption and corrupt leadership, and the general perceptions of a country linked to terrorism (Ferris 2010b).

Leadership and governance are two concepts very central to disaster management. The case studies in the previous section show that leadership in disasters is a multi-layered and a multi-dimensional phenomenon. Traditional understanding of leadership is limited to individuals that have authority according to their top hierarchical position. Traditional understanding of leadership is not totally undermined and overlooked in disasters, but it is certainly coupled with transformational and more collaborative forms of leadership in the form of central players within governance networks and leading coordinating bodies such as the UN in International humanitarian response.

Inspirational and transformational leadership is expected in disasters. Both in the case of Haiti and Pakistan, political leadership and national government response was largely weak. According to Ferris and Petz (2011), "Haitian President André Préval seemed dazed and withdrawn while Pakistani President Asif Zadari continued his travels in Europe" (p. 44). When we have to compare the political or government leadership of Haiti in response to the earthquake with Pakistan in response to the floods, we need to take into account how the events directly impacted the establishment and the government. The Haiti earthquake impacted the capital city and lead to the death of 26,000 civil servants, and the destruction of government ministries and organization headquarters of many agencies. Moreover, many government workers and officers lost their family and homes and there was an overall feeling of remorse and sheer disbelief which, most naturally, hindered effective government response (Ferris and Petz 2011).

Post-disaster effective decision-making and government leadership in Haiti was also confounded by uncertainty pertaining to when the competitive legislative and presidential elections scheduled for 2010 would be held. There were over a hundred political parties and a lot of protests, with so much uncertainty the parliament was dissolved in May. Finally elections were scheduled for November and many political leaders and ministries, instead of focusing their energies on disaster recovery, became involved in the competitive election. Even relief workers complained that government authorities were not keen to help as they were not certain whether they would continue to hold office after the elections. Thus, instead of being concerned about elections, the government should have been 'applying effective leadership to the recovery effort' (Ferris and Petz 2011, p. 45).

Similarly, Pakistan is as unstable in terms of its governance institutions and corrupt practices. But it does enjoy a strong military and civilian government that is capable of effectively respond to disasters. On the other hand, Haiti has no military and had to rely instead on the US military. The strong presence of the military and army fighting insurgents and Taliban's in northern Pakistan helped in mobilizing quick response for the flood victims in the north as well (Ferris and Petz 2011; Thomas and Rendon 2010). In Pakistan, when the government fails, everyone looks to the military for support. There is a general perception in the country that the military is the only sector and establishment which is organized and has the resources to deliver, especially in crises.

Developing countries like Haiti and Pakistan rely heavily on international humanitarian organizations and their leadership during crises events. The United Nations 'cluster system' is a popular response and relief system that is used in huge disasters through which relief agencies coordinate their efforts in clusters and sectors. The main goal of the cluster system is to provide timely and coordinated response in disasters. Each cluster is guided by a specific humanitarian service such as camp coordination and management. Each cluster is assigned a lead agency to oversee and coordinate efforts and also individuals that are referred to as cluster coordinators (Thomas and Rendon 2010). Although theoretically this system is an ideal collaborative governance arrangement or network, practically it is weak. The two case studies illustrate major weaknesses of this approach. One major weakness or problem with the UN cluster system is that while the UN's Emergency Relief Coordinator can deploy only personnel from OCHA, other UN organizations and NGOs which are working in the cluster system can go ahead and deploy their own staff. Therefore, while the UN tries to coordinate between agencies, each agency or actor is acting on its own accord, its respective organizational mandates and funding (Ferris and Petz 2011). Thomas and Rendon (2010) recommend that to improve the cluster system, cluster coordinators have to be "better trained in effective coordination leadership, protection practices and management; and require them to relinquish agency affiliation and responsibilities to allow for independence and a focus on cluster coordination activities" (p. 20).

In the Pakistan foods, the cluster system was criticized greatly due to inadequate staffing, resources and leadership. Some critics suggest that clusters did not focus on coordination task but were primarily concerned with generating funds and allocating them. Reports suggest that the UN and International NGOs have to work closely with the national government to ensure better development and sustainability patterns in recovery operations. Thomas and Rendon (2010) report that a senior UN official commented on the UN response saying, "The floods have demonstrated that whatever we have in place [to respond] is inadequate and whatever planning assumptions we made were insufficient (p. 5)". The response and relief efforts in southern parts of the country were weak since cluster coordinators and staff did not have pre-existing relationships with local officials and agencies (Thomas and Rendon 2010). Thus, better collaboration, pre-existing relationships, and strong partnerships should focus on improving the local capacity of national governments. Civil society actors and community actors need to be identified and trained to work with international organizations and their clusters, and similarly, cluster coordinators and staff need to build trusting relationships by being involved with local communities and agencies.

Moreover, in line with the same argument of developing local capacities, the general perception is that international teams are experts in search and rescue operations and have the equipment for effective response. However, reports show that most of the immediate rescue operations were carried by untrained civilians in Haiti and the 67 international rescue teams that reached Haiti were only able to rescue 132 people. The same holds true for the Pakistan floods, where most people got out of their homes in time and without the help of government authorities and international rescue teams. Thus, for effective immediate response, investment should focus on training and developing the capacity and capabilities of local authorities and community leaders instead of fully relying on deploying and seeking help from expensive international teams (Ferris and Petz 2011).

13.5 Conclusion

Our international response system needs revamping and improvements. It can be concluded that the leadership provided by the international humanitarian system is ineffective to deal with more than one huge crisis in a year (Ferris 2010b). Although emergency funding appeals were launched by the international system in both cases, twice the funding was raised for Haiti ($3.3 billion) when compared to Pakistan ($1.6 billion). The United States also contributed twice as much to Haiti than Pakistan. One reason for this difference is that the Haiti and Pakistan disasters were only 7 months apart, governments and the humanitarian community had already committed a majority of funds to Haiti (Ferris 2010b).

These two crisis cases show that the international humanitarian system is not as strong as it should be. They cannot function through the cluster approach which is theoretically sound but practically weak and need to follow more integrative approaches to respond. The government in both cases has to play a more responsible role. International agencies and the UN leadership is limited and should not only provide funds and services for response and recovery directly to the people or via the national governments, it should also closely partner with national governments to build a government's legitimacy by developing a sustainable infrastructure and institutional capacity of the country's government (Ferris 2010b). This is a key mitigation strategy and will improve resiliency and effective response within the country, relieve some of the mounting burden off the international humanitarian community and will shift the onus of responsibility to local and federal governments in less developed countries such as Pakistan and Haiti.

Moreover, comparative governance and leadership structures in the two countries show that the Pakistan military response and their leadership stand out for the commendable effort in the flood response and relief phase. On the other hand, US military played a leadership role in the Haiti earthquake but its role was limited in Pakistan, since Pakistan itself deployed around 60,000 of its own military troops for response (Ferris and Petz 2011; Ferris 2010a). Overall political leadership seemed to be very weak in the two countries. The issue of competitive elections in Haiti, the deaths of many government officials and the destruction of government offices especially in the capital city worsened their government response. International humanitarian leadership and its constraints were equally applicable to both cases.

As reflected through the discussions and cases, leadership in disasters needs to be multi-layered and reflect strong interdependence and collaboration. International humanitarian organizations such as the UN need to assign cluster leaders that work effectively with local community leaders, local NGOs and government officials that would ideally have effective leadership and collaboration/partnership skills. Moreover, presidential and political leadership has to work closely and responsibly with the national and international media, with the military leadership of their country and with foreign leaders and international agencies. Thus, the overarching aim of these collaborations and partnerships should be focused on building resiliency and developing the capacity of the disaster inflicted country.

Acknowledgement Sana Khosa assisted in this research to the end of the chapter.

References

Alexander, D. 2010. News reporting of the January 12, Haiti Earthquake: The role of common misconceptions. *Journal of Emergency Management* 8(6): 15–27.

Ansell, C., and A. Gash. 2008. Collaborative governance in theory and practice. *Journal of Public Administration Research and Theory* 18(4): 543–571.

Aon Benfield. 2010. *Event recap report: 1/12/10 Haiti earthquake*. Retrieved January 28, 2011. http://www.aon.com/attachments/reinsurance/201001_ab_if_event_recap_haiti_earthquake_impact_forecasting.pdf.

Archibold, R.C. 2011. Haiti: Quake's toll rises to 316,000. *The New York Times*, January 13.

Asian Development Bank (ADB). 2010. *Pakistan 2010 floods: Preliminary damage and needs assessment*. Retrieved February 2, 2011. http://www.gfdrr.org/gfdrr/sites/gfdrr.org/files/publication/Pakistan_DNA.pdf.

Atkinson, M., and W. Coleman. 1992. Policy network, policy communities and the problems of governance. *Governance* 5(2): 154–180.

Bevir, M., R.A.W. Rhodes, and P. Weller. 2003. Traditions of governance: Interpreting the changing role of the public sector. *Public Administration* 81(1): 1–17.

Bilham, R. 2010. Lessons from the Haiti earthquake. *Nature* 463: 878–879.

Bligh, M.C., J.C. Kohles, and J.R. Meindl. 2004. Charisma under crisis: Presidential leadership, rhetoric, and media responses before and after the September 11th terrorist attacks. *The Leadership Quarterly* 15: 211–239.

Boin, R.A., and P. 't Hart. 2003. Public leadership in times of crisis: Mission impossible. *Public Administration Review* 63(5): 544–553.

Bolton, M.B. 2011. *Human security after state collapse: Global governance in post-earthquake Haiti*, LSE Global Governance research papers, RP 01/2011. London: LSE Global Governance, London School of Economics and Political Science.

Burki, T. 2010. Aid slows to a trickle as Pakistan crisis enters a new phase. *Lancet* 376: 1041–1042.

Burns, J. 2010. Humanitarian aid v. National interests: The current dilemma in Pakistan. *The Review*, October 20. Retrieved May 2, 2011. http://www.gppreview.org/blog/2010/10/humanitarian-aid-v-national-interests-the-current-dilemma-in-pakistan/.

Cedar, O. 2011. A failure of conscience: How Pakistan's devastating floods compare to America's experience during Katrina. *Sustainable Development Law & Policy* 11(1): Article 16. Retrieved May 2, 2011. http://digitalcommons.wcl.american.edu/sdlp/vol11/iss1/16.

Charbonneau, L. 2010. Bill Clinton to coordinate Haiti relief efforts. *Reuters*, February 1. Retrieved April 2, 2011. http://www.reuters.com/article/idUSTRE6105NH20100201.

Derthick, M. 2007. Where federalism didn't fail. *Public Administration Review* 67(s1): 36–47.

Elderlein, H., S. Walti, and M. Zurn. 2010. *Handbook of multi-level governance*. Northamton: Edward Edgar.

Ferris, E. 2010a. *Earthquakes and floods: Comparing Haiti and Pakistan*. Brookings-Bern project on internal displacement, 26 August 2010. Retrieved May 2, 2011. http://www.brookings.edu/~/media/Files/rc/papers/2010/0826_earthquakes_floods_ferris/0826_earthquakes_floods_ferris.pdf.

Ferris, E. 2010b. *Natural disasters and human rights: Comparing responses to Haiti and Pakistan*. Brookings Institution, Presentation at Center for Human Rights and International Justice, Boston College, 3 November 2010.

Ferris, E., and D. Petz. 2011. *A year of living dangerously: A review of natural disasters in 2010*. The Brookings Institution – London School of Economics Project on Internal Displacement. Retrieved April 10, 2011. http://reliefweb.int/sites/reliefweb.int/files/resources/04_nd_living_dangerously.pdf.

Freeman, J. 1997. Collaborative governance in the administrative state. *UCLA Law Review* 45: 1–98.

Getha-Taylor, H., M.H. Holmes, W.S. Jacobson, R.S. Morse, and J.E. Sowa. 2011. Focusing the public leadership lens: Research propositions and questions in the Minnowbrook tradition. *Journal of Public Administration Research and Theory* 21(suppl 1): 83–97.

Goldfinch, S., and P. 't Hart. 2003. Leadership and institutional reform: Engineering macroeconomic policy change in Australia. *Governance* 16(2): 235–270.

Goldsmith, S., and D.F. Kettl (eds.). 2009. *Unlocking the power of networks: Keys to high-performance government*. Washington, DC: Brookings.

Huxham, C. 2000. The challenge of collaborative governance. *Public Management* 2: 337–357.

Independent Evaluation Group (IEG). 2010. *Response to Pakistan's floods: Evaluative lessons and opportunities*. The World Bank Group. Retrieved February 2, 2011. http://siteresources.worldbank.org/EXTDIRGEN/Resources/ieg_pakistan_note.pdf.

Johnston, E. 2010. Governance infrastructures in 2020. *Public Administration Review* 70: 122–128.

Jung, Y.D., D.A. Mazmanian, and S.Y. Tang. 2009. Collaborative governance in the United States and Korea: Cases in negotiated policymaking and service delivery. *International Review of Public Administration* 13(Special Issue): 1–11.

Kapucu, N. 2005. Interorganizational coordination in dynamic context: Networks in emergency response management. *Connections* 26(2): 33–48.

Kapucu, N. 2008. Collaborative emergency management: Better community organizing, better public preparedness and response. *Disasters* 32(2): 239–262.

Kapucu, N. 2010. *Governance reforms: Comparative perspectives*. Ankara: International Strategic Research Organization (ISRO).

Kapucu, N. 2011. Collaborative governance in international disasters: Nargis cyclone in Myanmar and Sichuan earthquake in China cases. *International Journal of Emergency Management* 8(1): 1–25.

Kapucu, N. 2012. *The network governance in response to acts of terrorism: Comparative analyses*. New York: Routledge.
Kapucu, N., and A. Ozerdem. 2013. *Managing emergencies and crises*. Boston: Jones & Bartlett Publishers.
Kapucu, N., F. Yuldhasev, and E. Bakiev. 2010. Collaborative public management and collaborative governance: Conceptual similarities and differences. *European Journal of Economic and Political Science* 2(1): 39–60.
Kim, P.S. 2010. Building trust by improving governance: Searching for a feasible way for developing countries. *Public Administration Quarterly* 34(3): 273–299.
Kowert, P.A. 2001. Leadership and learning in political groups: The management of advice in the Iran-Contra affair. *Governance* 14(2): 201–223.
Kronstadt, K.A., P.A. Sheikh, and B. Vaughn. 2010. *Flooding in Pakistan: Overview and issues for Congress*. CRS report for Congress. http://fas.org/sgp/crs/row/R41424.pdf.
Littlefield, R.S., and A.M. Quenette. 2007. Crisis leadership and Hurricane Katrina: The portrayal of authority by the media in natural disasters. *Journal of Applied Communication Research* 35(1): 26–47.
Lynn, L.E., C.J. Heinrich, and C.J. Hill. 2000. Studying governance and public management: Challenges and prospects. *Journal of Public Administration Research and Theory* 10(2): 233–261.
Margesson, R., and M. Taft-Morales. 2010. *Haiti Earthquake: Crisis and response*. CRS report for Congress.
Millner, D.A. 2010. Rebuilding Haiti: Making aid work better for the Haitian people. *Bread for the World Institute, Briefing Paper* 9: 1–8.
Moynihan, D. 2005. *Leveraging collaborative networks in infrequent emergency situations*. Washington, DC: IBM Center for the Business of Government.
Osborne, S.P. (ed.). 2010. *The new public governance? Emerging perspectives on the theory and practice of public governance*. New York: Routledge.
Oxfam. 2011. *Six months into the flood: Resetting Pakistan's priorities through reconstruction*. 144 Oxfam briefing paper. Retrieved February 2, 2011. http://www.oxfam.org.uk/oxfam_in_action/emergencies/downloads/oxfam_pakistan_6mth_briefing_note.pdf.
Peters, B.G., and J. Pierre. 2000. Governance without government: Rethinking public administration. *Journal of Public Administration Research and Theory* 8: 223–242.
Reinicke, W.H. 1999. The other World Wide Web: Global public policy networks. *Foreign Policy* 117: 44–57.
Rencoret, N., A. Stoddard, K. Haver, G. Taylor, and P. Harvey. 2010. *Haiti earthquake response: Context analysis*. Active learning network for accountability and performance in humanitarian action. Retrieved April 2, 2011. http://www.alnap.org/pool/files/haiti-context-analysis-final.pdf.
Schafer, W., J.M. Carroll, S. Haynes, and S. Abrams. 2008. Emergency management planning as collaborative community work. *Journal of Homeland Security and Emergency Management* 5(1): 1–17.
Schneider, S.K. 2005. Administrative breakdowns in the governmental response to Hurricane Katrina. *Public Administration Review* 65(5): 515–516.
Siddiquee, N.A., and M.Z. Mohamed. 2007. Paradox of public sector reforms in Malaysia: A good governance perspective. *Public Administration Quarterly* 31(3/4): 284–312.
Solberg, K. 2010. Worst floods in living memory leave Pakistan in paralysis. *Lancet* 376: 1039–1040.
Stoker, G. 1998. Governance as theory: Five propositions. *International Social Science Journal* 50(155): 17–28.
Streck, C. 2002. Global public policy networks as coalitions for change. In *Global environmental governance: Options and opportunities*, ed. Daniel Esty and Maria Ivanova. New Haven: Yale School of Forestry and Environmental Studies. http://environment.research.yale.edu/documents/downloads/o-u/streck.pdf.

Thomas, A., and R. Rendon. 2010. *Confronting climate displacement: Learning from Pakistan's floods*. Refugees International, November 2010. Retrieved April 4, 2011. http://www.refugeesinternational.org/policy/in-depth-report/confronting-climate-displacement.

Vangen, S., and C. Huxham. 2003. Enacting leadership for collaborative advantage: Dilemmas of ideology and pragmatism in the activities of partnership managers. *British Journal of Management* 14: 61–76.

Vigoda, E. 2002. From responsiveness to collaboration: Governance, citizens, and the next generation of public administration. *Public Administration Review* 62(5): 527–540.

Wallace, M., D. O'Reilly, J. Morris, and R. Deem. 2011. Public service leaders as 'change agents' – For whom? *Public Management Review* 13(1): 65–93.

Wallis, J., and R. Gregory. 2009. Leadership, accountability and public value: Resolving a problem in "New Governance"? *International Journal of Public Administration* 32: 250–273.

Warraich, H., A.K.M. Zaidi, and K. Patel. 2011. Floods in Pakistan: A public health crisis. *Bulletin of the World Health Organization* 89(3): 236–237.

Waugh, W.L., and G. Streib. 2006. Collaboration and leadership for effective emergency management. *Public Administration Review* 66(Special Issue): 131–134.

Naim Kapucu is a professor in the School of Public Administration at the University of Central Florida (UCF). He is also the founding director of the Center for Public and Nonprofit Management (CPNM) at UCF. He developed and coordinated Emergency Management and Homeland Security Programs. His main research interests are emergency and crisis management, decision-making in complex environments, and organizational learning and design. He is co-author of *Managing emergencies and crises* with Ozerdem (2011) and author of *Network governance in response to acts of terrorism: Comparative analyses* (2011). He is recipient of numerous excellence awards in research and teaching. He teaches emergency and crisis management, and collaborative public management.

Chapter 14
Structure, Process, and Agency in the Evaluation of Risk Governance

Urbano Fra.Paleo

For character too is a process and an unfolding.
(George Eliot, Middlemarch)

14.1 Introduction

When Burton et al. (1993) examined the patterns of societal behaviour regarding risk, they adopted a social–ecological perspective, and considered the type of human action driven by the attributes of change in the environment. Living with risk may lead to multiple modes of individual and collective adaptation to the hazard dimensions, which are determined by the disasters experienced in or ex situ, by the level of risk awareness and knowledge, and by the capacities and priorities of the community. Burton et al. (1993) recognize the incremental modes of coping (absorption, acceptance, reduction, and change) dissociated by processes of awareness, action, and eventually intolerance, which operate as influential thresholds. Progression in coping goes along with advancement in social integration, from individual to collective action and paired advancement in governance complexity. Absorption is a kind of human individual/household behaviour that incorporates low-level disruptions into daily life by adopting all kinds of strategies following a high-level awareness of the changes in the natural system and an intuitive process of adaptation. Acceptance implies a comparable capacity to adapt in an environment with larger disruptions that cause significant losses. Finally, although all modes of coping require change; reduction, land use, or location change call for higher levels of social organization, more resources, increased capacity, and a vision of future action. However, any human action is far from straightforward; instead, it is a complex combination of modes of coping with risk, and while responses are not necessarily exclusory, they are, according to Burton et al. (1993), a mix of adjustments.

U. Fra.Paleo (✉)
School of Social Sciences and Humanities, University of Extremadura,
Campus universitario, 10071 Cáceres, Spain
e-mail: upaleo@unex.es

U. Fra.Paleo (ed.), *Risk Governance*, DOI 10.1007/978-94-017-9328-5_14

The sociological perspective contends that "risk and safety exist in and through social organization rather than as objective conditions that individuals simply perceive either more or less accurately" (Stallings 1990: 80), and for this reason both conventional mass media and new social media are key mechanisms through which risk is socially created, filtered, debated, and distributed. Those groups with more power, with control or greater access to media, may be able to shape the message responding to their interests, notwithstanding the specific goals of media as corporations (news organizations according to Stallings (1990)). Therefore, risk is not currently being negotiated between groups with competing or aligned interests and perceptions, but imposed by certain societal groups who have more access to power and media, particularly through keynote discourse (Stallings 1990). Risk knowledge—beyond science and technology—is acquired and built by multiple societal groups and individuals as a combination of culture, past experience, information received, and action, creating a context of high ambiguity for policy-making. On this point, Renn (Chap. 2, this volume) states: "more knowledge does not necessarily reduce ambiguity."

Disasters are the result of past decisions made by societal actors such as administration, corporations, civil society groups, and citizens on hazard exposure and the escalation of vulnerability driven by a varied combination of root causes (needs/wants, transfer, or a lack of knowledge) (Fra.Paleo 2009). This makes all societal groups the protagonists of their own risks and those of society at large. Thus, risk is not just the product of social construction, but also the output of everyday individual and group decisions that may challenge observable hazards, the interests of other groups, and even those of society. Just as some risks are created by the media and powerful groups, the role played by individual actions and small communities should not be diminished. Therefore, although every social group depends on each other's decisions, the governance system does not provide instruments adapted to negotiate a more balanced and appropriate level of risk to prevent hazard exposure and reduce disaster risk.

If we consider that the objectives of evaluation may go beyond simply measuring performance and that it is not just experts who should be entitled or have a say in the elaboration of reports, then evaluation processes can be recognized as opportunities for increasing societal interaction between groups and for the social construction of risk if we are interested in the process as much as in the outcome. However, if we contend that measuring performance through evaluation ultimately seeks to increase the efficiency of practice (an identifiable outcome), we also can claim that processes of deliberation, consensus seeking, or negotiation are not futile exercises. Instead, they can be seen as estimable pathways to improve the design and implementation of policy decisions, and a less visible and direct avenue to increase the efficiency of public policies and public trust in policy making.

This chapter will examine the theory and practice of participatory evaluation to reach the goals of social learning and consensus about risk; explore the avenues for adopting this model of evaluation to increase the efficiency of disaster risk reduction (DRR) and public policies; and improve equality in the distribution of risk, by modifying key processes in policy making.

14.2 Reasons for Evaluation

The measurement of performance is the most common, and apparently single, objective of evaluation. From this perspective, it seeks to identify evidence of qualitative or quantitative advancement with respect to a point of reference in the past or to a pre-established baseline. In the first case, it allows for determining not just the current state but also any relative changes that have occurred, i.e., to measure progress. Opportunities for future development and improvement can be further identified while capacities and means for action can be determined through evidence-based policies. In the second instance, the process is designed to streamline action against standards established as a baseline.

In this way, evaluation conveys an implicit temporal dimension, some kind of agreement with monitoring. Wolter (2007) suggests monitoring has a diachronic nature and is "relevant to the achievement of a particular goal" (p. 62). It follows a systematic, continued process serving diverse goals. While monitoring places emphasis on time and continuity, in the case of evaluation the focus rests in examining quality. As Rauschmayer et al. (2009) state, "If processes are evaluated, those responsible for carrying them out will pay more attention to their quality" to the extent that "a timely evaluation has more influence on the governance process itself as a corrective device" (p. 169). Accordingly, this implies that the transformative power of policies and decision-making processes will only be recognized if it is recurrent, because "adaptive iterative evaluation is useful as an ongoing learning exercise" (p. 164). From this standpoint, evaluation and monitoring (although different notions) complement each other; although they have different perspectives, they share the same goal of measuring performance.

Accordingly, the distinction between outcome-oriented and process-oriented evaluation is relevant, because the focus is placed on when it is performed and at what point the impact in policy making can be anticipated. Worthen et al. (1997) adopt this pragmatic and utilitarian standpoint—evaluation to make intelligent political choice—when they say that, "policy makers need good information about the relative effectiveness of each program" (p. 4). Still, to a great extent, one single and separate approach hardly matches the goal of examining risk, a constant condition of human society, or of risk governance, a context where a succession of decisions regarding risk reduction or risk increase occurs on a continued basis. An evaluation approach that looks at objectives does not make sense unless we focus on a specific risk program, but risk governance is systemic and highly complex with multiple components, relationships, and processes.

Thus, a combined outcome–process evaluation fits better with the object of study, because it is more important to focus on what is pursued rather than when to evaluate. Weibelzahl (2005) articulates this idea reasoning that, "Often evaluation is seen as the final mandatory stage of a project. … However, when constructing a new adaptive system, the whole development cycle should be covered by various evaluation studies" (p. 58). In literature and practice, we also find a distinction between formative and summative evaluation, which seems very relevant for the evaluation of governance. While summative evaluation serves to determine whether the system has accomplished its objectives (Hamilton and Chervany 1981), formative evaluation is "designed to identify potential and actual influences on the progress and effectiveness of implementation efforts" (Stetler et al. 2006: S1) as well as strengths and weaknesses, when opportunities for adaptation and response to change are more likely to be within reach. Emphasis is thus placed on improvement when the "ways that existing organizational structures, processes, and cultures either facilitate or impede program implementation" (p. 8) are examined. However, beyond dichotomies, Chen (1996) interprets the two modes of classification as non-exclusionary and compatible, as he understands that evaluation in general has two functions (improvement and assessment) and two stages (process and outcome) that can be combined.

The choice of evaluation, either internal or external, exhibits a common understanding that considers the evaluated organization—particularly government—as an independent entity from society. However, "evaluators cannot be value neutral in these matters. Our conceptions and even our methodologies are value laden" (House 1991: 245). Fetterman (1994) is even more conclusive when he contends that, "The issue of objectivity is also a relevant concern. We needn't belabor the obvious point that science and specifically evaluation have never been neutral" (p. 10). After all, evaluation is in essence a question of values, "for at the heart of evaluation lies the question of whose values are driving the evaluation and whose standards are being met by the activities being undertaken and assessed or whose standards are being measured against" (Springett and Wallerstein 2008: 200). Evaluation reports convey both descriptive information that has been selected and filtered and assessments that distil subjective views based on individual and group knowledge, values, interests, or priorities. In such cases, both the evaluation outcome and process are not neutral, but dependent on how the latter is constructed. Thus, Rallis and Rossman (2000) contend that evaluation should be a learning process for social justice "to influence the fair and equitable distribution of social goods and to foster a more civil society" (p. 1). We might also add the equitable distribution of social bads when evaluating risk governance. MacDonald (1976) distinguishes three distinct approaches that link reasons to forms of evaluation and connect goals with processes: bureaucratic, autocratic, and democratic. The bureaucratic form of evaluation serves the purposes of and stays in the ownership of the agent whereas the evaluator does not have control over its political impact or the publication of results. In the case of autocratic evaluation, the evaluator retains the ownership of the report and can make the results public, but participation of non-institutional actors can be constrained at the discretion of the evaluator.

Alternatively, democratic evaluation recognizes the value of pluralism and access to information, challenging monopolies of "problem definition, of issue formulation, of data control, of information utilization" (Greene 2006: 120), and "helping all our peoples to choose between alternative societies" (MacDonald 1976: 12).

Response to change has also been argued as an aim of evaluation, or evaluation understood as adaptation. Risk and governance are very dynamic environments where the association between hazard and vulnerability may rapidly turn into disaster, with latest public policies or new developments creating a context of a very different risk class. While placing emphasis on achieving quality and measuring performance in a medical curriculum, Morrison (2003) understands the principal goal of evaluation is to push development, improvement, and adaptation in response to new conditions. Each evaluation creates a baseline that can be used to monitor future progress, and to define future goals and milestones (Fetterman 1994). Although traditionally seen as a feedback mechanism for policy making, evaluation can be indisputably understood as policy making; and not just because it is a political act but because of the observer effect, with adjustments made to the structure and process of the subject undergoing a particular evaluation. Observation drives change: Hyde (2007, 2010) documented a direct reduction in fraud, induced by international election observers visiting polling stations during electoral processes. Most commonly, learning from knowledge exchange can activate adaptation at both individual and organizational levels, although as Berkes (2009) reasons, learning does not always lead to adaptation. Accordingly, the responsive capacity of evaluation may develop into a driver of policy (structural and functional) reform and transition, as it draws attention to imbalances. Knowledge informs policy design, and evaluation, as a feedback mechanism, informs about policy implementation and gaps and helps to reorient policy making by establishing a link between knowledge and action.

In addition, Rauschmayer et al. (2009) claim that evaluation has multifaceted goals, and include the following: providing information about governance processes, the state of development (gaps and opportunities), monitoring compliance for policy formulation and implementation, learning from governance processes, increasing awareness of the need for action, the promotion of public participation with consensus-based standards instead of the sole judgment of experts, and to ultimately improve governance. However, its transformative capacity can only be achieved through democratic evaluation (MacDonald 1976).

14.2.1 Evaluation as Participation Beyond Measuring Performance

Just as the object of evaluation can be either outcomes or processes (Rauschmayer et al. 2009), or both, the process of evaluation itself may go beyond simply seeking to produce a report and act—in the best-case scenario—to guide future political action to also pay close attention to the course of evaluation. Risk assessment is commonly based on the knowledge, views, and interests of experts, following models of either bureaucratic or autocratic evaluation. Furthermore, the outcome of the assessment is sometimes used by policymakers to support aprioristic positions and views to assist short-term interests (political and economic, but rarely social), and infrequently contradicts designed or implemented policies. This can be likened to a DAD (decide, announce, and defend) strategy, where pretended participants are just mere spectators.

The number and type of roles played by participants in evaluations are not neutral matters, as evaluation results are very dependent on these types of decisions. The goals—and openness—of the evaluation promoter are a determinant that can limit participation or, in contrast, increase inclusiveness and, from the viewpoint of the government, become largely intrusive in the work of policymakers. An increase in the number of evaluators does not solely increase the quality of the process for the reason that the diversity and empowerment of accessible political players in the process are essential attributes that enhance the evaluation, and ultimately act to change the governance of evaluation. Thus, evaluation can be seen both as a means and as an end in itself, contributing to increase the quality of governance and democracy, because all citizens are agents and bearers of risk. Renn (2004) proposes a functional argument to support analytic–deliberative processes: the governance of complex, uncertain, and ambiguous matters in contemporary society requires input from all kinds of stakeholders, and is a normative justification in the enhancement of democracy as an opportunity for political equality and citizen sovereignty. As King et al. (1998) reason, the quest for "democracy seems to lie at the core of why authentic participation is important" (p. 321). Empowerment evaluation is, for Fetterman (1994), inherently political because it "can be used to help anyone with a desire for self-determination" (p. 12), including the disenfranchised, minorities, disabled individuals, and women. Participation in evaluation is thus understood as a right and not as a privilege (this is a reference to consensus-based policies as opposed to those that are evidence-based).

Participation in its multiple forms represents a change in the relationships between citizens and the state; it drives advancement characterized by transparency and accountability to maintain the trust and legitimacy of government, and does not solely serve to guide governing (Wolter 2007) through auditing and compliance. Stivers (1990) understands it to represent a leap toward active accountability because an "administrative legitimacy requires active accountability to citizens, from whom the ends of government derive. Accountability, in turn, requires a shared framework for the interpretation of basic values, one that must be developed jointly by bureaucrats and citizens in real-world situations, rather than assumed" (p. 247).

14.2.2 Evaluation as Social Learning

Formative assessment, also known as assessment for learning, is an approach that has been used in education to promote student learning via assessment (Black et al. 2002); in governance, evaluation can accomplish a similar goal. A number of authors support this argument: iterative evaluation may induce ongoing learning (Rauschmayer et al. 2009), can help to promote social justice (Rallis and Rossman 2000), and in some cases stimulate adaptation (Berkes 2009). The argumentation behind this theory considers that within participatory evaluation, knowledge sharing (the bringing together of different types of knowledge and the transfer of information and data), dialogue, and consensus are social interaction processes that fill knowledge gaps. Furthermore, they increase awareness of the complexity of the matter under discussion, of remaining uncertainties, and of the diversity of challenges, and help with the understanding of other parties' arguments, concerns, interests, and values, because "program participants are typically more in touch with the critical variables associated with their daily life and their effectiveness (or ineffectiveness) than any outside party" (Fetterman 1994: 10). "In essence, a deliberative process can handle the problem of ignorance, incompetence, and distorted perspectives if it succeeds in feeding the relevant knowledge and the full diversity of values and interest into the deliberation procedure" (Renn 2004: 316).

In addition, the evaluation practice proclaims that those evaluators who have no past experience or are not familiar with the specific process do become familiar with and learn about the assessment; in particular, they learn how to be systematically analytical about themselves, how to rate the different elements, and how to justify and document the basis of their assessment (Fetterman 1994). However, learning does not just occur within one group, especially regarding citizens, but also among other societal sectors. Administrators and scientists learn, in a two-sided (deLeon 1992), or more properly, a multi-sided learning process because "social learning is a process of iterative reflection that occurs when we share our experiences, ideas and environments with others" (Keen et al. 2005: 9). There are times when various types of knowledge—and worldviews—match, collide, or assist in identifying gaps or raise new questions, yet two knowledge types are dominant: local, indigenous, practical, tacit, or lay knowledge; and expert, scientific, technical, explicit, or codified knowledge.

Both knowledge and learning are components of the social–ecological system, as much as the social and the ecological (McCarthy 2009), and the political, because "evaluation is inherently a political activity with potential political influence" (Greene 2006: 119). Learning occurs at various stages and levels. Triple loop learning is revealed when correcting "errors by addressing or designing governance protocols and norms" (May and Plummer 2011: 47), including participatory evaluation, while the translation of the evaluation outcome to policy making responds to double loop learning, because it "corrects errors by making adjustments to values and policies" (May and Plummer 2011: 47).

Disasters bring loss, but they also trigger a learning gain both in local communities—at least in the short term—that is assimilated into local environmental knowledge, and in governments, that allows them to review either the impact of their policies and programs on risk reduction or the multiplied effects of inaction. But, opportunities for sharing, on a continued basis, information, knowledge, and concerns about gaps on the construction and distribution of risk among societal groups, as an instrumental component of risk communication, are lacking. The implementation of periodic participatory evaluation may serve to meet these goals.

14.2.3 Evaluation as Consensus Building

To go beyond simply measuring performance and to place social learning at the fore turns participatory evaluation into an exercise of political feedback where the pluralistic views of society regarding risk are displayed—it becomes an arena to confront the multiple and opposed interests, concerns, and individual and collective choices of societal actors. Heterogeneity and discordance cannot be used as an argument to avoid interaction, neither can the fact that expert judgment is often divided within committees who address risk assessment. Sure enough, experts do not always share positions, and complete scientific consensus is particularly harder to reach among social and natural scientists such as in climate change issues (Nordhaus 1994) or natural processes with a high degree of uncertainty (Woodward and Bishop 1997). Sometimes, scientific agreement is hidden under aggregate expert opinion (Roughgarden and Schneider 1999) or expressed as degrees of uncertainty (Iverson and Perrings 2012). Notwithstanding, disagreement among experts may or may not have a major impact in policy making (Morgan and Henrion 1990), a signal of the relative influence of scientific knowledge in public policy. As the issues of dialogue, agreement, and consensus are very relevant, if not central to evaluation, a short discussion on the issues involved is worthwhile.

Risk assessment neither has the goal of nor offers the proper environment to address the social distribution of risk, although it is an exercise of amalgamation between analysis and preferences. Thus, how are these preferences known or anticipated? Assessment does not provide mechanisms for social consensus. Agreement on how risk should be distributed (in a fair manner) and how it may be confronted (to address uncertainty) needs to be based on a double consensus. This consensus has to occur within and across the diverse players of the social–political system, including citizens and decision makers, experts and practitioners, businesses and civil society groups, and on an integrated basis rather than a paired basis, building opportunities for exchange, dialogue, and consensus. Participatory approaches result in encounters between societal groups that are predominantly or simultaneously cooperative and antagonistic and build a consensus regarding the distribution of risk.

Fetterman (1994) proposes a model of empowerment evaluation based on the principle of self-determination, in which the evaluator and subjects benefiting from the evaluation "are often on an even plane" (p. 10) because self-determination is "the ability to chart one's course in life" (p. 2). It includes "the ability to identify and express needs, establish goals or expectations and a plan of action to achieve them, identify resources, make rational choices from various alternative courses of action, take appropriate steps to pursue objectives, evaluate short- and long-term results (including reassessing plans and expectations and taking necessary detours), and persist in the pursuit of those goals" (p. 2). If applied to risk governance, participatory evaluation—as empowerment—implies that both government and citizens decide horizontally how a society negotiates the preferred level of risk.

14.2.4 Evaluation as a Process of Governance

The decisions regarding the implementation of the evaluation and its goals, and those concerning the choice of the model of evaluation, participant selection, the internal and external dissemination of the report, the use of the results and the implementation of recommendations, the verification of compliance, and the recursive evaluation of impact in disaster risk reduction, are a set and succession of actions that may temporarily change political processes. However, adjustments can only be transformative if the political structure also changes.

Government structure has a high level of organizational inertia and thus a high capacity to resist changes; this enables it to return to customary operations and can be viewed as a kind of resilience. Hannan and Freeman (1984) argue that "selection processes tend to favor organizations whose structures are difficult to change" and that "high levels of structural inertia in organizational populations can be explained as an outcome of an ecological-evolutionary process" (p. 149). Societal groups can push change but it is government that has the highest capacity to effectively transform the process and structure within and outside of the administration. In fact, adjustment takes place gradually and quietly because "changes have layered new challenges on top of the traditional institutions and their processes" (Kettl 2000: 488), with nongovernmental partners increasingly delivering services, and institutions being challenged by globalization and devolution (Kettl 2000). For example, in one case, markets set the economic rules to national institutions or they transfer power to supranational organizations—the European Union as the most conspicuous; and in a second case, "more responsibility for both making and implementing policy [flowing] to state and local governments" (Kettl 2000: 489) not only in the United States—as seen by this author—but also elsewhere, responding to the principle of subsidiarity or to other pragmatic requirements. However, the contribution of evaluation to the amplification of opportunities for co-action within the social–political system may be limited, not just by political resistance to incorporate evaluation processes but also by a poor or absent implementation of recommendations, driven by a lack of political will or clear, available instruments for the institutionalization of the evaluation results into mainstream policy making.

The gain from an evaluation based on a participatory multi-stakeholder approach can be lost if innovative instruments and practices are not put into operation. For example, co-action in evaluation is limited by the time/space concordance of participants—who are dispersed geographically—and by discordant agendas, but can be surpassed with the use of technological solutions for facilitating interaction

(Medsker et al. 1995). Thus, evaluation processes can be performed asynchronically, with each participant adapting to their own agenda, and remotely, without the need to displace evaluators by making use of information technologies. However, this occurs at the cost of not having face-to-face interaction or the ability to complete an assessment through a continuous task. The advantages of face-to-face communication include the formalisation of more cohesive and personal communication (Jonassen and Kwon 2001), including the availability of non-verbal signals with information of other participants' behaviour and attitude. Furthermore, computer-mediated communication facilitates a more task-oriented process and a clearer identification of own-role expectations. In the end "the effectiveness of groups problem solving is largely determined by how well the group's members communicate with each other" (Jonassen and Kwon 2001: 35) rather than the means of interaction.

Perhaps the most challenging and potentially conflicting issue in participatory evaluation is the transfer of power and legitimacy by governments to other societal groups to sustain the exercise. Thus, this implies a limited (but ultimately a loss of) control of the political process, because, according to Mathison (2000), "any deliberative democratic evaluation must by definition be empowering. If such an evaluation does not disrupt the balance of power with an eye to redistributing that power, then democracy will not be served" (p. 88). Furthermore, "a governance process involves multiple actors with multiple preferences leading to multiple goals" (Rauschmayer et al. 2009: 19).

On balance, public participation in political processes at large—excluding elections in representative democracies—has advanced little and the steps taken in the form of public involvement pervert the original concept because they only partially satisfy the goals of participation and governance. Moote et al. (1997) provide an example of what occurs in a wider context where "traditional public participation methods used by the public land management agencies have been criticized for allowing the agencies to nominally meet their statutory requirements for public involvement while effectively continuing to dispense predetermined management decisions" (p. 877). Few successful experiences can be drawn to date, however, Fung and Wright (2001) are able to give some examples of achievement in participatory budgeting, neighbourhood governance, habitat conservation, and the job market.

In their examination of hazard mitigation policies based on cooperation, and command and control, Burby and May (2009) show that both approaches can be equally effective because success was found to be basically dependent on the commitment of local governments. Effectiveness does not seem to be jeopardized by the adopted approach to governance—including participatory—but by the interplay of interests and values that are operating in any context and by the commitment of political leadership to change the course of action and surpass barriers to transform political structure and function. Systems based on coercion and control have been prevalent, but they have not been able to provide opportunities for less privileged societal groups to speak out.

In addition to principles, a risk governance evaluation system requires the criteria and procedures to be operational, and these issues will be discussed in the second part of the chapter.

14.3 Criteria, What Criteria?

14.3.1 Governance Indices and Indicators

Various indicators and indices—their use appears to change in literature—of governance have been proposed (Table 14.1) by different authors and organizations, and are widely used to measure the state of and progress in governance at either global or regional scales. Among the latter are, for instance, the Ibrahim Index of African Governance (IIAG), which focuses on African countries, or the Sustainable Governance Indicators (SGI), which only concerns OECD countries, whose use limits comparison among countries worldwide. Notwithstanding multiple criticisms regarding comparability, bias, and completeness (or its opposite, complexity), which could be formulated with regard to any other synthetic measurement, they allow for synchronic cross-country comparison, or monitor change over time towards desirable goals of better governance, whatever this means in each case. Changes in the nature of the indicators, owing to their evolutive character, and the variation in the number of variables according to the availability of information in different countries, brings support for the arguments of limitations in comparability. However, as we will see, governance indicators are only a recent development, and these allegations have not been sufficiently addressed.

These indicators share a conventional understanding of governance, made more explicit in their dimensions rather than in their own definition of the concept; this emphasizes the roles played by government and its operational performance. Notwithstanding, they portray and allow for a better understanding of the social–political context in which decisions are being made. Indices and indicators approach the measurement of governance with what de Boer et al. (2008) designate as the governance equalizer whereby, "the ways of governance are empirical combinations of the various dimensions of governance and these dimensions are independent and can be combined with each other in a variety of ways" (pp. 38–39). In a set of multiple dimensions, some are inevitably related or interdependent, and indicators within the same index show a certain degree of correlation (Kaufmann et al. 2010), or their distinction is blurry—Rice and Patrick (2008) state that this blurriness exists between effectiveness and legitimacy. Interrelatedness and fuzziness increases to a great extent with the number of variables used. Rice and Patrick (2008: 6) also argue that, "so many indicators may make it difficult for policymakers to identify priority sectors on which to focus attention and resources." Other weaknesses of indicators are not exclusive to this type of measurement, but to research in general, particularly the use of secondary data and its high level of dependence on the goals and methodologies of others.

Some combine qualitative and quantitative indicators, such as WGIr or SGI, although there is a preference for statistical sources. This type of indicator is perhaps less demanding during the process of data integration or is possibly oriented to satisfy the report reader or policymaker for its apparent simplicity, as if a single figure can completely synthesize the complexity of reality. In the same vein, Wolter

Table 14.1 Properties of selected indicators and indices used for the evaluation of governance

Indicator/index	Dimensions	Components	Strengths	Weaknesses
Ibrahim Index of African Governance (IIAG)	Safety and rule of law	24 indicators	Composite index of aggregated data	African countries
	Participation and human rights	88 variables		Dependent on information from other organizations
Mo Ibrahim Foundation 2007	Sustainable economic opportunity			Proxy measurement when data are missing
	Human development			Criteria capture the quality of services provided to citizens by governments
Sustainable Governance Indicators (SGI)	Quality of democracy	19 indicators	Qualitative and quantitative indicators	OECD countries
Bertelsmann Foundation 2009	Policy performance (economy and employment, social affairs, security, resources)	150 qualitative and quantitative criteria	Available data for evaluation	Dependent on expert evaluation Focuses on needs for reform and effectiveness of existing initiatives
Index of State Weakness in the Developing World	Economic Political Security	20 indicators	State weakness as a function of its effectiveness, responsiveness, and legitimacy	Focuses on state function but neglects other political players
Brookes Institution 2008	Social welfare			
Worldwide Governance Indicators (WGIr)[a]	Voice and accountability	6 composite indicators	Based on survey respondents	Based on perception
Brookings Institution, World Bank Development Research Group, and World Bank Institute 1996	Political stability and absence of violence/ terrorism Government effectiveness		Diverse group of respondents: citizens, entrepreneurs, and experts in the public, private sectors and NGO	The underlying principle associates good governance with economic development
	Regulatory quality Rule of law Control of corruption		It focuses on quality of governance	

(continued)

Table 14.1 (continued)

Indicator/index	Dimensions	Components	Strengths	Weaknesses
World Governance Index (WGIx)[a]	Peace/security	5 indicators	Synthetic index	Synthetic index
Forum for a new World Governance (FnWG) 2008	Democracy/Rule of law	37 indexes		Dependent on information from other organizations
	Human rights/ participation			
	Sustainable development			Only quantitative indicators
	Human development			
Global reporting initiative (GRI)	Economic	3 categories	Takes into account the economic, environmental and social dimensions	Oriented to reporting towards corporate social responsibility
Ceres and Tellus Institute 1999	Environmental	4 subcategories		
	Social	46 aspects		Any kind of organization, but particularly corporations

[a]Both indices call themselves WGI. Here, for clarification, they are differentiated with these non-standard acronyms

(2007) argues that, "purely statistical indicators produce governance information that is neither unequivocal nor complete" (p. 60). But dependence on other sources of information such as surveys has further shortcomings. This is the case of WGIr, where the composite is elaborated by integrating individual perceptions, which include a variety of assessments from both the public and private sectors.

14.3.2 Evaluation of Risk Management

The evaluation of performance in risk management is also a recent development and instruments to measure are not abundant. Some are very specific while others are comprehensive. Among the first, Parker (1999) developed a set of social, organizational, and institutional criteria for the evaluation of a tropical cyclone warning system, while Simpson and Katirai (2006) proposed the disaster preparedness index (DPi) to evaluate one component—but a decidedly important aspect—of risk governance, the capacity to efficiently respond in the case of disaster.

Another perspective on evaluation is to attempt comprehensiveness at the risk of being defiantly complex. The framework proposed by Mitchell (2003) to mainstream disaster risk reduction (DRR) into development policy includes 20 indicators with associated benchmarks, seeking to inform national policy making. Through an

Table 14.2 Dimensions and indicators in the DRR mainstreaming framework

Dimension	Indicator
Politics and legislation	Political commitment
	Regional linkages
	Legislation
	Emergency powers
	National disaster mitigation committee
Policy	Policy statements
	Participation
	Development plans
	National disaster administration
	National disaster planning
Knowledge	Risk and vulnerability
	Education
	Media
	Community networks
	Research
	Skills, capacity and motivation
Practice	Reconstruction/building codes
	Local community
	Insurance and finance
	Poverty reduction

Elaborated from Mitchell (2003)

inclusive, participatory process, indicators are graded using a scale ranging from 1 to 3 and extended to a fourth supplementary super goal, and are then weighted by evaluation participants. The framework has a normative perspective, providing guidance to "raise political will and commitment for disaster risk reduction" (Mitchell 2003: 26) as it can be seen (Table 14.2) in the combination of actions and goals and principles and institutions. A multi-sector and all-hazard focus—although sometimes apparently unrelated—is discernable, that speaks of the concern with an institutional context and with the measurement of advancement in DRR.

Similar principles inspire the methodology of the Hyogo Framework for Action Monitor (HFA Monitor): the review of progress—at national, regional and local levels—"in the implementation of disaster risk reduction and recovery actions … in accordance with the Hyogo Framework's priorities," performed by multiple stakeholders who measure levels of progress. Additionally, the HFA Monitor requires a motivation for evaluation through pre-established evidence, justification, and self-evaluation questions, and the identification of constraints that impede progress. Motivation increases the complexity of the process but ensures statements have to be ground based, discussed among the evaluation participants and agreed, and not be the result of a let-it-go, tolerated consensus. Levels of progress with descriptors in each indicator act as benchmarks. Sometimes, evidence requires not only additional information but also data, producing a dual qualitative–quantitative evaluation system under the HFA Monitor.

Cardona (2006) developed a set of composite indicators or indices to measure country performance in risk management: Disaster Deficit Index (DDI), the Local Disaster Index (LDI), the Prevalent Vulnerability Index (PVI), and the Risk Management Index (RMI). This framework was adopted by the Inter-American Development Bank (2010) for the evaluation of country performance in risk management in Central America, the Caribbean, and South America. The DDI measures the relationship between the losses in the case of a catastrophic event and the economic resilience of the country to cope with them. Governance is indirectly embedded in vulnerability and directly included in resilience, through the proper management of insurance, tax collection, credit, or reserve funds. In contrast, the LDI focuses on recurrent and chronic smaller events, which are very frequent in urban areas. It accounts for the persistence and number of victims and losses at the local level, and is more a measurement of exposure to these types of events as PVI accounts for vulnerability conditions. The level of development, weakness conditions, and their rate of change are computed in this second index using multiple variables, indirectly considering the impact of varied public policies. Finally, the RMI specifically focuses on risk management, integrating 24 indicators in 4 public policies: identification of risk, risk reduction, disaster management, and governance and financial protection. The RMI seeks to measure the risk management performance of disaster risk management and effectiveness against predefined qualitative targets or benchmark levels (Carreño et al. 2007).

In this approach, the evaluation process is reliant on expertise—scholars and administration officials—to determine both the distance between current policy conditions and the objective condition or benchmark, and to resolve the indicator weights; although this approach measures risk perception in the first dimension, it is solely dependent on science and governmental views. Moreover, RMI focuses on sets of public policies but not on individual indicators, which impedes evaluation results readdressing specific public policies. The set of indexes also involve calculations with a high level of mathematical complexity that makes them hardly transparent to the non-expert and difficult to be adopted in public participation processes and in processes of risk governance at large.

With the objective of addressing the Hyogo Framework for Action Priority Action 4, focusing on the reduction of underlying risk factors, and after identifying a lack of indicators to determine the conditions and capacities to manage disaster risk at national and subnational/local levels, DARA (2011) sought to develop indicators of underlying risk factors. Four factors have been identified: environmental degradation and natural resources, socio-economic conditions and livelihood, spatial planning, and governability. The first two deal with sets of conditions, while the third and fourth concern management, policy instruments, and effectiveness. Key informants from diverse sectors (national public sector, local public sector, private sector, NGO, civil society) were selected to evaluate and offer a range of perspectives, adopting a participatory approach. Evaluators at the subnational level were asked to respond to 118 questions grouped into the four dimensions, valuate them with a scale ranging from 0 to 9, and then assign weights. The process concluded with a workshop to share the results, identify ultimate causes, and to find solutions to disaster risk. At the national level a quantitative approach to evaluation was adopted, with 38 indicators based on data from national and regional databases.

As opposed to governance evaluation, a higher variety of approaches is found to address the assessment of risk management, although some of the instruments have incorporated public participation to be more sensitive and capture the variety of perspectives and complexity of the decision-making processes.

14.3.3 Risk Policy Principles

The Hyogo Framework for Action (HFA) (ISDR 2005) identifies five priorities for strategic political action, which stem from gaps and challenges in disaster risk reduction previously identified in the Yokohama Strategy (1994). The priorities propose internationally agreed targets for disaster risk reduction to be adopted by countries:

- Ensure that disaster risk reduction is a national and local priority with a strong institutional basis for implementation
- Identify, assess, and monitor disaster risks and enhance early warning
- Use knowledge, innovation, and education to build a culture of safety and resilience at all levels
- Reduce underlying risk factors
- Strengthen disaster preparedness for effective response at all levels

These priorities feature the five structural pillars of DRR that can be synthesised in governance: knowledge, awareness, avoidance, resistance and resilience, and response and recovery (Fig. 14.1). Broad policy principles dealing with specific phases of DRR go hand-in-hand with the recognition of the roles of information and knowledge in planning and risk awareness, and with implementing measures to avoid exposure to hazards and to reduce vulnerability.

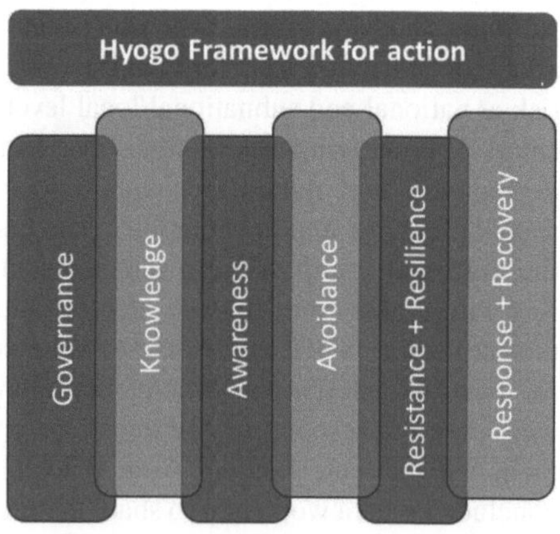

Fig. 14.1 The pillars of the Hyogo Framework for Action

In 2010, UNISDR (2010) launched the campaign Making Cities Resilient (MCR), which gained impetus with the 2011 Global Platform for Disaster Risk Reduction and the support of attending local policy makers. The ten essentials checklist (Table 14.3) of the MCR, aligned with the five priorities of the HFA, act as guiding principles for risk reduction at the local government level. Both the HFA and the MCR ten essentials adopt a combined proactive–reactive approach to disaster risk reduction that helps set priorities and streamline sectoral policies; they include the phases of the disaster cycle and go beyond disaster to incorporate the multiple dimensions of risk.

From an evaluation perspective, the principles and essentials may be used as a benchmark to contrast the realization of DRR policies. Thus, in the same way the principles are interpreted in the process of policy elaboration and implementation by governments, reversely, the alignment of those policies with the principles may be judged and clarified by risk governance evaluators.

Table 14.3 Dimensions derived from the ten essentials of the UNISDR campaign making cities resilient

Essential	Dimension
Institutional and administrative framework	Reform political structure and function
	Coordination of government departments
	Public participation
Financing and resources	DRR-oriented budget
Multi-hazard risk assessment	Risk information (hazard and vulnerability)
	Risk assessment
	Plan according to information
	Transparency and access to information
Infrastructure protection, upgrading, and resilience	Risk reduction infrastructure
	Incorporate the effects of climate change
Protect vital facilities: education and health	Assessment of critical infrastructures (schools and health centres)
	Risk-proofing of critical infrastructures
Building regulations and land use planning	Apply and enforce regulation
	Building coding
	Land use planning
	Upgrade housing
Training, education and public awareness	Education and training on DRR
	Raise risk awareness
Environmental protection and strengthening of ecosystems	Ecosystem conservation for DRR
Effective preparedness, early warning, and response	Early warning system
	Emergency management
	Involvement of communities in preparedness
Recovery and rebuilding communities	Victim-oriented disaster response and recovery
	Involvement of communities in response and recovery

14.4 The Evaluation of Risk Governance

14.4.1 Criteria for the Evaluation of Risk Governance

The broad-spectrum nature of the MCR essentials and HFA principles—though more specific in the case of the HFA Monitor—and their global scope are functional advantages that help to advance the structure and criteria of a comprehensive risk governance evaluation system, as well as the implicit rendering of the phases of the disaster cycle. Notwithstanding, the identification and generation of operational evaluation criteria of risk governance is challenged by six identifiable structural and functional conditions of both governance and risk: (i) the number of policies interested in risk that go beyond those related to disaster, and include economic, development, and social public policies that particularly involve the production of vulnerability; (ii) the intricate interactions between the multiple policies, direct and identifiable or indirect and subtle, sometimes in the form of overlapping responsibilities; (iii) the interplay between the multiple levels of administration, macro-regional, national, regional, and local; (iv) the interactions between societal players, including the strain and connection between governments, civil society organizations, and the market; (v) governments occasionally operate internally in terms of informal procedures and institutions, which are dominant in civil society interactions and abundant in the private sector as well as in the interactions between governments and other societal players; and (vi) the dynamic interactions within and between the social and natural systems, including the interplay between diverse hazards and between natural and technological hazards, not successfully translated into public policies. Factors intervening in governance and risk can thus be interpreted dialectically, as in the case of governmental and societal actors, policies of risk reduction and production, formal and informal processes, and proactive and reactive policies, which in the social–political context operate simultaneously, sometimes cooperatively, and occasionally antagonistically as agents or subjects of political action (Fig. 14.2). The upper part of the figure shows the observed political factors commonly taken into consideration in evaluation; informal arrangements, societal interactions, risk production factors, and proactive actions are usually overlooked.

The proposed DCC (Dimensions-Components-Criteria) framework for the evaluation of risk governance (Fig. 14.3) places criteria at the lowest level on the hierarchy of the evaluation system. Criteria are distinctive means accessible to participants to perform evaluations, grouped in families at a higher level as risk governance components, and are finally arranged as dimensions of risk governance. If we consider the number of elements in each level and their position in the categorisation, they fashion a multi-level pyramid of the evaluation system.

The framework is further developed in Fig. 14.4, starting with the seven dimensions at the first level: risk analysis, risk planning, risk reduction, emergency preparedness, emergency response, and disaster recovery. Dimensions capture

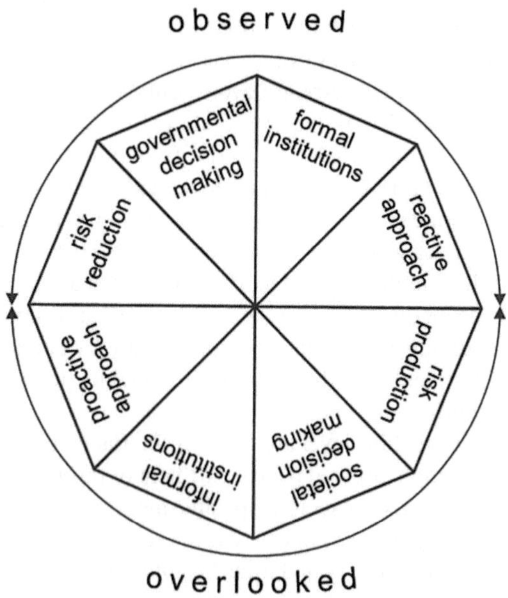

Fig. 14.2 Dialectic agency and subjecy in risk governance

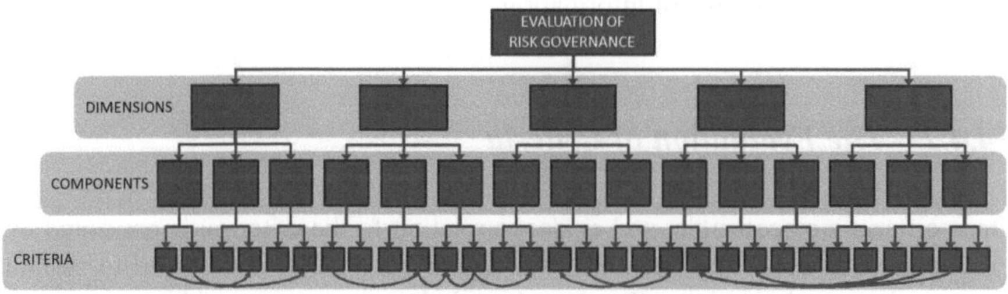

Fig. 14.3 Multi-level structure of the framework for the evaluation of risk governance

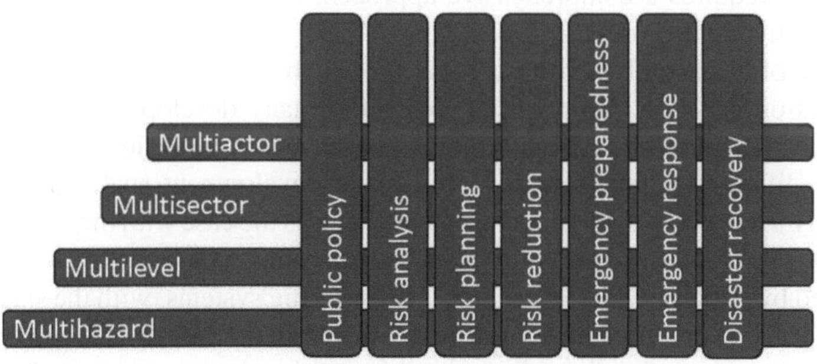

Fig. 14.4 Dimensions as pillars of evaluation and cross-cutting issues

strategically essential actions to suitably address the reduction of disaster risk, and include the phases of the disaster cycle (Quarantelli 1991), which is also known as the emergency response cycle (Cutter 2003), disaster risk cycle (Birkmann and von Teichman 2010), hazards cycle (Mileti et al. 1975), and the disaster management cycle (Weichselgartner 2001). In addition, it integrates other public policies that impact on risk production or reduction, and categorises mitigation and adaptation into dimensions of risk analysis, planning, and reduction.

At the second level, the components (Fig. 14.5) are identifiable policies, no matter how multi-sectoral they might be, that are associated with multiple governmental levels or departments responsible for their design and implementation. Once criteria pertaining to a certain component are evaluated, feedback is given to the related department or administration to tackle the recommendations of the evaluation output.

At the third level are the criteria. Each criterion singles out and describes a specific action targeted by a department or through concerted action with societal actors and the private sector. Specific measures can be taken to avoid and reduce risk, including hazard control and the reduction of vulnerability, in the context of either interdisciplinarity and transdisciplinarity. Thus, the evaluation of a criterion is inevitably related to the review of other criteria. The examination of policies, measures, and actions at this level during evaluation leads to the subsidiary review of the higher levels—components and dimensions—but their integration is hardly operational owing to the volume of information.

14.4.2 The Evaluation of Criteria

Once criteria are accessible, and evidence relating to the criterion is presented by evaluators and supporting organizations, arguments should be provided to sustain distinct views and positions. Benchmarking helps to compare evidence of policy making with pre-established measures and accomplishments to support evaluation outputs. With a view on facilitating participation, Mitchell (2003) proposes a "more qualitative approach using best practices to measure progress against benchmarks" (p. 7). This requires a comprehensive approach, including all possible measures, and exhaustive in their description.

A scale of levels with their descriptors for benchmarking is the basic element of an evaluation criterion. However, adopting too many development levels for the criterion results in difficulty in reaching a consensus about the level, while too few obscures differentiation. Qualitative five-stage development scales are the most commonly used owing to their manageability. This is the case with the RMI (Carreño et al. 2007)—low, incipient, significant, outstanding, and optimal—or the scale elaborated by Parker (1999) to evaluate early warning systems, with the stages ranging from basic to optimum development. However, in the case of DARA (2011), the system includes a scale of conditions and capacities that ranges from 0 to 9. The evaluation of sustainability in higher education with AISHE 2.0 (Roorda et al. 2009)

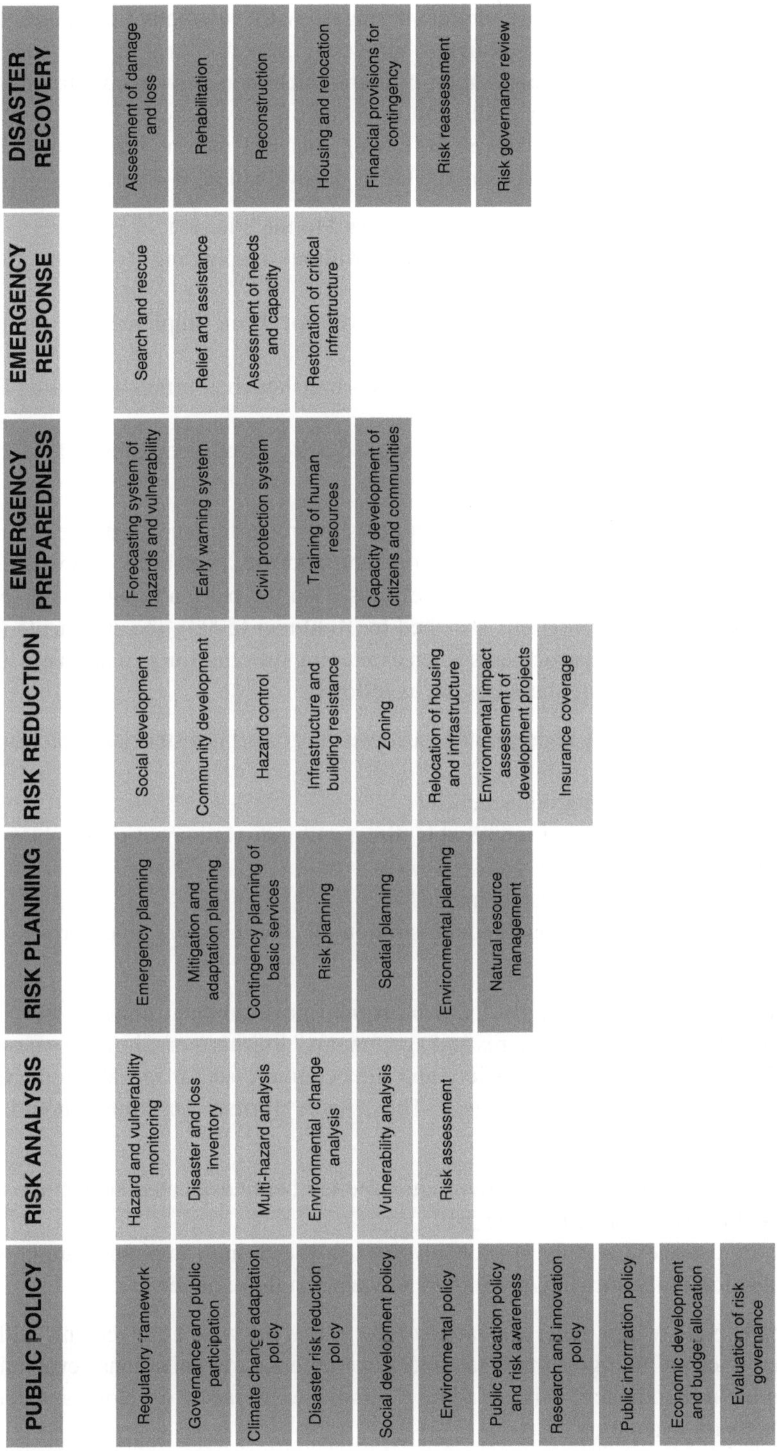

Fig. 14.5 The DCC (Dimensions-Components-Criteria) framework for the evaluation of risk governance

distinguishes between five development stages, with cumulative attributes and higher systemic complexity (activity-, process-, system-, chain-, and society-oriented). Combining the examination of hierarchical and networked attributes in policy measures enables the evaluation to focus on how policies relate to other policies and are linked to players. Following this last pattern Fra.Paleo (2014) proposes a scale of development applicable to the evaluation of criteria:

- No advances have been identified with regard to the criterion;
- Isolated measures have been adopted without responding to a structured program;
- Programs have been elaborated but have not been implemented through projects;
- Programs have been elaborated and projects translate their principles. Both plans and projects are occasionally evaluated;
- Programs and projects respond to a designed policy. Both programs and projects are regularly evaluated.

However, sound arguments to support decisions regarding the existence of measures or programs, their level of development, and the availability of evaluation need to be provided and assigned a level. Based on the basic attributes of ex-post evaluation of political programs sketched by Kuhlman (2003), the arguments that could be considered by participants when examining the criterion to reach a position about formal policies and measures are as follows:

- Is the program/policy appropriate to achieve the goal of disaster risk reduction or adaptation?
- Is a vulnerable target group reached?
- Is the reduction of risk a foreseeable direct and indirect impact in the short- or long-term?
- Are the formulated targets attainable?
- Is the implementation and administration of the program appropriate and efficient?

These arguments essentially focus on formal governmental policies, but how do we evaluate informal measures and agreements? Furthermore, how do we evaluate the role played by civil society and the economic sector in risk production and ultimately in risk governance? Thus, some further questions should be considered:

- Are the impacts of the measure/program/policy being evaluated against the targets of disaster risk reduction?
- Are civil society and the market involved in the design, implementation, and evaluation of the measure/program/policy against those targets?

Finally, with regard to the evaluation of the role of informal arrangements within the activities of non-governmental and governmental organizations, evaluation should include specific criteria to address this issue instead of developing specific arguments for evaluation.

14.4.3 Deliberation as Participation

Public participation in political processes may adopt multiple forms. In their evaluation of participation methods, Rowe and Frewer (2000) examine various modes such as referenda, public opinion surveys, public hearings, negotiated-rule making, consensus conferences, citizens' panels/juries, citizen advisory committees, and focus groups. Evaluation is however not considered, and its multiple forms ignored.

An advantageous point of view regarding the participation of social actors in risk management and political processes at large, is provided by Renn (2004) through the concept of cooperation in analytic–deliberative processes. Following Jürgen Habermas, deliberation is understood as a style and procedure of decision making that relies "on mutual exchange of arguments and reflections rather than on decision making based on the status of the participants, sublime strategies of persuasion, or social-political pressure." (Renn 2004: 303). Deliberation ought to occur among participants of different groups placed at the same level, in a neutral environment, minimiszing interferences and pressures by other players, generating dialectical exchange (Fischer 1993) and mutual influence through communication—"communication is interaction, sharing, influencing, and being influenced in turn" (Balkin: 2004: 43)—among participants rather than simply obtaining citizens' input (King et al. 1998). Transparency is thus a *sine qua non* condition to conform to the goals of participatory evaluation, which places greater responsibility on government than on citizens. The most important responsibilities (for which the government should be additionally accountable) are to steer the process, provide available information to participants, and to facilitate equal and extensive participation of societal groups. Transparency is attainable by augmenting the visibility of the evaluation processes using information technologies, such as Internet-based social media, giving voice to dissenting views, and enabling traceability of the evaluation. This would allow understanding of who, when, and what steps were taken and the specific contributions made by players.

But why deliberation? Renn (2004) argues this mode of societal interaction brings new opportunities, although it also poses three challenges: to value the relevance of both local and scientific knowledge, to identify the most appropriate way to deal with uncertainty, and to share the concerns of each social group. These challenges highlight "the importance of procedures, routines, and learning experiences for creating links or networks between the major systems of society" (Renn 2004: 308). According to Renn, deliberation drives a common understanding of the issue based on various types of knowledge, and mutual—empathetic (Mathison 2000)—learning of their respective position, helping society to increase its awareness of the diversity of interpretations and options available. Deliberation also highlights the reasons behind disagreement, as well as the limits of the problem, and with an openness to review one's position if contradicted by evidence (Mathison 2000). In addition, according to Renn (2004), although not evident at first, agreement may possibly arise from deliberation in diverse forms, such as true consensus, tolerated consensus, or compromise. True consensus implies a win-win solution that "serves the 'common good' and each participant's interests and values better than any other solution" (Renn 2004: 309), while a tolerated consensus is reached at the expense of other participants' loss. In contrast, compromise is the result of bargaining a less preferred solution "that will result in an acceptable if not perfect outcome" (Mathison 2000: 87). Ultimately, deliberation is the lubricant (Renn 2004) for consensus.

Deliberation involves the encounter of two rationalities (Renn 2004), the scientific and analytic approach with the communicative or deliberative mode. This implies the tense coexistence of different forms of knowledge, interpretations of information, values, interests and preferences regarding goals, strategies and action—as in resource management (Mitchell 2004)—and engenders both uncertainty and opportunities. Similarly, Mitchell (1997, 2004) distinguishes two fundamental themes in natural resource management: uncertainty and conflict. Notwithstanding, Renn (2004) contends decisions about risk are inescapably based on and driven by knowledge, values, and economic rationale. However, "Does it make sense to replace best institutional knowledge with intuition and personal interest? Indeed, research shows that public perception of probabilities and risks differs considerably from professional analysis" (Renn 2004: 311). It may be true that at times risk perception is refuted by scientific evidence but it is also reasonable to think that failure to implement public policies originates from ignoring and acting against public concerns, by not having displayed channels for deliberation and mutual learning.

14.4.4 Whose Reality Counts? or Who Counts Reality?

Robert Chambers, a strong proponent and similarly critic of standard participatory rural appraisal (PRA), titled his book with the provocative question *Whose Reality Counts? Putting the First Last* (1997) to question the distribution of power between players in rural development, putting the focus on the differences of views and interests between local residents and practitioners. Taking this question further, Estrella and Gaventa formulated a distinct but related question in their work *Who Counts Reality? Participatory Monitoring and Evaluation: A Literature Review* (1998). While Chambers' interest lies in the ownership of reality (a descriptive approach), Estrella and Gaventa speculate about which players should be (a prescriptive approach) defining reality through monitoring and evaluation to help make better decisions, that is, "whose voices and knowledge are used to define success?" (Gonsalves et al. 2000: xi).

Therefore, who is participating in evaluation? And who should be participating? In most processes it is scientists or practitioners (managers in the private sector) performing within a restricted environment, which Brand and Karvonen (2007) identify as the ecosystem of expertise—although understood in a wider perspective—where issues of epistemology, ontology, and power arise. Accordingly, evaluation is seldom interdisciplinary, and therefore does not readily enable understanding the multifaceted complexity of a problem, rarely is performed within an ecosystem of renovated expertise (Brand and Karvonen 2007), or is transdisciplinary and participatory. When a panel of experts is established with a diversity of expertise and experience, the issue of sympathy of interests with the contractor remains. "Hence, the adequate involvement of experts, stakeholders and the public in the risk governance process is a crucial dimension to produce and convey adaptive and integrative capacity in risk governance institutions" (Renn et al. 2011: 237).

Renn (2004) identifies four central actors or systems entitled to participate in risk management: governments, economic players, scientists, and civil society. Worthen et al. (1997) understand evaluation is not something strange to common human choice and agency because, after all, "everyone is, in his or her own way, an evaluator engaged in evaluation of a sort" (p. 6). Renn also extends these classes when recognizing experts do not necessarily need to be solely scientists, and that in government we may distinguish between the knowledge and interests of public officials and policy makers.

The identification and selection of participants in policy processes is not an easy matter, particularly if seeking to be inclusive, for there are no clear boundaries among groups; their interests are dynamic and dependent on the issue, and most of the time participant systems overlap. A multifaceted setting that integrates the main types of knowledge (experts and citizens), the diversity of interests and values (citizens and the market), and viability within the legal framework (public officials), can be operational. Any of the sectors may possibly have other forms of knowledge, interests, values, or understanding of the legal framework. The early involvement of civil society (through neighbourhood and environmental groups) and industry (through local, horizontal and sectoral, and vertical organizations) in risk governance facilitates the achievement of the goals of disaster risk reduction, i.e., the awareness that it is not just the action of government that builds risk but also that of individual and collective action.

Participation is associated with or based on the concept of political legitimacy, whereby not only do governors have the right and duty to make decisions as single lawful representatives, but theirs is the primary voice heard in political processes. Thus, demonstrations in contemporary democracies against liberal reforms and claims for a deeper and more participatory democracy (Glasius and Pleyers 2013) sometimes are not considered legitimate by the government. It is argued that legitimation only emerges from popular mandate (Buck 2013), a statement that is particularly aired when there is a sufficiently strong parliamentary majority. It is the government—as a legitimate institution that comes from broad consensus, power, historical continuity and stability—who dispenses legitimation. Therefore, other societal players who are not granted this attribute are excluded from policy making and barriers to other forms of participation can be easily raised.

Post-democratic forms of governance (Carlarne and Carlarne 2006) seek to generate legitimacy and extend the influence of government by establishing closer relationships with other societal actors based on the mutual exchange of credibility. Carlarne and Carlarne (2006) have documented this new development between intergovernmental and governmental entities in the developed world and local entities. With the progressive erosion of the legitimacy of the modern state, this responds to the adoption of various policy strategies of compensatory legitimation (Weiler 1983)—or the exchange of credibility as Carlarne and Carlarne (2006) claim: an increased "invocation of legal norms and institutions" (p. 262) (legitimation by legalisation), the use of experts in the policy-making process (legitimation by expertise), particularly by using experimentation and planning; and the development and introduction of participatory forms of decision making (legitimation by participation). However, this occurs at the expense of other possible avenues, such as reform and change. Weiler (1983) claims that when the state tries to "regenerate its own legitimation [it may act at] tolerating or actually instituting various schemes for citizen participation" (p. 272), particularly for those affected by the outcomes of the decision-making process.

According to Renn (2004), analytic–deliberative processes are an opportunity for representatives of the scientific, social, political, and economic subsystems to take part in the decision-making process. These four groups embody an acceptable representation of the diversity of public interests and values; they also provide anecdotal and systematic knowledge, secure economic balancing, and facilitate integration into established legal procedures of public policy processes. Renn (2004) also identifies three forms of participant engagement in the deliberation process: the selection of interested representatives, volunteers, and random representative selection. The advantages and disadvantages of each approach are relative to the representativeness, autonomy, availability, leadership, or motivation of the evaluation participants. Any of the selection processes can be internal or external, with or without the intervention of the identifiable groups or the government, although in this case it may lead to the co-option of outside interests into the machinery and particular interests of government (Saward 1992). To avoid interference, the self-selection of groups and representative identification of participants should be preferred. If participation in the evaluation process is extended to include interested groups from the societal system, this would increase equality, representativeness, and available knowledge, but at the same time amplify the complexity of input and interaction, and probably incorporate groups not preferred by the government. Demand-based participation can be supplemented with invitation if key groups are perceived to have been excluded. However, the principles of inclusiveness and legitimacy may clash if the legitimate entity finds itself allocating legitimacy only to politically favoured groups. As Mathison (1996) indicates "there remains, of course, contention about who counts as a stakeholder, just what inclusion means, and the consequences for evaluation when stakeholders are included in novel ways" (p. 86). By all means, access to evaluation should be neutral, and to guarantee this condition, the political promoter of evaluation should make the inclusion/exclusion process transparent.

14.4.5 Roles and Responsibilities

Evaluators should be mediators, asserts Kushner (2001), who maintains there should exist "some distance between those who judge and those who are judged" to preserve a politically neutral territory. Thus, the question arises: is neutrality reachable? The same author replies, stating that all individuals are co-opted into political ambitions. According to Fetterman (1994), as participants progress in their involvement in the process, they exhibit independence and the capacity to be critical. However, powerful groups have more options to influence, co-opt, or neutralise others according to their interests. Thus, facilitators can play a critical role in evaluation, particularly when the number of participants increases sharply as in participatory evaluation. In Fetterman's (1994) words "Evaluators can serve as coaches or facilitators to help others conduct their evaluation" (p. 4) and may become "the agents who give them voice and help them move forward" (p. 5).

When facilitators receive input and information openly provided by a number of participants in the process—as in the case of national or sub-national evaluations—they are required to identify what is relevant and what is not. Furthermore, they must confront differences and integrate regularities "through a continuous and repeated dialogue with stakeholders" (Mathison 2000: 86), according to their subjective judgment, to advance towards final evaluation. This mediated process can help both reconcile divergent views and offset individual subjectivity, but only if transparency is guaranteed by displaying both the jointly agreed consensus and dissenting arguments that have not been included in the course of the evaluation. If the outside facilitator is accountable to the evaluators, and there are quality controls throughout the process that help to show how the evaluation was made, it can be charged with the responsibility of making an evaluation credible and useful, and enable transformative learning (Fetterman 1994).

14.4.6 Deliberation in the Information Society

Mathison (2000) contends deliberation is based on three principles, reciprocity, publicity, and accountability, for it is based on an exchange of reasons and knowledge between participants that transcends private environments and that challenges "positions, evidence, and justification in ... public forums" (p. 87). In recent decades, the Internet and digital technologies have changed the social conditions in which people dialogue (Balkin 2004), and not just the means that citizens use to speak, facilitating "widespread cultural participation and interaction that previously could not have existed on the same scale" (p. 2). However, he warns against misuse, for "at the same time, it creates new opportunities for limiting and controlling those forms of cultural participation and interaction" (p. 2). The impact of the digital revolution on social conditions shifts the focus from democracy to freedom of speech (Balkin 2004), placing it before deliberation and even representative institutions, because the "individual's ability to participate in the production and distribution of culture" (p. 3) has been altered. Thus, it is not possible to put forward deliberation using digital technologies as an avenue for consensus without taking into consideration the progression of individualisation and the challenges to the individual's right to freely speak online and interact in public or restricted environments, either in a one-to-one or one-to-many mode.

Web-based evaluation offers various advantages as compared to on-site evaluation. Above all, it allows increasing the diversity and number of participants—essential in a comprehensive participatory process—although it may challenge the management capacity of the facilitators. Furthermore, web-based evaluation enables the evaluation to be conducted in real-time. Groupware, or collaborative software, allows for accomplishing an array of processes that serve the purposes of computer-supported cooperative work (CSCW), and satisfy the basic requirements of public participation. Andriessen (2003) identifies five CSCW accessible processes that can be grouped into three families: personal interchange processes (communication), group interchange processes (social interaction), and task-oriented processes (cooperation, coordination, and information sharing). Online participatory evaluation may particularly benefit from task-oriented processes, because these applications provide means for knowledge exchange, social interaction, and reach group consensus.

Based on attributes that stem from the time and place of social interaction and exchange, Johansen (1988) distinguishes between four types of CSCW software: face-to-face interaction, continuous interaction, synchronous distributed interaction, and asynchronous distributed interaction (Fig. 14.6). Geographically dispersed participants, with discordant agendas, can be gathered and remotely coordinated in the evaluation process using CSCW software that allows for asynchronous distributed interaction and input to reach consensus. In a society with increased spatial mobility, the specific advantages of collaborative software can enhance the feasibility of participatory evaluation, providing participants the opportunity to flexibly contribute and eliminate the need to gather or synchronize schedules.

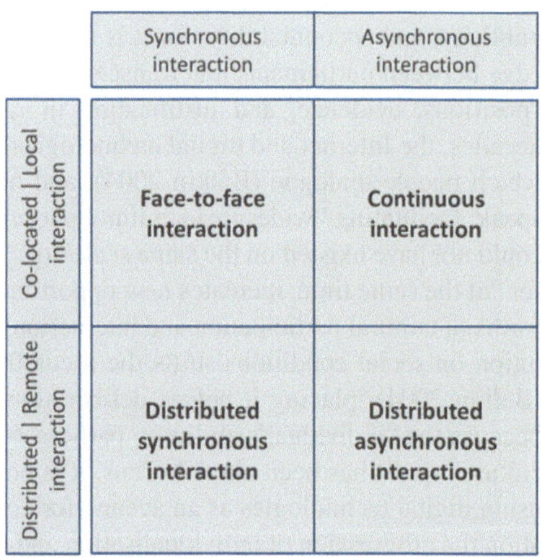

Fig. 14.6 Time-space matrix of computer-supported cooperative work, based on Johansen (1988)

14.4.7 The Continuity of Evaluation

Evaluation that occurs at one time and in one place is almost irrelevant; it is essentially an exercise with limited transformative power and impact on the structure and processes of a political system, and one that may disappoint societal participants. In contrast, a synchronic—cross-territorial—and diachronic comparison provides increased knowledge about the mode of governance as well as the capacities and pathways to transition to more developed risk governance modes. Disappointment also occurs when incomplete information about the limits of the evaluation output is delivered to participants, because higher—or beyond goal—expectations of the evaluation's influence may be created (McKnight and Sechrest 2004).

In diachronic evaluation, each process is a point of reference to measure progress or identify policy barriers that led to stagnation. From this standpoint, stable and durable evaluative criteria for measurement are required. However, if measurement is not seen as the principal goal of evaluation, but a collateral—although essential—objective, stability is not such a valuable property. Moreover, if risk governance is to be adaptive and responsive to change, dynamism should be a preferred property. Adjustment and evolution can therefore be essential properties of criteria or indicators. However, how do we compare? Or better yet, are comparison and learning compatible objectives? Some of the indicators and indices of governance cited above have evolved and changed throughout time, and this has not prevented their implementation as long as the adopted approach has continuity. Evaluation is necessarily dynamic if we understand two important points: that the concept of risk governance is changing and that the practice and modes of governance are continuously adapting.

14.5 The Limits of the Evaluation of Governance

Some of the limitations of participatory evaluation have already been reviewed but it may be wise to re-examine them as, according to Fetterman (1994) in his reference to empowerment evaluation, the participatory evaluation of risk governance is not a panacea. The principal underlying principle of this mode of evaluation can be debilitated or fully impaired with the intervention of unequivocal, but mostly inexplicit constraints, as Newman et al. (2004) state:

Public participation initiatives are sites in which processes of co-option and containment may be present. They are sites in which inequalities of power—between officials and the public, between statutory and voluntary/community organizations—are negotiated. They are uncomfortably situated in a dynamic field of changing relationships between central and local governance and between representative and participative democracy. In these conditions any move towards collaborative governance is likely to encounter difficulties (p. 219).

King et al. (1998) describe various barriers to effective public participation that have been identified by interviewees from both administration and citizenry. They range from the practical realities of daily life (work, mobility, family life) to administrative structure and processes, and techniques of participation, what Rowe and Frewer (2000) identify as the process of participation. Informants reported to King et al. (1998) that forums such as public hearings or meetings are inoperative, because they impede fruitful exchange, while citizen advisory councils and panels are usually biased in composition. Even the most participatory-prone political systems—such as that of the US—erect barriers to prevent "a too-active citizenry" (King et al. 1998: 318). However, "there is also a growing recognition on the part of administrators that decision making without public participation is ineffective" (King et al. 1998: 319). It must be noted that the opposite is also true for other authors, with excessive deliberation immobilising the regulatory system (Renn 2004). As bureaucrats are afraid of—and yet are also in need of—public participation, there is a high risk that any form of citizenry involvement is to be tamed, domesticated, or, in other terms, co-opted, to serve the interests of the player that has the legitimacy of making decisions. Apparently, governance resembles a two-faced Janus that requires and favours participation while concurrently its components, attributes, and functions raise barriers that either block intruders to access policy making or encourages them to remove themselves from the interaction. In the latter case, if governmental motivation for advancement in participation is genuine, then the following principle is applicable: "one change may be to go where the citizens are rather than asking citizens to come to them" (King et al. 1998: 323). As one administrator told the researchers, "We've got to go to them" (King et al. 1998: 323). These authors propose to work in three fronts: empower and educate community members, and "the central issue is one of access", re-educate administrators, and enable administrative structures and processes, perhaps the most difficult to change because "new forms of governance, then, do not displace the old but interact with them, often uncomfortably" (Newman et al. 2004: 218).

Evaluation in the information society addresses specific challenges because information and communication technologies (ICT) are not completely accessible to all citizens. Technical barriers, such as computer ownership, Internet access and speed, and design barriers, will influence a participant's accessibility. Other possible barriers exist, for example, large societal sectors are essentially—or completely—compute or internet illiterate, and participants may be reluctant or resistant to use new technologies for participation. Self-exclusion or limited intervention can be anticipated, as well as the opposite: the self-selection of those who are proficient. Control of digital technologies, and particularly collaborative software, by government or agencies responsible for evaluation may raise suspicion and deter civil society groups from participation. However, some of these barriers can also be identifiable in standard methods of participation or evaluation. While the design of software interfaces can be satisfactorily addressed by, for example, adopting the *Design for All*[1] principles contained in the EIDD Stockholm Declaration 2004 (they

[1] http://www.designforalleurope.org/Design-for-All/EIDD-Documents/Stockholm-Declaration/.

aim to promote eAccessibility in the information society), other barriers have a structural character and have more to do with social and economic equality, as well as governance at large. Thus, the transformative capacity of evaluation can only be understood in the wider context of democratic reform.

Evaluation results neither provide a complete picture of the evaluated matter, nor is their interpretation unequivocal. Such results can be read by different political players according to their perception, knowledge, interests, and expectations, or with regard to the contingent political context. All participants—as citizens—are risk bearers. However, throughout evaluation other interests may be superimposed because although civil society groups will look at evaluation criteria in terms of absolute safety, managers will think in terms of the distributed responsibility of risk or randomness that may "create a paradox that neither of the parties can overcome through means of communication" (Renn 2004: 311). Every so often, disagreement cannot be overcome because the clash between the extremely diverging perspectives of, for example, scientists and laypeople cannot be reconciled (Renn 2004), and will manifest regarding some modern technologies (such as genetic engineering or nuclear energy). The role of knowledge in evaluation is bewildering. It is associated with availability, the accessibility of information for evaluation to and by all participants, and with constraints at social, group, and individual levels. At the first two levels arise relations of power, with democratic transparency attenuating imbalances, but the appropriation of information by some players commonly takes place. At the third level bounded rationality steps in, because the understanding of the problem by each and every party is limited by the human capacity for analysis and interpretation. Different languages are used in processes of dialogue, though the scientific language is dominant. And we should not forget that information sharing is not positive by itself, unless this is translated to a common language; otherwise *ignotum per ignotius* comes into play. In response, Renn (2004) is sceptical of deliberation and maintains that "bringing experts, stakeholders, and citizens to one table has been tried many times, but usually without much success because such a setting makes the deliberations overly complex and unmanageable" (p. 327). He also suggests the properties of complexity, uncertainty, and ambiguity should be treated separately. But how can we separate them if these are mixed attributes of risk governance and of risk, and the limits are indistinguishable? In the same way as complexity, uncertainty and ambiguity emerge in an undifferentiated mode in policy making; it is troublesome to treat and evaluate the dimensions of risk governance separately because of its multi-level, multi-sector, multi-actor and multi-hazard nature. For this reason Rauschmayer et al. (2009) are very sceptical and contend that "the theory of multi-level governance does not yet allow for an evaluation of cross-level interaction" (p. 169), making "it difficult to select the right level(s), sector(s) and time frame(s)" (p. 166). This is particularly so when policy targets are not sufficiently and explicitly defined (Kuhlman 2003) to ensure that they are clearly evaluated with evaluation criteria. Moreover, the different rationalities of the parties do not only enter into conflict during deliberation, tension may continue after evaluation or even be fuelled by the process of interaction with blame, erosion of mutual trust, and loss of credibility (Renn 2004).

In the light of extant limitations and constraints, the evaluation of risk governance should seek satisfactory solutions rather than optimal ones, based on the exchange of multiple limited understandings to reach consensus among viewpoints and perceptions of risk and governance. The conflation of rationalities may produce consensus, dissension, or noise—again, Janus-like—that persist in future interactions. The form of evaluation, the process itself, can thus be decisive to lessen the conflict between the parts, facilitate societal integration, and promote consensus towards the goal of disaster risk reduction.

14.6 Conclusions

The governance of risk has evolved from isolated forms of organization to address risk, such as disaster response, to more integrated modes such as risk management, by continuously adapting policy making to environmental and societal processes. This evolution also includes the integration of sectoral policies, societal players, and levels of government to become more comprehensive, integrated, and functional (Fra.Paleo 2013). This development enables the appropriate responses to the complexity and diversity of the social system (Renn 2004), and to the changes of the natural system.

With the purpose of increasing efficiency and to make a qualitative leap in risk governance within the context of an increasing number of victims and losses caused by disasters, a novel framework for the evaluation of risk governance has been proposed. The DCC (Dimensions-Components-Criteria) framework is assisted by participatory evaluation to encourage the involvement of society in governance, to increase learning, help raise risk awareness, and to advance equality in the exposure to hazards and vulnerabilities. The framework is based on a hierarchical-networked system of criteria that seeks to capture the whole range of public policies (directly or indirectly) that deal with disaster risk reduction. It also includes the complex formal and informal interactions between sectoral policies, levels of government, and societal actors. The dimensions, evaluation pillars, stem from inclusive HFA principles and include—and even go beyond—the key components of the cycle of disaster.

The evaluation system is completed with a participatory process that seeks to be socially diverse and inclusive—including governments, private sector, scientists and civil society organizations—and acts to empower and legitimise different non-governmental groups. Furthermore, it aims to increase the efficiency of public policies dealing with disaster risk reduction and to promote multi-sided learning through mutual influence and communication of different types of knowledge. In seeking a common ground in planning traditions, Friedmann (1987) describes their essence to be: "knowledge should be properly linked to action" (p. 74). In the case of the evaluation of risk governance, the reverse also becomes true: future action is linked to the understanding of how current governance conditions are understood and respond to the expected or required reduction of disaster risk. Deliberation performed within an ecosystem of renovated expertise promotes consensus on the preferred level of risk and a more equitable distribution of risks in a more neutral environment. Despite problems of accessibility, information technologies can play a key role in facilitating interaction among participants within a context of geographical dispersion and discordant agendas. Although the interpretation of evaluation results is rarely unequivocal, governments receive political feedback that pushes improvement and the adjustment of the structures and processes of governance and policies. Although participatory evaluation is not a panacea that can radically change political processes, its transformative capacity and corrective character can be reinforced by recurrent evaluation.

The democratic governance of risk requires that societal groups exchange knowledge, reduce the level of ambiguity, and negotiate both the distribution of risks and the appropriate structure and process of decision making to reduce disaster risk. This mode of governance requires higher levels of social interaction and organization based on the principles of transparency, participation, accountability, effectiveness, and coherence (EC 2001) to which the participatory evaluation of risk governance may contribute.

Acknowledgement Copy editing of this chapter has been made possible with the financial support of the Government of Extremadura and the European Regional Development Fund (ERDF) through the grant GR10071 awarded to the research group on Geospatial Engineering and Urban Heritage of the University of Extremadura.

References

Andriessen, J.H.E. 2003. *Working with groupware. Understanding and evaluating collaboration technology*. New York: Springer.

Balkin, J.M. 2004. Digital speech and democratic culture: A theory of freedom of expression for the information society. *New York University Law Review* 79(1): 1–55.

Berkes, F. 2009. Evolution of co-management: Role of knowledge generation, bridging organizations and social learning. *Journal of Environmental Management* 90(5): 1692–1702.

Birkmann, J., and K. von Teichman. 2010. Integrating disaster risk reduction and climate change adaptation: Key challenges—Scales, knowledge, and norms. *Sustainability Science* 5: 171–184.

Black, P., C. Harrison, C. Lee, B. Marshall, and D. Wiliam. 2002. *Working inside the black box: Assessment for learning in the classroom*. London: GL Assessment.

Brand, R., and A. Karvonen. 2007. The ecosystem of expertise: Complementary knowledges for sustainable development. *Sustainability: Science, Practice, & Policy* 3(1): 21–31.

Buck, T. 2013. Spain's indignados switch focus to workaday struggles. *Financial Times*, 17/05/2013. Available at: http://www.ft.com/cms/s/0/342f1cee-bcaf-11e2-9519-00144feab7de.html#axzz2iXkInUGC.

Burby, R.J., and P.J. May. 2009. Command or cooperate? Rethinking traditional central governments' hazard mitigation policies. In *Building safer communities. Risk governance, spatial planning, and responses to natural hazards*, ed. U. Fra.Paleo, 34–43. Amsterdam: IOS Press.

Burton, I., R.W. Kates, and G.F. White. 1993. *The environment as hazard*, 2nd ed. New York: Guilford Press.

Cardona, O.D. 2006. A system of indicators for disaster risk management in the Americas. In *Measuring vulnerability to natural hazards: Towards disaster resilient societies*, ed. J. Birkmann, 189–209. New York: UNU Press.

Carlarne, C., and J. Carlarne. 2006. In-credible government: Legitimacy, democracy, and non-governmental organizations. *Public Organizations Review* 6: 347–371.

Carreño, M.L., O.D. Cardona, and A.H. Barbat. 2007. A disaster risk management performance index. *Natural Hazards* 41: 1–20.

Chambers, R. 1997. *Whose reality counts? Putting the first last*. London: Intermediate Technology Publications.

Chen, H.T. 1996. A comprehensive typology for program evaluation. *Evaluation Practice* 17(2): 121–130.

Cutter, S.L. 2003. GI science, disasters, and emergency management. *Transactions in GIS* 7(4): 439–445.

DARA. 2011. *Análisis de capacidades y condiciones para la reducción del riesgo de desastres. Costa Rica, El Salvador, Guatemala, Honduras, Nicaragua, Panamá y República Dominicana. Informe sumario*. Madrid: DARA.

de Boer, H., J. Enders, and U. Schimank. 2008. Comparing higher education governance systems in four European countries. In *Governance and performance of education systems*, ed. N.C. Soguel and P. Jaccard, 35–54. Berlin: Springer.

deLeon, P. 1992. The democratization of the policy sciences. *Public Administration Review* 52(125–1): 29.

EC (European Commission). 2001. *European governance: A white paper*. Available at: http://ec.europa.eu/governance/white_paper/en.pdf.

Estrella, M., and J. Gaventa. 1998. *Who counts reality? Participatory monitoring and evaluation: A literature review*, IDS working paper no. 70. Brighton: Institute of Development Studies, University of Sussex.

Fetterman, D.M. 1994. Empowerment evaluation. *American Journal of Evaluation* 15: 1–45.

Fischer, F. 1993. Citizen participation and the democratization of policy expertise: From theoretical inquiry to practical cases. *Policy Sciences* 26(3): 165–187.

Fra.Paleo, U. 2009. On exposure to natural hazards: Revisiting a neglected primal action. In *Building safer communities. Risk governance, spatial planning, and responses to natural hazards*, ed. U. Fra.Paleo, 61–78. Amsterdam: IOS Press.

Fra.Paleo, U. 2013. A functional risk society? Progressing from management to governance while learning from disasters. ISC and UNESCO. In *World social science report 2013. Changing global environments*, 434–438. Paris: OECD/UNESCO.

Fra.Paleo, U. 2014. Principles for the measurement and evaluation of risk governance. In *Earthquake hazard impact and urban planning*, ed. M. Bostenaru, I. Armas, and A. Goretti. Berlin: Springer.

Friedmann, J. 1987. *Planning in the public domain*. Princeton: Princeton University Press.

Fung, A., and E.O. Wright. 2001. Deepening democracy: Innovations in empowered participatory governance. *Politics and Society* 29(1): 5–41.

Glasius, M., and G. Pleyers. 2013. The global moment of 2011: Democracy, social justice and dignity. *Development and Change* 44(3): 547–567.

Gonsalves, J.F., IIRR, and J. Gaventa. 2000. Preface. In *Learning from change: Issues and experiences in participatory monitoring and evaluation*, ed. M. Estrella, xi–xiii. Londres: IDS. Institute of Development Studies.

Greene, J. 2006. Evaluation, democracy, and social change. In *The Sage handbook of evaluation*, ed. I.F. Shaw, J. Greene, and M. Mark, 118–140. Thousand Oaks: Sage.

Hamilton, S., and N.L. Chervany. 1981. Evaluating information system effectiveness—Part I: Comparing evaluation approaches. *MIS Quarterly* 5(3): 55–69.

Hannan, M.T., and J. Freeman. 1984. Structural inertia and organizational change. *American Sociological Review* 49(2): 149–164.

House, E. 1991. Evaluation and social justice: Where are we? In *Evaluation and education: At quarter century*, ed. M.W. McLaughlin and D.C. Phillips, 233–247. Chicago: University of Chicago Press.

Hyde, S.D. 2007. The observer effect in international politics: Evidence from a natural experiment. *World Politics* 60(01): 37–63.

Hyde, S.D. 2010. Experimenting in democracy promotion: International observers and the 2004 presidential elections in Indonesia. *Perspectives on Politics* 8(02): 511–527.

Inter-American Development Bank (IDB). 2010. *Indicators of disaster risk management. Program for Latin America and the Caribbean Summary Report*. Washington, DC: IDB.

ISDR (International Strategy for Disaster Reduction). 2005. *Hyogo framework for action 2005–2015: Building the resilience of nations and communities to disasters*. A/CONF.206/6. Available at: http://www.unisdr.org/2005/wcdr/intergover/official-doc/L-docs/Hyogo-framework-for-action-english.pdf.

Iverson, T., and C. Perrings. 2012. Precaution and proportionality in the management of global environmental change. *Global Environmental Change* 22(1): 161–177.

Johansen, R. 1988. *Groupware. Computer support for business teams*. New York: The Free Press.

Jonassen, D.H., and H.I. Kwon. 2001. Communication patterns in computer mediated versus face-to-face group problem solving. *Educational Technology Research and Development* 49(1): 35–51.

Kaufmann, D., A. Kraay, and M. Mastruzzi. 2010. *The worldwide governance indicators: Methodology and analytical issues*, World Bank policy research working paper no. 5430. Washington, DC: The Brookings Institution – World Bank.

Keen, M., V.A. Brown, and R. Dyball. 2005. *Social learning in environmental management: Towards a sustainable future*. London: EarthScan.

Kettl, D.F. 2000. The transformation of governance: Globalization, devolution, and the role of government. *Public Administration Review* 60(6): 488–497.

King, C.S., K.M. Feltey, and B.O.N. Susel. 1998. The question of participation: Toward authentic public participation in public administration. *Public Administration Review* 58(4): 317–326.

Kuhlman, S. 2003. Evaluation of research and innovation policies: A discussion of trends with examples from Germany. *International Journal of Technology Management* 26(2/3/4): 131–149.

Kushner, S. 2001. Culture, standards, and program qualities. In *Vision of quality: How evaluators define, understand and represent program quality*, ed. A.P. Benson, D.M. Hinn, and C. Lloyd, 121–134. Bingley: Emerald.

MacDonald, B. 1976. Evaluation and the control of education. In *Curriculum evaluation today: Trends and implications*, ed. D. Tawney, 125–136. London: Macmillan.

Mathison, S. 1996. *The role of deliberation in evaluation*. Annual meeting of the American Evaluation Association, Atlanta, November 1996.

Mathison, S. 2000. Deliberation, evaluation, and democracy. *New Directions for Evaluation* 85: 85–89.

May, B., and R. Plummer. 2011. Accommodating the challenges of climate change adaptation and governance in conventional risk management: Adaptive collaborative risk management (ACRM). *Ecology and Society* 16(1): 47.

McCarthy, D. 2009. *A critical systems approach to socio-ecological systems: Implications for social learning and governance*. Saarbrücken: VDM Verlag.

McKnight, K.M., and L. Sechrest. 2004. Program evaluation. In *Comprehensive handbook of psychological assessment, Vol. 3: Behavioral assessment*, ed. S.N. Haynes and E.M. Heiby, 246–266. Hoboken: Wiley.

Medsker, L., M. Tan, and E. Turban. 1995. Knowledge acquisition from multiple experts: Problems and issues. *Expert Systems with Applications* 9(1): 35–40.

Mileti, D.S., T.E. Drabek, and J.E. Haas. 1975. *Human systems in extreme environments: A sociological perspective*. Boulder: Institute of Behavioral Science.

Mitchell, B. 1997. *Resource and environmental management*. Toronto: Longman.

Mitchell, T. 2003. *An operational framework for mainstreaming disaster risk reduction*, Disaster studies working paper 8. London: Benfield Hazard Research Centre, University College London.

Mitchell, B. (ed.). 2004. *Resource and environmental management in Canada: Addressing conflict and uncertainty*, 3rd ed. Don Mills: Oxford University Press.

Moote, M.A., M.P. McClaran, and D.K. Chickering. 1997. Theory in practice: Applying participatory democracy theory to public land planning. *Environmental Management* 21(6): 877–889.
Morgan, M.G., and M. Henrion. 1990. *Uncertainty: A guide to dealing with uncertainty in quantitative risk and policy analysis*. New York: Cambridge University Press.
Morrison, J. 2003. ABC of learning and teaching in medicine: Evaluation. *British Medical Journal* 326: 385–387.
Newman, J., M. Barnes, H. Sullivan, and A. Knops. 2004. Public participation and collaborative governance. *Journal of Social Policy* 33(2): 203–223.
Nordhaus, W.D. 1994. Expert opinion on climate change. *American Scientist* 82: 45–51.
Parker, D. 1999. Criteria for evaluating the condition of a tropical cyclone warning system. *Disasters* 23(3): 193–216.
Quarantelli, E. 1991. *Patterns of sheltering and housing in American disasters*, Preliminary paper no. 170. Newark: Disaster Research Center, University of Delaware.
Rallis, S.F., and G. Rossman. 2000. Dialogue for learning: Evaluator as a critical friend. *New Direction in Program Evaluation* 86: 81–92.
Rauschmayer, F., A. Berghöfer, I. Omann, and D. Zikos. 2009. Examining processes or/and outcomes? Evaluation concepts in European governance of natural resources. *Environmental Policy and Governance* 19: 159–173.
Renn, O. 2004. The challenge of integrating deliberation and expertise. Participation and discourse in risk management. In *Risk analysis and society: An interdisciplinary characterization of the field*, ed. T.L. McDaniels and M.J. Small, 289–366. Cambridge: Cambridge University Press.
Renn, O., A. Klinke, and M. van Asselt. 2011. Coping with complexity, uncertainty and ambiguity in risk governance: A synthesis. *Ambio* 40(2): 231–246.
Rice, S.E., and S. Patrick. 2008. *Index of state weakness in the developing world*. Washington, DC: The Brookings Institution.
Roorda, N., C. Rammel, S. Waara, and U. Fra.Paleo. 2009. *AISHE 2.0 Manual. Assessment instrument for sustainability in higher education*. http://www.slideshare.net/NRoorda/aishe-20-manual.
Roughgarden, T., and S. Schneider. 1999. Climate change policy: Quantifying uncertainties for damages and optimal carbon taxes. *Energy Policy* 37: 415–429.
Rowe, G., and L.J. Frewer. 2000. Public participation methods: A framework for evaluation. *Science, Technology & Human Values* 25(1): 3–29.
Saward, M. 1992. *Co-optive politics and state legitimacy*. Aldershot: Dartmouth.
Simpson, D.M., and M. Katirai. 2006. *Indicator issues and proposed framework for a disaster preparedness index (DPi)*, Working paper 06–03. Louisville: Center for Hazards Research and Policy Development, University of Louisville.
Springett, J., and N. Wallerstein. 2008. Issues in participatory evaluation. In *Community based participatory research for health: Process to outcomes*, 2nd ed, ed. M. Minkler and N. Wallerstein, 199–220. San Francisco: Jossey Bass.
Stallings, R.A. 1990. Media discourse and the social construction of risk. *Social Problems* 37(1): 80–95.
Stetler, C.B., M.W. Legro, C.M. Wallace, C. Bowman, M. Guihan, H. Hagedorn, B. Kimmel, N.D. Sharp, and J.L. Smith. 2006. The role of formative evaluation in implementation research and the QUERI experience. *Journal of General Internal Medicine* 21(2): S1–S8.
Stivers, C. 1990. Active citizenship and public administration. In *Refounding public administration*, ed. G.L. Wamsley, R.N. Bacher, C.T. Goodsell, P.S. Kronenberg, J.A. Rohr, C.M. Stivers, O.F. White, and J.F. Wolf, 246–273. Thousand Oaks: Sage.
UNISDR. 2010. *The 10 essentials for making cities resilient*. Available at: http://www.unisdr.org/campaign/resilientcities/toolkit/essentials.
Weibelzahl, S. 2005. Problems and pitfalls in evaluating adaptive systems. In *Fourth workshop on the evaluation of adaptive systems*, ed. S. Weibelzahl, A. Paramythis, and J. Masthoff, 57–64. Edinburgh.

Weichselgartner, J. 2001. Disaster mitigation: The concept of vulnerability revisited. *Disaster Prevention and Management* 10(2): 85–95.

Weiler, H. 1983. Legalization, expertise and participation: Strategies of compensatory legitimation in educational policy. *Comparative Education Review* 27(2): 259–277.

Wolter, S.C. 2007. Purpose and limits of a national monitoring of the education system through indicators. In *Governance and performance of education systems*, ed. N.C. Soguel and P. Jaccard, 57–84. Berlin: Springer.

Woodward, R.T., and R.C. Bishop. 1997. How to decide when experts disagree: Uncertainty-based choice rules in environmental policy. *Land Economics* 73(4): 492–507.

Worthen, B.R., J.R. Sanders, and J.L. Fitzpatrick. 1997. *Program evaluation. Alternative approaches and practical guidelines*, 2nd ed. New York: Addison Wesley Longman.

Urbano Fra.Paleo is Professor in Human Geography at the University of Extremadura, Spain. He is also Visiting Professor at the UN-mandated University for Peace (UPEACE), Costa Rica. He worked at the United States Geological Survey in Denver (1995) and Hawaii (1999) and was Research Associate at the Environment Institute of the University of Denver (1996). Fra.Paleo was Fellow of the American Geographical Society Library, University of Wisconsin-Milwaukee, in 2005. In 2009, he received the Award for Sustainable Actions in Social Entrepreneurship and the Innovation Award by the University of Santiago de Compostela. Since 2011, he is a member of the Scientific Committee of the Integrated Risk Governance (IRG) Project, an initiative within the International Human Dimensions Programme on Global Environmental Change (IHDP), and of the Disaster Risk Reduction Thematic Group within the International Union for Conservation of Nature (IUCN). Urbano Fra.Paleo has edited the book *Building Safer Communities. Risk Governance, Spatial Planning and Responses to Natural Hazards* in 2009, a volume that addresses the adoption of spatial planning as a proactive approach to deal with risk exposure. His interests lie in the collaborative evaluation of risk governance and the development of proactive strategies of mitigation and adaptation to risk from natural hazards.

Weichselgartner, J. 2001. Disaster mitigation: The concept of vulnerability revisited. *Disaster Prevention and Management* 10(2): 85–95.

Weiler, H. 1983. Legalization, expertise and participation: Strategies of compensatory legitimation in educational policy. *Comparative Education Review* 27(2): 259–277.

[illegible] 2003. Proposal for the [illegible] of a national monitoring of the educational system [illegible]

[illegible]

[illegible]

[illegible]

Part II
Governance in Regions and Domains of Risk

Chapter 15
Climate Governance and Climate Change and Society

Geoff O'Brien and Phil O'Keefe

15.1 Introduction

There is a great irony to the climate debate. When the United Nations Framework on Climate Change (UNFCCC) came into force it was clear that global business interests, particularly in the energy sector, were not too convinced by the underlying assumptions in the Convention. Firstly, they did not like the idea of common but differentiated responsibility, meaning that the developed world would carry the greater burden for emission reductions. Secondly, they were dismayed by the lack of any real financial incentive. This left global business interests cold. Yes, there was some scope for technology transfer, but essentially there was little in it for business. For governments, there was the opportunity to make political capital as green issues appeared to be moving up the public agenda. Politicians could parade themselves on this global platform as visionary leaders ready to stretch out a hand to help the weak and the vulnerable, whilst simultaneously saving the world from the horrors of climate change. This soon became a charade. Substantial action has not appeared.

Fortunately for both business and government the Intergovernmental Panel on Climate Change (IPCC) effectively trashed the idea of Annex I nations cutting emissions below the 1990 level by 2020. This led to the Kyoto Protocol and gave both government and business some of what they wanted. For government, it gave flexible targets and for business a market based approach as an incentive. This should have produced real momentum in UNFCCC. Not so.

In UNFCCC, within developed countries there were divided opinions in the United States, together with Australia, Canada and Russia, and Europe. Japan sat rather on the sidelines. Within developing countries, the divisions were even greater.

G. O'Brien (✉) • P. O'Keefe
Department of Geography, Faculty of Engineering and Environment,
School of Built and Natural Environment, Northumbria University,
Ellison Place, Newcastle upon Tyne NE1 8ST, UK
e-mail: geoff.obrien@northumbria.ac.uk; phil.okeefe@northumbria.ac.uk

U. Fra.Paleo (ed.), *Risk Governance*, DOI 10.1007/978-94-017-9328-5_15

China and India, essentially two rapidly industrialising nations with one third of the world's population, wished to have similar opportunities to that already enjoyed by developing countries to use hydrocarbons. Arab members of OPEC sought compensation for any loss of oil or gas markets. Brazil, Malaysia and Indonesia fought against UNFCCC because they found common cause, not in the climate issue but because they objected to the parallel Convention on Biological Diversity, commonly known as the biodiversity Convention, that was being opened at the same Earth Summit and which would restrict their export of tropical hardwoods. Then there was the Alliance of Small Island States (AOSIS), an intergovernmental organization of low-lying coastal and small island countries, including Bangladesh and Egypt, established in 1990, to consolidate the voices of Small Island Developing States (SIDS) to address global warming. And then there were the rest of the developing countries organized by G77. It came as no surprise that UNFCCC did not specify which problem should be faced, or how. That required the Kyoto Protocol.

The UNFCCC is an international environmental treaty with the goal of achieving the stabilization of greenhouse gas concentrations at a level that would prevent dangerous anthropogenic interference with the climate system. The Kyoto Protocol is a protocol to UNFCCC, aimed at fighting global warming.

The Protocol was initially adopted on 11 December 1997 in Kyoto, Japan, and entered into force on 16 February 2005. As at September 2011, 191 states have signed and ratified the protocol. Under the Protocol, 37 countries (Annex 1 countries), essentially the developed world, committed themselves to a reduction of four greenhouse gases (GHG) (carbon dioxide, methane, nitrous oxide, sulphur hexafluoride) and two groups of gases (hydrofluorocarbons and perfluorocarbons) produced by them: all member countries give general commitments. The protocol allows for several 'flexible mechanisms', such as emissions trading, the Clean Development Mechanism (CDM) and Joint Implementation (JI) to allow Annex 1 countries to meet their GHG emission limitations by purchasing GHG emission reduction credits from elsewhere.

It is clear that initially the focus of UNFCCC was on mitigation. This is at odds with the overall aims of the Convention. The Convention acknowledges sustainable development as the context for dealing with the climate problem, in that it must address both current consequences and future risks. In short it must not jeopardise opportunities today or in the future. We need to protect ourselves against nature (adaptation) whilst protecting nature from us (mitigation) (Stehr and von Storch 2005). There needs to be a twin track approach with equal emphasis on mitigation and adaptation. Mitigation is essentially a process of 'stopping' the destruction of nature and is usually a technological response: it is capital investment favoured by both business and governments. Adaptation is adjusting to what nature has already done and is doing: it is essentially about changing livelihoods and is a recurrent cost: adaptation is favoured by developing countries and NGOs.

It is clear that the drafters of the Convention were more focused on mitigation. Adaptation to climate change is recognized within the Convention but was specifically argued as a requirement for the Least Developed Countries (LDCs). The LDCS were then grouped with the OPEC countries as they argued they would require help to adjust their economies as mitigation measures would lead to a downturn in demand for fossil fuels. There are two problems with this grouping. Firstly, it makes little sense to group some of the poorest nations in the world with rich oil producing states. Poorer nations typically do not have export led economies and their livelihood systems, typically agriculture and pastoralism, would be threatened by climate change. Excessive use of hydrocarbons is the driver of climate change. Secondly, demand for hydrocarbons had been increasing up to the introduction of the Convention and was projected to continue increasing, militating against any immediate need for structural adjustment to the economies of OPEC members.

The governance of the Convention is realised through the Convention of the Parties (COP). This body comprises ministerial representatives of member states. The COP meets annually and the first meeting was held in 1996 in Berlin. Because of the link between oil producing states and LDCs little progress on adaptation was made until the 2004 COP10 meeting in Buenos Aires. COP10 (sometimes referred to as the Adaptation COP) separated adaptation under the Convention (adverse effects) from the impact of response measures on oil producing countries (Ott et al. 2005). However progress on adaptation has been slow. The meeting of the Durban COP in 2011 finally agreed an adaptation fund, though there is still uncertainty about the source of these funds. Progress has been slow and has not been helped by what appears to be a concerted campaign by climate deniers. This is despite economists such as Stern arguing that it is more cost effective to address climate change now than to deal with the consequences (Stern 2006).

15.2 Climate Change Denial

There appears to have been a concerted attack on climate scientists by sceptics, allied to fossil fuel interests and industries which use fossil fuels. This campaign is extraordinary, going well beyond issues of peer review and reasonable doubt. This attack has been designed to undermine credibility of both climate scientists and UNFCCC processes. As opposed to scepticism, the attacks have been based on the view that climate change is some sort of global conspiracy. Michael Mann and colleagues, who worked on the 'hockey stick graph', have, however, been repeatedly vindicated, as his work clearly presents evidence of climate change. In 2011, he fought off an attempt by a pro-industry and industry funded think-tank, American Tradition Institute, to gain access to thousands of private emails. Other prominent climate scientists, for example James Hansen, have suffered similar attacks (Goldenberg 2011).

Political lobbyists have also been active. In the UK, the former Chancellor of the Exchequer, Lord Lawson, who heads a group, Global Warming Policy Foundation, which offers a range of disinformation about climate has been challenged by Freedom of Information request to reveal the foundation funders (Hickman 2012).

Lobbying in the US has stopped progress on climate change legislation and the current frontrunners for the Republican Party Presidential nomination have shied away from claiming that climate change is produced by human actions. What we have seen is a failure of the US government as it buckles under the weight of lobbyists. The outcome was the weak agreement at Copenhagen, no doubt influenced by the ClimateGate email hacking saga. Even the EU has been knocked off course when, in 2011, a group of UK Conservative Members of the European Parliament (MEPs) planned to derail the EU plan to raise the emission reduction target for greenhouse from 20 to 30 % by 2020. Of course, the usual claims were made that the higher cuts would damage economic prospects particularly by those industries in the generation sector and those intensive users of energy. They neglected to mention that many large energy users and producers are sitting on mountains of carbon credits that could be used to sustain a transition period to low carbon sources. Remarks made by the Polish EU Budget Commissioner Janusz Lewandowski, that there was a question mark over climate change, effectively halted the move to a higher emission reduction target (Euractiv 2011).

There is evidence that the climate community is winning its way back. The Fourth Assessment Report (FAR) by IPCC gave the strongest message that there is a discernible signal of human interference with the climate signal (IPCC 2007). This has gained support. A recent study by the Berkeley Earth Group, partly funded by climate sceptics, agrees with climate change modellers that there is a discernable increase in temperature caused by human actions (Berkley Earth Temperature Group 2011. There are three broad reasons.

Firstly, politics and economics. Getting all signatories to agree, let alone getting them to start adapting the global energy system to low carbon sources, is fiendishly difficult. At present the rich nations are transfixed by the economic crisis and they appear to have little appetite to tackle the climate crisis. This is despite warnings from Lord Stern and others that it is more cost effective to deal with the problem now as opposed to dealing with the consequences later (Stern 2006). Governments are fearful of being first. They are caught in what is known as the Giddens' paradox -that is governments won't act until something actually goes wrong- by then it may be too late (Giddens 2009).

Secondly, distorted markets. Renewable energy is all but unlimited, but transformation into usable energy supplies appears more costly than fossil fuel polluting alternatives. This hides the fact that fossil fuel industry received $409 bn in handouts in 2010, compared with $66 bn for renewable technologies. The external costs of fossil fuel use are not reflected in price mechanisms. The market is distorted. This is evident in Cap and Trade, the mechanism lauded by many as a powerful tool to address climate change. But there is little evidence of its effectiveness. The EU Emissions Trading System (ETS) is claimed to have contributed 2–5 % carbon reductions in the pilot phase (2005–2007) (Ellerman and Buchner 2008; Ellerman et al. 2010). Others claim that any reductions were realised through carbon offsetting, meaning that no actual carbon reductions in the EU can be attributed to carbon trading (Gilbertson and Reyes 2009). The value of the traded volume in the EU ETS was US$50bn in 2007 and this rose to just under US$120bn in 2010 (World Bank 2011). This raises the question of effectiveness. Would it not be more effective to use some of this money to invest in alternatives and adaptation? Mol (2012) argues that the commodification of carbon can contribute to mitigation providing that state and non-state actors strongly advocate climate change mitigation. Given the track record of the financial sector there is a danger that without proper safeguards and regulation that carbon credits could become entangled in dubious financial schemes, thus undermining any credibility they may have.

Thirdly, technological lock-in. We are still trapped in thinking about energy systems where high density fuels are transformed into energy services, often at a gigantic scale. Carbon Capture and Storage (CCS) is hailed as the new future for coal. However, it is only experimental at present and there are huge uncertainties, both technical and legal, about storage. Shale gas appears to be the next plentiful source and is predicted to provide one fifth of global energy supplies by 2035. But this will also require CCS. This still leaves the issue of fracking, the extraction method for shale gas, and its adverse impacts on groundwater sources and geological stability. In a densely populated country like Britain, where mining subsidence occasionally affects housing, there is reason to be cautious about introducing fracking.

15.3 Climate Change and Disaster Management

Some new areas are beginning to emerge in the disaster management field. Disaster management can be characterised as being reactive, that is responding after an event. In the developed world it is embedded within government structures and is increasingly professionalized. It typically has a Command and Control structure (O'Brien 2006). Throughout much of the developing world the approach is similar with some exceptions where the focus is on community based disaster management. However, through the work of UNISDR, there is a shift from reaction to prevention.

Internationally, thinking about disaster management has shifted from 'managing' disasters to prevention. It is now recognized that hazards have a significant impact on societies and that natural hazards have a disproportionate impact on the developing world. Between 1991 and 2005 almost a million deaths and tens of billions $US in damages resulted from the impact of natural hazards (Schipper and Pelling 2006).

The international response to this growing trend was the declaration that the 1990s would be known as the International Decade of Disaster Reduction (IDNDR) by adopting resolution 44/236. A conceptual shift was signalled by a mid-term review of IDNDR, the Yokohama Strategy in 1994, which requested practitioners and organizations to adopt a new approach that emphasized prevention rather than just reactive action. This changed disaster management to disaster risk management, often described as preparedness practice (Sperling and Szekely 2005).

In 2000 UNISDR was established to respond to the challenges identified by the Yokohama Strategy. The mission of UNISDR is to build disaster resilient communities by promoting increased awareness of the importance of disaster risk reduction (DRR) as an integral component of sustainable development, with the goal of reducing human, social, economic and environmental losses due to natural hazards and related technological and environmental disasters (UNISDR 2004). UNISDR places prevention at the heart of disaster management and articulates the view that Disaster Risk Reduction (DRR) should determine how we conceptualise disaster management. UNISDR defines DRR as:

> The concept and practice of reducing disaster risks through systematic efforts to analyse and manage the causal factors of disasters, including through reduced exposure to hazards, lessened vulnerability of people and property, wise management of land and the environment, and improved preparedness for adverse events (UNISDR 2009).

UNISDR identifies the two components of risk: hazard and vulnerability. Further it acknowledges that adverse events will occur by advocating a focus on preparedness. This affects the policy framework. Perhaps the most significant shift is the focus on people. There is acknowledgement that governments are the most appropriate body to promote and manage DRR, but this cannot be done without the participation of people and communities. Essentially this marks a shift from a Command and Control approach to an approach that can be described as a partnership between government and people.

The mid-term review of UNISDR produced the *Hyogo Framework for Action (HFA) 2005–2015: Building the Resilience of Nations and Communities to Disasters*, which was adopted at the World Conference on Disaster Reduction (UNISDR 2005). This framework sets out a plan of action for developing national platforms for DRR. What can be discerned is the shift to self-reliance or resilience where the emphasis is placed on the local. Adaptation measures can be viewed as being unique to location and socio-economic conditions. It is clear that poorer communities will have less capacity to adapt.

The issues of climate change and disaster risk reduction raise issues of environmental governance. From a governance perspective, there is a parallel to military practice where small units are scattered throughout the operational theatre and exercise control of their locale and are able to coordinate through communications equipment. This can be likened to a top-down enabling environment that encourages local level action.

In 2006 UNISDR was invited by UNFCCC to discuss areas of common interest (UNISDR 2006). UNISDR has many ideas, but no money, about how people and communities can prepare for disruptive events. UNFCCC has access to considerable funding but lacks skills in people preparedness. The outcome of this is the *Special report on managing the risks of extreme events and disasters to advance climate change adaptation*. The *Summary for Policy Makers* has been published and the full report became available in February 2012 (IPCC 2012). This report epitomises DRR.

There are parallels with Community Based Disaster Management (CBDM), sometimes known as Community Based Disaster Risk Management (CBDRM), that has emerged in South East Asian countries. A good example of this is Bangladesh. In 1970 and 1991 cyclones resulted in deaths of 500,000 and 138,000 respectively. Following the 1970 disaster the government and other agencies began to implement Bangladesh Cyclone Preparedness Programme, a bottom-up programme aimed at communities reducing their vulnerabilities and enhancing resilience.

The national government worked in partnership with other agencies to develop a community based approach to disaster management. They developed a system of early warnings and cyclone shelters. These measures are not fully implemented throughout Bangladesh, but where they have been implemented the impact on death rates has been significant (Schultz et al. 2005; Akhand 2003).

15.4 Issues in Climate Governance

The equity issues of climate change are broadly twofold, namely the right to 'grandfather' -equal access to pollution of the commons over time- regimes and the issue of global sharing of the carbon sink. The issue of 'grandfather' relates to developing states having the same rights to carbon and other emissions as the developed world which has built its wealth on an intensive use of energy resources: this particularly applies to the BRICS countries (Brazil, Russia, India, China and South Africa) which have significant levels of hydrocarbon resources, especially coal. The global sharing of the carbon sink implies not just a reduction from carbon emissions by individual nation state but the idea that, rather than nation state allowances, a per capita allowance be generated giving every person equity in access to the carbon commons.

Disaster risk reduction is increasingly framed in a rights agenda with the push being to draft international law effectively, so that individual and community rights to not suffer risk. Drafting negative laws is always problematic and 'soft' laws, where causal responsibility is difficult to 'prove', are not justiciable. There may be better reason, however, to rely on the Universal Declaration of Human Rights and subsequent conventions which have entrenched these in domestic law, in order to confront the problem of vulnerability. This would avoid the difficulty of a burden of proof for showing that no harm has been done; and it would, also avoid the difficulty of a standard of proof, i.e., the question of what goes far enough towards eliminating any doubt that the case has been proved. Technological disasters, such as Bhopal, have much to teach us: it is much more difficult to apportion blame in the recent Japanese tsunami with its mix of technological and natural causes. It is almost impossible in natural disaster such as the Asian tsunami. Most legal advice suggests that a human rights stance on risk would not make good operational law. With regard to climate change, UNFCCC requires individual nation states to report their emissions and their broad approaches to mitigation and adaptation. There are common methodologies for calculating emissions and model exemplars of mitigation and adaptation strategies. While never perfectly accurate, these efforts at reporting show a commitment to sharing an information base. The difficulties arise because, unlike the Montreal Protocol, there are too many pathways to carbon equivalents pollution and too many technologies, of different environmental value, to judge impact accurately, especially at regional or national level. There are questions of modelling scale.

The real problems of environmental governance emerge with the Kyoto Treaty itself, the vehicle designed to implement UNFCCC. There are difficulties over the non-signatories to the original Kyoto agreement, the withdrawal (Canada) from the original agreement and then the failure to put in place a replacement which reflects conflicting national interests. At present there is an intention to sign, by 2020, a global agreement but it is a long road ahead. What makes a long road is the US will not bind itself to international agreements and the BRICS (Brazil, Russia, India, China and South Africa) show no commitment to global solutions.

The UNFCCC adopts a principle of 'common but differentiated responsibilities'. The parties agreed that the largest share of historical and current global emissions of greenhouse gases originated in developed countries, that per capita emissions in developing countries are still relatively low, although not trivial, and that the share of global emissions originating in developing countries will grow to meet social and development needs. But 'common and differentiated responsibilities' enshrines difference that makes future agreement difficult.

The existing Kyoto Protocol has several 'flexible mechanisms', such as carbon trading, the CDM and JI that allow Annex I countries, the largely developed world, to meet their GHG emission limitations by purchasing GHG emission reductions credits from elsewhere, through financial exchanges, projects that reduce emissions in non-Annex I countries, from other Annex I countries, or from Annex I countries with excess allowances.

There are criticisms of carbon trading. Carbon trading encourages business-as-usual as expensive long-term structural changes will not be made if there is a cheaper source of carbon credits. Cheap 'offset' carbon credits are frequently available from the developing countries, where they may be generated by local polluters at the expense of local communities.

The CDM allows industrialized countries to invest in emission reductions wherever it is cheapest globally. Between 2001, which was the first year CDM projects could be registered, to 2012, the end of the Kyoto commitment period, the CDM is expected to produce some 1.5 billion tons of carbon dioxide equivalent (CO_2e) in emission reductions. Most of these reductions are through renewable energy, energy efficiency, and fuel switching. In short, CDM can allow industrialized countries to avoid action at home.

One of the most vexing issues is that of land use, land-use change and forestry (LULUCF). This is defined by the United Nations Climate Change Secretariat as "A greenhouse gas inventory sector that covers emissions and removals of greenhouse gases resulting from direct human-induced land use, land-use change and forestry activities." (UNFCCC undated). The baseline measurement of existing conditions is difficult, especially the vegetative matter that is underground. More importantly, this agenda relates to the Biodiversity Convention which itself is problematic between countries. There is, therefore, much which is debatable in regard to making any international agreement on climate change or measures to cope with it. Environmental governance needs law to support its efforts; but law needs certainty before it can be effective.

In many senses, as we have already noted, the adaptation agenda is stronger in the Disaster Risk Reduction area than in the climate change one. The DRR agenda, with its emphasis on local deployment of resources, is not, however, beyond criticism. Emphasising the local allows the nation state to deny responsibility for the local while simultaneously claiming it has taken action. This is pernicious as there is a wealth of material, including a national survey (Aryal 2012) over the last century in Nepal showing that the overall impact of small scale local disasters is greater than that of the large scale disasters; it is the latter, however, that guide most policy making.

15.5 Towards a Kind of Progress for Climate Governance

What has emerged from this analysis is that governance of climate change has undergone change. Globally UNFCCC led the debate but the withdrawal of the USA from the Kyoto Protocol signalled the end of an international consensus. Though President Bush cited economic reasons for withdrawal it is clear that there has been a growing wave of climate scepticism in the USA. It is probable that this scepticism has been fostered by industrial interests. The rise of the right following the election President Obama through the vehicle of the Tea Party has seen increasing scepticism amongst potential Republican Congressional and, more recently Presidential candidates. The first clear signs of increasing scepticism were seen at the Copenhagen COP, where hopes of a successor to the Kyoto protocol were dashed when President Obama failed to secure Congressional agreement for a climate change deal. The outcome of Copenhagen was a limp form of words in the Copenhagen Accord agreeing that efforts should be made to maintain global average temperature increase below two degrees Celsius (Copenhagen Accord 2009). Some progress was made on adaptation. Nevertheless, this should be seen as a failure of international policy.

It is not clear how effective the efforts of the climate sceptics have been in influencing this shift. The hacked e-mails of the ClimateGate affair offered an ominous portend of what was to come. But it should be noted that there has been a history of lobbying by the energy sector. In addition they have funded groups promoting climate scepticism which have continued to attack the claims of IPCC and consequently the work of UNFCCC. This is a failure of business practice. But it was the economic crash of 2008 and the current economic crisis that seems to have preoccupied political leaders and, as evidenced by the Durban COP, there seems to be little interest in taking a path toward a low carbon economy. This is institutional failure in the case of the financial sector and political failure of many of the bigger players in UNFCCC.

Prior to Copenhagen there were a number of significant publications; e.g., *IPCC Fourth Assessment Report*, the *Stern Review*, *The Politics of Climate Change and World at Risk* (IPCC 2007; Stern 2006; Giddens 2009; Beck 2009). The IPPC report brings up to date the scientific evidence on climate change. Stern investigates the economics of climate change, arguing that it would be beneficial economically to act in response. Giddens explores the politics of climate change, highlighting the great danger in carrying on with the usual forms of behaviour; and Beck explores world risk. Each of these works is important for the climate debate. They all offer, in a number ways, thoughts on what may happen and how to make a transition from a high to a low carbon society.

The IPCC report clearly states that the evidence of human induced warming cannot be denied. It goes on to say that with 'business as usual', that is without effective action to reduce greenhouse emissions, the stock of these gases could treble. By the end of this century there would be a 20 % risk of a more than 5 °C increase in average global temperature. This projection was confirmed by a study by Anderson and Bowes (2008) which concluded that it was highly unlikely that the current policy framework would be able to keep the average global temperature rise below 2 °C. There is now a consensus that we should be preparing for a rise of at least 4 °C.

Such a rise would mean that adaptation as currently framed is totally inadequate. Such a temperature would transform our world and not necessarily in a positive way. It would mean severe disruption, as the *Copenhagen Diagnosis* indicates. This report was produced by scientists who had contributed to previous IPPC assessment reports. Its purpose was to inform the Copenhagen COP of the latest scientific evidence on climate change. It should focus all policy-making and decision-making minds. Tipping points, where positive feedbacks, such as the melting of the Arctic ice-cap, acts to reinforce climate change will severely disrupt ecosystems to the point where we may experience irreversible change. The impact on global agriculture, animal husbandry, flora and fauna and people would be immense. Such dislocations are likely to lead mass migration (Copenhagen Diagnosis 2009). It is reminiscent of Lovelock's vision of ragged groups of survivors trekking towards the poles in search of favourable living conditions (Lovelock 2006).

IPCC was courageous to highlight the consequence of inaction. Yet it is amazing that political leaders fail to take any relevant action. Neo-liberal capitalism seems to have locked political leaders into the view that we can expand indefinitely along the same path in a world where there are limits and severe consequences if we exceed those limits. This is clearly an example of the political classes being unable to think differently. It is also an example of the ill effects of globalization, in that these limited views have been distributed by the mass media so that 'the ordinary voter' everywhere is kept moving on the same path of consumption economics, regardless of the costs which he is incurring for his children.

Stern places emphasis on technology as both the cause and the cure. Technology is neutral. The question is how we use technology. There must be a shift in public attitudes towards a genuine concern for the world in which we live. Stern offers no recipe. What Stern does argue for are policy interventions to correct market failures. This is the nub of the issue. The basic premise of our market-based economy is that humans make rational decisions. Nothing could be less true. We are influenced by trends and fashions. We can be compulsive. We may follow someone who promises a better future with little evidence of how it may be attained. Part of the strength of climate change denial is that scepticism about the future envisaged by climate science allows the sceptics to say 'we know what we are talking about. They are giving you less evidence of reality than we are giving you.' There are plenty of people peddling 'get-rich-quick' schemes. The failure of economic practice is at the core of the economic crisis. It has allowed the financial sector to engineer financial models that gave the illusion of minimising risk when in fact all they did was *to conceal risk.* The financial sector, driven by greed and a lack of regulation, has been able to portray itself as the engine-house of the world economy. The so-called 'Masters of the Universe' soon proved to be a chimera, an illusion. Their incompetence has plunged the world economy into crisis. One of the dangerous consequences is that the obvious responsibility of the U.S. economy for the crisis produces a robust self-protective denial by the Americans. This merely strengthens the denial of climate change science in these same quarters where continued reliance on fossil fuels causes much of the risk.

It is ironic that the Kyoto Protocol is underpinned by a market model based on emissions trading. The irony is that this was pushed by the same institutions that were responsible for the 2008 crash. There is little evidence that Cap and Trade has had any impact on emission levels. In the EU there are now moves at reforms. Perhaps the flaw is in the model of economics that underpins our market economy.

Beck's 'world risk society' sees greenhouse gases as being dangerous, global in their effects, and invoking fear. Beck believes that this global risk has unknown outcomes, because it will cause events which are progressively unlimited in time and space. He argues that pervasive fear could generate a politics of resistance. It is also possible that this pervasiveness of fear could lead to re-trenchment, a reversion back behind the walls of the state.

15.6 Climate Governance and Society

Giddens provokes thought with his paradox, which says that people will not respond, that is change behaviour, until something happens. That is partially the reason why governments do not seem willing to act. Although there have been climate related disasters, such as hurricane Katrina, they have not been on a sufficient scale. The fact that Kiribati is considering acquiring land for its entire population in Fiji does not seem to affect world leaders. Further, governments are aware that the IPCC scenarios all seem to agree that nothing really bad will happen for at least 10 years. It is a problem for the future. Thus, policy and decisions can be deferred; procrastination can prevail. Giddens argues that the state is central in shifting from high to low carbon societies. There would need to be new forms of collaboration and technology exchange but nonetheless the state will be the driver. States are powerful but in general they react to the demands of their citizens and to date there has been little public outcry in Europe and North America for change. Some may argue that the outcry has been marginalized by the enormous power of the mass media.

At present we appear to have an international process that hopes by 2020 to have some kind of an agreement on greenhouse gas mitigation, but there is little indication that it will provide the radical architecture for mitigation. We are likely to retain the market base approach to mitigation which, so far, has been futile. Although there were some suggestions of a more equitable and just approach based on the principles of Contraction and Convergence, this seems very unlikely. The governance of climate change at the international level is in a mess.

It is evident that the top-down nature of UNFCCC has proved to be problematic. In addition there is the immense challenge of trying to forge an agreement between so many countries with so many different interests. The nation state is the next level down where climate actions can be realized. This is dependent upon the view of government and it is fair to say that there are many shades of green, as well as any other colour except green, of national governments. In the EU there are examples of governments taking positive actions, for example Sweden. In this case the government has committed to de-carbonising the energy system and its vision is to have a sustainable and resource-efficient energy supply and no net emissions of greenhouse gases in the atmosphere by 2050 (Government of Sweden 2009). Although the government has set out a series of policy instruments it will be reliant upon civil society to help meet its goals. Legislation, of itself, will be insufficient.

There are a range of actors in both the public and the private sectors, non-governmental organizations (NGOs), communities, households and individuals who will all need to be part of any drive to a low carbon future. In Denmark innovation within the business sector has supported the wind energy sector. In Germany the Passive House is re-defining the domestic construction sector. The Transition Town movement with its focus on localism and self-sufficiency has effectively communicated its message that it is possible to do things differently. There is a form of climate governance emerging from local initiatives with the state acting through non-state actors but the boundaries of climate governance are blurred (Bulkeley and Newell 2010). There are questions about the states' ability to harness such initiatives towards a climate target. This is a function of the level of trust the state places in people and local adaptive capacity. The governments of Denmark and Germany have actively encouraged community involvement and ownership of renewable technologies. In contrast, the UK has not. Yet some might think that the voluntary principle so well established in British social and economic history that if there were government encouragement now for this transition, the results would be exemplary.

NGOs play significant roles at the local, regional and national level as lobbyists, educators and community organisers. NGOs can work across national boundaries, for example the International Council for Local Environmental Initiatives (ICLEI) which is a network of cities that are promoting sustainable development including climate mitigation and adaptation. The plethora of transnational actors, that can include both public and private actors raises questions about how, why and where governance is taking place (Bulkeley and Newell 2010). But as the role of the state remains a fundamental fact, the idea of 'governance' may be attenuated in many cases.

One further point to note is the deliberate attempts by climate deniers to undermine the climate debate. Good science needs scepticism in the sense that its findings need to be questioned and verified. Climate denial is quite different and is often malicious. Take the case of Michael Mann, referred to earlier, and the problems he has had to deal with. Denial and the phenomenon of 'astroturfing' (giving misleading information that may influence someone's understanding of an issue) are shoddy practices. This has nothing to with good science or good governance.

So what form of governance is likely to emerge? That is almost impossible to predict. The main actions currently are the grassroots levels; individuals, households, NGOs and local government. There are uncoordinated actions by many through such acts as installing renewable capacity at the local level, raising concerns, usually after events of how they can best protect their properties or business. This is not governance but it manifests the failure of national governments to take the problem seriously. But where governments have encouraged different behaviour, there is evidence of success. In Germany the Feed-in Tariff has encouraged household take up of renewable technologies. The 20/20/20 programme provides the underpinning for climate mitigation in the EU. Other countries are rapidly implementing renewable technologies, for example China and the USA; although these choices are offset by a commitment to fossil fuels, and that is worsening the risks. From the standpoint of the manufacturing sector, the automotive industry is investing in alternative approaches to the conventional internal combustion engine. There is also widespread interest in 'new materials' including nanoparticles. This is a piecemeal approach that lacks an overall strategic direction, targets or timetables.

People should be at the heart of climate governance. The current debate is dominated by physical scientists who are principally focused on impacts and economists that are looking for technologies in market based solutions. Understanding climate change processes is critical to understanding how we should respond. That response must be social, but this requires brokerage of ideas to individuals, and the mass media are interested in trivia and not in the reality of the current transition. The challenge is both to adapt the global energy system to new non-fossil resources and to adapt many of our human support systems to a changing climate. The future might be very different particularly if we reach tipping points that lead to irreversible change.

Economic analysis is predicated on the view that people make rational choices. Nothing could be less true. At times we are swayed by trends and fashions. At times we are predictable. We are often conditioned by our social worlds we occupy such as faith, ethnicity, gender, family and class. The type of goods and services used are often a function of our social worlds. We will need to find ways of transforming our social worlds to a low carbon pathway. Many goods such as appliances have inbuilt obsolescence. Others, such as housing, do not, but there is no political movement to retrofit housing for the age of energy conservation and low carbon. The challenge will be to use the variety of communications channels to embed the importance of low carbon living within these social worlds. Social scientists should focus on the kinds of society which can be sustained by various energy base assumptions and habits, if we are to move to a low carbon path (Urry 2012). We are both the end user and end polluter.

This analysis shows that governance of climate change is fractured. Although UNFCCC is flawed this has been used as an excuse by national governments for inaction. The Durban COP has agreed to have an agreement by 2020 to reduce greenhouse gases. This is some time off and many things can happen to subvert this hope. National governments should develop an enabling framework of legislation and incentives that will encourage bottom-up innovation. This may then begin to give some structure to the myriad of individual and collective actions that will bring social worlds into the fold. Perhaps only a catastrophe related obviously to global warming can change the indifference of governments to action for the future. If such a horror were to occur, let us hope it would produce co-ordinated and far-sighted action.

References

Akhand, M.H. 2003. Disaster management and cyclone warning systems in Bangladesh. In *Early warning systems for natural disaster reduction*, ed. J. Zschau and A.N. Kuppers, 49–64. Berlin: Springer.

Anderson, K., and A. Bowes. 2008. Reframing the climate change challenge in light of post-2000 emission trends. *Philosophical Transactions of the Royal Society A* 366(1882): 3863–3882. doi:10.1098/rsta.2008.0138.

Aryal, K.R. 2012 *Getting down to local level. Exploring vulnerability to improved disaster management systems in Nepal*. Ph.D. thesis, Northumbrian University.

Beck, U. 2009. *World at risk*. Cambridge: Polity Press.

Berkley Earth Temperature Group. 2011. http://berkeleyearth.org/index.php.

Bulkeley, H., and P. Newell. 2010. *Governing climate change*. Oxford: Taylor and Francis.

Ellerman, A.D., and B. Buchner. 2008. Over-allocation or abatement? A preliminary analysis of the EU ETS based on the 2005–06 emissions data. *Environmental and Resource Economics* 41(2): 267–287.

Ellerman, A.D., F.J. Convery, and C. de Perthuis. 2010. *Pricing carbon: The European emissions trading scheme*. Cambridge: Cambridge University Press.

Euractiv. 2011. *Poland's EU commissioner in surprise climate denial move*. Euractiv, 23 June 2011. http://www.euractiv.com/climate-environment/polands-eu-commissioner-surprise-climate-denial-move-news-505869.

Giddens, A. 2009. *The politics of climate change*. Cambridge: Polity Press.

Gilbertson, T., and O. Reyes. 2009. *Carbon trading. How it works and why it fails*, Critical currents no.7. Uppsala: Dag Hammarskjöld Foundation.

Goldenberg, S. 2011. Climate change scientist Michael Mann fends off sceptic group's raid on emails. *Guardian,* November 2, 2011. http://www.guardian.co.uk/environment/2011/nov/02/climate-change-michael-mann-emails.

Government of Sweden. 2009. *A sustainable energy and climate policy for the environment, competitiveness and long-term stability*. http://www.sweden.gov.se/content/1/c6/12/00/88/d353dca5.pdf.

Hickman, L. 2012. Climate scientists back call for sceptic think tank to reveal backers. *Guardian*, January 23, 2012. http://www.guardian.co.uk/environment/2012/jan/23/climate-sceptic-lawson-thinktank-funding.

IPCC. 2007. *Climate change synthesis report: Summary for policymakers*. Geneva: IPCC. http://www.ipcc.ch/pdf/assessment-report/ar4/syr/ar4_syr_spm.pdf.

IPCC. 2012. Summary for policymakers. In *Managing the risks of extreme events and disasters to advance climate change adaptation*, A Special Report of Working Groups I and II of the Intergovernmental Panel on Climate Change, ed. V. Barros, T.F. Stocker, D. Qin, D.J. Dokken, K.L. Ebi, M.D. Mastrandrea, K.J. Mach, G.-K. Plattner, S.K. Allen, M. Tignor, P.M. Midgley, and C.B. Field, 1–19. Cambridge: Cambridge University Press.

Lovelock, J. 2006. *The revenge of Gaia*. London: Penguin.

Mol, A.P.J. 2012. Carbon flows, financial markets and climate change mitigation. *Environmental Development* 1: 10–24.

O'Brien, G. 2006. UK emergency preparedness: A step in the right direction? *Journal of International Affairs* 59(2): 63–85.
Ott, H.E., B. Bouns, W. Sterk, and B. Witneben. 2005. It takes two to tango. Climate policy at COP10 in Buenos Aires and beyond. *Journal for European Environmental and Planning Law* 2: 84–91.
Schipper, L., and M. Pelling. 2006. Disaster risk, climate change and international development: Scope for, and challenges to, integration. *Disasters* 30(1): 19–38.
Schultz, J.M., J. Russell, and Z. Espinel. 2005. Epidemiology of tropical cyclones: The dynamics of disaster, disease and development. *Epidemiologic Reviews* 27: 21–35.
Sperling, F., and F. Szekely. 2005. *Disaster risk management in a changing climate*. Informal discussion paper prepared for the world conference on disaster reduction on behalf of the Vulnerability and Adaptation Resource Group (VARG), Washington, DC.
Stehr, N., and H. von Storch. 2005. Introduction to papers on mitigation and adaptation strategies for climate change: Protecting nature from society or protecting society from nature? *Environmental Science and Policy* 8: 537–540.
Stern, N. 2006. *Stern review on the economics of climate change*. London: UK Treasury.
The Copenhagen Accord. 2009. http://unfccc.int/files/meetings/cop_15/application/pdf/cop15_cph_auv.pdf.
The Copenhagen Diagnosis. 2009. *Updating the world on the latest climate science*, ed. I. Allison, N.L. Bindoff, R.A. Bindschadler, P.M. Cox, N. de Noblet, M.H. England, J.E. Francis, N. Gruber, A.M. Haywood, D.J. Karoly, G. Kaser, C. Le Quéré, T.M. Lenton, M.E. Mann, B.I. McNeil, A.J. Pitman, S. Rahmstorf, E. Rignot, H.J. Schellnhuber, S.H. Schneider, S.C. Sherwood, R.C.J. Somerville, K. Steffen, E.J. Steig, M. Visbeck, and A.J. Weaver. Sydney: The University of New South Wales Climate Change Research Centre (CCRC).
UNFCCC. undated. *Glossary of climate change acronyms*. http://unfccc.int/essential_background/glossary/items/3666.php#L.
UNISDR. 2004. *Living with risk: A global review of disaster reduction initiatives*. New York: United Nations International Strategy for Disaster Reduction Secretariat (UNISDR).
UNISDR. 2005. *Hyogo framework for action 2005–2015: Building the resilience of nations and communities to disasters*. Geneva: UNISDR.
UNISDR. 2006. *Statement to the twenty-fifth session of the subsidiary body for scientific and technological advice (SBSTA 25)*. Nairobi, 6 November. Statement prepared and delivered by the ISDR Secretariat. http://unfccc.int/files/adaptation/sbsta_agenda_item_adaptation/application/pdf/isdr_statement.pdf.
UNISDR. 2009. *UNISDR terminology on disaster risk reduction*. http://www.unisdr.org/eng/library/UNISDR-terminology-2009-eng.pdf.
Urry, J. 2012. *Climate change and society*. Cambridge: Polity Press.
World Bank. 2011. *State and trends of the carbon market 2011*. Washington, DC: World Bank.

Geoff O'Brien is a Senior Lecturer in environmental management and sustainable development (undergraduate) and disaster management and sustainable development (postgraduate) at Northumbria University. Geoff's current research interests are in resilient approaches to disaster management with particular reference to accelerated climate change and increased variability. His writing has focused on response in the developed world. Geoff is actively involved in public life and serves as a City Councillor in Newcastle upon Tyne. Some examples include the co-authored papers *UK emergency preparedness – A holistic local response?* (2008) and *Approaching disaster management through social learning* (2010).

Phil O'Keefe is a Professor of Economic Development and Environmental Management at Northumbria University. Phil's research interests focus on the political economy of environmental issues with particular reference to Eastern and Southern Africa. He has written extensively on environmental risk and hazard which, over the last 15 years, has increasingly focused on the delivery of humanitarian assistance after natural disaster and in complex emergencies. He is actively involved in broader debates about the production of nature, sustainability, vulnerability and poverty alleviation, contributing to major global policy documents as well as polemically challenging what this policy means in practice.

Chapter 16
Climate Change and the Politics of Uncertainty: Lessons from Iraq

Michael Heazle

16.1 Introduction

The issue of values, and how they colour perceptions of uncertainty and risk, is of central importance in understanding the limits of both expert advice and rationalist expectations of evidence-based policy making. As March and Olsen have noted, the concept of knowledge informing and largely defining a rational, apolitical process of policy and decision making is made particularly dubious by the separation of knowledge and politics that it requires, a separation that is "impossible to sustain, either conceptually or behaviourally":

> Expert knowledge clearly rests on values that regulate the way knowledge is organised and validated. … Like other people, experts seem to find facts and theoretical implications consistent with their policy preferences and forget facts and theoretical implications inconvenient for their purposes. For their part policy makers often seem to use advice from specialists as an excuse for doing what is unpopular with some groups. They often seem to be inattentive to the cautions and fine details of expertise. … When experts disagree, policy makers often seem to view the disagreement as justification for accepting whatever advice is convenient. Agreement among specialists, on the other hand, is likely to be treated as a sign of conspiracy. (March and Olsen 1989, pp. 30–31)

Exposing the falsity of this separation and the ways in which it inhibits rather than promotes agreement on "legitimate" policy is an essential part of attempting to bring politics back into not only our understanding of policy decision making, but also the process by which one decision rather than another is legitimized. Broader and more rigorous public debate on complex, high stakes policy issues – as envisaged by liberal notions of a "marketplace of ideas" – is the mechanism by which the "political" is given life and power in liberal democratic societies (Crick 2000, p. 18). Moreover, it is their essentially political character that defines them as free

M. Heazle (✉)
School of Government and International Relations, Griffith University,
Nathan Campus, Brisbane, QLD 4111, Australia
e-mail: m.heazle@griffith.edu.au

U. Fra.Paleo (ed.), *Risk Governance*, DOI 10.1007/978-94-017-9328-5_16

and open. But policy debates are unable to be framed in the context of the values and uncertainties that drive disagreement when competing knowledge claims take centre stage. The values underpinning particular positions on an issue are instead made obscure by an ostensibly science-driven discourse where different knowledge claims are used by various protagonists to support their policy preference and discredit the competing knowledge claims supporting other normative positions. Thus, the fundamental challenge, in terms of building broader and more effective policy debates, is for us to recognize that our values and interests not only shape the way we look at science and expert advice, but that they also drive disagreement over how problems should be identified, prioritized, and responded to.

The sooner these values can be openly debated, the sooner competition between goals and priorities can be resolved, thereby allowing science to concentrate on the task of how best to achieve, as opposed to expecting it to somehow determine, what is politically acceptable. If we are to conduct policy debates of the kind used to argue the superior policy making ability of liberal democracies, which in the context of wicked policy problems are infrequent at best, it is the competing values behind the various expert advice on offer that must be openly debated so that the *actual* sources of disagreement can be identified and then negotiated on *both* the merits of the competing normative judgments that drive them, and the expert advice/knowledge claims invoked to support them. In such a context, the limits of specialist advice and its susceptibility to differing uncertainty claims, regardless of the type of advice on offer, are much more likely to become clear, allowing debate to move beyond the pointless claim and counter-claim over who has the "real" science/knowledge, which occupies much of what has come to be accepted as public policy debate. Moreover, and contrary to rationalist claims, the problem is that policy debate is *under* "politicized," since politicization means openly stating and contesting values rather than attempting to manipulate or pervert the quest for objective, evidence-based policy in order to achieve self-serving ends. Indeed, the "post-normal" character of complex policy challenges like determining the level of threat posed by a WMD[1] armed Iraq or an appropriate response to the potential consequences of a changing climate makes politicization of the policy decision process both essential and inevitable.

16.2 Evidenced-Based Policy Making: The Rationalist Approach

By the time of the Intergovernmental Panel on Climate Change (IPCC) Third Report in 2001, international scientific opinion was largely in agreement that anthropogenic generated greenhouse gas emissions were exerting a warming influence on the

[1] Weapon of mass destruction.

so-called "global climate."[2] But despite claims of a growing consensus around the IPCC's findings, debate and controversy continued to grow, especially over the scale, nature, and likelihood of future climate change impacts. Numerous disagreements over the reliability of the global warming scenarios produced by Global Circulation Models (GCMs), and in particular the data and assumptions fed into these models, have continued to rage, thereby allowing differing interpretations of the available evidence and the many uncertainties it involves to be used by governments and various groups to substantiate either their support for or opposition to calls for immediate emission reductions. The Howard Government in Australia and the Bush Administration in the USA, and various industry groups, initially chose to amplify the uncertainties in climate change science and its predictions as a means of justifying their opposition to emission reductions, while also arguing that the levels of uncertainty – over future emissions levels, their environmental impacts, and the effectiveness of the Kyoto Protocol approach – did not warrant the significant economic costs that they believed implementation of the Kyoto agreement would "certainly incur."

Shortly before taking office in 2001, the then Governor George W. Bush already had made his doubts over the conclusions of global warming science, and the need for more certainty, clear during his presidential campaign:

> I – of course there's a lot – look, global warming needs to be taken very seriously, and I take it seriously. But science, there's a lot – there's differing opinions. And before we react, I think it's best to have the full accounting, a full understanding of what's taking place. (*Commission on Presidential Debates Transcript* 2000)

Then, in 2001, in a letter to Republican Senators, President Bush highlighted both energy security and the economy in explaining his rejection of mandatory emission reductions:

> At a time when California has already experienced energy shortages, and other Western states are worried about price and availability of energy this summer, we must be very careful not to take actions that could harm consumers. This is especially true given the incomplete state of scientific knowledge of the causes of, and solutions to, global climate change … (Fuller 2001)

Still focused on energy and the economy, Bush later informed reporters that:

> We are now in an energy crisis. And that's why I decided to not have mandatory caps on CO_2, because in order to meet those caps, our nation would have had to have had a lot of natural gas immediately flow into the system, which is impossible. … We'll be working with our allies to reduce greenhouse gases, but I will not accept a plan that will harm our economy and hurt American workers. (PBS *Online NewsHour*, 29 March 2001)

In December 2003, Australian Prime Minister John Howard explained his ongoing opposition to the Kyoto Protocol, having informed the Australian Parliament in

---

[2] The notion of a "global climate" is an abstract derived from the concept of a "global average temperature." Climate behavior and temperature are regional phenomena that vary significantly and can be influenced by a very broad range of both local and more global factors. Reliable modeling of regional climate behavior remains beyond the capability of contemporary climate models and our understanding of the climate system.

2002 of his government's intention not to ratify the Kyoto Protocol, by saying: "I'm not going to be a party to something that destroys jobs and destroys the competitiveness of Australian industry" (*The Age*, 2 December 2003). Three years later, in August 2006, Howard continued to downplay the risks of climate change relative to the economic risks of reducing emissions. In an interview with the Australian Broadcasting Corporation's *Four Corners* program, the Prime Minister stated that:

> I accept that climate change is a challenge. I accept the broad theory about global warming. I am sceptical about a lot of the more gloomy predictions. I also recognise that a country like Australia has got to balance a concern for greenhouse gas emissions with a concern for the enormous burden to be carried by consumers through much higher electricity prices, higher petrol prices, falls in GDP of too dramatic an imposition of what you might call an anti-greenhouse policy. It's a question of balance. (*ABC Four Corners*, 28 August 2006a)

The "prudent, and rational, response" was, therefore, according to President Bush and Prime Minister Howard, to resolve the uncertainties through more research before committing to any particular policy response – especially one likely to affect economic and energy security. Underpinning such a course of action, or in this case non-action, was the Rationalist policy model and its faith in science's ability to inform policy decisions by producing reliable knowledge, which in turn will be used to arrive at legitimate policy decisions based on science. This approach to uncertainty has remained bi-partisan, with the only real difference being disagreement over various policy issues on how much certainty is needed (or how much uncertainty can be tolerated) before acting on the specialist advice and evidence at hand. The U.S. Global Change Research Program, for example, has spent billions of USD in government funding since 1989 on research that, according to then-Senator Al Gore's justification of the program, has been aimed at building political consensus on environmental challenges through the reduction of uncertainty. In Gore's view, the solution to complex policy problems is obvious: "More research and better research and better targeted research is absolutely essential if we are to eliminate the remaining areas of uncertainty and build the broader and stronger political consensus."[3]

But, as already indicated, the Rationalist model can also legitimize decisions *not* to act, that is, to wait and see until the evidence is in – which is precisely what the Bush and Howard governments were advocating despite their acknowledgement of global warming as an important policy issue. For those wanting a policy response sooner rather than later, the obvious strategy is to downplay the uncertainties involved by arguing that failing to act introduces unacceptable risk; rising oceans, increasingly frequent extreme weather events, and depleted bio-diversity are, after all, more fundamental challenges to human societies and their economies than price increases, unemployment, and energy shortages – even if they are in the distant future. Global warming's credentials as a major security issue deserving of the same kinds of policy action taken to protect against terrorism, external military threats, and economic recession thus are imbued with a sense of urgency – vis-à-vis the

[3] Quoted in Pielke and Sarewitz (2002/2003). By 2002, the U.S. had spent more than US$20 billion on funding research under this project.

logic of the precautionary principle – that makes taking a "wait and see approach" more difficult to justify. Such a precautionary approach, however, suffers from the same challenge faced by proponents of the Iraq invasion – that is, how to demonstrate that currently imagined future threats will be, or are, likely to occur. Policy debates, framed in these terms, become dominated by competing perceptions of future outcomes and the nightmares invoked to illustrate them, and provide only a choice between either a "look before you leap" (or, wait until we *know* more) or a "he who hesitates is lost" (or, take preventative action before it is too late) approach to managing uncertainty and its associated risks.

The alternative suggested here is to further develop and adopt a "post-normal science" approach to uncertainty and policy making, and the specialist advice that informs decision makers' perspectives on both – as has been pioneered by Jerry Ravetz and others.[4] We should accept that, for a variety of compelling epistemological reasons (that are certainly not new), the uncertainties surrounding major policy issues, including climate change, are not reducible beyond a certain point. Maintaining that certainty, or something resembling it, is attainable provides, as is argued herein, nothing other than the opportunity for policy protagonists to use uncertainty and competing knowledge claims as a disguise for the interests that underlie the policies they support. But, if we accept from the outset that certainty is not available, and that it is competing political agendas rather than simply objective knowledge that set the parameters of policy making and debate, it may be possible to produce debates that, rather than being sidetracked by disputes over who has "the real science," focus instead on the *actual* political and interest-based obstacles to agreement and compromise. Doing so would produce more transparent, credible, and consistent policy decisions and, therefore, also better international policy coordination and cooperation. As Daniel Sarewitz has argued:

> No longer able to hide behind scientific controversy, politics would have to engage in processes of persuasion, reframing, disaggregation, and devolution, to locate areas of value consensus, overlapping interests, or low-stakes operations (e.g., "no regrets" strategies) that can enable action in the absence of a comprehensive political solution or scientific understanding. In particular, the abandonment of a political quest for definitive, predictable knowledge ought to encourage, or at least be compatible with, more modest, iterative, incremental approaches to decision making that can facilitate consensus and action. (Sarewitz 2004)[5]

[4] For post-positivist related perspectives on science and technology and the role of expert advice in policy making, see Funtowicz and Ravetz (1992, 1993); Ravetz (1999); Daston (2005); Hajer (2003); and Goldenberg (2006).

[5] Daniel Sarewitz's call for more "incremental approaches to decision making" on the basis of consensus being achieved, not on scientific or knowledge disputes but on values and interests, invokes Lindblom's ideas about the need to abandon the quest for "synoptic" or complete knowledge as the basis of policy and instead to face up to our inability to know very much at all about complex policy issues, or even to reach shared understandings of policy goals. Lindblom's famous description of policy making and analysis as "a science of muddling through," made necessary by the sheer complexity of policy problems, represents what he later stated "is and ought to be the usual method of policy making." See Lindblom (1959, p. 517).

16.3 Scientising Politics: Climate Change and the "Power of Nightmares"

The executive elites of governments elected to 3 of 4 year terms before facing re-election are necessarily more focused on the short term, as are the majority of voters striving to manage today's financial and family related pressures and concerns. Thus, it is hardly surprising for national governments and societies to be less concerned with speculation over threats in the distant future than with those they feel are much more likely to occur in the shorter term. This point is supported by a 2009 Gallup Poll in the U.S. showing higher levels of concern for the economy than the environment in the wake of the so-called global financial crisis. For the first time since the 1980s, levels of environmental concern fell below a sharply increased level of concern for the economy in the U.S. and elsewhere, indicating the extent to which environmental priorities may be a product of economic good times (Saad 2009; Newport 2009). In the case of global warming, adopting a strong precautionary approach today as protection against future (and unknown) warming impacts will not guarantee a safer future but will guarantee a politically difficult, perhaps even fatal, set of problems for governments that must be addressed in the short term. From a policy perspective, such calculations are unavoidable; but they are seldom, if ever, made explicit in contemporary policy discourses where values prioritizing the importance of maintaining biological diversity and an unchanged natural environment often compete against human-centric priorities such as development and economic growth. As Sarewitz has noted, governments seeking to justify or delay policy in the public sphere choose to avoid the politics of competing values and interests where possible, preferring instead to rely on the credibility of "scientific authority" and invoking it as the source of objective rationality behind the decision taken:

> If you were a policy maker, would you rather participate in a debate about the scientific aspects of a controversy, or about the interests and values that underlie the controversy? Arguing about science is a relatively risk free business; in fact, one can simply mobilise the appropriate expert to do the talking, and hide behind the assertion of objectivity. But talking openly about values is much more dangerous, because it reveals what is truly at stake. (Sarewitz 2000).

On the other side of the political equation, we see alternative policy positions supported by groups with very different interests who in turn derive and put forward very different scientific interpretations of the data and evidence in order to objectify and rationalize their own particular policy preference. Environmental groups, alternative fuel lobbies (including the nuclear lobby),[6] and segments of the media, for

[6]The argument put forward by Howard and others that nuclear energy should be an option for reducing greenhouse gas emissions also serves as an example of the precautionary principle dilemma. Is the risk of a nuclear accident more or less acceptable than the still unknown but possible risks posed by greenhouse gas emissions? Who decides and on what basis?

example, are unrestrained by electoral accountability and view global warming through an entirely different prism of values and interests. For these players, the short term political and economic consequences of taking action today are of little concern in comparison to the long term risks they invoke as justification for a precautionary policy response. Moreover, the debate over competing knowledge claims has spilled over from climate-related science into the realm of economics where very different arguments have been made over both the current cost of acting against climate change and the future cost of not acting.

The Bush and Howard governments cited sometimes sparse or incomplete economic analysis[7] warning of the high costs emission reductions will impose on their respective economies, while other analyses, most notably the Stern Report, advised that the potentially high future costs of not reducing emissions can be avoided if we accept the relatively small cost of reductions now. Stern's approach, which controversially values future benefits and consumption equally with the present as a means of undermining the economic dangers of action today, clearly illustrates the normative dimension of climate change debate. As one economist noted, "the strong, immediate action on climate change advocated by the authors [of the Stern Report] is an implication of their views on intergenerational equity; it isn't driven so much by the new climatic facts the authors have stressed" (Dasgupta 2006).[8] The kinds of risks cited both by Nicholas Stern and the IPCC were indeed extreme enough to capture the public imagination, despite their place in the more distant future, and compete for priority against the short term social and economic costs of emission reductions that governments and other groups (e.g., business and industry) have advocated avoiding. Climate science and its many associated uncertainties then, as Stephen Bocking argues, have become "deeply embedded in political debates" (Bocking 2004, p. 118), and are regularly invoked in the service of specific interests:

> [T]hose opposed to action on climate view it as more effective to question the science than to defend their interests directly. On the other hand, those advocating climate action tend to minimise scientific uncertainties .... In effect science serves as a surrogate for political and economic conflict, imparting authority to positions on either side, but at the expense of becoming fully embroiled in these conflicts. (Bocking 2004, p. 126)

---

[7] See, for example, Baker (2007) and Nordhaus (2006).

[8] Nicholas Stern's cost-benefit analysis of the present versus future costs of acting and not acting to mitigate anthropogenic climate change attracted criticism from many economists due to (1) its employment of a pure rate of time preference or social discount rate of almost zero (0.1 %) and an overall discount rate of only 1.4 % as opposed to the 4–6 % realm normally used when discounting future costs and benefits; (2) the report allegedly downplaying the real cost of spending 1 % of annual world GDP on mitigation and the sacrifice this will require, particularly among developed countries where it will more likely amount to 1.8 % of *each* economy's GDP annually; and (3) Stern's 'cherry picking' of worst case scenarios and damage estimates as the basis of his cost-benefit analysis. See also, for example, Neumayer (2007); Nordhaus (2006); and Pielke (2006).

The interpretation of uncertainty in political debate then is mostly about values, particularly in big issues like climate change where uncertainty is high (preventing science from providing the kinds of specific information policy requires) and values are hotly disputed (which are more important, jobs or polar bears?).[9] But because the rationalist model still survives, despite the sustained scholarly criticism it has received over the years,[10] people still think in terms of good policy simply being based on the best knowledge, as demonstrated by the ongoing faith of politicians in so-called "evidenced based" policy making, which of course means "scientifically derived" knowledge. So what happens when science can't give us definitive, testable answers? Policy elites and people in general nevertheless claim that the science is on their side in order to make their values and priorities seem more legitimate than other people's (who are using different scientific opinion to do the same thing), which causes what Jurgen Habermas, writing in the late 1960s, called the "scientisation of politics" (an inversion of what rationalists call the "politicisation of science"): dead-end debates over who has the "real" science that obscure the values actually in dispute.[11]

The fundamental conflict in values underpinning the distinction between short-term economic and energy security and long term "climate security" becomes, as a consequence, obscured as protagonists use competing scientific claims to gain advantage in what are actually political disputes over competing values (what Sarewitz (2000) also has referred to as "an excess of objectivity"). And as the scientific disputes, fuelled by uncertainties, intensify, political debate is subsumed into what soon becomes a zero-sum contest for the mantle of scientific objectivity and the policy legitimacy it provides. The result is highly polarized debate that: (i) excludes all but the more extreme policy options (future risk warrants drastically reducing emissions versus economic consequences of reductions and uncertainties over future impacts too great to justify any significant short term reductions); and (ii) employs scientific advice and uncertainty as a smoke screen to hide the core values and interests actually driving disagreement between the various actors. In contrast to the Rationalist complaint of politics distorting what otherwise could be rational policy making informed by science – popularly referred to as the

[9]Princeton physicist and climate skeptic Freeman Dyson also sees conflicts over fundamental values as underpinning the competing scientific arguments over global warming's causes. Nicholas Dawidoff (2009) writes that: "Beyond the specific points of factual dispute [over global warming science], Dyson has said that it all boils down to 'a deeper disagreement about values' between those who think 'nature knows best' and that any gross human disruption of the natural environment is 'evil,' and 'humanists,' like himself, who contend that protecting the existing biosphere is not as important as fighting more repugnant evils like war, poverty and unemployment."

[10]One of the best known critiques of the rationalist model is Charles Lindblom's "science of muddling though." See works by Lindblom (1959, pp. 79–88; 1979, pp. 517–526). More contemporary examples include Jasanoff (1990); Stone (2002); Pielke (2007); and Nieman and Stambough (1998).

[11]According to Habermas (1971, p. 75), scientisation causes values-based issues (i.e., political issues) to be redefined as technical issues that can be rationally solved or managed by scientific enquiry. See also Sarewitz (2000).

"politicisation of science" – highly complex, or "wicked,"[12] policy problems like climate change most often create debates where the "scientisation of politics" becomes the main obstacle to not only the development of any broadly supported policy, but also to policy responses made *rational* by their attempt to manage rather than vanquish uncertainty issues.

The United Nations Framework Convention on Climate Change (UNFCCC), for example, frames climate change in the positivist, human-centric language of anthropogenic global warming and betrays an ongoing subscription to the linear good science equals good policy orthodoxy of the rationalist model. By doing so, the convention has effectively encouraged the scientisation of policy debate from the outset. The UNFCCC, unlike the much broader IPCC definition, defines climate change as an issue of policy and decision making *only* in terms of human activity having a "dangerous" influence on climate and by doing so automatically excludes natural climate change or variation as irrelevant in addition to causing confusion over what is or is not conclusively known about current climate change's causes.[13] Making such a distinction, however, assumes that not only is the distinction between human and naturally caused climate change knowable on the basis of science, but also that the question of what constitutes "dangerous climate change" can be resolved scientifically (Pielke 2005, pp. 553–555). Thus, according to the UNFCCC, climate change can only qualify as a policy issue if it can be shown to be both human induced and dangerous. As a consequence, the climate change debate has become extremely narrow in its scope, limited to unresolvable disagreement over one set of potential causes and outcomes (greenhouse gases, or GHGs, in particular carbon, emissions are the primary cause of future climate change threats), and one set of unimplementable policy proposals in response to those highly uncertain outcomes and causes (global reduction of carbon emissions is necessary to mitigate future climate change threats).

This narrow framing of the climate debate to date is well illustrated by the 2009 announcement of a conference convened by University of Oxford, the Tyndall Centre for Climate Change Research, and the U.K. Met Office entitled "4 Degrees and Beyond: Implications for people, ecosystems, and the earth system." Ignoring ongoing uncertainties over climate sensitivity, feedback mechanisms, and future human behaviour that so far have prevented the IPCC from asserting anything more than a *possible* temperature range over this century, adjusted in 2007 to between 1.1

[12]Wicked policy problems are broadly understood as policy issues of great complexity involving systems within systems, which not only defy any uniform definition but also are highly resistant to analysis and resolution due to the numerous system uncertainties (epistemic and variability) and multi-causal factors involved.

[13]Former Australian delegate to the IPCC John Zillman (1997) writes that: "According to the [UNFCCC] Convention, 'climate change' is that which is due to human activity and is in addition to natural variability. The IPCC WG I, on the other hand, regards 'climate change' as including natural variations. Thus, when the IPCC says 'climate has changed over the past century,' it is simply saying the climate now is not the same as it was a century ago (whatever the cause) whereas the UNFCCC listener will reasonably interpret such a statement as the scientific community affirming that *human* influence has changed climate over the past century."

and 6.4 °C, the conference's call for participants asserts that global temperature will increase "well beyond" 4 °C and limits its focus only to the consequences and policy options relevant to such a major increase:

> Despite 17 years of political negotiations since the Rio Earth Summit, global greenhouse gas emissions have continued to rise, which presents the global community with a stark challenge: Either instigate an immediate and radical reversal in existing emission trends or accept global temperature rises well beyond 4 °C.[14]

Thus, according to the conference organizers, societies and policy makers alike are confronted only with a simple choice between accepting the costs of radical action to dramatically reduce emissions immediately or accepting the implied catastrophic consequences of an extreme increase in global temperatures somewhere beyond 4 °C. Such framings of the climate change/global warming issue, relying as they do on a misrepresentation of the complexities and uncertainties involved in order to invoke future nightmare scenarios as the cost of not following one particular course of action (and indeed with no regard for the costs of following that course of action), serve only to further delay policy responses and cooperation. The choice presented here is for the vast majority of people and no doubt all governments completely unacceptable without a good deal more certainty that such a "stark" proposition is what we must now face, which is of course not available. So rather than discussing what kind of policies may be adopted and implemented, given the different values, interests, and unknowns, debate remains locked within an all or nothing conflict driven by political competition over values but fought in the language of science.

16.4 The Climate Consensus and Policy-Making: New Rhetoric but No Paradigm Shift

Through the selective use and interpretation of sometimes vague or unsubstantiated tactical intelligence, proponents of the 2003 Iraq war were able to introduce a range of difficult to dismiss potential threats as justification for military action while also obscuring the partisan values and policy priorities that actually informed the decision to go to war. Investigations into the intelligence assessments and evidence used to justify the U.S.-led invasion of Iraq in the U.S., the U.K., and Australia – in addition to the failure of coalition forces and the Iraq Survey Group to uncover any evidence of chemical, biological, or nuclear weapons in Iraq – indicate that the coalition case for going to war largely was based mostly on speculative thinking in an environment where, as a joint Australian parliamentary investigation concluded in 2004, "policy was running strong" (Parliament of the Commonwealth of Australia 2004; United States Senate 2006). Moreover, the largely speculative conclusions

[14] Email call for conference participants received 3 April 2009 from Mary Mansfield via the Climate Change Info Mailing List; sent by bounce-875604-330596@lists.iisd.ca.

drawn concerning Iraq's threat capability and potential often reflected only some of the many intelligence assessments available during the lead up to the Iraq invasion – much of which was based on dubious sources and occurred largely in isolation from the strategic assessments on offer. *Jane's Intelligence Digest* (30 October 2003), for example, noted that much of "the often flawed intelligence cited by both the U.S.A. and U.K." came from "outside the usual channels," in particular the U.S. State Department funded Iraqi National Congress.

In understanding the Bush and Howard governments' opposition to the Kyoto Protocol and the Blair government's contrary support for it, it is necessary to look again at the already existing policy priorities, political circumstances, and values base from which the executive policy elite (i.e., those with the executive authority to publicly state what is or is not government policy)[15] in each government made calculations of the "national interest" and the Kyoto Protocol's potential for either helping or hindering the pursuit of established policy goals. Among all three governments, policy was again "running strong" in the treatment of specialist advice, and again the policy debate became dominated by competing knowledge claims and assertions as protagonists arguing for and against the implementation of the Kyoto Protocol looked to specialist advice and uncertainty for evidence and arguments that would "legitimise" their policy preference. When comparing the treatment of uncertainty issues in the global warming/climate change debate with their role in the arguments and specialist advice used by Bush, Blair, and Howard to justify military action in Iraq, the most obvious difference is that uncertainty was used as a basis *for* acting, or as insufficient reason for *not* acting, in one case (Iraq) but then interpreted as a basis *for not* acting, or as insufficient reason for immediate action *beyond* existing policy, in the other (the Kyoto Protocol). A less apparent distinction concerns changes in policy settings and instruments versus a more fundamental paradigm shift in policy goals and priorities as per Peter Hall's (1993) often cited analysis of the U.K. government's shift from a Keynesian economic philosophy to what has since become known as neo-liberalism under Margaret Thatcher.

The decision to invade Iraq essentially represented a change first in policy settings (more frequent U.N. inspections and stronger warnings of "dire consequences" for Iraqi non-compliance) and then a shift in the policy instruments used (military force and occupation rather than containment). The use of military force to topple Saddam Hussein from power was a change in the choice of policy instruments

[15] Whether or not executive decision makers are reacting to a policy challenge or proactively pursuing ideological preferences, justified in the "national interest," the imperative that they publicly state what they intend to do, and how they intend doing it, and why is a major determinant of the policy process; it is an aspect of policy making that, in addition to defining policy, clearly reflects the strong link between policy and its justification. According to former U.S. Secretary of State Warren Christopher, "in any given week as Secretary, I received dozens of memoranda advocating various particular policy directions. However persuasive their contents, they did not constitute U.S. policy unless they were incorporated into a speech, public statement or formal government document. The challenge of articulating a position publicly compels leaders to make policy choices. Often decisions on what to do and what to say publicly are made simultaneously." Quoted in Chollet and Goldgeier (2002, p. 170).

rather than a paradigm shift in policy thinking since the need to remove Saddam from power had been part of publicly stated U.S. policy since at least 1997.[16] And although this particular outcome did not become "policy" in the U.K. or Australia until after 9/11, when the Bush administration made its intention to remove Saddam sooner rather than later clear, their respective commitments to U.S. foreign policy and goals – upheld in the interests of strengthening their security and economic relations with the U.S. – were longstanding, bipartisan pillars of nationalist interest perceptions in both countries. The Iraq debate then was not over the question of whether Saddam should go, but rather the question of by what means and how quickly he should go and in particular whether the alleged risks his regime posed justified the kind of change in policy instrument advocated by the U.S. and its allies (i.e., from containment to invasion). In contrast, the climate change debate quickly became polarized over more than simply questions of the appropriate settings and instruments. The response mandated by the Kyoto agreement effectively required, in the eyes of the U.S. and Australian governments at least, a paradigm shift in policy priorities and goals within an area of policy where policy makers are highly risk averse: the economy.

Because the policy elites within the U.S., U.K., and Australian governments saw military action against Iraq as fitting within their existing policy paradigm – either in terms of Saddam's removal or the importance of security relations with the U.S. – they downplayed uncertainties over the need for, and outcomes of, using military force. In the case of the Kyoto Protocol, however, only the Blair government regarded this agreement's mandatory carbon emission reductions, and the changes they required in fossil fuel usage and cost, as being compatible enough with existing policy to allow their implementation through further adjustment of policy settings and instruments. For George W. Bush and John Howard, no such accommodation within existing policy and calculations of the national interest was possible – even under Australia's entitlement under Kyoto to a limited increase in emissions until 2012[17] – and a policy shift in economic and energy thinking of the scale required by the Kyoto agreement was unacceptable. For Bush and Howard, it was the implementation of the Kyoto Protocol's emission reductions and the paradigmatic shift in policy priorities and thinking required, rather than the unknown potential for

[16] Speaking at Georgetown University in March 1997, Secretary of State Madeleine Albright (1997) made future U.S. policy on Iraq contingent on Saddam's removal from power: "We do not agree with the nations who argue that if Iraq complies with its obligations concerning weapons of mass destruction, sanctions should be lifted. Our view, which is unshakeable, is that Iraq must prove its peaceful intentions. It can only do that by complying with all of the Security Council Resolutions to which it is subject. Is it possible to conceive of such a government under Saddam Hussein?… The evidence is overwhelming that Saddam Hussein's intentions will never be peaceful…. Clearly, a change in Iraq's government could lead to a change in U.S. policy. Should that occur, we would stand ready, in co-ordination with our allies and friends, to enter rapidly into a dialogue with the successor regime." Quoted in "The Iraq Crisis" (1998).

[17] The Howard government attracted considerable criticism, especially from the E.U., when it demanded and received an 8 % *increase* on its 1990 emission levels during negotiations for the Kyoto Protocol. Prime Minister Howard later announced Australia would not ratify the Kyoto agreement in 2003.

long term climate change consequences cited by the protocol's supporters, that represented the clearest and most immediate threat to the security of their countries – despite the efforts by some scientists, economists, NGOs, and governments to make anthropogenic global warming a more compelling "security" issue than traditional state-centric notions of threats to the economy or the integrity of the state. Furthermore, support for the Kyoto Protocol among even its most vocal government supporters like the U.K. government ultimately proved to be dependent on the extent to which emission reductions could be reconciled within established policy priorities and objectives.

Opposition to the emission reductions called for in Kyoto on the grounds of the harm it would cause to the U.S. economy was made clear in the Republican dominated U.S. Senate as early as 1997 by the 95–0 vote supporting the Byrd-Hagel Resolution's (1997) rejection of U.S. participation in any emissions reduction agreement that excluded developing countries. In addition to concerns expressed by both the Bush administration and Howard government over the uncertainties surrounding claims of anthropogenic climate forcing and its future impacts, the other major reason cited by both governments was that the Kyoto agreement did not require emission reductions or limits from developing economies as well, in particular China and India. This often cited source of opposition to ratifying the Kyoto Protocol's planned emission reductions illustrated the kind of state pre-occupation with self-interest and concerns about international cooperation creating the kinds of unequal benefits that realist theorists had long warned of as an obstacle to international agreements. But given the increasing domestic pressure within both countries – encouraged by the IPCC and various media reports and NGOs – for their governments to recognize global warming as a serious threat, it was also becoming clear by late 2005 that if Kyoto was off the list of policy options, then some kind of surrogate response was needed to show that the issue at least had the government's attention. Thus, President Bush and Prime Minister Howard found themselves in the kind of bind described by Robert Putnam's (1988) "two level game" depiction of how policy elites are forced to reconcile tensions between foreign policy ambitions shaped by their own notions of the national interest and the international system on the one hand, and the domestic plurality of competing interests that elected governments are ultimately accountable to on the other hand. The solution Bush and Howard adopted for their two level game dilemma over how to appear engaged with global warming as a policy issue while minimizing any risks to their fossil fuel dependent economies was to propose a new international forum with a very different strategy, technology based strategy for controlling emissions: the Asia Pacific Partnership Group (AP6).

Buttressing the need for an alternative approach to global warming were the numerous critics of the Kyoto Protocol that had emerged by this time. In addition to the now regularly cited issue of the Kyoto agreement's failure to require any emission reductions in the developing world, some argued that the Kyoto strategies were undermined by too many questionable assumptions in relation to the likely costs involved. Meanwhile debate raged – even among those supporting the anthropogenic global warming consensus – over how effective, if at all, the protocol's reductions would be even if full international co-operation and implementation were possible, which was by now looking increasingly unlikely. Indeed, by late 2005 even Tony Blair was becoming critical of the approach drawn up in Kyoto, drawing accusations that he was again falling into line with Washington, as he had over Iraq. At a climate change conference convened by former U.S. president Bill Clinton, Blair's criticism of the Kyoto agreement and endorsement of the technology focused response to global warming proposed by his allies in Washington and Canberra attracted a storm of protests and accusations of backtracking on his formerly strong pro-Kyoto rhetoric:

> I'm changing my mind about this … no country is going to cut its growth or consumption substantially in the light of a long term environmental problem. To be honest, I don't think people are going, at least in the short term, to start negotiating another major treaty like Kyoto…. How do we move forward and ensure that, post-Kyoto, we do try to get agreement? I think that can only be done by the major players in this coming together and finding a way for pooling their resources, their information, their science and technology. (*The Independent*, 25 September 2005).

Interestingly, Tony Blair's publicly stated reluctance to sacrifice economic security in the pursuit of "climate security" and emphasis on a "market-based" response prefaced a softening of climate policy in Washington and Canberra the following year in late 2006. The Bush and Howard governments had been gradually toning down their sceptical position on global warming impacts since 2002 as public fears in both countries over climate change intensified, especially in the wake of the release of Al Gore's documentary *An Inconvenient Truth* and *The Stern Review Report on the Economics of Climate Change*. Prime Minister Howard made his new found enthusiasm for acting against climate change clear at the Asia Pacific Economic Co-operation (APEC) summit in Hanoi in November 2006, where he actively promoted discussions on regional measures against climate change, giving it the same priority as Iraq and global trade (*The Sydney Morning Herald*, 17 November 2006). The Howard government, however, continued to reject policy it claimed would harm Australia's economy and fossil fuel interests, but, nonetheless, made global warming a government priority, particularly in relation to the energy debate that was emerging at home over his government's calls to end Australia's long running ban on nuclear power. President Bush, too, was undergoing a Gestalt switch of sorts on global warming policy that was remarkably similar to the new line of thinking espoused by his Australian ally, as demonstrated by the softening of the Bush administration's position on the need for international policy action on global warming witnessed at both the APEC meeting and later at the 2007 United Nations Climate Change Conference in Bali. In his 2007 State of the Union speech, President Bush already had indicated a shift in his position on climate change, and in particular America's oil dependency, was underway when he said:

> America's on the verge of technological breakthroughs that will enable us to live our lives less dependent on oil. And these technologies will help us become better stewards of the environment, and they will help us to confront the serious challenge of climate change. (*The Washington Post*, 23 January 2007)

Howard's and Bush's rhetorical shift on global warming most likely was motivated by increasing public concern and the greater weight anthropogenic global warming had acquired as a policy issue, particularly with federal elections looming in both countries, since no new or more compelling evidence on the likelihood of global warming's most serious threats being realized appeared prior to either government adopting a less sceptical position. But despite the greater acknowledgement by both leaders of global warming's importance and potential consequences, their priorities were still very much about avoiding any economic disruption at home.

At the inaugural meeting of the AP6 in January 2006, George W. Bush and John Howard, two of the Kyoto Protocol's biggest critics, talked up the importance of developing renewable energy sources as a way of combating global warming threats without incurring potentially crippling economic penalties; little actually had changed, however, in terms of the policy priority given to short term economic growth over the longer term and still largely speculative challenges posed by global warming. Prime Minister Howard, for example, also made it quite clear the Australian government remained committed to fossil fuels with his endorsement of the AP6 view that fossil fuels "will be an enduring reality for our lifetime and beyond" (Peatling and Frew 2006). According to figures reported in *The Australian*, of the A$100 million the Howard government had dedicated to the partnership over the next 5 years, only A$5 million per year was for developing renewable energy projects. This, according to a government AP6 press release, was in addition to the A$200 million the Howard government claimed it already had invested in developing renewable energy (A$500 million meanwhile had been "invested" in so-called "low emission technologies" such as carbon sequestration). For its part, the U.S. government, which spends more than US$350 billion on its military each year, committed a meagre US$52 million from its 2007 budget, subject to approval by Congress ("expected" to grow to US$260 million by 2011) (Hodge and Maiden 2006). Moreover, both governments essentially used the AP6 as a cover for dodging the global warming issue entirely by announcing their intention to hand the job of developing and implementing new energy technology over to the private sector.

The ongoing focus in climate change policy discussion on often vague and complex market-based responses to climate change, such as the various carbon trading scheme models, is perhaps the clearest indication that it still is not regarded as a "security" issue by many governments, despite efforts and claims to the contrary within the European Union. The U.K. government in 2007, for example, attempted to promote climate change as a security issue in the U.N. Security Council (U.K. Concept Paper, 23 April 2007) while the European Commission has sought to use climate change as a catalyst for new energy security measures and targets for renewable energy among member states (Trombetta 2008a). But even among governments supportive of emission reductions and targets, there has been no direct government policy response of the kind normally seen when an area of "national security" has been at risk. Put simply, the policy rhetoric on climate change and its threat potential has not been matched by the kinds of measures and policy shifts (e.g., the rapid introduction of carbon taxes by governments) one would expect to see if such threats were indeed being taken seriously:

> Whether someone is serious about tackling the global-warming problem can be readily gauged by listening to what he or she says about the carbon price. Suppose you hear a public figure who speaks eloquently of the perils of global warming and proposes that the nation should move urgently to slow climate change. Suppose that person proposes regulating the fuel efficiency of cars, or requiring high-efficiency light bulbs, or subsidizing ethanol, or providing research support for solar power – but nowhere does the proposal raise the price of carbon. You should conclude that the proposal is not really serious and does not recognise the central economic message about how to slow climate change. (Nordhaus 2008, p. 22)[18]

Unlike the transnational threats posed by terrorism, failing states, pandemics, and the ongoing economic fall-out from the sub-prime mortgage meltdown in the U.S., for example, where governments have directly intervened using state resources on the grounds of national security, policy responses to climate change have relied almost entirely on market mechanisms and the private sector for their implementation and funding. And, in so far that governments have taken the lead in implementing climate change responses, these measures have focused much more on the issue of energy security than climate security. Moreover, they are motivated and informed primarily by traditional state-centric notions of security and relative gains in contrast to the more multi-lateral, "global public good" framing of "climate security" characterizing both domestic and international policy debates (Trombetta 2008b). And for the economies of Australia and the U.S., and many developing states, the cure may well represent a bigger threat than the disease. Thus, the Janus-like nature of emission reductions as either a threat or threat response, depending on one's circumstances, perceptions, and priorities, is the main obstacle to anthropogenic climate change becoming securitized to the extent that its potential threats are directly linked by states to their own national security.

The US and Australian governments' change in policy rhetoric indicated their acceptance of global warming as a "political fact" that no longer could be ignored or downplayed, regardless of the many uncertainties that still surrounded its causes and possible impacts, but only in terms that maintained the economy as the *central* referent of security. Bush and Howard also shared very similar views and values on security and the economy as the fundamentals of the "national interest," and in particular on the need to avoid sacrificing any aspect of the national interest on the altar of multi-lateral engagement and commitment (Elliot 2007). The economic costs that both governments believed major emission cuts would have in their respective societies, despite the public mood towards the potential dangers of global warming, remained the dominant influence on the kind of specialist advice the policy elites in these two governments were prepared to accept as the basis of policy. In contrast, the positions on the need for an international response to global warming, especially

[18] It is also worth noting that U.K. emissions began increasing in the late 1990s under the Blair government.

on the kind of role their countries should play in that response, taken by Bush's and Howard's Democratic and Labor Party predecessors stemmed from a national interest perspective more accepting of the potential for environmental threats and also far more supportive of engagement with international institutions as a response to international issues and threats. But as ongoing domestic opposition to pricing carbon in both Australia (currently a Labor-led coalition government) and the U.S. (Democrats) demonstrates, the economic costs of implementing Kyoto were always going to make it an unlikely proposition under either side of politics in Australia and the U.S., given the ongoing fossil fuel dependence both economies share.

One conclusion to draw from the apparent about-face on climate change by President Bush and Prime Minister Howard is that a combination of expert consensus, international pressure, and domestic public opinion finally proved effective in influencing U.S. and Australian policy, giving some weight to rationalist notions about how knowledge can and does play a significant role in policy decisions and also pluralist claims over the ability of the "truth" to finally win out in the course of public debate and then, on the basis of its validation in the marketplace of ideas, become accepted by governments as the legitimate basis of policy. Constructivists and liberal institutionalists interested in the ability of international regimes and emerging norms to influence the policies of member governments and facilitate cooperation among states on international issues also might point to the U.S. and Australian policy shifts as evidence that new norms can evolve and influence state notions of national interest and that cooperation in international politics need not be hamstrung by fears of unequal relative gains and an unwavering commitment to state self interest. According to these perspectives, the case of Iraq would be an aberration in policy terms; a blatant example of political manipulation or ideology perverting the idealized rationalist policy making schema and further proof of the need, therefore, to get politics *out* of policy making.

The explanation presented here, however, paints a very different picture of the role specialist advice and uncertainty played in the policies adopted by the Bush, Blair, and Howard governments – one that looks not only to how established policy priorities and the world views of policy elites shape the interpretation of uncertainties in policy advice, but also to the role such interpretations play in justifying the adoption of some specialist advice over other specialist advice. Moreover, this perspective also questions the extent to which the kinds of explanations outlined above are useful in understanding the pro-Kyoto policies adopted by some governments. In contrast to the Australian and U.S. positions, it also seems clear that those governments that supported the Kyoto Protocol, such as the Blair government, had judged that the political benefits of doing so (e.g., a positive commitment to environment values) could be realized without incurring unacceptable levels of job losses, price increases, and GDP impact.[19] In the U.K. significant emission

[19] Some critics of the science and economic advice used to support the Kyoto reductions even went so far as to suggest that European governments initially supported the protocol's implementation only because they believed that it would never come into force due to the opposition of the U.S. and other likeminded governments. See, for example, Singer (2000).

reductions had already been achieved during the 1990s due to the phasing out of Britain's coal-fired power plants in favour of cheaper and cleaner natural gas powered generators. And in contrast to Australia and the U.S., climate change already had a history of bi-partisan support as a major policy issue – due in no small part to the complementary role climate change threats played within ongoing government plans for privatizing the U.K.'s electrical providers and making the country more energy secure – which dated back to a mix of energy and pro-market reform policies that began with Margaret Thatcher's confrontation with Britain's coal miners and her goal of expanding Britain's supply of nuclear generated electricity.

For some governments, like the Blair government and many of its E.U. partners, the implications of such a paradigm shift had largely been minimized by already implemented industry and energy reforms that had, rather ironically, been deemed necessary for reasons of economic security and growth. In the U.K., the move away from high coal and oil dependency towards cleaner options in the form of natural gas and "renewables" had been underway since the early 1980s. U.K. emissions had been falling since the late 1980s thanks largely to the "dash for gas" and increased emissions regulation that followed Margaret Thatcher's liberalization of the U.K.'s energy industry, at the expense of the country's coal mines, and the discovery of large oil and gas deposits in the North Sea.[20] Moreover, these domestic energy reforms meant that the timing of the U.K.'s reductions meshed nicely with the Kyoto agreement's 1990s baseline for emission reductions. As John Howard remarked in a 2006 interview: "… with very great respect to my good friend Tony Blair, Kyoto was in a sense designed to suit the Europeans because of the starting date and which happened to coincide fairly neatly with some very significant emission-reducing events that took place in Europe" (*Four Corners*, 12 September 2006b).[21] But for the still coal and oil dependent economies of the U.S. and Australia, where emission levels steadily had been increasing, emission reductions of the scale and nature being called for by the Kyoto Protocol and its supporters represented a shift in policy thinking that not only would be difficult to implement but also would certainly involve significant short term economic risk and pain with no more than the prospect of highly uncertain long term benefit in return.

16.5 Conclusion

What has become apparent since the Kyoto Protocol, and its attempt to compel societies and their governments to prioritize "climate security" over economic and energy security, is that the "scientific" consensus, plagued as it is by un-testable

[20] Michael Grubb notes that approximately half of the U.K.'s emissions reduction in the 1990s can be attributed to the switch from coal to natural gas electrical generation. See Grubb (2002, p. 142).

[21] William Nordhaus makes a similar point in criticizing the use of base year emission targets: "Base year emissions have become increasingly obsolete as the economic and political fortunes of different countries have changed. The 1990 base year penalizes efficient countries (like Sweden) or rapidly growing countries (such as Korea and the United States). It also gives a premium to countries with slow growth or with historically high carbon-energy use (such as Britain, Russia, and Ukraine)." See Nordhaus (2005).

assumptions and knowledge gaps, is not on its own a sufficient basis for "political" consensus when important or entrenched values and interests are at stake. The "science informing policy and making it rational depiction" of how policy should be made relies on science being able to provide certainty, or something very close to it, in its explanations of phenomena and cause and effect – at least to the point where all are sufficiently compelled by logic and reason to choose only the "right" response as illuminated by science. But science is seldom able to provide such guarantees and guidance in issues of policy where complexity and uncertainty abound and the making of one decision rather than another can result in heavy losses and consequences, both known and unknown. In post-normal policy issues, Rationalist expectations of science encourage political disagreements to be played out under the guise of scientific debate (since in post-normal issue areas no-one can claim a decisive victory on the basis of science) until one set of values finally becomes dominant and further debate becomes untenable. The scientific advice aligned with the most widely embraced values (e.g., save the whale, anti-smoking, anti-GMF) subsequently becomes the "consensus view" (read "only" scientific view), while all remaining dissenting scientists become contrarians and are tainted by suspicion (e.g., self-interested links with industry and/or government). Thus, the process becomes self-perpetuating since all the while the values that actually drive the debate remain out of the spot light and are therefore never explicitly debated outside the realm of competing knowledge claims. And because a scientific consensus is ultimately declared in support of those values (even though the values are likely to be informing the science, rather than vice-versa), science is seen to have won the day by producing the rational policy!

It is also worth noting that while it is often regular, normal scientific endeavour that identifies the potential problem or risk (as with climate change or ozone depletion for example), the issue then becomes open to being more or less hijacked by whoever has an interest in pursuing it. At this point the pressure for scientific consensus begins to build depending on how well the issue bites politically, causing the normal practice of science, particularly when it is forced to operate in the post-normal realm, to produce ongoing controversy rather than "consensus." There exist, then, fundamental tensions and incompatibilities between what science can actually do and what policy makers, interest groups, the general public want it do (i.e., prove to everyone else that the policy response that supports and reflects their values is the appropriate one!). And therein lies the rub: policy making is necessarily about the future whereas science is restricted to the present and the past for the evidence it relies on to theorize and test knowledge claims. The business of determining policy responses to climate change impacts, for mitigation and adaptation strategies alike, is an entirely political process that must manage competing values, choices, and preferences. And although some scientists appear to think otherwise, policy advocacy is beyond the realm of scientific expertise. While most physical scientists no doubt believe their task is about uncovering the realities of the natural world, policy is about the reality of what is acceptable and therefore achievable in the political world. Even if, for example, everyone accepted the consensus on human induced "global warming" as the most compelling explanation for what is happening today, as most governments now have, we would be no closer to understanding either what global warming means in terms of what will happen tomorrow or reaching agreement on how best to respond to it.

So what should acceptance or rejection of one or another of the various global warming scenarios (the IPCC has produced some 40 "scenarios," not "predictions"), which range from minor to catastrophic climate change consequences, be based upon? Hard evidence, guesswork, ideology, faith, or all of the above? And, most importantly, how does our confidence in such scenarios actually occurring stack up against the costs of taking precautions today against the possible (but unknown) costs of global warming tomorrow? Given the numerous uncertainties that characterize our understanding of the global climate and the effects of our interaction with it,[22] it is not surprising that many governments, especially those in developing countries, are unwilling to accept a high risk of significant economic cost and hardship today – despite the Stern Report's relatively optimist assessment on this point – in order to limit only one of the many variables that may or may not cause future global warming catastrophes. The central policy question then should not be all about who has got it right; we should also be thinking about how we can develop a strategy that best manages the risks involved with getting the causes and potential effects of climate change wrong, at least until we are in a position to more confidently discuss what may or may not happen and adjust our policy responses accordingly.

There are, for example, many policy initiatives that not only would contribute directly to managing the potential for future climate change impacts, but could be justified independently of climate change science. Reframing climate change mitigation to focus on phasing out fossil fuel use rather than reducing carbon emissions would make uncertainty over human influence on the climate largely redundant in policy terms since eliminating industrial carbon output in this way would, in addition to *certainly* reducing carbon levels, also help address a number of other uncontroversial, broadly recognized policy challenges ranging from energy security concerns over future energy supply and competition to the health and environmental impacts of air pollution. Adaptation measures – such as more energy efficient buildings and infrastructure and more careful planning of where and how they are built – offer numerous benefits that would apply regardless of whether carbon emissions will cause sea levels to rise or breed more destructive weather events. And, because the benefits of such measures are not dependent on the accuracy of climate science and its predictions, adopting such a change in focus also would provide us with some insurance against being wrong in our current assessment of how and the extent to which human activity alters climate behaviour. Climate change research in the meantime should continue – not as a justification for delaying policy action until the "facts are in" or in the belief that the certainty policy makers demand can be achieved – but rather in the hope that it will at least explain how some of the many uncertainties we necessarily face in policy making might be better managed.

[22] Not to mention our inability to *know* what either the climate or we might be doing in 50 or 100 years time.

References

ABC Four Corners. 2006a. What price global warming? Australian Broadcasting Corporation broadcast on August 28, 2006. Transcript available at: http://www.abc.net.au/4corners/content/2006s1726376.htm.

ABC Four Corners. 2006b. Interview with John Howard. Australian Broadcasting Corporation, September12. Transcript available at: http://www.abc.net.au/4corners/content/2006/s1738726.htm.

Albright, M. 1997. *Preserving principle and safeguarding stability: United States policy towards Iraq*. Speech by U.S. Secretary of State Madeleine Albright at Georgetown University, Washington DC, March 26, 1997.

Baker, R. 2007. PM feels the heat. *The Age*, February 25. http://www.theage.com.au/news/national/pm-feels-the-heat/2007/02/24/1171734063333.html.

Bocking, S. 2004. *Nature's experts: Science, politics, and the environment*. New Jersey: Rutgers University Press.

Byrd-Hagel Resolution. 1997. 105th Congress, 1st Session, S.Res. 98. United States Senate, July 25.

Chollet, D.H., and J.M. Goldgeier. 2002. The scholarship of decision-making: Do we know how we decide? In *Foreign policy decision-making (revisited)*, ed. R. Snyder, H.W. Bruck, B. Sapin, V.M. Hudson, D.H. Chollet, and J.M. Goldgeier. New York: Palgrave Macmillan.

Crick, B. 2000. *In defence of politics*, 5th ed. London: Continuum.

Dasgupta, P. 2006. *Comments on the Stern review's economics of climate change*. Seminar on the Stern review's economics of climate change, Foundation for Science and Technology at the Royal Society, London, November 8, 2006. http://www.econ.cam.ac.uk/faculty/dasgupta/STERN.pdf.

Daston, L. 2005. Scientific error and the ethos of belief. *Social Research* 72(1): 1–28.

Dawidoff, N. 2009. The lonely prophet. *The Courier Mail Weekend*, April 18–19, 23.

Dodd, T., and T. Youngs. 1998. *The Iraq crisis*. House of Commons Library, Research paper 98/28, February 16, 1998. http://www.parliament.uk/commons/lib/research/rp98/rp98-028.pdf.

Elliot, L. 2007. Pragmatism, prosperity, and environmental challenges in Australia's foreign policy. In *Trading on alliance security: Australia in world affairs*, ed. J. Cotton and J. Ravenhill, 2001–2005. South Melbourne: Oxford University Press.

Fuller, J. 2001. U.S. officials cite serious energy shortage. *Washington File*, U.S. Department of State, April 2, 2001. http://usinfo.org/wf-archive/2001/010402/epf111.htm.

Funtowicz, S., and J. Ravetz. 1992. The good, the true, and the post-modern. *Futures* 24(10): 963–976.

Funtowicz, S., and J. Ravetz. 1993. Science for the post-normal age. *Futures* 25(7): 739–755.

Goldenberg, M. 2006. On evidence and evidence-based medicine: Lessons from the philosophy of science. *Social Science and Medicine* 62(11): 2621–2632.

Grubb, M. 2002. Britannia waives the rules. *New Economy* 9: 139–142.

Habermas, J. 1971. *Toward a rational society*. London: Heinemann Educational Books. First published in German in 1968.

Hajer, M. 2003. Policy without polity? Policy analysis and the institutional void. *Policy Sciences* 36(2): 175–195.

Hall, P. 1993. Policy paradigms, social learning, and the State: The case of economic policymaking in Britain. *Comparative Politics* 25(3): 275–296.

Hodge, A., and S. Maiden. 2006. Ferguson splits left on Kyoto. *The Australian*, January 13.

Jane's Intelligence Digest. 2003 *Iraq: A war without intelligence*. October 30. http://articles.janes.com/articles/Janes-Intelligence-Digest-2003/Iraq-a-war-without-intelligence.html.

Jasanoff, S. 1990. *The fifth branch: Science advisers as policy makers*. Cambridge, MA: Harvard University Press.

Lindblom, C.E. 1959. The science of muddling through. *Public Administration Review* 19: 79–88.

Lindblom, C.E. 1979. Still muddling, not yet through. *Public Administration Review* 19: 517–526.

March, J., and J. Olsen. 1989. *Rediscovering institutions: The organizational basis of politics*. New York: Free Press.

Neumayer, E. 2007. A missed opportunity: The Stern review on climate change fails to tackle the issue of non-substitutable loss of natural capital. *Global Environmental Change* 17: 297–301.

Newport, F. 2009. *Americans: Economy takes precedence over environment*. March 19. http://www.gallup.com/poll/116962/Americans-Economy-Takes.

Nieman, M., and S. Stambough. 1998. Rational choice theory and the evaluation of public policy. *Policy Studies Journal* 26(3): 449–465.

Nordhaus, W. 2005. *Life after Kyoto: Alternative approaches to global warming policies*. Yale University, December 9. http://www.econ.yale.edu/~nordhaus/kyoto_long_2005.pdf.

Nordhaus, W. 2006. *The Stern review on the economics of climate change*. November 17. http://qed.econ.queensu.ca/pub/faculty/milne/872/SternReviewD2.pdf.

Nordhaus, W. 2008. *A question of balance: Weighing the options on global warming policies*. New Haven: Yale University Press.

Parliament of the Commonwealth of Australia. 2004. *Intelligence on Iraq's weapons of mass destruction*, Parliamentary paper (Australia. Parliament), no. 47. Canberra: Parliamentary Joint Committee on ASIO, ASIS and DSD.

PBS Online NewsHour. 2001. *Bush and the environment*. March 29. http://www.pbs.org/newshour/bb/environment/jan-june01/bushenv_3-29.html.

Peatling, S., and W. Frew. 2006. Greenhouse battle handed to industry. *The Sydney Morning Herald*, January 13.

Pielke Jr., R. 2005. Misdefining "climate change": Consequences for science and action. *Environmental Science and Policy* 8: 548–561.

Pielke Jr., R. 2007. *The honest broker: Making sense of science in policy and politics*. New York: Cambridge University Press.

Pielke Jr., R. 2006. Stern's cherry picking on disasters and climate change, *Prometheus*, October 30. http://sciencepolicy.colorado.edu/prometheus/archives/climate_change/000973sterns_cherry_picki.html.

Pielke Jr., R., and D. Sarewitz. 2002/2003. Wanted: Scientific leadership on climate. *Issues in Science and Technology* 19(2): 27–30.

Putnam, R. 1988. Diplomacy and domestic politics: The logic of two level games. *International Organization* 42(3): 427–460.

Ravetz, J. 1999. What is post-normal science? *Futures* 31: 647–653.

Saad, L. 2009. *Increased number think global warming is "exaggerated"*. March 11. http://www.gallup.com/poll/116590/Increased-Number-Think-Gl.

Sarewitz, D. 2000. Science and environmental policy: An excess of objectivity. In *Earth matters: The earth sciences, philosophy, and the claims of community*, ed. R. Frodeman, 79–98. Upper Saddle River: Prentice-Hall.

Sarewitz, D. 2004. How science makes environmental controversies worse. *Environmental Science and Policy* 7(5): 385–403.

Singer, S.F. 2000. *Climate policy – From Rio to Kyoto: A political issue for 2000 and beyond*. Stanford: Stanford University: Hoover Institution on War Revolution and Peace.

Stone, D. 2002. *Policy paradox. The art of political decision making*. New York: Norton & Company.

The Age, I won't ratify Kyoto Protocol: PM. December 2, 2003. http://www.theage.com.au/articles/2003/12/02/1070127416251.html.

The Independent. 2005. Blair falls into line with Bush view on global warming. September 25. http://www.independent.co.uk/environment/blair-falls-into-line-with-bush-view-on-global-warming-508336.html.

The Second Gore-Bush Presidential Debate. 2000. *The commission on presidential debates*, October 11.

The Sydney Morning Herald. 2006. Howard talks up breakaway climate group at summit. November 17. http://www.smh.com.au/news/environment/howard-talks-up-breakaway-climate-group/2006/11/16/1163266712683.html.

The Washington Post. 2007. President Bush's 2007 State of the Union Address (as delivered in the House Chamber). January 23. http://www.washingtonpost.com/wpdyn/content/article/2007/01/23/AR2007012301075.html.

Trombetta, M.J. 2008a. Environmental security and climate change: Analysing the discourse. *Cambridge Review of International Affairs* 21(4): 585–602.

Trombetta, M.J. 2008b. Environmental security and climate change: Analysing the discourse. *Cambridge Review of International Affairs* 21(4): 599–601.

U.K. Concept Paper. 2007. *Energy, security, and climate.* U.K. Mission to the U.N., April 23. http://unfccc.int/files/application/pdf/ukpaper_securitycouncil.pdf.

United States Senate. 2006. *Postwar findings about Iraq's WMD programs and links to terrorism and how they compare with pre-war assessments.* Select Committee on Intelligence, 109th Congress. September 8. Washington, DC: U.S. G.P.O. http://intelligence.senate.gov/phaseii-accuracy.pdf.

Zillman, J. 1997. *The IPCC: A view from the inside.* Melbourne: Australian APEC Study Centre. http://www.apec.org.au/docs/zillman.pdf.

Michael Heazle is an Associate Professor with the Griffith Asia Institute and the Griffith University School of Government and International Relations where he teaches international relations and politics. Dr. Heazle taught in the faculties of Economics and Business Administration at Ritsumeikan University, Kyoto, Japan. He has researched and published in the areas of energy, human, and environmental security; policy making and the treatment of specialist advice; and China-Japan relations. He is the author of *Scientific Uncertainty and the Politics of Whaling* (2006). His most recent books are *Uncertainty in policy making: Values and evidence in complex decisions* (2010), and *Understanding Australia's neighbours: An introduction to East and Southeast Asia,* (2011) with Nick Knight.

Chapter 17
Water Risk Management, Governance, IWRM and Implementation

Bruce Mitchell

17.1 Introduction

Most societies are at risk, and therefore vulnerable, because of inability to deal well with multiple and inter-related natural and human systems characterized by ongoing change, complexity and uncertainty, all of which enhance risks. Organizations usually do not perform well when, to be effective, each needs to function harmoniously in a network of stakeholders with varying needs, interests and priorities. Instead, too often, individual organizations emphasize 'local optimization', with negative consequences for 'system optimization', and resultant conflict.

Integrated water resource management (IWRM) is one approach or means to resolve the challenges arising from the 'silos' that often characterize water management structures, mechanisms and processes. The intent of IWRM is to achieve effectiveness and efficiency through systematic coordination and collaboration (Mitchell 1990. However, IWRM is not without challenges (Molle 2008; Chéné 2009; Butterworth et al. 2010). Critics argue that IWRM has not been defined adequately, unrealistically assumes equitably distributed power among stakeholders, and has frequently led to ineffective implementation (Biswas 2004, 2008a, b; Rahman and Varis 2005).

Implementation challenges are endemic in planning and management (Pressman and Wildavsky 1971; Mazmanian and Sabatier 1989; Goggin et al.

B. Mitchell (✉)
Department of Geography and Environmental Management, University of Waterloo, 200 University Avenue West, Waterloo, ON N2L 3G1, Canada
e-mail: mitchell@uwaterloo.ca

U. Fra.Paleo (ed.), *Risk Governance*, DOI 10.1007/978-94-017-9328-5_17

1990; Weale 1992; Joseph et al. 2006). Therefore, they are not unique to IWRM. Nevertheless, to reduce vulnerability from water security or hazard problems, more effort is needed to overcome the 'implementation gap'. A specific way to resolve implementation challenges regarding IWRM is to improve governance capacity.

The purpose here is to explore how an integrated approach and innovative governance can improve implementation of strategies and plans, and thereby reduce risk and vulnerability, related to water problems and management.

17.2 Water Risk Management

17.2.1 Risk Defined

Smith (1996, p. 5) has considered the relationship between hazards and risks. He notes that risk is often treated as interchangeable with hazard, but observes that risk incorporates an additional dimension: the probability of a particular event occurring. Thus, risk combines attention both to exposure to hazard and to likelihood of loss. He highlights the significance of these two aspects with an example. Two people set out to cross an ocean, one on a liner and the other in a rowing boat. The most significant hazards (deep water, high waves) are present for both individuals. However, the risk (likelihood of drowning) is much higher for the individual in the rowing boat. Thus, a key conclusion is that people and their choices plus behaviour contribute in a major way to the types and magnitudes of risk to which they become exposed. This position is re-iterated by Blaikie et al. 1994, p. 23) who observe that, "…our view is that the risk faced by people must be considered as a complex combination of vulnerability and hazard."

17.2.2 Implications for Governance and Management

Given that Smith (1996, p. 5) argues, "People, and what they value, are the essential point of reference for all risk assessment….", the implications of risk for governance and management are nontrivial. As noted in the next section, if governance focuses on how decisions are implemented, including mobilization of resources and taking action, then governance is a key component for societies striving to reduce risk and vulnerability triggered by threats to water, whether it be too much or too little water, or water of unsuitable quality. In a more general way, and as highlighted by Smith (1996, p. 54): "There are great difficulties in deciding what is an acceptable level of risk, who benefits from risk management, who pays, and what constitutes success or failure." Such matters are often at the core of governance and management.

17.3 Governance

17.3.1 Governance Defined

Various interpretations exist regarding governance related to water (Global Water Partnership 2002; Rogers and Hall 2003; Tortajada 2010. Bakker (2007, p. 16) differentiates between governance and management in the following way,

> Simply put, '… governance' refers to the decision-making process we follow, whereas '… management' refers to the operational procedures we adopt. Governance refers to how we make decisions and who gets to decide; management refers to the models, principles, and information we use to make those decisions. Obviously the two are interrelated; however, management is often the focus of debate, whereas governance is often overlooked.

Olsson (2007, p. 269) extends Bakker's comments by distinguishing among governance, management and monitoring:

> … governance is the process of resolving trade-offs and providing a vision and direction for sustainability, management is the realization of this vision, and monitoring provides feedback and synthesizes the observations to a narrative of how the situation has emerged and might unfold in the future.

With regard to water and governance, the World Water Assessment Programme (2006, p. 47) stated that:

> … The governance of water in particular can be said to be made up of the range of political, social, economic and administrative systems that are in place, which directly or indirectly affect the use, development and management of water resources and the delivery of water services at different levels of society. Governance systems determine who gets what water, when and how and decide who has the right to water and related services and their benefits. The representation of various interests in water decision-making and the role of politics are important components in addressing governance dynamics. …. In short, governance is about making choices, decisions and trade-offs.

And, Lautze et al. 2011, p. 7) defined water governance and identified desirable qualities for it. In their words,

> Water Governance consists of the processes and institutions by which decisions that affect water are made. Water governance does not include practical, technical and routine management functions such as modelling, forecasting, constructing infrastructure and staffing. Water governance does not include water resource outcomes.
>
> Good Water Governance Qualities can be proposed as: openness and transparency; broad participation; rule of law (predictability); and ethics, including integrity (control of corruption).

17.3.2 Significance of Governance

The World Water Assessment Programme (2006, p. 47) concluded that governance usually has not been given as much attention as technical matters. It further highlighted that governance requires attention to relationships between organizations and social groups involved in water decision making, horizontally across sectors and between urban and rural areas, and vertically, from local to international levels. It also stated that key operating principles for best practice in governance are downward and upward accountability, transparency, participation, equity, rule of law, ethics and responsiveness.

Finally, it argued that governance includes much more than water policies. It must include them, but in addition must consider

> .... the exercise of power in policy-making and whether or not to implement particular policies. Which actors were involved in influencing the policy in question? Was the policy developed in a participatory and transparent fashion? Can revenues and public and bureaucratic support be raised to implement the policy? These are just some of the important questions involved, but they indicate that governance is about the process of decision-making, its content and the likelihood of policies and decisions to be implemented. To be able to understand why water is allocated in different ways, it is necessary to look into the dynamics of policy and decision-making, informal and formal legislation, collective action, negotiation and consensus-building, and how these interact with other institutions (World Water Assessment Programme 2006, p. 48).

17.4 Integrated Water Resource Management

17.4.1 IWRM Defined

The Global Water Partnership (GWP) (2000, p. 22) has defined integrated water resource management (IWRM) as "a process which promotes the coordinated development and management of water, land and related resources, in order to maximize the resultant economic and social welfare in an equitable manner without compromising the sustainability of vital ecosystems." Although the GWP statement is the most frequently cited interpretation of IWRM, others exist. For example, Jaspers (2003, p. 91) stated that integrated river basin management "can be understood as the management of all surface and subsurface water resources of the river basin in its entirety with due attention to water quality, water quantity and environmental integrity." He added that "a participatory approach is followed, focusing on the integration of natural limitations with all social, economic and environmental interests". This latter observation is reinforced by Braga's (2001, p. 581) view that its multidisciplinary nature requires "... many professionals to deal with the different aspects of the planning process".

17.4.2 Three Levels of IWRM

Garcia (2008, pp. 28–29) argued that it is desirable to differentiate among three levels of IWRM. First, at a *constitutional* level, attention should focus upon actions to draft, approve and apply basic water legislation and IWRM policies and strategies to create the necessary enabling environment. And, although not mentioned by Garcia, a precondition for establishing and applying legislation, policies and strategies is sustained commitment from key political and management leaders. Second, at an *associative* level, the need is to establish the means to facilitate a river basin or watershed approach, including creation of river basin organizations which can allocate water flows, determine assimilative capacity, and facilitate ecosystem maintenance. Third, at an *operational* level, consideration is given to financing such aspects as water supply and sanitation, irrigation and drainage and hydropower projects in order to meet demands and needs. Garcia (2008, p. 31) concluded that most effort has been at the constitutional level, and, even when that has been done well, "few actions meaningful for the end-user have resulted so far. There is a tendency to be heavy on diagnostics but weak in solutions …."

Several of the definitions of 'governance' in Sect. 17.3.1 above, the three levels of IWRM identified by Garcia (2008) in this section, and the governance arrangements identified by Jaspers (2003) in Sect. 17.4.3 below have also been applied to analysis of experience with IWRM in CHINA (Mitchell 2013).

17.4.3 IWRM and Governance Arrangements

The connections between IWRM and governance are attracting increasing attention (Conca 2005; Mollinga et al. 2006; Kemper et al. 2007; Lautze et al. 2011). In terms of governance arrangements to facilitate IWRM, Jaspers (2003, p. 83) concluded the following are essential: (1) management based on hydrological boundaries, (2) an organizational structure involving river basin and sub-basin authorities with their respective by-laws to incorporate decision making at the lowest appropriate level, (3) a planning system oriented to production of integrated river basin plans, (4) mechanisms to facilitate stakeholder involvement in decision making, and (5) a system of water pricing and cost recovery.

Regarding the river basin as the spatial unit for management, Jaspers (2003, p. 81) argued that "the need for water management on hydrological boundaries is mainly triggered by the growing competition for water or by the need to co-operate in an upstream-downstream relation for flood control or both". Furthermore, he argued that "The setting of priorities for water allocation in an equitable and efficient way other than on hydrological boundaries is, by definition, physically impossible".

For organizational arrangements, Jaspers (2003) stated that three ways exist for the public sector to create regional organizations: (1) *de-concentration*, in which executive tasks and competencies are assigned to regional offices of a central authority or to lower levels within a central authority structure. Basic authority and responsibility remain with the central institution; (2) *delegation*, in which executive tasks and competencies are given to some other public or private organization, with transfer of responsibility but without irreversible transfer of authority; and, (3) *devolution*, in which executive tasks and competencies are given to other administrative organizations on an ongoing basis and with a complete shift of both responsibility and authority. He observed that while examples of devolution are infrequent, delegation is very common and increasing examples are found of de-concentration.

And, it should be highlighted that without solid political support for IWRM none of the three options is likely to be successful.

The specific tasks and functions of river basin authorities can differ significantly. However, Blanco (2008) suggested that a river basin organization will function effectively in allocating water if it is given a clear right to distribute water, if it does not have to serve too many water users, if it is located in close proximity to its users, and if it has sufficient information about the natural system and its human users. Jaspers (2003) also suggested that normally a river basin authority emphasizes collective choice functions whereas a sub-basin authority or a water users' association focuses on operational matters.

With reference to an integrated planning system, a key output of IWRM should be river basin plans to facilitate horizontal and vertical integration of initiatives for water quantity, water quality, environmental services, and ecosystem integrity. Ideally, lower level sub-basin plans are completed first and then are drawn upon to develop a plan for the entire river basin. In Ontario, Canada, four levels of plan are used, moving from the greatest detail and smallest area to less detail and larger area: site, tributary, sub-basin and basin (Mitchell and Shrubsole 1992; Ontario 1993, 1997).

Regarding a participatory approach, Jaspers (2003, p. 83) observed that "A crucial issue is the process of stakeholder participation. It has become very clear that water resources planning without the participation of stakeholders in decision making is highly ineffective". He further argued that participation of stakeholders in decision making or in other management functions is essential (see also Veale 2010).

Water pricing and cost recovery initiatives can be controversial, especially if charges make the poorest people in a river basin unable to access sufficient water quantity and quality to meet basic needs. Nevertheless, a key principle should be that the price for being serviced with potable raw water, being protected through water treatment services, or being protected against flooding should be paid by the user or polluter or beneficiary. However, there needs to be capacity to understand the ability and willingness to pay, and to decide explicitly whether subsidies will be provided to some sectors or groups.

It is self-evident that many of the above characteristics require systematic information and understanding about the behaviour of natural and human systems. It is also indisputable that managers often do not have all desired knowledge and insight on which to base decisions. Consequently, all management decisions involve potential for unintended negative outcomes, surprises, and mistakes, highlighting that attention must be given to determine how aspects of risk should be addressed.

17.4.4 Best Practice

Based on the literature and practice, elements of best practice for IWRM include:

- A vision, policy and strategy for IWRM, including a clear definition of IWRM.
- A legislative basis to provide credibility and authority for IWRM. A precondition for a statutory foundation is political commitment and endorsement for IWRM,

along with concomitant policies and strategies. This and the previous element address the constitutional level identified by Garcia (2008).

- A river basin organization to coordinate development and management in a river basin. This element reflects Garcia's associative level, and will be most effective if based on the delegated or devolution model (Jaspers 2003).
- Explicit specification of functions, responsibility and authority of a river basin organization, to address Garcia's operational level. Functions, responsibility and authority should reflect the following (Blanco 2003; Jaspers 2003): and facilitate application and coordination of initiatives regarding:
  - upstream and downstream needs and interests related to water use and discharge of wastes, through planning and managing at river basin and sub-basins scales;
  - water supply, water pollution and flooding;
  - administrative (permits for water use and waste discharges): economic (charges for use and pollution) and planning (regional land and water use plans) instruments;
  - protection and enhancement of both economic development and ecological integrity;
  - interactions among water, land and other resource systems;
  - ensuring data are valid and reliable relative to various aspects of the hydrological system, and are accessible to all stakeholders in a river basin;
  - explicit means to address risk and uncertainty, in order to acknowledge imperfect understanding and knowledge, as well as evolving contexts and conditions; and,
  - issues associated with vertical and horizontal administrative structures;
- Public participation and community involvement, to facilitate transparency and accountability; and,
- Creation and updating of an IRWM strategy and plan.

17.5 Overcoming Implementation Barriers

"Implementation failure is like original sin; it is everywhere and it seems ineradicable" (Weale 1992, p. 43). In planning and management, what is often referred to as the 'implementation gap' is well recognized, reflecting the difficulty in moving from visions and plans to action.

17.5.1 Challenges for Effective Implementation

Water management encounters implementation challenges due to complexities arising from inter-connections among water, land and other resources, surface and ground water, and upstream and downstream areas of basins, as well as the role of water related to economic development and ecosystem integrity (Hartig and Law 1994; Gurtner-Zimmermann 1995; Jones and Taylor 1999; McLaughlin and Krantzberg 2011). By explicitly addressing such inter-connections, it should be possible to be effective, efficient and equitable. Nevertheless, we should recall that "There is no simple solution for implementing the concept of IWRM; it depends on the particular framework and institutions related to the water resources of a country" (Blanco 2008, p. 91).

17.5.2 Best Practice

For effective implementation, experience suggests at least the following need to be addressed (Mitchell 2009, pp. 8–9, 2011):

- Appreciation of the importance of context or local conditions, leading to custom-designed solutions and avoidance of a standardized, 'one size fits all' approach.
- A long-term perspective. Many resource and environmental problems evolve over decades and most will not be resolved in the period between the usual election cycles. Decades often are required to slow down, stop and reverse degradation or to resolve scarcity problems.
- A vision outlining the future desired condition. Different values, needs and interests are often major obstacles to develop a vision. But, without one, it is difficult to know which road to follow.
- Creating legitimacy or credibility for IWRM, best achieved through a combination of ongoing commitment from elected leaders, statutory foundation based on laws and regulations, endorsement through policies, structures and processes associated with governance arrangements, and sufficient resources (financial and human) to facilitate necessary work.
- One or more leaders or champions in place. Presence of a committed and talented leader is often the key variable for successful implementation.
- Willingness to share or redistribute power and authority to facilitate desired and positive change – difficult to achieve when a public agency has statutory responsibility for specified functions or issues.
- A multi-stakeholder approach so that various values, interests and needs are incorporated.
- Appreciation that turbulence and surprises will be encountered. Such conditions require capacity for flexibility, resilience and adaptability, in order to cope with risk and uncertainty.
- Up-front commitment to monitor and assess results of implementation, accompanied by willingness to learn from experience.
- High quality communication – in 'plain language', understandable to all participants.
- Demonstration projects to provide tangible evidence of action and progress in the short term.

- Celebrating accomplishments, with credit openly acknowledged. Such occasions help people to carry on during the difficult times when they can become discouraged.

Attention to the above considerations does not guarantee implementation but experience suggests attention to them leads to a higher probability of success in implementation.

17.6 Example, Water Management in Ontario, Canada, Related to Flooding and Water Quality

The Ontario conservation authorities in general, and the Grand River Conservation Authority (GRCA) in particular, are the focus here to examine the way in which governance, integration, risk management and implementation are addressed by them.

17.6.1 Ontario Conservation Authorities

Thirty-six conservation authorities exist in Ontario. Established by legislation in 1946, they illustrate integrated water resources management in action. Two authorities, the Grand River Conservation Authority (GRCA) and the Lake Simcoe Region Conservation Authority (LSRCA), received the Thiess International River*prize* in 2000 and 2009, respectively, in Brisbane, Australia. The Thiess International River*prize*, first awarded in 1999 and worth $350,000 in 2012, is awarded "to those who have developed and implemented outstanding, visionary and sustainable programs in river management".[1] The main considerations in selecting the award recipient are "measurable outcomes demonstrating maintenance of or improvements in whole of basin aquatic ecosystem health." Special attention is given to "actions to achieve aquatic ecosystem sustainability, focusing on ecological, social and economic values." Furthermore, organizations are required to provide evidence of delivering high levels of effective and efficient program outcomes, along with a sustained track record of value for money, public accountability, inclusiveness (engaging all relevant interest groups, genuine participation): and innovation (use of technology, program design, research, planning, management practices, policy development, institutional arrangements).

The GRCA was recognized for its long-term restoration and management plan and related initiatives for the Grand River basin, significant improvement of water quality due to changed agricultural practices, enhancement of fish habitat and wildlife, and increasing use of the river system for recreation. The LSRCA documented

[1] International RiverFoundation http://www.riverfoundation.org.au/riverprize_about.php.

that, by 2009, through IWRM, it had reduced phosphorus in Lake Simcoe by 35 % relative to the worst conditions in the early 1990s, while dissolved oxygen in deep waters of the lake had been increased by more than 200 %. The improved dissolved oxygen levels are essential for survival of the lake trout fishery, a defining feature of the aquatic system. Such results and recognition suggest that these two conservation authorities have learned how to deal with risk management and overcome 'implementation gaps'.

Given the track record and external recognition outlined above, the conservation authorities provide an opportunity to examine interrelationships among governance, integration and implementation in the context of managing water and related resources to deal with risk and vulnerability (Shrubsole 1996). This view is shared by others. Almost 50 years ago, Lord (1962, p. 28) observed that the conservation authorities had attracted "worldwide attention". Shortly thereafter, Hamilton (1971, p. 110) concluded that on a global scale they had come the closest to achieving "unified action in water management". In reflecting on the first 25 years of their experience, Lord (1974, p. ix) suggested "the success of the conservation authorities has resulted from sound basic thinking and shrewd planning in the formative years. The initial concepts were simple, but so fundamentally right that only minor adjustments have been necessary." And, more recently, Watson (2004, p. 244) observed that "Today, the Ontario Conservation Authorities are internationally recognized as leading examples of an integrated approach to watershed management."

This case study draws from Mitchell and Shrubsole (1992): with updating (Plummer et al. 2005) and ongoing research.

17.6.1.1 Structural and Non-structural Adjustments

The GRCA uses both structural and non-structural measures to reduce the risk of flood damage. Structural measures include a network of seven dams and reservoirs to hold water and reduce peak flows. The dams serve multiple purposes, however, as water is released during summer months to augment flow and thereby enhance water quality. In this way, both the risk of flooding and of poor quality water is reduced. In addition to the dams, the GRCA owns and maintains dyke systems, especially in urbanized areas, which complement the upstream dams to protect areas on or adjacent to floodplains. To guide decisions related to reduction of flow peaks and enhancement of summer flows, the GRCA monitors weather conditions and river flows through a network of gauging stations.

Regarding non-structural measures, the GRCA has a flood forecasting system, and each winter conducts a flood warning system test to ensure messaging (advisories and/or warnings) reaches intended audiences. Such testing is a critical component of a risk reduction strategy, as it is critically important that information about potential flooding conditions reaches those who could be affected, as well as for them to know what reactions are appropriate to reduce the risk. It also has a reforestation program throughout the watershed, with the objective to stabilize terrestrial systems and reduce surface run off. The GRCA further controls development in flood-prone lands through land-use zoning and development controls, in collaboration with municipalities. Following a serious flood in 1974, the GRCA acquired properties in vulnerable areas adjacent to shorelines of the Grand River in Cambridge. Buildings on such properties, with the exception of selected heritage buildings, were then torn down, and the properties became part of a river-side dyke and park system.

By combining structural and non-structural measures, the GRCA addresses inter-relationships between land and water systems, is attentive to upstream and downstream characteristics, and uses a multiple-purpose approach to address various management objectives: flood damage reduction and water security (both quantity and quality): all of which is consistent with IWRM. These initiatives are undertaken to reduce risks associated with flooding and poor water quality, based on a belief that a mix of complementary measures is needed. A governance model facilitating cooperation and coordination between the GRCA and municipalities is a critical variable allowing systematic implementation.

17.6.1.2 Source Water Protection

In mid May 2000, 0157LH7 strain *E. coli* bacteria entered the municipal water supply system of Walkerton, a town of about 5,000 people in southwestern Ontario. Seven people died, and more than 2,300 became ill. A public inquiry followed. In his report, Justice Dennis O'Connor (2002a, b) concluded that the *E. coli* entered the supply system through one well due to spreading of manure on a field adjacent to the well; the farmer had followed proper procedures when spreading the manure; chlorination equipment, which if operating would have prevented the problem, was not functioning due to repair work; the supply system operators had a history of errors; the general manager withheld information from public health officials, making the situation worse; approvals and monitoring systems of the provincial government were not adequate; and, budget cutbacks had reduced capacity of government laboratory testing services (see also Perkel 2002). In combination, all except the first two aspects above exacerbated the risk of contamination of the drinking water for the community.

A key recommendation of Justice O'Connor was that, to ensure safe drinking water, and reduce risk, a multi-barrier approach should be used. In other words, a series of measures should be used to protect drinking water quality, so that if any one did not function then others still would. Such an approach involves deliberate redundancy in protective measures. Justice O'Connor also stated that a comprehensive approach to manage all aspects of watersheds was essential. Subsequently, he recommended that watershed-based source protection plans should be created. To ensure a local perspective and buy in, source protection planning should be undertaken wherever possible at a local level and be led by those most affected (i.e., municipalities and other local groups). With regard to how a strong source protection program can reduce risks, Justice O'Connor (2002b, p. 8) remarked that:

> It [a strong source protection program] lowers risk cost-effectively: keeping contaminants out of drinking water sources is an efficient way of keeping them out of drinking water. This

> is particularly so because standard treatments cannot effectively remove certain contaminants. And protecting drinking water sources can in some instances be less expensive than treating contaminated water so that it meets required safety standards.

The government of Ontario passed the Ontario Clean Water Act, 2006, and then introduced an Ontario Drinking Water Source Protection Program. To implement the source protection approach, Source Protection Committees were established (Ontario Ministry of Environment 2008, pp. 1–2). Their members come from municipal councils, and their focus is at a watershed scale. Eleven source protection regions and eight stand-alone source protection areas have been designated in the province. For each source protection region, one conservation authority coordinates the initiatives of all committees within the region. The committees are responsible to develop terms of reference, prepare an assessment report with attention to risks regarding drinking water, and create a source protection plan.

The membership includes municipalities, conservation authorities, business people, farmers, and other stakeholders. Municipalities prepare and implement policies to reduce risks to water supply systems under their responsibility. An example would be requiring property owners to remove threats to drinking water located on their properties. Conservation authorities have a complementary role. As the Ontario Ministry of Environment (2008, p. 2) states, based on their watershed perspective, the authorities "… will help source protection committees develop source protection plans by gathering and sharing information, facilitating cooperation and coordination among communities and stakeholders, and providing technical support and advice to the source protection committees."

Preparation of the Grand River basin Source Protection Plan is led by the Lake Erie Region Source Protection Committee, with responsibility for the Grand River basin and three other adjacent basins. The GRCA is providing the support outlined above. The first step is preparation of an assessment report to identify possible threats or risks to water quality and supplies. In addition to identifying vulnerable areas in the watershed related to drinking water sources, the assessment specifies number and types of threats, and ranks them as low, moderate or significant. The second step is creation of a Source Protection Plan to outline actions to eliminate or reduce significant existing threats or to prevent new threats. This plan was to be completed by 2012; it would be reviewed and approved by the provincial government.

Worte (2010) observes that the source protection approach should be but one component of IWRM. In his view, although the Clean Water Act specifies a comprehensive watershed approach, its focus is on municipal drinking water sources. While drinking water safety is important, Worte argues it is only one aspect of risk and vulnerability for society related to water and land management (e.g., floods, drought). Worte noted that Justice O'Connor (2002b, p. 9) recognized the need to move beyond a single-issue perspective when writing that, "I want to emphasize that a comprehensive approach is needed and should be adopted by the Province. Source Protection plans should be a subset of broader watershed plans."

Worte went on to state that the desirable, broader approach can be achieved through Integrated Watershed Planning which recognizes the important interconnections between land and water systems, the foundation for Ontario's conservation authorities' approach to managing on an ecosystem basis. In Worte's (2010, p. 7) words,

> Integrated Watershed Management is the process of managing human activities and natural resources on a watershed basis, taking into account social, economic and environmental issues, as well as community interests, in order to manage water resources sustainably.

Worte (2010, p. 7) also observed that IWM is increasingly being adopted in Canada, and that most provincial water strategies include it as a fundamental principle. In his view, the benefits of IWM are compelling:

> Effective IWM ultimately leads to better decision making, smarter priority setting, opportunities to pool existing resources, and increased efficiency between a variety of stakeholders such as municipalities, residents, provincial agencies and businesses.

17.7 Conclusions and Implications

17.7.1 Water Risk Management

Risks and hazards triggered by water can rarely be totally eliminated. Instead, risks need to be identified, and then explicit decisions taken related to what vulnerability is acceptable, what amount and kinds of resources (financial, human) will be allocated to address them, and what capacity needs to be created to support individuals and communities which suffer from the consequences of being exposed to hazards associated with risks.

Experience is being gained related to effective governance approaches related to Integrated Water Resources Management to ensure a holistic approach underlies management decisions related to water risk. In addition, improvements are being made to ensure effective transition from strategies and plans to implementation of specific initiatives. Challenges continue to be encountered, and setbacks occur. Nevertheless, accumulated experience by organizations such as the Ontario Conservation Authorities indicates that capacity to identify and manage risks and hazards is steadily improving. In the following subsections, more specific observations are presented.

17.7.2 Integrated Water Resources Management and Governance

The Ontario conservation authorities (CAs), and the GRCA in particular, exemplify the principles related to IWRM identified at the beginning of this chapter (Jasper 2003; Garcia 2008). Garcia stipulated a constitutional foundation, satisfied by the Conservation Authorities Act, 1946, and the Grand Strategy. The associative requirement for a river basin agency is met through the 36 watershed-based conservation authorities and their integrated approach to manage water, land and other related natural resources. The GRCA and other conservation authorities further reflect Jasper's (2003) devolution model, in which executive tasks and competencies are allocated to them on an ongoing basis and with a complete shift of both responsibility and authority to them. The operational requirement for financing is met through the CAs receiving levy-based funding from municipalities in the catchment. Jasper's (2003) other principles are met because CAs mange on the basis of hydrological boundaries, prepare and update strategic river basin plans, actively engage stakeholders, and have capacity through user fees for cost recovery.

In terms of other aspects of best practice identified earlier in this chapter, the GRCA has an explicit vision and strategy, and receives legitimacy through the Ontario Conservation Authority Act as well as other statutes (e.g., Clean Water Act) and programs (e.g., Ontario Drinking Water Source Protection Program). Vertical and horizontal coordination is sought through partnerships with provincial government departments (e.g., Agriculture, Environment, Natural Resources) as well as local governments. In addition, cooperation is frequent with federal agencies (e.g., Environment Canada, Fisheries and Oceans Canada).

The CAs deal with multiple management issues, including water security, pollution, flooding, ecological protection, recreation and education. The CAs apply administrative, financial and planning instruments to facilitate implementation, and explicitly seek to balance economic development and ecological integrity. Development of the source protection plan has been based on extensive public participation and community involvement. There is a commitment, and has been explicit action, to update and renew the basin-wide IWRA plan.

Finally, the GRCA collects weather and hydrological data compatible with province- and nation-wide data systems. In that regard, it should be highlighted that part of the success of the GRCA can be attributed to use of substantial information acquired over decades related to hydrological processes, water quality, water uses, and land-use impacts to inform decisions related to risk management. Furthermore, such data have been used in science-based modeling to help identify potential issues and to forecast trends as well as to develop strategic responses to risks. And, ongoing monitoring of watershed conditions and processes provides information about trends and thresholds that help in anticipating possible risks.

17.7.3 Implementation

Regarding aspects to facilitate systematic and effective implementation not already addressed in the above subsection, each CA is attentive to the significance of local context by preparing basin-specific integrated plans, while simultaneously reflecting principles and meeting expectations of the provincial government. The leadership role of the chair of a conservation authority is recognized as very important. The GRCA has chosen chairs who have arrived with or cultivated effective networks at local, provincial and national scales, have often taken on leadership roles in provincial or national organizations, and, in some cases, have led major international initiatives.

The GRCA shares power with local governments. For example, the floodplain regulations of the GRCA must be coordinated with land use zoning regulations of the local governments if that non-structural measure is to be effective. The GRCA applies a multi-stakeholder approach, given that it is in a formal partnership with the provincial and local governments. It also must work with a First Nations community which occupies significant land on both sides of the Grand River in the lower central part of the basin.

The GRCA has a number of tools or means to help it manage risks which emerge because of the reality of multiple jurisdictions being involved in management decisions for water and land in the Grand River basin. Thus, the Board of Directors, by having members representing local municipalities, allows a direct link between the GRCA and municipalities in the watershed. In addition, sophisticated strategies are used to facilitate communications within the basin and to monitor the impacts of decisions implemented. In addition to being tightly linked to municipalities in the watershed, the Board works diligently to maintain effective links with provincial ministries which have responsibility and authority related to management of land and water. These relationships with municipal and provincial partners contribute to a collaborative approach to development of policies, as well as coordination of initiatives related to risk from both land and water issues

Monitoring and assessment of experience is demonstrated through watershed report cards which track activities and progress, and are accessible in print or electronic form. Annual conservation awards are given to acknowledge landowners, business people and others who have taken initiatives to improve conditions in the watershed. In this manner, the GRCA both celebrates successes and allocates credit. Such awards are celebrated when they are presented to recipients, and are publicized in both electronic and print newsletters.

17.7.4 Limitations

While CAs exemplify an IWRM approach, innovative governance and effective implementation, they have limitations. First, a basic foundation of the CAs has been a close provincial–municipal partnership. During the mid 1990s, however, a provincial government with a strong neo-conservative ideology reduced funding from partnerships such as the CAs, de-stabilizing their financial base. This problem has been partially addressed by introducing more fees-for-service, not always viewed enthusiastically by local governments which also pay a per capita levy to the CAs.

Second, CAs, as with all organizations with board members drawn from elected local officials, can be vulnerable to lack of continuity when key people are not re-elected to local councils. Third, many of the 36 CAs are small organizations, and often struggle to have all staff competencies needed for an interdisciplinary approach. To resolve the staffing challenge, some smaller CAs have formed regional groupings in order to share the cost of specialized expertise. In addition, to strengthen the human resource base, some large and small CAs partner with universities and colleges in their watersheds in order to influence research programs by faculty and graduate students and/or to become active participants in such programs.

Fourth, the CA model has not been replicated in other parts of Canada. The main reason has been that all local government partners in the watershed with a CA must contribute financially, and with extra contributions for benefits specific to a municipality. This principle requires local governments to have a sufficient tax base to fund their contribution, meaning CAs are only feasible in areas with a relatively large population base. Southern Ontario is one of the few regions in Canada satisfying this requirement.

Other regions meeting this precondition are the St. Lawrence River valley in Quebec, and the lower reach of the Fraser River basin in British Columbia. In the last decade, watershed-based organizations have been created in Quebec under the Quebec Water Policy of 2002. In British Columbia, the Fraser Basin Council was established in 1997 to provide leadership for the entire Fraser River basin, extending the work started by the Fraser River Management Board in 1992 (Dorcey 1997; Calbick et al. 2004; Watson 2004, 2005).

Another benefit of the financial capacity requirement is that local governments normally only promote initiatives for which they have capacity to help fund, creating desirable discipline for decisions. A different financial-related challenge emerges, however, when any CA enters into a cost-sharing agreement with provincial and/or federal government departments. When such agreements involve what should be ongoing activity, such as monitoring river flows, if priorities of one of the other partners changes, its funding may be reduced or withdrawn. Or, for matters such as monitoring river flows, if a change is initiated regarding the protocol or standards for such data, incremental expenses may be created for the CA and it may not have a ready way to cover them.

Acknowledgements I am very grateful for suggestions provided by Urbano Fra.Paleo and one anonymous reviewer. Their suggestions were constructive and have been incorporated into this chapter.

References

Bakker, K. 2007. Introduction. In *Eau Canada: The future of Canada's water*, ed. K. Bakker, 1–20. Vancouver: University of British Columbia Press.

Biswas, A.K. 2004. Integrated water resources management: A re-assessment. *Water International* 29(2): 248–256.

Biswas, A.K. 2008a. Integrated water resources management: Is it working? *Water Resources Development* 24(1): 5–22.

Biswas, A.K. 2008b. Current directions: Integrated water resources management – A second look. *Water International* 33(3): 274–278.

Blaikie, P., T. Cannon, I. Davis, and B. Wisner. 1994. *At risk: Natural hazards, people's vulnerability, and disasters*. London/New York: Routledge.

Blanco, J. 2003. Integrated water resources management in Columbia: Paralysis by analysis? *International Journal of Water Resources Development* 24(1): 91–101.

Blanco, J. 2008. Integrated water resources management in Columbia: Paralysis by analysis? *International Journal of Water Resources Development* 24(1): 91–101.

Braga, B.P.F. 2001. Integrated urban water resources management: A challenge into the 21st century. *Water Resources Development* 17(4): 581–599.

Butterworth, J., J. Warner, P. Moriarity, S. Smits, and C. Batchelor. 2010. Finding practical approaches to integrated water resources management. *Water Alternatives* 3(1): 68–81.

Calbick, K.S., R. McAllister, D. Marshall, and S. Litke. 2004. *Fraser river basin case study, British Columbia, Canada*. Washington, DC: The World Bank.

Chéné, J.-M. 2009. Integrated water resources management: Theory versus practice. *Natural Resources Forum* 33(1): 2–5.

Conca, K. 2005. *Governing water: Contentious transnational politics and global institution building*. Cambridge: MIT Press.

Dorcey, A.H.J. 1997. Collaborating towards sustainability together: The Fraser Basin Management Board and program. In *Practising sustainable water management: Canadian and international experiences*, ed. D. Shrubsole and B. Mitchell, 167–199. Cambridge: Canadian Water Resources Association.

Garcia, L.E. 2008. Integrated water resources management: A 'small' step for conceptualists, a giant step for practitioners. *Water Resources Development* 24(1): 23–36.

Global Water Partnership. 2000. *Integrated water resources management*, Technical Advisory Committee background paper no. 4. Stockholm: Global Water Partnership.

Global Water Partnership. 2002. *Introducing effective water governance*, GWP technical paper. Stockholm: Global Water Partnership.

Goggin, M.L., A.O.M. Bowman, J.P. Lester, and L.J. O'Toole. 1990. *Implementation theory and practice: Toward a third generation*. Glenview: Scott, Foreman/Little, Brown Higher Education.

Gurtner-Zimmermann, A. 1995. A mid-term review of remedial action plans: Difficulties with translating comprehensive planning into comprehensive action. *Journal of Great Lakes Research* 21(2): 234–247.

Hamilton, L.S. 1971. Concepts in planning for water resources development and conservation – The American experience. *Biological Conservation* 3(2): 107–112.

Hartig, J.H., and N. Law. 1994. Institutional frameworks to direct development and implementation of Great Lakes remedial action plans. *Environmental Management* 18(6): 855–864.

Jaspers, F.G.W. 2003. Institutional arrangements for integrated river basin management. *Water Policy* 5: 77–90.

Jones, M.L., and W.W. Taylor. 1999. Challenges to the implementation of the ecosystem approach in the Great Lakes basin. *Aquatic Ecosystem Health and Management* 2(3): 249–254.

Joseph, C., T.I. Gunton, and J.D. Day. 2006. Implementation of resource management plans: Identifying keys to success. *Journal of Environmental Management* 88(4): 594–606.

Kemper, K.E., W. Blomquist, and A. Dinar (eds.). 2007. *Integrated river basin management through decentralization*. Berlin: Springer.

Lautze, J., S. de Silva, M. Giordano, and L. Sanford. 2011. Putting the cart before the horse: Water governance and IWRM. *Natural Resources Forum* 35(1): 1–18.

Lord, G.R. 1962. A conservation authority and water management. *Journal of Soil and Water Conservation* 18(1): 28–30.

Lord, G.R. 1974. Introduction. In *Conservation by the people: A history of the conservation authority movement to 1970*, ed. A.H. Richardson. Toronto: University of Toronto Press.

Mazmanian, D.A., and P.A. Sabatier. 1989. *Implementation and public policy*. Lantham: University Press of America.

McLaughlin, C., and G. Krantzberg. 2011. An appraisal of policy implementation deficits in the Great Lakes. *Journal of Great Lakes Research* 37(2): 390–396.

Mitchell, B. 1990. Integrated water management. In *Integrated water management: International experiences and perspectives*, ed. B. Mitchell, 1–21. London/New York: Belhaven Press.
Mitchell, B. 2009. Implementation gap. *IWRA Update* 22(3): 7–12.
Mitchell, B. 2011. The implementation gap: From policy to implementation. In *Water innovation forum report: A competitive and innovative agricultural sector*, ed. D. Cunningham, L. Coates, and M. Harris, 56–61. London: Lawrence National Centre for Policy and Management.
Mitchell, B. 2013. Governance and water: Integrated water resource management. In *Perspectives on governance and society: Essays in honour of professor OP Dwivedi*, ed. N. Trivedi, T.A. Mace, D. Vajpeyi, and S.P. Dwivedi, 143–159. Jaipur: Rawat.
Mitchell, B., and D. Shrubsole. 1992. *Ontario conservation authorities: Myth and reality*, Department of Geography publication series no. 35. Waterloo: University of Waterloo.
Molle, F. 2008. Nirvana concepts, narratives and policy models: Insights from the water sector. *Water Alternatives* 1(1): 131–156.
Mollinga, P.P., A. Dixit, and K. Athukorala (eds.). 2006. *Integrated water resource management: Global theory, emerging practice and local needs*, vol. 1. London: Sage Publications.
O'Connor, D.R. 2002a. *Report of the Walkerton Inquiry: Part One: The events of May 2000 and related issues*. Toronto: Ontario Ministry of the Attorney General.
O'Connor, D.R. 2002b. *Report of the Walkerton Inquiry: Part Two: A strategy for safe drinking water*. Toronto: Ontario Ministry of the Attorney General.
Olsson, P. 2007. The role of vision in framing adaptive co-management processes: Lessons from Kistianstads Vattenrike, southern Sweden. In *Adaptive co-management: Collaboration, learning and multi-level governance*, ed. D. Armitage, F. Berkes, and D. Doubleday, 168–185. Vancouver: University of British Columbia Press.
Ontario Ministry of Environment. 2008. *The clean water act: Roles and responsibilities*. Toronto: Ministry of Environment.
Ontario Ministry of Environment and Energy and Ontario Ministry of Natural Resources. 1993. *Watershed management on a watershed basis: Implementing an ecosystem approach*. Toronto: Queen's Printer for Ontario.
Ontario Watershed Planning Implementation Project Management Committee. 1997. *An evaluation of watershed management in Ontario: Final report 1997*. Toronto: Ministry of Environment and Energy, and Ministry of Natural Resources.
Perkel, C.N. 2002. *Well of lies: The Walkerton water tragedy*. Toronto: McClelland and Stewart.
Plummer, R., J. FitzGibbon, A. Spiers, and J. Imhoff. 2005. The expanding institutional context for water resources management: The case study of the Grand River watershed. *Canadian Water Resources Journal* 30(3): 227–244.
Pressman, J.L., and A. Wildavsky. 1971. *Implementation: How great expectations in Washington are dashed in Oakland*. Los Angeles: University of California Press.
Rahman, M.M., and O. Varis. 2005. Integrated water resources management: Evolution, prospects and future challenges. *Sustainability: Science, Practice and Policy* 1(1): 15–21.
Rogers, P., and A.W. Hall. 2003. *Effective water governance*, TEC background paper no. 7. Stockholm: Sweden Water Partnership.
Shrubsole, D. 1996. Ontario conservation authorities: Principles, practice and challenges 50 years later. *Applied Geography* 16(4): 319–335.
Smith, K. 1996. *Environmental hazards: Assessing risk and reducing disaster*, 2nd ed. London/New York: Routledge.
Tortajada, C. 2010. Water governance: Some critical issues. *International Journal of Water Resources Development* 26(2): 297–307.
Veale, B.J. 2010. *Assessing the influence and effectiveness of watershed report cards in watershed management: A study of watershed organizations in Canada*. Ph.D. dissertation, University of Waterloo, Geography Department, Waterloo..
Watson, N. 2004. Integrated river basin management: A case for collaboration. *International Journal of River Basin Management* 2(4): 243–257.
Watson, N. 2005. Partnerships for integrated land and water management: Lessons from the Fraser basin. In *Sustaining our futures: Reflections on environment, economy and society*, Department

of Geography publication series no. 60, ed. D. Shrubsole and N. Watson, 409–446. Waterloo: University of Waterloo.
Weale, A. 1992. Implementation failure: A suitable case for review? In *Achieving environmental goals: The concept and practice of environmental performance review*, ed. E. Lykke, 43–63. London: Belhaven.
World Water Assessment Programme. 2006. *Water, a shared responsibility, The United Nations world water development report 2*. Paris/New York: United Nations Educational, Scientific and Cultural Organization/Berghahn Books.
Worte, C. 2010. 10 years after Walkerton: Protecting municipal drinking water sources. *Municipal World*, March, 5–7.

Bruce Mitchell is a Professor of Geography and Environmental Management, at the University of Waterloo in Ontario, Canada. His research focuses on policy and governance arrangements regarding resource and environmental management, with particular focus on integrated water resource management. He has authored or edited 31 books and written 165 articles and chapters. He is a Fellow of both the Royal Society of Canada and the International Water Resources Association, and also an honorary professor at four Chinese universities. He edited the volume *Resource and environmental management in Canada: Addressing conflict and uncertainty* (1995), and with Philip Dearden has recently co-authored the book *Environmental change and challenge: A Canadian perspective* (2012).

Chapter 18
The Emergence of Landscape Governance in Society-Environment Relationships

David J. Brunckhorst

18.1 Introduction

This book explores the key principles and multiple dimensions of risk governance in order to improve understanding towards systemic and conceptually integrated policy approaches. These multiple dimensions include and are shaped by a variety of interacting elements inherent in 'landscapes'. When we look out from an elevated vantage point, an office window, or aeroplane we generally visually observe such regional and local patterns that are in essence, social-ecological responses, reflected in 'landscapes'. The manifestations of the interdependencies between human and biophysical elements across the dimensions of society, space, time and ecological systems are reflected in the mosaic of patterns we observe at various scales close to home, across regional landscapes and from images recorded from higher observation posts such as satellites (Forman 1995; Brunckhorst 2000).

Landscapes are not simply the creative product on the canvas of a painter. Likewise, geological and biological structures and function do not provide a complete interpretation of regions and landscapes, or of how they change. Humans and their societies shape, interpret, re-shape and re-interpret their surroundings. Individually, collectively, and from different policy perspectives we interpret the patterns we see in our surroundings, and further afield, in a variety of ways and for different purposes. For example, in terms of risk or potential, different actors might see fire hazard, landslide hazard, or the potential for mining, human settlement or environmental conservation. While landscapes synthesize and reflect human and ecological interactions, they are a social construct, whether imagined or understood, or constructed inadvertently or deliberately (Fig. 18.1).

D.J. Brunckhorst (✉)
UNESCO Centre for Bioregional Resource Management, University of New England, Armidale, NSW 2351, Australia
e-mail: dbrunckh@une.edu.au

U. Fra.Paleo (ed.), *Risk Governance*, DOI 10.1007/978-94-017-9328-5_18

Fig. 18.1 Human shaped landscapes are created through human interpretation, imagination and interactions with 'nature'

> The human region … is a complex of geographic, economic, and cultural elements. Not found as a finished product in nature, not solely the creation of human will and fantasy, the region, like its corresponding artefact, the city, is a collective work of art. (Lewis Mumford 1938: 367)

Human influence and capabilities for large scale landscape and environmental change and exploitation of natural resources has escalated exponentially since the industrial revolution. Socially, large scale environmental change, was further fuelled through the eighteenth and nineteenth centuries by an increasing population and notions of 'Taming the Frontier' through resource capture and exploitation. In North America and Australia in particular, civil liberty became associated with private land ownership. Increasing proximity of individualist and private land management can create and transfer impacts to others or society generally. There may be no need for systems of governance for resource and risk management as long as the spatial patterns of ecosystems are able to buffer or insulate the activities of individuals from each other and from 'natural' hazards. However, when the progression of resource use by individuals and society results in the loss of the buffering capacity of ecosystem services and hazardous or harmful externalities begin to be transmitted, perhaps via ecosystems or the atmosphere, the need for governance systems becomes important. In the post-industrial era, human activity has been the major force shaping landscape change. Governments are increasingly focused on adjusting to the accelerating pressures of climate change along with landscape change. Landscape change, including expanding human settlements with changing social circumstances, can increase hazard exposure and risk in terms of future climate change perturbations and natural disasters.

Landscape governance deals with spatial contexts shaped by and given meaning, through the interactions of ecological functions and social relationships and constructions of places. These regional spaces may change over time but retain meaning for community interest, civic collective action and formal and informal governance institutions. People and communities, resource production and related industries, economies and political institutions, biodiversity and ecological systems all have relationships with a place in space – a locatable area – and so can be represented spatially as a landscape component. These components operate at various scales and interact at a variety of levels. While constant change is the normal state, an understanding of the pattern and form of multi-scale and long-term landscape mosaics is required to work towards more resilient, sustainable, and adaptable futures. Understanding current and future landscape change can provide valuable insights into future hazard and risk scenarios, risk governance and preparedness and forward design and planning for lower risk alternative future landscapes and governance arrangements.

18.2 Systems Interactions: Co-evolution of Landscapes and Institutions

Environmental history mirrors human institutions. Reactions to landscape change are reflected in new policies, planning and activity that create new landscape change (Gunderson et al. 1995). Government and policy responses might include land and resource tenure adjustments such as conversion of wetland or agricultural land to urban development, increased regulation, biodiversity conservation or reallocation of resource rights for higher economic value.

Landscape ecology theory provides a useful regional approach to understanding social-ecological systems interactions to assist the design of institutional arrangements towards increased adaptive capacity. Landscapes internalize many of the interactions amongst ecosystem and social elements, which we observe as patterns – forest, clearings, wheat fields, towns, a web of roads, meandering streams, dry river beds. Mosaics of changing landscape patterns reflect responses and feedbacks of social-ecological interactions, which drive change in natural resource capacity and ecosystem health. Institutions and landscapes evolve together over time via feedback and response loops (Fig. 18.2) either increasing or decreasing adaptive capacity in relation to resilience to future change or natural hazards (Berkes et al. 2003). Over time feed-back and feed-forward loops drive the non-linear co-evolution of landscapes and institutions within and across geographic spaces and produce an array of emergent conditions (Brunckhorst 2010; Fig. 18.2).

Patterns or processes that emerge from interdependent interactions occurring across landscapes are uniquely different from the individual ecosystem elements that created them. Conditions that emerge from society-environment interactions

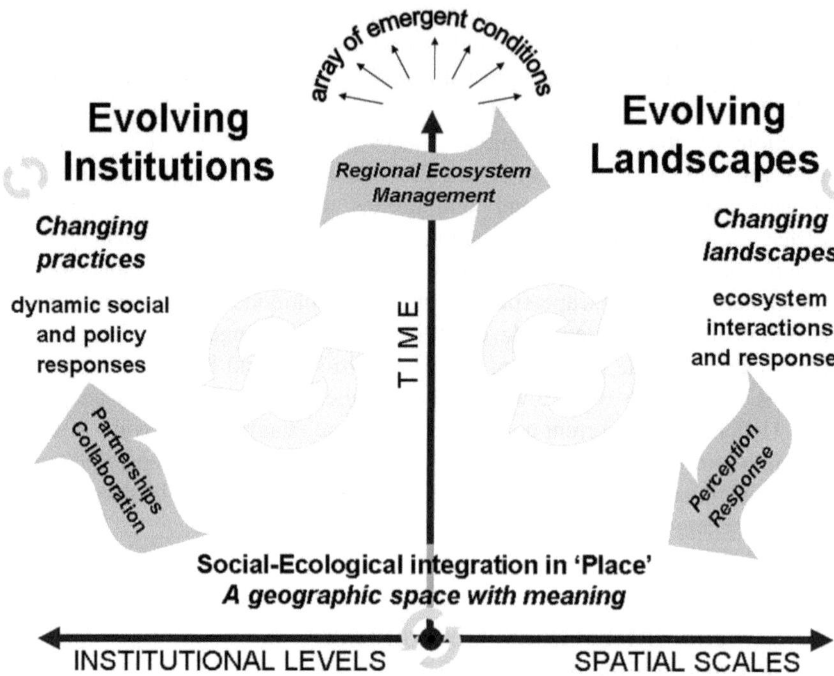

Fig. 18.2 The co-evolution of changing institutions and landscape change creates meaningful 'contexts' at nested scales and levels of ecosystems and institutions through which governance systems might arise that are adaptable and responsive to issues of resource sustainability and natural hazards (Adapted from Brunckhorst 2010: 18)

are often at the heart of sustainability and risk management issues and may involve interactions of fast and slow moving variables, feedbacks, threshold effects and reorganization. A fusion of systems interactions can lead to manifestation of surprises, including possible systems crash, such as collapse of viable species populations, ecosystems or whole social-ecological systems (Diamond 2005; Walker and Salt 2006; Duit and Galaz 2008). Alternatively, new institutions for governance might arise from social-ecological systems interactions, leading to improved resilience and adaptive capacity. Complex-adaptive systems are characterized by the possession of self-organizing capacities which are receptive to pressures of change. Such adaptive capacity of natural and human systems is an important responsive mechanism for dealing with risk, vulnerability and buffering pressures of change. Landscapes are therefore emergent responsive conditions, manifest as spatial patterns from feedback cycles that reflect social-ecological systems interactions (Fig. 18.2), which in turn define the multi-scale geographical context of interactions and interdependencies including institutional elements of governance that will shape responses to natural hazard and sustainability issues (Ostrom et al. 2002; Berkes et al. 2003; Brunckhorst et al. 2006, 2008). The challenges for improving the governance of social-ecological systems include understanding how various social actors can make flexible policy decisions and take action in the light of uncertainty, the collective circumstances pointing to (adaptive) change, the conditions for sustaining cooperation and co-management, and how various scales and levels of interactions between social and biophysical elements affect governance (Brown and Raymond 2007; Duit and Galaz 2008).

Within an explicit spatial location of similar patterns and human attachment, 'context' is created (Cheng et al. 2003; Brunckhorst 2010). Structuring of landscapes and regions through social-ecological systems interactions defines operational contexts in which to integrate cross-scale interactions of resource use, property rights, agency jurisdictions and ecological patterns and processes. Bioregional planning is similarly a landscape context focused approach based on local resident's collective identity with a 'place' (McGinnis 1999; Johnson et al. 1999; Brunckhorst 2000). From landscape structure and change emerges 'sense of community' and place attachment, resource knowledge, preferred land uses, and creation of various externalities of interactions which may in turn influence (or ameliorate) risk and vulnerability to natural hazards.

Identity with and attachment to a place is often reflected in how local people invest in shaping the landscape and their place in it over time (Cheng et al. 2003; Fabricius et al. 2007). These interactions and responses effect positive and negative change on social-ecological systems, shaping 'sense of place' contexts in a landscape space (Brown and Raymond 2007; Brunckhorst et al. 2008). Meaningful places and spaces are valuable concepts for understanding sensitivity (or adaptive capacity) to disasters in relation to risk governance in society. Investing in place shaping of landscapes, resources, livelihoods and the space called 'home' (not just the physical house but the area identified as home and/or community) create important meaning for residents. Interwoven with social creation of community and land use, the landscapes' natural ecological features also have special meaning to local residents. It is where the main local community of interest exists, where residents interact, have networks of trust and have an interest in local civic affairs. In turn, this collective shared history develops social capital frameworks to forge collaboration and integration of local level governance, policy and community co-management initiatives and responsive adaptive capacity to disaster management. Place and community are important in local cooperation and observance of formal and informal rules, facilitating and motivating self-organization for collective action (Berkes et al. 2003; Armitage et al. 2007). There exists 'built in', local knowledge about the landscape, its resources and residents in and about that particular geographic context that is valuable to risk governance and management. At higher levels of policy development, governments continue to struggle to understand the importance of place making and community attachment in their policy and planning, despite a histories of communities 'rallying' in natural disaster circumstances.

It follows that systems of governance need to be more seamlessly integrated at all levels of social and environmental management and institutional arrangements to match landscape scales of social-ecological interdependencies. At broader scales however, how can policy makers, communities and scientists better understand the local regional landscape (social-ecological) context? How can context be used to integrate intra-scalar requirements for governance – including governance aimed at preparedness and responsive capacities to deal with increasing natural hazards and disasters? In addition to some identified design and operational rules for successful governance institutions (Ostrom et al. 2002), several other principles are considered essential for the successful design of ecologically-sustainable cross-scale interactions of social-ecological systems (Armitage et al. 2007; Stedman 2003). To understand a regional landscape context to integrate multi-jurisdictional resource governance, three further principles are considered important. Firstly, that the ecosystems of the landscape context possess a relatively high level of homogeneity. Secondly, that the regional boundaries maximize the area that residents consider important for civic engagement and reflect their local to regional communities of interest. The third principle is a nested multi-scaling capacity for dealing with externalities of conservation and resource use. These principles have been used to develop techniques for the definition of nested spatial frameworks for governance integrating natural resource management, planning and government administration; a technique now known as **'eco-civic' regionalisation** (Brunckhorst et al. 2006, 2008; Davenport and Anderson 2005; Brown and Raymond 2007). Nested frameworks of spatially, socially and ecologically meaningful contexts provide an appropriate theatre for multiple actors to collaborate. The sharing of responsibilities and power between government, local resource users and/or other resident stakeholders, often referred to as co-management, is likely to be essential for creation of effective governance. Co-management mobilizes resources and knowledge across different levels, bridges organizations (e.g., voluntary and government agency disaster response organizations), builds trust and coordination and assists co-learning for the future (Berkes et al. 2003; Armitage et al. 2007).

In summary, landscape theory and principles of landscape ecology provide a useful basis within which to understand **context** – a concept of critical importance to 'governance', in particular risk governance. Issues of future sustainability, natural hazard and risk management are interconnected and need to be considered in the design of governance frameworks, organization and response capabilities. Human shaping and perceptions of landscapes create sense of community, attachment and identity to those places and landscapes, a blend of human community and ecological interrelationships defining that place and attachment to it. A strong basis has been established for a considerable local community role in participatory governance (Kearney et al. 2007; Shaw et al. 2009). The success of sustainability and risk governance arrangements will depend on their structural coupling to the social-ecologically characterized human-landscape context in which it is applied. Landscape governance refocuses multi-level policy and decision making on spatial contexts that reflect the greatest extent of internalisation of interdependencies of social-ecological relations. Characterizing and understanding contextual circumstances relevant to a place will be valuable to the design of governance systems at appropriate scales (or nested arrangements for scaling). The achievement of coordination and response is likely to occur through co-management and subsequent joint knowledge building from collectively shared experience (Berkes et al. 2003; Williamson et al. 2003). The concept of landscape governance contributes enhanced understanding towards participatory governance and will prove to be a useful tool to provide innovative direction to risk policy, planning and response.

Eco-Civic Regionalisation: Understanding scale and geography of local contexts

The following is an example from the State of New South Wales, Australia.

Residents' identity and affiliations with 'place' and each other, their 'communities of interest' can be mapped from various question framings in mapped social surveys. In spatial form, individual residents shared identity with a place and community may spatially oincide. Like a 'stack of pancakes', areas of shared interest create a social surface or topography in which hills represent the greatest coincidence of shared identity and interest in a place, and the valleys indicate little overlap of areas of shared identity or civic interest.

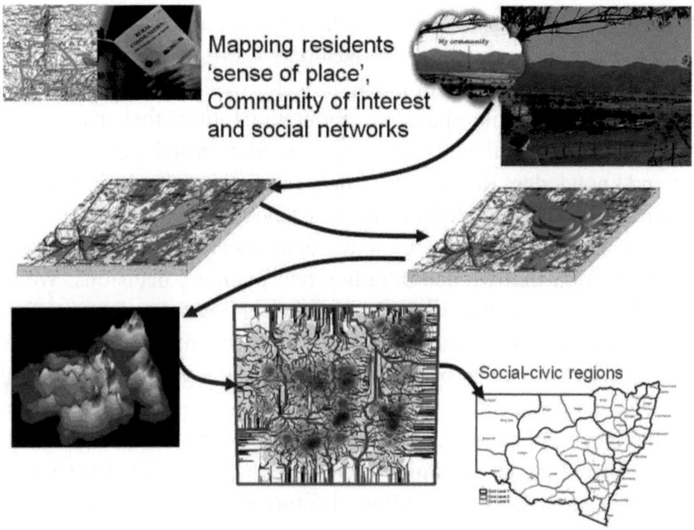

The depth and width of the social valleys provide room to optimize the integration of ecological landscapes, and nested scales.

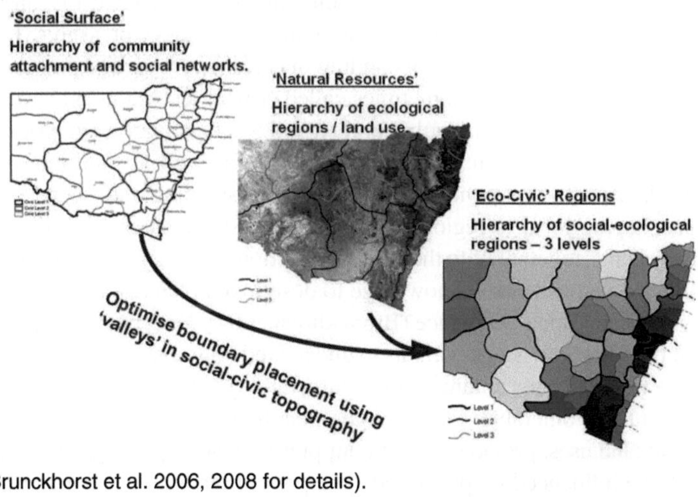

(see Brunckhorst et al. 2006, 2008 for details).

18.3 Future Landscapes: Forward Planning for Risk Governance

While it has been said that the only things one can be certain of are death and taxes, it might be rejoined that even these have elements of uncertainty. One thing we can be certain of, however, is **change**. Change is a continuous process; change begets change and, creates uncertainty. Uncertainty about future change (except we know change will happen) creates additional complexity for policy, planning and governance at all levels of society. Nevertheless, forecasts for future conditions are made each day, depicting a variety of environmental, social, and economic interests.

Predictions are made regarding the weather, sports events, stocks and shares, housing markets, ecosystem health (such as water quality or pollution levels), and other interests. When people speculate about the future, they use models either formally derived and described, or developed within an individual's framework of experience and knowledge. Regardless of how predictions are made, they are usually wrong in some way. Any modelling process tends to complicate or hide implicit uncertainties or natural variability. Despite some uncertainty, these models provide important information that we use in policy and planning decisions. We still make judgments and choices about what to do based on the weather forecast. It is important, therefore, not to focus so much on the details or advantages of particular models, but rather to focus on what can be identified as something that might be valuable in making choices about the future. In terms of landscape governance, natural hazard and risk management, planning and policy choices need to shape local and regional landscape and their human communities towards adaptive, resilient futures in the light of (some but limited) understanding of future hazards and risks.

The past history of policies, planning and land use have created the landscape components, patterns and processes of the present (Antrop 2005). It is from these that future landscape patterns of land use, ecosystems and processes will evolve in response to policy, planning and human activity along with other change pressures (Figs. 18.2 and 18.3). A regional landscape understanding of current patterns and processes along with change pressures, drivers (sometimes called critical uncertainties) is valuable to understanding future change and vulnerability, and what alternative (long-term) plans and actions (adaptation) might reduce vulnerability as well as contribute different changes to landscape components. Through understanding the social-ecological interactions shaping local to regional context, the past pressures of change in these interacting systems, together with the current directions of policies and circumstances, it is possible to synthesize new knowledge to design and plan towards more resilient conditions for landscape governance (Brunckhorst 2004). Understanding past change and current pressures of change, help us to understand how and where present policies or decisions might create or ameliorate vulnerability, and how various landscape elements might be vulnerable. Past change, including past climate change, policies, planning, and land uses, provides us with our present landscape (Fig. 18.3). Although still likely to be influenced by past trends, the present landscape is our starting point towards understanding future change, enabling us to assess positive or negative change impacts and/or vulnerability across ecosystems and human communities, making it

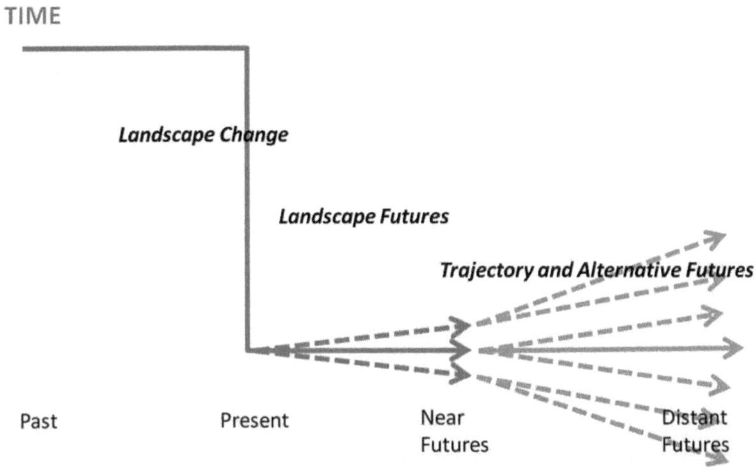

Fig. 18.3 Landscape change of places and patterns over time creates a trajectory of change and influences options for future change. Design of plausible alternative landscape futures could provide greater adaptive capacity and options for risk governance

possible to see how society might adapt to such change, using thoughtful design to move towards more sustainable and resilient futures. Many climate change or natural hazard vulnerability and risk analyses of ecosystems, human settlements and infrastructure are done using the configuration of the current landscape. Future hazard impacts (for example 2050 or 2070 climate change predictions of increased intensity weather events) will be occurring on a different future landscape, most likely with expanded or higher density human settlements. A present-future landscape analysis approach allows a more integrated analysis of "whole" landscape parameters that might change, or identification of components that might become more or less vulnerable. Also, the approach supports the examination of adaptation benefits of alternative landscape futures that need to be planned and achieved to increase resilience and reduce negative impacts (Brunckhorst et al. 2010).

18.4 Conclusions

Landscapes are constructed physically and mentally by people and human society. Human shaping and perceptions of landscapes create a sense of community, attachment and identity to those places and landscapes. The blend of human community and ecological interrelationships can define place and human attachment to it across space and time (referred to as 'eco-civic' landscapes or biocultural regions). Thus within an explicit spatial location of similar patterns and human attachment, 'context' is created. Such context, as I refer to it, is a geographical, emergent property of social-ecological relationships. Structuring of landscapes and regions through social-ecological systems interactions defines contexts meaningful to local communities – governance contexts relevant to planning, policy, natural resource management and natural hazard management. A context of mutually shaped and shared social, civic and ecological variables provides knowledge and insights about hazards, risk and potential vulnerability. Such contexts are useful for planning over different time scales from present to medium and long term horizons. Longer term policies and planning might be considered towards the design of less vulnerable future landscapes and their human settlements. Examining likely future landscapes and alternatives can include scalable capacity to capture and deal with externalities at appropriate levels of local to regional to national governance. Assessing future landscapes can also assist understanding of loss of buffering capacity to natural hazards and possible pathways to increase buffering and flexibility which might provide further adaptive capacity in the future. These same landscape contexts, as nested spatial arrangements, similarly provide frameworks for natural hazard risk management, such as disaster relief deployment logistics, temporary local to regional control centres, and their constituent networks of emergency services.

Changing landscapes alter future hazard impacts and resultant risk and vulnerability. Vulnerability and risk assessments should consider what the future landscape might be when a particular natural hazard might eventuate – such as climate change enhanced storm impacts in say 2050 when the landscape pattern of human settlements and ecosystems is likely to be vastly different to what it is today. Analysing future landscapes to understand risk governance can incorporate many variables; including for example, the future development of intensive human settlements, changing socio-economic circumstances, young or aging populations of suburbs, enhanced water catchments, mono-cultures, and reduced natural buffers which influence exposure and vulnerability to natural and other hazards. Design and analysis of alternative landscape futures can guide policy, planning and governance arrangements towards more resilient and adaptable landscapes and human settlements (e.g., Brunckhorst et al. 2010).

Landscape governance refocuses multi-level policy and decision making on spatial contexts that reflect the greatest extent of internalisation of interdependencies of social-ecological relations; that is, between socially constructed spaces and the ecosystem function and conditions of places. The concept of landscape governance contributes enhanced understanding towards participatory governance and will prove to be a useful tool to provide innovative direction to risk policy, planning and response.

References

Antrop, M. 2005. Why landscape of the past are important for the future. *Landscape and Urban Planning* 70: 21–34.

Armitage, D., F. Berkes, and N. Doubleday (eds.). 2007. *Adaptive co-management: Collaboration, learning and multi-level governance*. Vancouver: University of British Columbia Press.

Berkes, F., J. Colding, and C. Folke (eds.). 2003. *Navigating social-ecological systems – Building resilience for complexity and change*. Cambridge: Cambridge University Press.

Brown, G., and C. Raymond. 2007. The relationship between place attachment and landscape values: Toward mapping place attachment. *Applied Geography* 27: 89–111.

Brunckhorst, D.J. 2000. *Bioregional planning: Resource management beyond the new millennium*. London/Amsterdam: Taylor & Francis.
Brunckhorst, D. 2004. Turning points towards sustainability: Integrative science and policy for novel (but real) landscape futures. *Ethics in Science and Environmental Politics* 2004: 83–91.
Brunckhorst, D.J. 2010. Using context in novel community based natural resource management: Landscapes of property, policy and place. *Journal of Environmental Conservation* 37(1): 1–7.
Brunckhorst, D., P. Coop, and I. Reeve. 2006. 'Eco-civic' optimisation: A nested framework for planning and managing landscapes. *Landscape and Urban Planning* 75: 265–281.
Brunckhorst, D., I. Reeve, P. Morley, and K. Bock. 2008. Strategic spatial governance: Deriving social–ecological frameworks for managing landscapes and regions. In *Landscape analysis and visualisation: Spatial models for natural resource management and planning*, ed. C. Pettit, W. Cartwright, I. Bishop, K. Lowell, D. Pullar, and D. Duncan, 253–276. Berlin: Springer.
Brunckhorst, D., I. Reeve, P. Morley, M. Coleman, E. Barclay, J. McNeill, R. Stayner, R. Glencross-Grant, J. Thompson, and L. Thompson. 2010. *Case studies to support a 'first pass' national climate change coastal vulnerability assessment. Case study 6: Hunter and central coasts*. Refereed report to Australian Government. Canberra: Department of Climate Change, Australian Government. Retrieved January 20, 2012, from http://www.climatechange.gov.au/publications/adaptation/hcc-report.aspx.
Cheng, A.S., L.E. Kruger, and S.E. Daniels. 2003. 'Place' as an integrating concept in natural resource politics: Propositions for a social science research agenda. *Society & Natural Resources* 16(2): 87–104.
Davenport, M.A., and D.H. Anderson. 2005. Getting from sense of place to place based management: An interpretative investigation of place meanings and perceptions of landscape change. *Society and Natural Resources* 18: 625–641.
Diamond, J. 2005. *Collapse: How societies choose to fail or succeed*. New York: Viking Penguin.
Duit, A., and V. Galaz. 2008. Governance and complexity: Emerging issues for governance theory. *Governance* 21(3): 311–335.
Fabricius, C., C. Folke, G. Cundill, and L. Schultz. 2007. Powerless spectators, coping actors, and adaptive co-managers: A synthesis of the role of communities in ecosystem management. *Ecology and Society* 12(1): 29. Retrieved November 29 2011, from, http://www.ecologyandsociety.org/vol12/iss1/art29/.
Forman, R.T. 1995. *Land mosaics: The ecology of landscapes and regions*. Cambridge: Cambridge University Press.
Gunderson, L., C. Holling, and S. Light. 1995. *Barriers and bridges to the renewal of ecosystems and institutions*. New York: Columbia University Press.
Johnson, N., F. Swanson, M. Herring, and S. Greene. 1999. *Bioregional assessments: Science at the crossroads of management and policy*. Washington, DC: Island Press.
Kearney, J., F. Berkes, A. Charles, E. Pinkerton, and M. Wiber. 2007. The role of participatory governance and community-based management in integrated coastal and Ocean management in Canada. *Coastal Management* 35: 79–104.
McGinnis, M.V. 1999. *Bioregionalism*. London: Routledge.
Mumford, L. 1938. *The culture of cities*. New York: Harcourt Brace.
Ostrom, E., T. Diez, N. Dolsak, C. Stern, S. Stonich, and E. Weber (eds.). 2002. *The drama of the commons*. Washington, DC: National Academy Press.
Shaw, A., S. Sheppard, S. Burch, D. Flanders, A. Wiek, J. Carmichael, J. Robinson, and S. Cohen. 2009. Making local futures tangible – Synthesizing, downscaling, and visualizing climate change scenarios for participatory capacity building. *Global Environmental Change* 19: 447–463.
Stedman, R.C. 2003. Is it really just a social construction? The contribution of the physical environment to sense of place. *Society & Natural Resources* 16(8): 671–686.
Walker, B., and D. Salt. 2006. *Resilience thinking*. Washington, DC: Island Press.
Williamson, S., D. Brunckhorst, and G. Kelly. 2003. *Reinventing the common: Cross-boundary farming for a sustainable future*. Sydney: Federation Press.

David J. Brunckhorst was the inaugural Director of the Institute for Rural Futures (2000), University of New England, Australia. David has worked as a Scientific Editor for the Australian Biological Resources Study, Director of land and marine conservation management programs with Federal Environment and Conservation Agencies, where he also established the National Reserve System Program for Australia. He has also been principal scientific advisor to the Federal Parliamentary Standing Committee on the Environment, USA Secretaries of the Interior and Defense. His research and capacity building projects focus on integrating the big picture of conservation, ecological services and production in the milieu of human society. He was recipient of a UNESCO Medal in 2000 and the George B. Fell award (2009).

Chapter 19
Risk Complexity and Governance in Mountain Environments

James S. Gardner

19.1 Introduction

Mountain regions, many of which have been subject to a disproportionate number of disastrous events (Hewitt 1997a, b), contain complex geo- and social-ecological systems that pose challenges for managing risk from natural and other hazards. Physical, socio-economic and political isolation of these regions have complicated risk governance and management, exacerbating losses and damage in many instances. While the potential for hazardous processes is inherently high, the social-ecological systems in mountain areas have rapidly evolved from relatively discrete and isolated pockets of population, subsistence land use and commodity exchange to larger, diversified economies and populations linked to national and global interests. This has magnified exposure to hazardous processes and conditions and increased vulnerability to damage and destruction, while eroding physical and political isolation and broadening the need for pro- and reactive risk governance. Natural processes in mountain environments also may contribute to disasters beyond the mountain margins and the increased economic, social and political integration has magnified this effect.

The increasing proximity of hazardous processes and growing populations and expanding infrastructure in mountains regions, such as the European Alps (Fig. 19.1), South American Andes and the Himalayan and Trans-Himalayan ranges of South Asia and southwest China, have called for close attention to disaster and accident risk reduction (Brundl et al. 2009; Fuchs 2009). Likewise, global media coverage of disaster events, such as the 2004 South Asian earthquake and tsunami, 2005 Kashmir earthquake, 2008 Wenchuan earthquake, 2010 Haiti earthquake and Pakistan floods and 2011 Fukushima earthquake and tsunami, have focused public and professional attention generally. In mountain regions, risk governance must

J.S. Gardner (✉)
Natural Resources Institute, University of Manitoba, Winnipeg, MB R3T2N2, Canada
e-mail: gardner@cc.umanitoba.ca

U. Fra.Paleo (ed.), *Risk Governance*, DOI 10.1007/978-94-017-9328-5_19

Fig. 19.1 Chamonix-Mt. Blanc, France, showing the altitudinal geo-ecological transitions and the interface between the inherently hazardous steep slopes and the Glacier Bossons and Glacier Taconnaz, and the expanding residential, commercial and transportation land uses, creating a high level of risk. Some structural and non-structural measures to control floods and avalanche runouts, protect structures and infrastructure and limit building are evident, as is the portal of the Trans-Alpine Mt Blanc Tunnel in the *left-center* (James Gardner)

account for high-magnitude disasters, such as those that disrupt all aspects of civil society, as well as chronic hazards that produce accidents focused on individuals or small groups and are not widely disruptive, but over time, may impose significant costs on society. The latter arise from on-going processes such as snow avalanches, flash floods, rock falls, debris flows and extreme weather events. Significant advances in knowledge and practice, particularly those related to the incorporation of social, economic, historical and political factors (Wisner et al. 2004), along with improved data on and understanding of bio-physical processes, have contributed to more effective prevention, mitigation and recovery, at least in terms of human life if not economic costs (*The Economist* 2012).

The objectives in this chapter are to describe the evolution of disaster risk in mountain environments and the governance response to the hazards and vulnerabilities present. Two components, the extremely variable and changing mountain geo-ecological systems and the evolving social-ecological systems (Berkes et al. 2003), give rise to risk complexity. The concept, risk complexity, is used to convey the notion that the extreme spatial and temporal variability and scale of influence characteristic of mountain geo-ecological systems, plus the varied and rapidly changing social-ecological systems in and interconnected with mountain regions pose significant challenges to governing and managing risk. The method used in choosing the cases was not a controlled representational sampling. Rather, it was a subjective sampling from the author's fieldwork experience meant to illustrate the components of risk complexity in mountain environments. The two components of risk, hazard and vulnerability, are discussed in turn, followed by a consideration of the resulting implications for risk governance.

19.2 Hazard Complexity in Mountain Environments

Hazardous processes and conditions are shaped by large-scale contextual factors and the inherent geo-ecological characteristics of the mountain environment. The former include global and regional crustal tectonics, atmospheric and oceanic circulations and solar energy variations. The latter include topographic, climatic, hydrological, ecological and anthropogenic variables. Their interactions produce many hazard types, hazard cascades and highly variable spatial dimensions of hazard processes, all of which are influenced by long-term environmental change and variability and anthropogenic factors. This mix of factors produces complexity in the hazard component of risk.

19.2.1 Hazard Types

Hazards such as earthquakes and volcanic eruptions are common in mountain areas and their geography reflects the broad context of crustal tectonics and crustal plate boundaries. Though not restricted to mountain areas, earthquakes and volcanic eruptions historically have contributed to numerous disasters in and adjacent to the mountains of south and east Asia, Europe, Latin America, North America and New Zealand. Disasters arising from the 2008 Wenchuan earthquake in Sichuan, China and 2005 Kashmir earthquake in northern Pakistan and India, illustrate the importance of crustal plate boundary locations and the inherent complexity of mountain hazards. The island of Java, Indonesia, a product of ancient and on-going volcanic processes, provides examples of disastrous eruptions, while contributing to fertile volcanic soils that support large populations. Some eruptions produce ash clouds of regional and global scale that alter energy balances, causing earth-surface cooling and catastrophic crop failures.

The proximity of mountain areas relative to large-scale atmospheric and oceanic circulations also influences hazardous processes and conditions. Examples include location relative to: westerly atmospheric circulation (North American Cordillera, Alps, northwest Himalaya, New Zealand Alps), monsoonal processes (Himalaya of South and East Asia), warm and cold ocean currents (Alps, Scandinavian mountains, central and southern Andes) and oscillatory ocean surface events such as *el Niño* and *la Niña* (Andes, North American Cordillera, southern Himalaya). Energy and moisture balances and their inter-annual, annual and longer variability are affected. Lower summer monsoon intensity in South and East Asia is linked to *el Niño* events in the western Pacific, translating into drought disasters (Davis 2002), whereas the opposite produces high precipitation (floods, landslides, etc.) in the mountains and their margins. While the coastal and interior ranges of Alaska, British Columbia and the Pacific Northwest of the U.S. are inherently high precipitation areas due to proximity to Pacific waters, *la Niña* events produce cooler temperatures and higher snowfall, leading to severe snow avalanche outbreaks and higher probabilities of spring floods.

Variations in solar energy output, as occurs through sunspot cycles, have a global influence on energy and moisture balances that may translate into short-term altitudinal shifts in temperature, snowline, glacier mass balance and biological processes leading to more hazardous conditions. The effects in mountain environments may be magnified by the compact altitudinal zonation of inherent geo-ecological properties (Fig. 19.1).

The inherent properties of mountain environments, in combination with contextual factors, produce conditions favorable to hazard types that range from the geological to the biological and vary in frequency, magnitude, speed of onset, duration and spatial extent. Topography, weather and climate, hydrology, earth materials and plant and animal life are important environmental properties. For example, steep and long slopes, heavy precipitation, easily mobilized earth materials, freezing and thawing, erosional undercutting, among other factors, contribute to rock falls, rock avalanches, debris flows, landslides, flash floods and snow avalanches. Steep slopes magnify the gravitational effect and therefore, downslope movement. Mountain areas capture rain- and snowfall and serve as important water sources locally and regionally. Variations of rain- and snowfall below and above resource thresholds produce hazardous droughts and floods respectively.

Extreme weather, such as high intensity precipitation, deep frosts, wind, droughts and blizzards, may trigger disasters in mountain environments and surroundings.

The processes they create (e.g. snow avalanches, flash floods, debris flows, landslides, blizzards, killing frosts, etc.) are common hazards to transportation, agriculture/horticulture, recreation and tourism requiring some form of risk management. They are a product, and are reflective, of the variable weather and the geo-ecological diversity of high mountain environments. In moving from lower to higher altitudes, one can be transposed from mild weather into episodes of cold and snow. Or, at higher elevations summer-like weather can change quickly to winter-like blizzard conditions. The probability of such events increases in the spring and autumn transitions. Historically, unexpected snow events (blizzards) plagued the trans-Himalayan trade, nomadic and transhumance processes, killing people and livestock and destroying trade goods (Rizvi 1999). Today, blizzards, cold and snow avalanche outbreaks disrupt the trans-Himalayan tourism, road and air transport and continue to plague traditional nomadic and transhumance processes. Elsewhere in the mid-latitude mountain regions, these and other extreme weather events act as routine and disaster-generating hazards for road transport, agriculture and horticulture and tourism and recreation (e.g. see McClung and Schaerer 2004).

Biological hazards involve organisms and their related processes and conditions. Few are unique to mountain environments but some have had profound and disastrous impacts therein. Insects and many micro-organisms historically have posed a hazard to agriculture/horticulture, animal husbandry and human life. The introduction of new cash crop monocultures into suitable mountain settings has improved incomes for locals and outsiders but also has opened entry points for plant- and animal-specific hazardous organisms and micro-organisms. Thus, the introduction of vineyards, orchards, vegetable cash crops, common in all mountain areas and often replacing long-standing subsistence crops, has encouraged new biohazards and elevated economic vulnerability.

Indigenous mountain people have suffered from exotic infectious diseases like smallpox, measles and influenzas and, more recently, from tuberculosis and HIV/AIDS. Resulting from European-based economic and colonial expansion post-fifteenth century, epidemics produced catastrophic collapses of social-ecological systems and disasters of unprecedented proportions (Mann 2006). Early examples in mountain areas coincided with the arrival of the Spanish in the central Andes in the sixteenth century where spread of infection was magnified by the prior concentration of large numbers of workers and their families in high altitude mining settlements, such as Potosí, under the pre-Inca and Inca forced labour systems. Lack of natural immunity, crowded living conditions, poor health due to malnutrition and high altitude and social erosion due to relocation made people highly susceptible to infection. The physiological stresses imposed by high altitude, including hypoxia (oxygen deprivation) and persistent cold, were geo-ecological factors magnifying hazardousness. Isolation had ensured a lack of natural immunities to infection and delayed the onset of catastrophic epidemics in some areas. Examples are found in isolated pockets of the Himalaya, Karakoram and Hindu Kush where smallpox epidemics took their toll in the twentieth century. Infectious diseases, including tuberculosis, drug-resistant TB, HIV/AIDS and gastro-intestinal infections, remain the most dangerous hazards among mountain residents.

Forest fires are common in mountain environments. The contributing geo-ecological factors include: extensive forest cover, fire-susceptible/dependent species, weather that encourages periodic drought and wind, and topography that facilitates rapid upslope spread of fire. Social-ecological factors related to forest use and management practices and expansion of settlements into susceptible forested areas have been equally important but not unique to mountain environments. The escalation of forest fire risk is extremely important in the mountain regions of North America (e.g. Filmon 2004) and it is used later to illustrate risk escalation due to anthropogenic factors.

19.2.2 Hazard Cascades

Hazard cascades occur when a primary hazard leads to a secondary and different hazard which may, in turn, produce a tertiary hazard and so on. Such cascades, which are common in mountain regions, magnify the temporal and spatial dimensions and impacts of disaster, adding to risk complexity. For example, forest fires may lead to increased surface water runoff, floods, soil erosion and snow avalanches over a time period measured in years, if not decades. Earthquakes provide a problematic example of hazard cascades and examples in Peru 1970, Pakistan 1974, Taiwan 1999, Kashmir 2005 and Sichuan 2008 are illustrative. Each produced significant ground shaking, structural damage, deaths and injuries. Ground shaking in the Peru event triggered a rock, ice and snow avalanche on Huascarán, the highest peak in the Cordillera Blanca. The avalanche fell into a pro-glacial lake, producing a water wave that over-topped and eroded a moraine dam, resulting in a large and growing debris flow that destroyed the town of Yungay and its approximately 20,000 inhabitants (Morales 1970; Bode 1990). Similar disastrous cascades have occurred repeatedly in the Cordillera Blanca and Rio Santa valley since 1941 (Carey 2010). The 1974 Pakistan earthquake at Pattan in the northwest Himalaya, produced numerous rockfalls/slides and other forms of slope failure that added to the toll of death, injury and destruction (Hewitt 1976). The 1999 Taiwan, or Chi-Chi, earthquake was notable for the cascade of landslides and debris flows that followed. Research from this event has added significantly to knowledge of earthquake-generated hazard cascades and their longevity (Shieh et al. 2009). The 2005 Kashmir earthquake impact was substantial, particularly in Pakistan, where direct structural damage, deaths and injuries have had a lingering effect (Halvorson and Hamilton 2010). Over 4,000 landslides were triggered (Owen et al. 2008), several potentially dangerous landslide-dammed lakes were created and elevated landslide frequency continued for 2 years following the event (Saba et al. 2010).

The 2008 Wenchuan earthquake unleashed a disastrous cascade that has unfolded over several years and may not yet be concluded (Cui et al. 2011). The main shock (8.2 Mw) occurred on May 12, 2008 accompanied by 9 m horizontal surface displacement and extreme ground shaking at its epicentre on the Beichuan Fault at the Indo-Eurasian Plate boundary. The mountain margin is densely populated and

Fig. 19.2 The Minjiang valley near Yingxiu, Sichuan, showing the destruction of a national highway by the 2008 Wenchuan earthquake and landslides and 2010 debris flows. The bridge in the background carries the new route of the highway and was constructed within months of the earthquake (James Gardner)

intensively used as are the valleys leading into and through the mountains. About 89,000 fatalities and 374,216 injuries, >5 million homeless, 26,888,000 destroyed or damaged dwellings and extreme damage to road, rail and other infrastructure resulted (Cui et al. 2009). Approximately 3,700 landslides left large deposits of loose rock and sediment precariously perched on steep slopes (Yin et al. 2009; Chigira et al. 2010). The landslides contributed substantially to the disaster by destroying and damaging highways and bridges, thus delaying emergency disaster response operations (Fig. 19.2).

The hazard cascade continued with 32 landslide-dammed lakes formed and in danger of failing and producing outburst floods (Cui et al. 2009) (Fig. 19.3). Such landslide-dam outburst floods have resulted in horrific flood disasters in and beyond the mountains in this area previously (Dai et al. 2005). A large rescue, recovery and reconstruction program, supported by the central and provincial governments, the Peoples Liberation Army and private enterprise ensued, including mitigation of the landslide-dam outburst flood threat. In mid-August 2010, heavy rain mobilized earthquake debris into destructive debris flows that damaged newly reconstructed villages, towns, highways and bridges, power installations, etc. (Tang et al. 2011;

Fig. 19.3 An earthquake-generated landslide-dammed lake from the 2008 Wenchuan earthquake. The dam has been hardened with concrete and the outflow is controlled to prevent erosion of the dam and outburst flooding. The presence of the lake, however, carries the threat of a future landslide-generated flood overtopping the dam (James Gardner)

Xu et al. 2012) (Fig. 19.4). Debris flow forecasting and warning systems kept fatalities and injuries to a minimum.

The Wenchuan earthquake has initiated to date, a multiyear hazard cascade of destructive landslides, landslide dams, and debris flows resulting in a prolonged or delayed disaster with enormous impacts on life, communities and economies (>50 billion$). These few examples demonstrate that risk governance systems must account for mountain hazard cascades and disastrous consequences that may extend to years, decades or more.

19.2.3 Spatial Dimensions of Hazard

The spatial dimensions of hazards in mountain regions range from local to regional. Individual landslides, debris flows and snow avalanches generally reoccur in specific locations. Such sites in aggregate may cover, and the hazards may occur coincidentally when and where conditions are conducive, extensive areas. Whether at individual sites or in aggregate, these hazards are relatively easy to

Fig. 19.4 2010 debris flow destruction, by earthquake and landslide debris, of newly (post-2008 earthquake) constructed and reconstructed hotels and shops in the Longxijiang valley, Sichuan, China (James Gardner)

anticipate in terms of location and timing which makes hazard mapping and forecasting important tools in risk governance and management. Earthquakes, droughts and wildfires, on the other hand, generally are regional in extent. Their spatial dimensions of impact and their timing are more difficult to anticipate and forecast, though wildfire hazard forecasting over large areas has become a routine tool in risk management (Pyne 2007).

Some hazards affect areas beyond the mountains. As catchments and storage places for water, snow and ice, mountain areas are sources of floods that may do their primary damage outside the mountains. The front ranges of the Himalaya and Meghalaya, with copious monsoon precipitation, are generators of flood disasters throughout the Indo-Gangetic Plain. The extended 2010 flood disaster on the Indus River and its tributaries in Pakistan is an example. Thus, mountains while serving a significant 'water towers' for regional water supply, also serve as regional flood disaster engines. Risk governance and management under these conditions must therefore involve cooperation and collaboration across many jurisdictions, including international jurisdictions.

Landslide-dam and glacier-lake outburst floods, in addition to their local impacts, have affected downstream areas outside mountain source areas. Historic, disastrous floods on the Dadu He and Minjiang in Sichuan resulting from landslide-dam bursts are examples (Dai et al. 2005). The 1841 catastrophic flood on the Indus River, which extended through the northwest Himalaya to the plains, originated as an earthquake-generated landslide dam outburst. More recently, exceptionally large floods on the Indus extending into the plains have originated, at least in part, from glacier-lake outbursts in the Karakoram (Hewitt 1982). With the advent of glacial shrinkage and the build-up of pro- and sub-glacial lakes in the Himalaya, Karakoram, Tien Shan, Alps, Andes and the North American Cordillera, the threat of local and regional outburst floods has become an important concern in risk governance and management across jurisdictions.

Volcanic eruptions produce two types of hazardous processes and conditions that have regional and global dimensions. Firstly, the 1985 eruption of Nevado del Ruiz in the Andes of Colombia triggered a volcanic debris flow (*lahar*) that traveled down the Rio Gualí, away from the mountains, overwhelming agricultural land and villages and the unsuspecting population of 23,000 in Armero (NRC 1991). Extra-mountain debris flows are a significant concern in the Pacific Northwest of the U.S.A. and southwestern British Columbia where glacier-covered Mts. Rainier, Hood, Baker and Meager, all with a record of lahars, pose a threat to the densely populated urban and suburban areas of Portland, Seattle and Vancouver (e.g. Clague and Turner 2003). Again, the dimensions of hazard require risk governance and management that crosses several jurisdictions.

Secondly, explosive volcanic eruptions eject fine-grained ash high into the atmosphere where it is distributed regionally and globally. The fallout of ash can destroy and damage crops, agricultural land and ecosystems, kill and injure people and livestock, and damage infrastructure and machinery. Damaging ash fallout is normally greatest near the source but can extend great distances away from the source. Volcanic dust high in the troposphere can have a global reach and reduce incidence of solar radiation at the Earth surface, resulting in ground surface and lower atmospheric cooling which triggers climate change, glacial expansion, altered water balances and ecosystem change (e.g. Miller et al. 2012), all or some of which has disastrous consequences for social-ecological systems.

19.2.4 Environmental Variability and Change

Environmental variability and change, especially climate change, are of growing concern in regard to hazards and disasters (Shaw et al. 2010). Shrinkage of mountain glaciers and snow cover are obvious manifestations of climate warming (e.g. Orlove et al. 2008). Glacier shrinkage is one factor in an increase of glacial lake outburst floods (GLOFS), especially in the Himalaya and the Peruvian Andes. Carey (2010) has described this phenomenon in the latter, and the accompanying evolution of risk governance. In regions where water demands already exceed local supplies, as in the southwest U.S.A. and the Indo-Gangetic Plain, a decrease in snow cover in tributary mountain watersheds elevates the risk of greater water shortages and drought disasters.

Glacier shrinkage and permafrost degradation due to warming destabilize valley-side rock and debris slopes, causing increased rock avalanche, rock fall, landslide and debris flow hazards. This is evident in the Alps (Gruber and Haeberli 2007; Ravanel and Deline 2010).

Change and long-term variability in temperature and precipitation are driving shifts in energy and moisture balances and other inherent properties of mountain environments. These may be manifest in extended droughts, wet periods and shifts in plant distributions and types (e.g. changes in treeline), altering the types, frequencies, magnitudes and distribution of hazardous processes.

Cooler and moister conditions have, in the past, caused expansion of glaciers which in some cases were damaging to settlements and land uses. Best documented is the expansion of glaciers into settlements and cropland in the western Alps during the seventeenth through nineteenth centuries (Vivian 2005).

19.2.5 Anthropogenic Change

Mountain environments have been modified by human use, altering hazard frequency, magnitude, location and impact. Understanding of such modifications is central to risk governance. Forest use, expansion of agriculture, extraction of mineral resources, use and alteration of streams and rivers and construction of buildings, roads and other infrastructure have been the primary modifiers.

Forests on mountain slopes modulate processes, stabilizing soil, water and snow and resisting erosion. Agricultural land clearing and timber harvesting has occurred on a large scale in mountain areas for about three millenia (Williams 2006). Forests have regenerated naturally or by replanting in some areas. In others, as in parts of Europe and the Andes, forests have not returned. The process has been repeated several times in some parts of the Himalaya, including southwest China (Elvin 2004). At present, widespread timber harvesting is denuding large areas of the North American Cordillera. Tree removal exposes the ground surface to increased water and wind erosion, causes snow to be more unstable and decreases the stability of soil and other material provided by roots. Soil erosion, floods, landslides, debris flows and snow avalanches may increase in frequency, magnitude and spatial distribution. Eckholm (1975) proposed the *Himalayan Theory of Environmental Degradation*, claiming that widespread forest clearing by mountain agriculturists was responsible for elevated flooding, erosion and sedimentation in the Gangetic Plains region of India and Bangladesh. This claim of widespread deforestation and its influence on flooding was debunked (Ives and Messerli 1989; Hofer 1993), but at a scale of individual slopes and small watersheds, the relationship between deforestation and increased erosion, flood and snow avalanche hazard has been demonstrated widely.

Forest protection practices, well-intended to counteract the foregoing effects, have had unintended hazard-elevating consequences in some mountain areas. The vast Cordilleran area of Canada and the United States is managed through a mosaic of federal, provincial/state, regional, First Nations, municipal and private jurisdictions. Early in the twentieth century, at a time when forest fires were taking a significant toll on standing timber, settlements and infrastructure and the parks and protected areas movement was becoming well-established, fire suppression became a common practice across jurisdictions (Pyne 2007). The goal was to extinguish all fires as quickly as possible and instill a public culture of fire prevention. By 1980, large areas had become reforested with even-age (60–80 years) monocultures of a few species, often dominated by fire-tolerant pine species. Research was revealing that complete fire suppression was hampering species and biodiversity progression and raising wildfire hazard. Some agencies (U.S. National Parks Service and Parks Canada) adopted controlled burning to replicate frequent, low magnitude fires. The cataclysmic Yellowstone fire of 1989 that started from several controlled fires, altered the geo-ecology of a third of Yellowstone National Park and demonstrated that high hazard conditions were encouraged by fire suppression. Mature, even-age pine forests also proved to be ideal incubators for Mountain Pine Bark Beetle populations (Nikiforuk 2011). Large areas of mountain pine forests were ravaged by exploding beetle populations. Warmer than normal winter temperatures also reduced the usual beetle larvae winter kill-off. Left standing were large areas of dead timber, adding to fuel load and resulting in widespread wildfire outbreaks in warm and dry summers such as in 2003. The key event of 2003 was the Kelowna Mountain Park Fire which badly damaged the Park and the outskirts of Kelowna, British Columbia, negatively impacted the local and regional economy and drew national and international attention to mountain wildfire hazard, particularly that at the interface between forests and settlements (Filmon 2004). A mix of natural and anthropogenic factors increased the hazard while population and economic growth elevated exposure and vulnerability. The elevated risk remains in many mountain areas as a testimony to ill-informed fire risk governance (Nikiforuk 2011).

Construction has been an anthropogenic instrument of changing hazard and risk in mountain environments. Construction of settlements, roads and other infrastructure has both elevated vulnerability and assisted with hazard prevention, control and mitigation. Here we focus on two examples of construction that change hazard processes: road construction and construction of dams and reservoirs. Roads ranging from temporary roads for resource extraction to multi-lane highways destabilize slopes because most use cut-and-fill construction across the slope fall line. Upslope is undercut, magnifying gravitational pull by removing support and slope steepening. Downslope of the road is destabilized by added loose material and local steepening (Fig. 19.5). Roads, unless carefully designed and maintained, also act as conduits and storage places for water. Soil erosion, landslides, debris flows and debris torrents increase in frequency and magnitude and elevated sediment loads are added to streams (Barnard et al. 2001; Sidle et al. 2011).

Dams of all sizes are used to impound water on mountain streams for water supply, power generation, flood and low flow control and recreation purposes. Today, dams and reservoirs are seen as important sources of hydro-electric power and they are used for such in most mountain regions. They both elevate and reduce hazard, and have been opposed and supported on the basis of each and on the basis of their

Fig. 19.5 Road relocation and construction required by the construction of the Tehri Dam and reservoir on the Bagirathi (Ganges) River, Uttarakhand, India. Slope destabilization along, above and below the road is clearly evident in the numerous scars and deposits from landsliding, soil erosion and debris flows and torrents (James Gardner)

various social-ecological costs and benefits. Some potential hazards associated with mountain dams and reservoirs include: catastrophic flooding due to dam failure, destabilization of reservoir shorelines and slopes, dam overtopping and catastrophic flooding due to landslides into reservoirs and crustal destabilization leading to heightened earthquake activity. Evidence for each does exist. Most dam failures have involved small installations. Shoreline and slope destabilization is common in large valley-bottom reservoirs but the impacts are usually local, affecting agricultural land and roads. The 1963 Vaiont Dam disaster in the Italian Alps resulted from a large rockslide into the reservoir that produced dam over-topping and a downstream flood wave that took 2,500 lives and destroyed agricultural, residential, industrial and transportation facilities for 30 km down the Piave valley (Superchi et al. 2010) (Fig. 19.6). Minor crustal destabilization associated with large reservoirs has been demonstrated and the 2008 Wenchuan earthquake raised the spectre of contributory destabilization by the Ziping Dam and Reservoir (Ge et al. 2009) which is located within 5 km of the Beichuan Fault rupture zone and epicentre. Some evidence indicates that a significant water level draw-down prior to the earthquake may have contributed to crustal instability in an already stressed fault zone. Large fluctuations in

Fig. 19.6 The Vaiont Dam in the Italian Alps, now decommissioned as a hydro-electric facility following the 1963 landslide (deposit shown to the *left* of the photo), impact wave and flood disaster in the Piave valley. The photo is taken from the reservoir area overlooking the dam to the narrow opening to the Piave River valley and the city of Longarone (James Gardner)

water levels are characteristic of hydro-electric power generation reservoirs, especially in environments where natural water yield is seasonal, and growing evidence suggests that the fluctuations trigger shore, slope and crustal instability.

19.3 Social-Ecological Systems and Change

Social-ecological systems (Berkes et al. 2003) are comprised of people, families, communities, settlements, states and nations, coupled with their many institutions and relationships with place, habitats and resources. Often portrayed as physically, socially, economically and politically isolated, most mountain people are now integrated with wider economies and jurisdictions. This both has reduced and elevated exposure and vulnerability to hazardous processes and represents the second broad area of risk complexity (Folke et al. 2005). Risk reduction has occurred through hazard prevention, control and mitigation, modification of exposure and vulnerability, enhancement of resilience and improvements in disaster response, recovery and reconstruction, all facilitated by new knowledge and technologies, including traditional and local knowledge (Gardner and Dekens 2007). Elevated exposure and vulnerability, and therefore risk, relates to growth and concentration of population, changes in the socio-economic characteristics of mountain social-ecological systems, changes in land uses, livelihoods, technologies and activities, and erosion of local knowledge.

19.3.1 Population and Social Change

The population of mountain areas has been well-studied (e.g. Funnell and Parrish 2001). About 10 % of the global population, or 650 million people, live on a permanent basis in mountain areas with the majority concentrated in developing regions in Nepal, Peru, India and China for example. High annual growth rates of 3 % occur in, but are not limited to, these areas. Reduced infant mortality, increased life expectancies and in-migrations have contributed to the population increase. In addition to permanent residents, there are millions who have semi-permanent residency (Moss 2006) and even larger numbers who are transients and tourists in mountain areas. They all are relevant to risk governance.

Growth in population has been accompanied by changes in the social structure, driven by new economic opportunities in resource extraction, construction, horticulture, administration, and recreation and tourism, many of which also encourage in-migration. The presence, numbers, distributions and activities have elevated exposure to hazards in most mountain regions, particularly in the developed and emerging economies of Europe, the Americas and east and south Asia. The isolated and independent mountain people and communities of the past are no more, with a few exceptions.

Urbanization and other forms of population concentration are rampant in mountain areas (Perlik and Messerli 2004). Ribbon development along major highways, recreation communities at ski, golf and amenity sites, company towns at resource extraction and construction sites and settlements around manufacturing, processing and transportation are common throughout the mountain world (Fig. 19.1). Increasing numbers of people, facilities and infrastructure are crowding into sites that may be exposed to hazards and onto sites that have traditionally been avoided due to hazards. Risks are increasing as are efforts to reduce risk through many structural and non-structural means.

19.3.2 Livelihoods and Activities

Livelihoods are the means by which people 'make a living', which in mountain environments now range from traditional agro-pastoral work, gathering and hunting, to employment in a multitude of industries and services. Some livelihoods do cause greater exposure to hazards than others. Resource extraction, particularly timber harvesting and mining, and some recreation livelihoods, guiding and driving have high degrees of exposure to chronic or routine hazards. These risks must be managed just as disaster risks are managed. Regulations governing conditions of employment, required safety training and measures and insurance coverage are risk management tools. Many contemporary mountain livelihoods are at risk when the commodities, means of production, markets and security are subject to highly destructive hazards and conditions, and large-scale regional disasters. For instance, biohazards can wipe out cash crops; roads and railroads can be closed by landslides and earthquakes; epidemics, violence and civil disturbance can compromise the tourism industry (e.g. Gardner et al. 2002). Earthquake damage and disruption of security typically throw large numbers of people into temporary or permanent unemployment. Added to population growth and change, diversification of livelihoods has elevated exposure to hazards and raised vulnerability to natural, economic and political crises and shocks.

19.3.3 Linkages

Linkages include trails, roads, highways, railroads, telecommunications and information technologies all of which facilitate movements of people, materials, goods, services, information and ideas. Linkages have been primary facilitators of all the changes apparent in mountain social-ecological systems. They have been the means by which isolation has been overcome and through which outside interests have exerted power and influence in mountain areas. Roads and railroads were instrumental in opening the Alps to modern mass tourism and fundamental to widespread resource exploitation in the North American Cordillera. Trails and roads facilitated the pre-Inca, Inca, Spanish and now, national and international exploitation and use of the Andes. Frantic road building in the Himalaya and Trans-Himalaya in Pakistan, India and China has been an instrument for socio-economic development, nation building and national security (e.g. Kreutzmann 2004). Trans-Andean road construction is facilitating resource exploitation, tourism and international trade in South America. Pipelines for the transport of fossil fuels are highly controversial links that have been built through mountains in central Asia, Europe and the Cordillera of the Americas.

All land-based links in mountain areas are highly vulnerable to damage, destruction, closure and delays by many hazards. This is due in part to their linearity perpendicular to the direction of most slope hazards. The social and economic impacts of closures are widespread, extending well beyond the mountains. Closures also compromise disaster assistance in mountain regions. Today, air transport by fixed wing aircraft and helicopters address some of the shortcomings of land-based transport in supporting day-to-day life and hazard and disaster management. Telecommunications and information technologies have been instrumental in supporting and driving many of the changes that both encourage and reduce vulnerability in mountain social-ecological systems.

19.4 Risk Governance in Mountain Environments

Policies and procedures that reduce exposure and vulnerability to hazards and that facilitate recovery from accidents and disasters are important parts of risk governance. Our consideration of governance refers to the ways decisions are made through individual and/or collective action and, as such, defines the distribution of influence, power, authority and responsibility. Ideally, risk governance is sensitive to the local knowledge, traditions, rules, laws and institutions embedded in social-ecological systems. In mountain regions, risk governance has evolved from local systems based on local knowledge to a combination of local, regional and international systems guided by knowledge from physical and social sciences and engineering within legal and political contexts. Strictly speaking, risk governance authority and responsibility are embedded in the legal framework originating at the level of the nation state and devolved to the regional and local levels. International organizations, such as the UN and its agencies, NGOs and ICIMOD, provide guidelines, knowledge and many forms of assistance but they lie outside the strict governance framework. The tendency in risk governance has been towards a technocratic and top-down approach at the expense of local traditions (Hewitt 1983), though this is changing with renewed interest in 'indigenous' knowledge (Mercer et al. 2010).

Prior to the establishment of imperial, colonial and state systems of governance in mountain areas, management of risk from hazards largely was the responsibility of individuals, families and communities. Knowledge about hazards was gained through experience and shared mainly by oral transmission. In disaster events, transmission of news and outside assistance was often slow and difficult. Risk reduction focused on place-specific adaptations to reduce exposure/vulnerability, such as avoidance of dangerous locations adjacent to streams, gullies, glaciers and unstable and snow avalanche slopes (De Scally and Gardner 1994). Transhumance and migratory practices were used to access resources while avoiding dangerous conditions. Diversification of crops and planting locations were common adaptations to biohazards, floods, debris flows and landslides. Community fragmentation and migration were adaptations to regional hazards such as earthquakes and droughts (Fagan 2008). Where hazards and vulnerabilities were not understood, as with infectious diseases, disasters ensued. When newcomers, such as traders, migrants and refugees, arrived in a mountain area, they may have been exposed through lack of local knowledge and/or unavailability of safe locations (Gardner 2002).

The arrival of imperial, colonial and state forms of governance saw a shift away from local forms of risk management, with varying intent and success. The Pre-Inca and Inca empires in the Central Andes were early, pre-sixteenth century, examples of regional governance with a primary emphasis on resource exploitation in support of monarchy and elites. Forced labour migration resulted in large concentrations of population that, at high altitudes, were highly susceptible to exotic infectious diseases introduced during the sixteenth century Spanish colonization. The disaster inflicted on indigenous peoples suggests that the shift to regional governance did more to increase risk than to mitigate it. Parts of the Himalaya fell under Gurkha (Nepalese), Tibetan, Sikh and British authority at various times starting in the eighteenth century. Eventually, British authority extended over the largest area, coming to an end in 1947. Evidence from the 1905 Kangra earthquake indicates that the civil administration, with military support, attempted to address emergency response and recovery in the affected mountain areas by re-establishing lines of communication, especially trails, roads and bridges which were badly damaged by ground shaking, rockfalls, landslides and snow avalanches (Gardner 2002). In doing so, the value of coordinated risk management and governance by a higher or regional level authority was demonstrated, as it continues to be in more recent disasters of regional scale. Nonetheless, first response rescue of earthquake victims remains largely in the hands of neighbors, friends, family members and local police, fire-fighters and medical personnel. The British administration in India was less effective in addressing the infectious disease and famine disasters in the decades prior to 1905. Indeed, that administration has been held responsible for creating the conditions of disaster in those cases (Davis 2002).

The evolution of risk governance in the Rio Santa valley of the Cordillera Blanca, Peru (Carey 2010), provides an excellent example of cross-scale and intra-community tensions and inequalities that can arise in risk governance. In this situation, hazards arising from climate warming and glacier shrinkage have increased in frequency. Since the 1941 destruction of Huaraz by a glacial lake outburst flood, a series of similar events has caused destruction in neighboring locations, the most famous being the 1970 landslide/debris flow destruction of Yungay (Bode 1990). Local, national and international scientific interest and understanding of the hazards, vulnerabilities and risks developed following 1941. An important step in risk governance occurred in 1950 with the establishment of the Andean Lakes Commission, shifting the locus of control to the national level. Coupled with the interests in disaster prevention, were competing interests in water supply and power generation at the local and national levels. Risk reduction measures involving structural controls at the lakes and along the stream courses and non-structural measures such as land use zoning in towns and villages were established or proposed. Local elites opposed land use restrictions and various national governments, ranging from authoritarian to democratic, prevaricated in support or not of risk reduction measures. Economic liberalization in the 1990s further eroded risk reduction efforts. International interest and resources, such as that from the World Bank in 2008, driven by growing concern about climate change, have now come to bear on risk management in the central Andes. The case well illustrates the tensions in risk governance across scales from the local to the international and between rational, science-based planning and that driven by economics, elitism and inequalities.

The 2005 Kashmir and 2008 Wenchuan earthquakes, as compared with the 1905 Kangra earthquake, illustrate the evolution of risk governance through the twentieth century as applied to regional-scale disasters in mountain areas. The nature of the hazard has changed little. The earthquakes, were similar in magnitude (8+Mw), had many foreshocks and aftershocks, affected areas similar in geo-ecological conditions and produced hazard cascades of landslides, etc. The relative severity of impacts was similar, though the absolute impacts in terms of deaths, injuries and property and infrastructure damage were magnified in recent cases due to population growth and concentration, growth and diversification of economies and construction of new settlements, industries and infrastructure. In 1905, local and regional administrations took responsibility for rescue, recovery and redevelopment. Pre-event preventive efforts amounted to traditional location avoidance and construction methods and materials. The disaster itself became known beyond the locally affected areas only after many days and weeks.

The 2005 and 2008 earthquakes were known around the world immediately via seismic monitoring, telephone, internet, television, radio and other media. Cities were badly damaged, Manshera, Balakot and Muzzaferabad in 2005 and Dujiangyan, Yingxu and Beichuan in 2008 for example. Access to and from mountain villages was hampered by earthquake and landslide damage and inclement weather. Deaths, injuries and damage were magnified by poor construction and by the timing of the events during the work/school day when large numbers of people were in public buildings and spaces, schools, work places, etc. The location of the Kashmir Earthquake astride the LOC (line of control) between Indian and Pakistani controlled territories added to the jurisdictional complexity. Apart from local first response efforts to rescue victims, national and international attention from governmental and non-governmental organizations was immediate and sustained in the Kashmir case (Zimmermann and Issa 2009). The 2008 Wenchuan event was unique in disaster history in China as international media and expertise were permitted access to information and some locations following the earthquake. Disaster response, recovery and redevelopment was managed through the substantial human, financial and technical resources of local, municipal, provincial and central governments of China, the Peoples Liberation Army, businesses and service clubs, along with some international assistance.

In most mountain areas, it is now recognized that routine hazards and the possibility of disasters are present by virtue of inherent properties of the environment and the types of exposures and vulnerabilities present. Nations, where mountains cover large areas and/or make up a large part of national territory, are sensitive to the challenges of mountain environments and have in place multi-scale risk governance frameworks with defined authority and responsibility across jurisdictional scales. Mitigation, control and prevention of routine hazards, such as those posed by local landslides, debris flows, rock falls and snow avalanches, are often managed at the municipal or district levels utilizing general research and development, national and international resources and local knowledge of hazards and conditions of vulnerability. Hazard forecasting, preventive interventions, warning systems, building codes, land use zoning, activity regulation, structural controls, first-response measures, insurance and public education are among the common tools. Widespread, coincident outbreaks of otherwise site-specific hazards can lead to disasters that require intervention at the national and international scales. Large-scale regional hazards, especially those that generate hazard cascades and impact large areas within and beyond the mountains, such as earthquakes, explosive volcanic eruptions, wildfires, regional floods and biohazard epidemics, require risk governance frameworks that span jurisdictions from the local to the international. While risk reduction is greatly enhanced by preparatory and preventive action at the local scale, disaster response and recovery generally requires the financial, material and human resources at the national level, usually with international assistance in less-developed jurisdictions.

19.5 Conclusions

Risk governance and management in mountain environments has evolved from the domain of individuals and communities to that of regional and national governments, international agencies and non-governmental and private sector organizations. Early isolation from centers of population and control and difficult access and communication inhibited the flow of information from and attention to disasters and accidents in mountain environments. Imperial, colonial and commercial expansion was accompanied by an expanded domain of responsibility and authority for protection of people, though it did so in a faltering manner until the latter half of the twentieth century. More often than not imperial, commercial and colonial interests focused on protection of assets in mountain areas, with less regard to the people. Lack of knowledge about the types and complexities of mountain hazards challenged effective risk governance and management. Local knowledge was mostly ignored or discounted. Improvements in access and information technologies facilitated external dominance and use of mountain environments, usually with increases in vulnerabilities and thus risk. The same improvements ensured that news of even minor accidents as well as major disasters could become widely and quickly known, generating public pressure and the option for remedial action. Recognition of the importance of natural resource and other assets and services of mountain environments to societies at large, plus a growing global concern for human rights and humanitarian assistance, further advanced risk governance in mountain areas.

The challenges for risk governance in mountain areas today are in creating a system that:

- incorporates local needs and resources with those at the national and international scales,
- recognizes the complexities and spatial and temporal dimensions of hazards and hazard cascades proactively,
- anticipates the impact of social-ecological and environmental variability and change on risk,
- co-ordinates the actions of local authorities and governmental, non-governmental, community-based, private sector and international organizations,
- builds resilience in affected social-ecological systems by ensuring financial and other resources are in place to complete recovery and reconstruction at least to their prior state.

References

Barnard, P.L., L.A. Owen, M.C. Sharma, and R.C. Finkel. 2001. Natural and human-induced landsliding in the Garhwal Himalaya of northern India. *Geomorphology* 40: 21–35.

Berkes, F., J. Colding, and C. Folkes (eds.). 2003. *Navigating social-ecological systems: Building resilience for complexity and change*. Cambridge: Cambridge University Press.

Bode, B. 1990. *No bells to toll: Destruction and creation in the Andes*. New York: Paragon House.

Brundl, M., H.E. Romang, N. Bischof, and C.M. Rheinberger. 2009. The risk concept and its application in natural hazards risk management in Switzerland. *Natural Hazards and Earth System Sciences* 9: 801–813.

Carey, M. 2010. *In the shadow of melting glaciers: Climate change and Andean society*. New York: Oxford University Press.

Chigira, M., X. Wu, T. Inokuchi, and G. Wang. 2010. Landslides induced by the 2008 Wenchuan earthquake, Sichuan, China. *Geomorphology* 118: 225–238.

Clague, J.J., and B. Turner. 2003. *Vancouver, city on the edge: Living with a dynamic geological landscape*. Vancouver: Tricouni Press.

Cui, P., Y.-Y. Zhu, Y.-S. Han, X.-Q. Chen, and J. Zhuang. 2009. The 12 May Wenchuan earthquake-induced landslide lake: Distribution and preliminary risk evaluation. *Landslides* 6: 209–223.

Cui, P., X.-Q. Chen, Y.-Y. Zhu, F.-H. Su, F.-Q. Wei, Y.-S. Han, H.-J. Liu, and J.-Q. Zhuang. 2011. The Wenchuan earthquake (May 12, 2008), Sichuan Province, China and resulting geohazards. *Natural Hazards* 56: 19–36.

Dai, F.C., C.F. Lee, J.H. Deng, and L.G. Tham. 2005. The 1786 earthquake-triggered landslide dam and subsequent dam-break flood on the Dadu River, southwestern China. *Geomorphology* 65: 205–221.

Davis, M. 2002. *Late Victorian holocausts: El Nino famines and the making of the third world*. New York: Verso.

De Scally, F., and J.S. Gardner. 1994. Characteristics and mitigation of the snow avalanche hazard in Kaghan Valley, Pakistan Himalaya. *Natural Hazards* 9: 197–313.

Eckholm, E.P. 1975. The deterioration of mountain environments. *Science* 189: 764–770.

Elvin, M. 2004. *The retreat of the elephants: An environmental history of China*. New Haven: Yale University Press.

Fagan, B. 2008. *The great warming: Climate change and the rise and fall of civilizations*. New York: Bloomsbury Press.

Filmon, G. 2004. *Firestorm 2003: A provincial review*. Victoria: Government of British Columbia.

Folke, C., T. Hahn, P. Olsson, and J. Norberg. 2005. Linkages between vulnerability, resilience and adaptive capacity. *Global Environmental Change* 16: 293–303.

Fuchs, S. 2009. Susceptibility versus resilience to mountain hazards in Austria – Paradigms of vulnerability revisited. *Natural Hazards and Earth System Sciences* 9: 337–352.

Funnell, D., and R. Parrish. 2001. *Mountain environments and communities*. New York: Routledge.

Gardner, J.S. 2002. Natural hazards risk in the Kullu District, Himachal Pradesh, India. *The Geographical Review* 92: 282–306.

Gardner, J.S., and J. Dekens. 2007. Mountain hazards and the resilience of social-ecological systems: Lessons learned in India and Canada. *Natural Hazards* 41: 317–336.

Gardner, J.S., A.J. Sinclair, F. Berkes, and R.B. Singh. 2002. Accelerated tourism development and its impacts in Kullu-Manali, H.P., India. *Tourism Recreation Research* 27: 9–20.

Ge, S., M. Liu, N. Lu, J. Godt, and G. Luo. 2009. Did the Zipingpu reservoir trigger the 2008 Wenchuan earthquake? *Geophysical Research Letters* 36: L20315. doi:10.1029/2009GL040349.

Gruber, S., and W. Haeberli. 2007. Permafrost in steep bedrock slopes and its temperature-related destabilization following climate change. *Journal of Geophysical Research* 112: F02S18. doi: 10.1029/2006JF000547.

Halvorson, S.J., and J.P. Hamilton. 2010. In the aftermath of the Qa'yamat: The Kashmir earthquake disaster in northern Pakistan. *Disasters* 34: 184–204.

Hewitt, K. 1976. Earthquake hazards in the mountains. *Natural History* 85: 30–37.
Hewitt, K. 1982. Natural dams and outburst floods of the Karakoram Himalaya. Hydrological aspects of high mountain areas. *International Association Scientific Hydrology Publication* 138: 259–269.
Hewitt, K. (ed.). 1983. *Interpretations of calamity: From the viewpoint of human ecology*. London: Allen and Unwin.
Hewitt, K. 1997a. *Regions of risk: A geographical introduction to disasters*. Harlow/Essex, UK: Addison Wesley Longman.
Hewitt, K. 1997b. Risk and disasters in mountain lands. In *Mountains of the world: A global priority*, ed. B. Messerli and J.D. Ives. London/New York: Parthenon Publishing Group.
Hofer, T. 1993. Himalayan deforestation, changing river discharge and increasing floods: Myth or reality? *Mountain Research and Development* 13: 213–233.
Ives, J.D., and B. Messerli. 1989. *The Himalayan dilemma: Reconciling development and conservation*. New York: Routledge.
Kreutzmann, H. 2004. Accessibility in high Asia: Comparative perspectives on northern Pakistan's traffic infrastructure linkages with its neighbours in the Hindukush-Karakoram-Himalaya. *Journal of Mountain Science* 1: 193–210.
Mann, C.C. 2006. *1491: New revelations of the Americas before Columbus*. New York: Random House.
McClung, D., and P. Schaerer. 2004. *The avalanche handbook*. Seattle: The Mountaineers.
Mercer, J., I. Kelman, L. Taranis, and S. Suchet-Pearson. 2010. Framework for integrating indigenous knowledge with scientic knowledge for disaster risk reduction. *Disasters* 34: 214–239.
Miller, G.H., A. Giersdottir, Y. Zhong, D. Larsen, B.L. Otto-Bliesner, M.M. Holland, D. Bailey, K.A. Refsnider, S.J. Lehman, J.R. Southon, C. Anderson, H. Bjornsson, and T. Thordarson. 2012. Abrupt onset of the Little Ice Age triggered by volcanism and sustained by sea-ocean feedbacks. *Geophysical Research Letters* 39: L02708. doi:10.1029/2011GL050168.
Morales, B. 1970. El día más largo en el Hemisfero Sur. *Revista Peruana de Andinismo y Glaciología* 9: 63–71.
Moss, L.A.G. (ed.). 2006. *The amenity migrants: Seeking and sustaining mountains and their cultures*. Oxford: CABI.
Nikiforuk, A. 2011. *The empire of the beetle: How human folly and a tiny bug are killing North America's great forests*. Vancouver: Greystone/D & M Publihers.
NRC (National Research Council). 1991. *The eruption of Nevado del Ruiz volcano, Columbia, South America, November 13, 1985*, Committee on Natural Disasters, Division of Natural Hazards, Natural Disaster Studies, A Series. Washington, DC: National Academic Press.
Orlove, B., E. Wiegandt, and B.H. Luckman (eds.). 2008. *Darkening peaks: Glacier retreat, science, and society*. Berkeley: University of California Press.
Owen, L.A., U. Kamp, G.A. Khattak, E.L. Harp, D. Keefer, and M.A. Bauer. 2008. Landslides triggered by the 8 October 2005 Kashmir earthquake. *Geomorphology* 94: 1–9.
Perlik, M., and P. Messerli. 2004. Urbanization in the Alps: Development processes and urban strategies. *Mountain Research and Development* 24: 215–219.
Pyne, S.J. 2007. *Awful splendour: A fire history of Canada*. Vancouver: University of British Columbia Press.
Ravanel, L., and P. Deline. 2010. Climate influence on rockfalls in high-Alpine steep rockwalls: The north side of the Aiguilles de Chamonix (Mont Blanc massif) since the end of the "Little Ice Age". *The Holocene* 21(2): 357–365.
Rizvi, J. 1999. *Trans-Himalayan caravans: Merchant princes and peasant traders in Ladakh*. New Delhi/New York: Oxford University Press.
Saba, S.M., M. van der Meijde, and H. van der Werff. 2010. Spatio-temporal landslide detection for the 2005 Kashmir earthquake region. *Geomorphology* 94: 419–437.
Shaw, R., J.M. Pulhin, and J.J. Pereira (eds.). 2010. *Climate change adaptation and disaster risk reduction: Issues and challenges*. Bingley: Emerald.

Shieh, C.L., Y.S. Chen, Y.J. Tsai, and J.H. Wu. 2009. Variability in rainfall threshold for debris flows after the Chi-Chi earthquake in central Taiwan, China. *International Journal of Sdeiment Research* 23: 177–188.

Sidle, R.C., T. Furuichi, and Y. Kono. 2011. Unprecedented rates of landslide and surface erosion along a newly constructed road in Yunnan, China. *Natural Hazards* 57: 313–326.

Superchi, L., M. Floris, M. Ghirotti, R. Genevois, M. Jaboyedoff, and D. Stead. 2010. Implementation of a geodata base of published and unpublished data on the catastrophic Vaiont landslide. *Natural Hazards and Earth System Sciences* 10: 865–873.

Tang, C., J. Zhu, J. Ding, X. Cui, L. Chen, and J. Zhang. 2011. Catastrophic debris flows triggered by a 14 August 2010 rainfall at the epicentre of the Wenchuan earthquake. *Landslides* 8(4): 485–497.

The Economist. 2012. Counting the cost of calamities. *The Economist*, January 14, 60–61.

Vivian, R. 2005. *Les glaciers du Mont-Blanc*. Montmelian: La Fontaine de Siloe.

Williams, M. 2006. *Deforesting the Earth: From pre-history to global crisis*. Chicago: University of Chicago Press.

Wisner, B., P. Blaikie, T. Cannon, and I. Davis. 2004. *At risk: Natural hazards, people's vulnerability and disasters*, 2nd ed. London: Routledge.

Xu, Q., S. Zhang, W.L. Li, and T.W.J. van Asch. 2012. The 13 August 2010 catastrophic debris flows after the 2008 Wenchuan earthquake, China. *Natural Hazards and Earth System Sciences* 12: 201–216.

Yin, Y., F. Wang, and P. Sun. 2009. Landslide hazards triggered by the 2008 Wenchuan earthquake, Sichuan, China. *Landslides* 6(2): 139–152.

Zimmermann, M., and S.S. Issa. 2009. Risk-conscious reconstruction in Pakistan-administered Kashmir. *Mountain Research and Development* 29: 202–210.

James S. Gardner is Professor Emeritus, Natural Resources Institute, University of Manitoba and Adjunct Professor, Department of Geography, University of Victoria in Canada. Formerly Provost and Professor at the University of Manitoba (1991–2001) and Professor and Dean of Graduate Studies at the University of Waterloo (1987–1991), he has pursued research and teaching in geomorphology, hydrology, glaciology and resources and hazards management with field studies in mountain environments in Canada, Europe, India, Pakistan and China. He has published widely on alpine geomorphology and resources and hazards. Now retired he continues to write, teach occasionally and assist in mountain and coastal research and in supervision of graduate students at the Universities of Manitoba and Victoria.

Chapter 20
On the Edge: Coastal Governance and Risk

Rhoda Ballinger

20.1 Introduction. Coastal Risks and Needs

Coastal zones require special consideration. Loosely defined as those zones at the interface between marine and terrestrial systems, not only in terms of natural biological and physical processes, but also in terms of their governance, they pose unique and complex issues for natural risk management. Such zones are also some of the most populous areas of the planet (Mee 2010), where most megacities are located (Nicholls 1995; Grimmond 2011) and population densities average at least three times the global average (Small and Nicholls 2003). Many areas face multiple challenges, being susceptible to hazards such as storms, flooding, erosion and tsunamis as well as increasing environmental degradation and development pressures, including land subsidence, coastal habitat degradation, fisheries decline and pollution issues (Charlier 1989; Li 2003; Hadley 2009). Alongside this, the world's coast has lost much of its 'natural' coastal defence capacity, with 50 % of wetlands having disappeared over the last century due to human interference (Creel 2003). This has left high concentrations of people and assets at risk, particularly in deltas and other low lying coastal areas (McGranahan et al. 2007).

Such risks are also likely to rise as global population continues to grow and climate change exacerbates risks. Estimates for population growth vary, but some suggest the number of people living within 60 miles of coastlines will increase by about 35 % by 2025 compared with 1995 figures. Climate change, inducing a range of secondary impacts, including increased flooding, erosion, salinity changes and degradation of habitats, is likely to expose billions more worldwide to such risks (Creel 2003). Within South East Asia and the Pacific alone millions are likely to become sea level refugees by the end of the century (Wetzel et al. 2012). Whilst

R. Ballinger (✉)
School of Earth and Ocean Sciences, Cardiff University,
Main Building, Park Place, Cardiff CF10 3AT, UK
e-mail: BallingerRC@Cardiff.ac.uk

U. Fra.Paleo (ed.), *Risk Governance*, DOI 10.1007/978-94-017-9328-5_20

there remain huge uncertainties regarding climate change impacts and associated sea level rise predictions, such implications require a detailed, critical review of coastal governance and adaptability.

The chapter commences by reviewing traditional approaches to the management of coastal hazards before considering recent advances towards a more integrated approach to the management of coastal risks. The rest of the chapter focuses on the challenges which the coastal zone poses, including those associated with building and maintaining coastal resilience within the context of significant environmental including climate change. Within such discussions, the potential of Integrated Coastal Zone Management (ICZM) as a new governance approach is considered.

Throughout the chapter, there is focus on North West Europe where the author has considerable first-hand experience. Lessons from this region should be relevant elsewhere as much of the coast, particularly around the southern North Sea, is low lying, densely populated, and faces a range of coastal hazards. Also, whilst the level of centralisation and formality of arrangements associated with coastal protection varies from country to country (O'Connor et al. 2010), the region includes some of the most advanced coastal defence practice in the world. On top of this, public expenditure dedicated to coast protection has risen significantly in recent decades and is projected to escalate over the next half century (EUROSION 2004). This is fuelling debate over the future of hard defences and the need for consideration of other adaptation options and governance arrangements, including the role of ICZM.

20.2 Changing Approaches to Managing Coastal Risk

20.2.1 Traditional Approaches

Traditionally, coastal communities across North West Europe have battled 'against' nature, constructing hard coastal defences, including sea walls and even tidal barriers, along highly populated low-lying shores, particularly in the southern North Sea region. This almost unquestioning reliance on technological fixes (Mee 2010) was perpetuated after the 1953 North Sea storm event which resulted in a significant death toll, particularly in the Netherlands (Hillen et al. 2010).

Consequently, the Development – Defend cycle has been a feature of much decision-making until recently (Fig. 20.1) (Ballinger et al. 2002; Milligan and O'Riordan 2007). This has been perpetuated by local populations who have expected 'hold the line' solutions (Milligan and O'Riordan 2007), feeling safer living behind hard, clearly visible sea walls. Local politicians, frequently not well versed in coastal processes and engineering, have appeased their electorate, taking short-term decisions to sanction schemes. However, as Fig. 20.1 shows, such decisions have often provided impetus for further development on land behind defences, sometimes of inappropriate type and density. In turn, this has left populations and assets vulnerable, leading to further pressure for even higher levels of protection and sometimes even more defences.

Fig. 20.1 The development–defend cycle

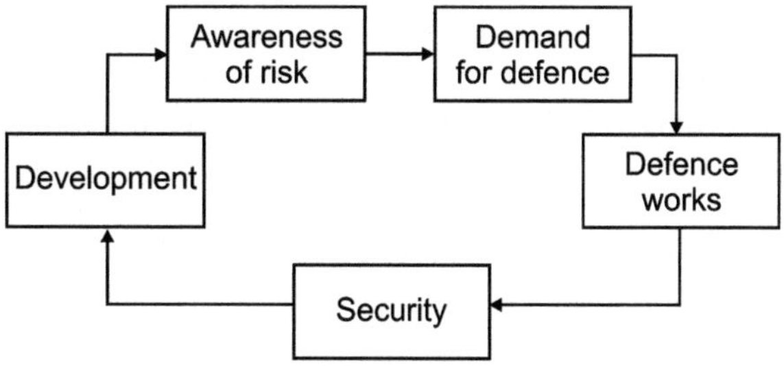

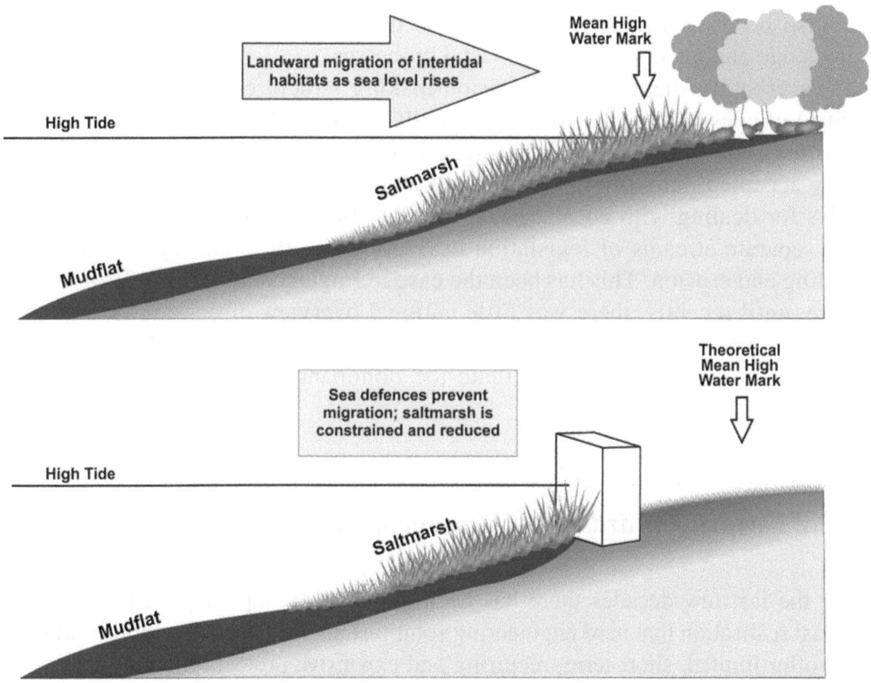

Fig. 20.2 Coastal squeeze

Driven by short-term, local needs, and based on the priorities of individual authorities, traditional hard engineering approaches have frequently been beset with problems. Parochialism has often exacerbated erosion issues down-drift of coastal protection schemes and other structures, as exemplified by the IJmuiden – Holland and Zeebrugge coasts (EUROSION 2004). There have also been issues associated with erosion of coastal intertidal habitats along low-lying coasts, particularly in estuarine areas where such habitats are frequently of high conservation value. As Fig. 20.2 shows, coastal squeeze occurs when intertidal habitats no longer can migrate landwards naturally and are 'squeezed' against fixed hard defences.

This leads to their subsequent degradation and erosion (Doody 2004), a significant concern given the EC Birds (2009/147/EC) and Habitats (92/43/EEC) Directives. Whilst these require the conservation interest and integrity of Natura 2,000 coastal sites to be maintained, the European Commission and European Court of Justice have yet to decide definitively whether this applies in the context of climate change and accelerated sea level rise. Issues have been compounded by an existing legacy of coastal infrastructure, including ports and harbours, promenades and even military defences, which have influenced coastal sediment budgets and processes, and, consequently, vulnerability to coastal hazards. Indeed, the re-allocation of some existing infrastructure, notably promenades, and in the case of Jersey former World War II military defences, to coastal defence usage, has also been particularly problematic as these were not designed with a full knowledge and understanding of coastal processes.

Exacerbating these issues, decision-making structures associated with the control of coastal hazards have traditionally been overly complex as legislation has often evolved piecemeal, reacting to individual hazard events. With no generally accepted definition of the coastal zone, fragmented institutional frameworks for dealing with coastal hazards have often developed, sometimes even with separate streams of legislation developing for different hazards, notably flooding and erosion. This has been the case in England and Wales (Pettit 1999), where, until recently, there was little national overview of coastal hazards and their management. At local levels, such complexity and the lack of a national framework, has tended to perpetuate the dominance of local, 'reactive' hard engineering responses.

20.2.2 Move Towards Risk Management

Over the last few decades there has been a change in approach, fashioned by a gradual realisation that hard engineering solutions provide only one option and may only offer limited, short-term, recurring and expensive protection (Charlier 1989). Globally, focus has turned to a wider range of responses, particularly in the context of climate change, which can also help provide other benefits, particularly for recreation and conservation. The register of extreme flood events across Europe over the last 10 years has certainly forced many governments to reconsider their positions. Even in the Netherlands, where flood protection is essential to two-thirds of the country and nine million people (Transport and Water Management Inspectorate 2006), there is incipient concern and public debate about flood risks in the wake of a recent national report which suggests risks from flooding in some locations far exceed that from other human-induced hazards (Klijn et al. 2008). Whilst the Netherlands largely holds on to its existing protectionist stance, other countries in the region, notably UK and France, have embraced a risk-based paradigm and promoted this through relevant measures at a national level. Realising that it is not possible to prevent all coastal flooding and erosion, a wide range of actions are suggested to manage risks and reduce impacts. Common elements of risk-based approaches include the need to:

- understand the nature of the risks, including their temporal and spatial extents
- communicate the risks appropriately with stakeholders including infrastructure providers
- take appropriate adaptive actions to reduce risks, damage and disruption

The National Flood and Coastal Erosion Risk Management Strategy for England (Environment Agency 2010) is typical, in which the onus on risk minimisation goes well beyond the province of the engineering community, forcing new shared, ways of decision-making, challenging existing working approaches and governance structures.

In terms of taking appropriate actions to reduce risk, Table 20.1 summarises the five generic shoreline policy options available to coastal managers and relates these to the three adaptation response strategies proposed by the Intergovernmental Panel on Climate Change's Coastal Zone Management experts (Gilbert and Vellinga 2005). These options are well recognised in Europe (see for example, MESSINA 2006 and EUROSION 2004) where they have been suggested as options for Coastal Sediment Management Plans, regional plans based on units defined by sediment cell boundaries (op. cit.). Similar options however, have been pioneered by and are already the cornerstone of the regional shoreline management plan process in England and Wales. Here, these non-statutory plans provide a strategic approach, supplying generic policies for the next hundred years for each management unit. As such, they direct local planning decisions and investments in coastal defence schemes. Whilst the realisation of these policies at local levels relies on appropriate funding, land availability and changing local priorities (Environment Agency 2012), these plans are becoming recognised as an important vehicle for the management of coastal risk, fostering engagement with a wider range of interests than merely the engineering community.

Table 20.1 also shows the extent to which the engineering and spatial planning communities need to be involved in decision-making related to each option. Clearly, spatial planning has a critical role as a gatekeeper of coastal change (Taussik 2000), preventing or restricting development in areas at potential risk. As indicated in the table, planning's input to ensure the sustainability of the 'no active' or 'limited intervention' options is essential. Its full involvement in retreat/managed realignment decisions, given the need to prevent development in areas at risk, is also vital. Whilst zonation of the coast in England and Wales has occurred within the shoreline management process and national planning guidance has been issued by the authorities on flood risk and coastal climate change adaptation, the non-statutory nature of the plans and indeed the guidance, threaten the sustainability and interpretation of the policies at local levels.

Indeed, at this level, where there are increasingly limited budgets and continued local development pressures and community concerns, some of the more unsavoury policies are already being challenged, sometimes unravelling deep-seated and protracted local coastal conflicts. Ballinger et al. (2002) and others (Greiving et al.

Table 20.1 Generic shoreline policy and adaptation options

Strategic shoreline policy		IPPC response strategy involved	Sectoral involvement	
Title	Description		Engineering	Planning
Do nothing/no active intervention	No planned investment for flood or erosion defence	**Accommodation**: Continued occupancy and use of vulnerable areas	×	✓
	This allows natural processes to 'take their course' (irrespective of whether or not an artificial defence previously existed)	*Requires advanced planning and accepting loss of some coastal value*		
Hold the line	To maintain the current shoreline position.	**Protection**: Defending vulnerable areas	✓	✓
	This will involve maintaining or improving standards of protection. It could include a variety of approaches including improvements to existing artificial defences and undertaking works in front of existing defences.	*Includes population centres, economic activities and natural resources*		
Advance the line	New defences built seaward of current shoreline	**Protection**: as above	✓	✓
	In practice the use of this is limited although there are opportunities associated with land reclamation, surfing and renewable energy generation.			

Managed realignment	Allowing natural shoreline movement, with some form of intervention, managing and directing the process in certain areas	**Retreat**: Abandonment of land and structures in vulnerable areas	✓	✓
	Provides opportunities for nature conservation	*Involves resettlement of inhabitants*		
Limited intervention	Reducing risks by working with natural coastal processes and change.	**Accommodation/Protection**: as above	(✓)	(✓)
	Includes measures attempting to:			
	Reduce rather than eliminate coastal erosion and cliff recession (e.g. nourishments)			
	Address public safety (e.g. flood warning systems, dune and forest maintenance, coastal building restrictions)			

Based on Environment Agency (2012), MESSINA (2006), and Gilbert and Vellinga (2005)

2006) have noted the rather disappointing and overestimated involvement of spatial planning in risk management across UK and many other countries of Europe, attributing this to a range of factors including the inadequate training and understanding of risk by planners, as well as the mere advisory nature of risk assessment and associated advice. However, there are some countries and regions where a stronger legal framework has provided a more effective approach. Article 8(2) of the ICZM Protocol to the Barcelona Convention states that contracting parties (i.e. Mediterranean countries) should establish a zone where no construction is allowed. This is not to be less than 100 m wide and should take account of climate change and natural risks. Similarly, the *Loi Littoral* in France demands a shoreline exclusion zone of 100 m (la bande littorale) where no construction is allowed (littorale non constructible) (EUROSION 2004). Over the last decade, this has been supplemented by Natural Risk Prevention Plans (Plan de Prévention des Risques) to control development within various risk zones, including coastal areas (Deboudt 2010). These plans are being prepared by county prefectures in consultation with local councils producing detailed spatial plans.

Whilst not specifically coastal, the Floods Directive (2007/60/EC) is promoting a risk-based rather than a flood management approach to both fluvial and coastal flooding. It is also encouraging a wider perspective and evaluation of human factors through a well prescribed statutory process, including mapping of potential flood extent, assets and humans at risk. It also requires adequate and coordinated measures to reduce areas at significant risk. Whilst some question the extent to which it will translate into the management of people, property and other human assets rather than flood control (Klijn et al. 2008), it certainly has been a significant catalyst for addressing coastal risk in some areas, such as Northern Ireland, where traditionally coastal flood and erosion have previously been low government concerns (Dodds et al. 2010). Whilst not explicitly related to coastal risk, the European Commission has recently devoted considerable effort into addressing and incorporating climate change into much of its legislation, particularly through it guidance on adaptation to climate change in water management and its subsequent 'A Blueprint to Safeguard Europe's Water Resources' (European Commission 2012). Similarly, the European Water Framework Directive (2000/60/EC) and the Marine Strategy Framework Directive (2008/56/EC), whilst not dedicated to risk management, provide opportunities for framing sustainable erosion and flood risk management practices within wider marine spatial and river basin planning systems.

20.3 Challenges for the Management of Coastal Risk

Despite moves towards a more risk-based management approach, a number of significant challenges hamper the management of coastal risks. These stem both from the unique characteristics of coasts and the distinctive institutional framework which has evolved to deal with coastal matters. The following paragraphs summarise those relevant to North West Europe.

The inter-connectivity of human, biological and physical coastal systems with associated complex feedback mechanisms provides particular challenges to risk management, especially in the context of the complex governance regimes of such areas, where contrasting land and marine institutions and policies come together. Indeed, as Moser et al. (2012) contend, coastal issues are often 'wicked' problems, resulting from this systemic complexity. They defy complete definition or understanding, which in turn negates against any simple solutions, given the limited time, discipline and spatial frames under which most coastal managers operate. Whilst many have called for a systems view to underpin coastal management (for example, van der Weide 1993) and associated modelling processes (Nicholls and Cazenave 2010), there is limited embedding of such systems into practice, albeit there have been some important pilots undertaken in Europe over the last few years to demonstrate the value of a 'systems approach' (Reis et al. 2014).

There are challenges resulting from the complexity of physical systems in coastal areas for coastal risk management, particularly given land-sea and catchment sediment flows and the nature of these dynamic and long term processes (Pethick 2001). Aspects of scale are particularly complex, with much debate and emerging evidence linking global and local processes, particularly in the context of storm incidence. This is well exemplified by a recent analysis of the role of the global circulation, notably the North Atlantic Oscillation, on storm tracks and severity along the coasts of south Wales (Phillips and Crisp 2010).

However, management continues, hampered by the limited knowledge and understanding of such processes (McFadden 2007), often compounded by inadequate monitoring systems and associated data. Good monitoring practice is patchy across Europe. Whilst the MESSINA project found that, in some parts, such as in England, the Netherlands and German Landers, LIDAR and other advanced coastal monitoring systems are regularly, routinely and comprehensively employed, in other countries, such as Ireland and France, coastal monitoring is confined to specific locations or is linked to experimental research projects (MESSINA 2006). Limited knowledge and understanding, however, are much more prevalent across Europe. For example, a recent European report revealed the limited investigation of coastal erosion and processes within many EIAs for projects where such matters should have received more in depth study (National Institute for Coastal and Marine Management of the Netherlands 2004).

Issues associated with poor understanding, monitoring and science give rise to high levels of uncertainty. These, in turn, may make management decisions harder to justify to communities who may expect 'simple' answers and solutions. This may be particularly an issue when unsavoury adaptation options are under consideration, involving conflicts for space in already congested coastal space and/or high levels of expenditure within budgets that are already under strain. Clearly, uncertainties abound when climate change and its associated secondary impacts, including accelerated sea level rise and increased flooding, are considered. Communities not even currently living on the coast are likely to have to engage with such debate too as natural systems and associated habitats attempt to migrate landwards (Pethick 2001), calling into question approaches to risk communication and associated science translation (van Aalst et al. 2008).

There is also a magnitude of challenges posed by the human system in coastal areas, which suggest that good governance and integrated policy making and implementation may be more difficult to achieve than elsewhere. This is particularly true given the complexity of property and other rights in the coastal zone as well as issues of changing access and distribution of resources, risk and social capital linked associated with coastal adaptation (Dolan and Walker 2006). Whilst some authors have highlighted the need to address economic issues (Cheong 2011) and called for holistic coastal resource assessment (Turner 2000), including economic and social aspects/consequences (O'Riordan et al. 2008) in relation to coastal risk, such matters are rarely adequately addressed. With the exception of the procedures in place in England, assessment of costs and benefits of coastal defence options at local, scheme and regional levels is rarely done systematically (MESSINA 2006). This is despite the considerable coastal defence expenditure in countries such as the Netherlands where between 30 and 40 millions Euros is annually devoted to beach and foreshore nourishment (op. cit.).

Of all the aspects of the human system, however, institutional and associated governance issues remain the most difficult challenge. There is no harmonisation of legislation on coastal erosion or flooding measures across the EU and so organisational structures vary from State to State. Generally, there are several tiers of administration involved, including local, regional and national bodies (Ballinger et al. 2008). Responsibilities are further frequently divided between bodies with off and onshore remits. Whilst the former tend to take a long term and more strategic, often national view, generally the latter have contrasting local and shorter-term priorities (O'Hagan and Ballinger 2010). In relation to the terrestrial environment, local government bodies dominate, taking key decisions relating to specific local coastal defence schemes and spatial planning. In contrast, Central Government oversees national offshore concerns such as shipping and renewable energy generation and, in the context of coastal defence, provides the steer for longer-term monitoring and some funding for local projects. The complexity of jurisdictions is particularly apparent in estuary areas, where boundaries between local administrations occur. This is the case in the Severn Estuary where recent devolutionary processes have led to a burgeoning of bodies with coastal interests, as government agencies and other bodies are duplicated on either side of the English-Welsh border (Fig. 20.3).

Government responsibilities in most countries are generally fragmented and are sectorally or issue-based, creating potential issues for engendering a more holistic approach to the management of coastal risks. The piecemeal evolution of legislation over decades in reaction to specific concerns (Ballinger 1999), has resulted in the delivery of functions and services being divided amongst Government departments and agencies (op. cit.). This has led to the perpetuation of a silo mentality as sectors and associated administrations work in relative isolation (Ballinger et al. 2002). Given these narrow windows of decision making, there is a possibility that 'win – win' scenarios remain unrecognised and future adaptation options are overlooked, particularly planned retreat. This has been the case in Australia, where Abel et al. (2011) suggest that a legacy of former planning decisions, development pressures and liability laws has 'squeezed out' managed retreat in favour of development. In North

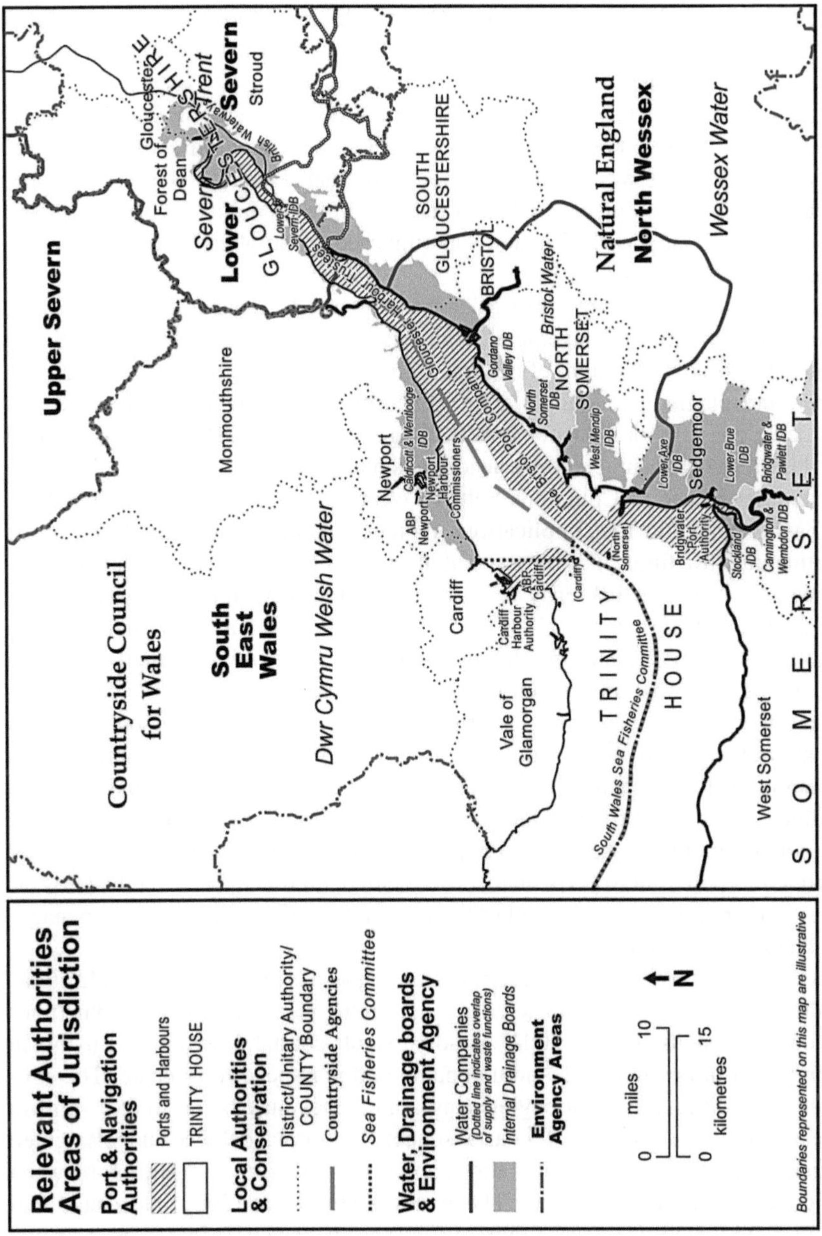

Fig. 20.3 Severn estuary bodies (2012)

West Europe, the plethora of laws, interests and sectors, alongside the lack of any overarching framework for resolving issues (Ballinger et al. 2008) inevitably results in conflicts between sectors and incompatibilities between coastal uses, as well as inefficiencies and short-term horizons. Coastal defence decisions and associated financing of coastal defence schemes 'compete' with those from other sectors. Indeed, the complexity of legislation and the lack of a clear hierarchy of coastal management objectives are frequent bugbears of practitioners. Whilst some recent European legislation, notably the Habitats Directive (92/43/EEC) and Water Framework Directive (2000/60/EC), does, however, demand a more coordinated implementation by competent authorities, there remains confusion related to the additional bureaucracies and plans brought about such legislation, superimposed on an already complex system (Ballinger and Stojanovic 2010).

In spite of some moves towards more integrated management approaches, through the shoreline management processes described above, the sectoral and somewhat artificial division of responsibilities between coast protection, sea defence and planning remains a significant impediment to fully integrated and sustainable coastal risk management (Ballinger et al. 2002). Problems associated with this are well documented by academics and policy makers for the UK (See for example: House of Commons Select Committee on Agriculture 1998; Pettit 1999) who point out that implications include inappropriate local government departmental structures and the associated, limited liaison between planners and engineers as well as a tendency to narrow, technocentric approaches (Ballinger et al. 2002). O'Hagan and Ballinger (2010) highlight a similar situation in the Republic of Ireland, illustrating this by reference to one of the councils in which the planning unit was, until recently, totally unaware of the council's committee on coastal erosion, even though both were housed in the same building. As O'Connor et al. (2010) note, with no national shoreline management policy, management tends to be reactive, responding to local demands, politics and economics. Given the sectoral fragmentation of responsibilities this becomes potentially even more problematic. Across Europe, the Floods Directive 2007/60/EC, referred to earlier, has potential to perpetuate a fragmented approach to coastal hazards, as coastal flood risk management rather than coastal flood and erosion risk management becomes targeted.

On top of these institutional issues, there are complex socio-economic considerations to address. As note above, climate change is forcing some difficult and sometimes unsavoury decisions (Lowry 2002) as some small coastal communities living in vulnerable locations may become untenable and may need to relocate. The economics of traditional coastal engineering approaches may be no longer valid in such cases forcing a new type of 'coastal squeeze' as coastal defence budgets tighten associated with the economic slow down and ever increasing competition between local government budget streams. The Wales Audit Office (2009) has stated that funding would need to increase threefold just to manage existing assets over the next 25 years. Such issues test coastal governance systems to their limit and pose questions of accountability, social justice, compensation and associated rights, much debated in the academic literature (see for example, Tompkins et al 2008; Cooper and McKenna 2008). These include questions not only over how to balance private interests and the common good but also about how to balance national and local interests (Stallworthy 2006). Within such discussions, there is a gradual recognition that public financial responsibility for coastal risk should be limited and that there needs to be a less piecemeal and more accountable approach to public intervention (EUROSION 2004).

Matters are coming to a head as policies for local areas, taking on board economic issues and climate change, advocate the withdrawal or non-maintenance of sea defences and managed realignment. Whilst recent European reports have highlighted the requirement for expropriation or compensation measures to comply with EC competition regulations (Bucx 2010; EUROSION 2004), such mechanisms are not uniformly in place across Europe. Whilst French Law facilitates expropriation of assets threatened by coastal natural hazards under the *Loi Barnier* (op. cit.), the situation differs in the UK. The recent Flood and Water Management Act 2010 redefines coast protection to include anything undertaken to maintain or restore natural processes and 'passive inaction that allows sea defences to be breached naturally would seem to be legitimate, provided that the authorities act reasonably' (Gibson 2011a). However, as authorities only having statutory duties to pay 'compensation for loss or disturbance due to the exercise of their powers' if their conduct could be deemed as nuisance or negligence in common law in terms of coast protection (Gibson 2011b), there is no compensation mechanism (Defra 2009). With fundamental human rights incorporated into UK law and well-being powers provided to Local Government under the Local Government Act 2000, however, there still remains much confusion about public liability, compensation and ethical considerations.

Faced with uncertainty on compensation, community action has escalated (Milligan and O'Riordan 2007). This has been compounded by community disquiet about perceived inadequate consultation on shoreline management plans and concern that birds are fairing better than humans under the provisions of the European Habitats Directive in parts of the UK. Here, Central Government has made significant efforts to engage with all this, exploring and promoting new coastal adaptation and funding options for local communities through its recent £11 million Coastal Pathfinder Programme (Defra 2012), but, this has not, as yet, resulted in any major changes in Government policy or strategy.

20.4 ICZM: A New Governance Approach

Given the well recognised failings of existing institutional structures in coastal areas, it has been suggested that ICZM may be the answer. ICZM has been purported to be able to provide a framework for managing competing resources and tackle 'wicked,' multi-dimensional coastal problems including climate change adaptation (Vellinga and Klein 1993).

This section explores the role and potential of integrated planning and management in facilitating a more balanced and sustainable approach to the management of coastal risks, focusing on Integrated Coastal Zone Management (ICZM), defined as:

> a dynamic process in which a coordinated strategy is developed and updated for the allocation of environmental, social, cultural and institutional resources to achieve the conservation and sustainable multiple use of the coastal zone. (Sorensen 1993)

20.4.1 ICZM Development

Whilst early attempts at ICZM date back several decades to initiatives such as the US 1972 Coastal Zone Management Act, it was the international prescriptions of the subsequent decades, which brought ICZM centre-stage and resulted in its global support by UN agencies, the World Bank and others (Cicin-Sain and Knecht 1998; WWF 1998). All proposed ICZM as a means of delivering sustainable development for coastal areas, helping facilitate multi-sectoral development and resolution of coastal conflicts alongside attempts to protect coastal habitats and coastal system integrity (Thia-Eng 1993). A range of tools were put forward, including many directly relevant to the management of coastal risk, including environmental impact assessment and information management. However, unlike coastal defence management at the time, ICZM encouraged states to manage the coastal zone and its watershed as an integral, single unit and to encompass all uses and users of the coastal zone within an integrated framework.

As Cicin-Sain and Knecht (1998) argued, ICZM represented a 'new paradigm of management for managers and a new way of thinking,' challenging existing management approaches, legal systems and administrative arrangements, particularly those being sectoral, discipline or problem-based. Within this new approach 'integration' has been viewed as a central concept with various dimensions of integration having been categorised (op. cit.). These include:

- intersectoral integration – amongst different coastal sectors;
- intergovernmental integration – amongst levels of government;,
- spatial integration – including land–ocean interaction;
- international integration – for transboundary issues; and
- science–management integration – between disciplines and between science and management/policy.

Driven largely by environmental problems, including depletion of resources, pollution and ecosystem damage, 380 ICZM efforts had been established by 2000 (Sorensen 2000). However, it was not until 2002, that the European Parliament and the Council adopted a Recommendation on Integrated Coastal Zone Management (ICZM) (European Parliament and Council 2002). Recognising that there was already significant ICZM practice at local levels across already, the Recommendation, a non-binding policy document, defined principles which together, it was suggested,

Table 20.2 The European ICM principles

Broad holistic approach
Long-term perspective
Local specificity
Working with natural processes
Adaptive management
A combination of instruments
Supportandinvolvementofallstakeholders-
Participatory approach

would help deliver better coastal governance (Table 20.2). It also sought to encourage rather than require European Member States to develop national strategies, based on the common principles and following national audits of coastal governance and associated institutional arrangements. Whilst there was little prescription in terms of the nature of the national strategies, the use of the principles has helped fashion some sort of European approach to ICZM (O'Hagan and Ballinger 2009). There have been calls for the principles to be clarified and prioritised (op. cit. and Ballinger et al. 2010). However, a preoccupation with developing a European Integrated Maritime Policy and associated Maritime Spatial Planning by DG MARE left ICZM policy development in Europe in limbo for some time. There, has also been European funding support for numerous short-term ICZM projects, but these have tended to perpetuate the short term, project-based nature of ICZM efforts across the region (Shipman and Stojanovic 2007).

20.4.2 ICZM Performance and the Management of Coastal Risk

Whilst ICZM could be considered to be still in its early stages within Europe, there is considerable evidence which suggests that ICZM can make a considerable contribution to the management of coastal risk. Clearly the ICZM principles are relevant with some, notably 'working with natural processes,' 'adaptive management', and 'the long term perspective' closely aligned to the needs of coastal risk management. Encapsulated within the first of these, for example, is a need to work within the carrying capacities and limits of coastal ecosystems and natural physical systems: within the second, the principle suggests adjustment to management should occur with increased knowledge and understanding of problems, implying the need for sound scientific evidence to underpin coastal management decisions. Similarly, the long-term perspective principle, with its links to the precautionary principle, has clear ramifications for the management of coastal risk and associated coastal defence decisions, although as Mee (2010) points out there are difficulties associated local ICZM projects making meaningful long-term priorities, given their limit remit and perspective.

Whilst there is much 'common sense' in the ICZM principles, it could be suggested that they are merely a repackaging of many of the principles of good environmental governance, commonly used in other spheres of environmental management. Their detailed interpretation at an operational level within an ICZM context has also been problematic and has led to much debated in the academic literature (for example, Cooper and McKenna 2008; McKenna and Cooper 2006; Ballinger et al. 2010). Even the widely accepted principle of 'working with natural processes, Cooper and McKenna (2008) note can be interpreted in various ways according to the time frame adopted. There are also problems of scale with a need for reconciliation between the need for 'local specificity' on the one hand and the need to take a 'broad approach' on the other (Ballinger et al. 2010). The participatory principle has also been criticised for its promotion of bottom-up' 'voluntary, powerless, under-funded and non-sustainable' approaches (McKenna and Cooper 2006) even though some evidence would suggest that some local coastal partnerships in the UK, whilst struggling with limited resources, have much to deliver (Stojanovic and Barker 2008).

In terms of other aspects of ICZM, Table 20.3 summarises the contribution of ICZM to addressing some of the key challenges currently facing the management of coastal risk, described in the previous section. Clearly, there is potential for ICZM to help alleviate some of these. ICZM can, for example, provide a neutral platform to bring together stakeholders from many backgrounds, disciplines and institutions to discuss coastal issues. This can build shared responsibility and understanding as well as fostering trust and respect. In turn this may and sometimes does lead to the 'win-win' situations, so much needed in the management of coastal risk. This is well demonstrated by the practical experiences of the Severn Estuary Partnership. This ICZM programme has provided multiple benefits over the last couple of decades (Ballinger and Stojanovic 2010), developing overarching estuary-wide policies to inform sectoral policy development. The Partnership has also provided a neutral platform for debate of coastal issues through regular multi-stakeholder engagement meetings including annual forums as well as providing significant assistance with the public consultation process on the recent shoreline management plan for the estuary. Whilst many ICZM efforts have been criticised for their limited linkage with their science base (McFadden 2007; Billé 2007; Mee 2010), the Partnership, through its close links with Cardiff University, has managed to address this. As well as science-based outputs, including a State of the Severn Estuary, associated report cards and education materials, have informed a range of audiences of the importance and characteristics of the estuary (Severn Estuary Partnership 2011), a science-policy forum has been established to address coastal adaptation matters for the whole estuary.

Whilst ICZM might appear like the panacea for coastal areas and indeed for the management of coastal risk, it frequently has not delivered as much as promised. However, there are some local success stories and achievements, particularly at local levels (Stojanovic and Barker 2008; Morris 2008) and sometimes in spite of a limited national supporting framework (O'Hagan and Ballinger 2010). Throughout much of Europe, apart from the Mediterranean where an ICZM Protocol to the

Table 20.3 The contribution of ICM to addressing key coastal risk management challenges

Challenge	ICM's contribution	
Inter-connectivity of human, biological and physical coastal systems	Limited/minimal but with significant potential	ICM programmes and principles recognise this but ICM's contribution in practice is limited due to resource issues
Complexity of physical systems	Limited/minimal but with significant potential	As above
Need for holistic resource assessment	Limited/minimal but with significant potential	Possible ICM programmes can promote this although rarely have resources to fully do so
Limited knowledge and understanding	Widespread and with further potential	Many ICM programmes play an important role in awareness raising and improving knowledge of stakeholders
Inadequate monitoring and data	Moderate but with further potential	Some ICM programmes are involved in monitoring and data gathering programmes
Complex property and other rights	Limited/minimal but with potential	ICM programmes could be used to help explain these
Institutional issues	Limited/minimal but with potential	Whilst ICM cannot solve these, ICM programmes could help explain institutional arrangements and responsibilities
Administrative issues & split responsibilities	Moderate but with significant potential	ICM programmes can bring administrations together
Complex jurisdictions	Limited but with further potential	ICM programmes could be used to help explain these
Silo mentalities	Moderate but with significant potential	The wide outlook of ICM programmes and multi-sectoral engagement can help reduce this
Competing priorities	Moderate but with significant potential	As above
Domination of narrow, technocentric approaches	Moderate but with significant potential	As above
Local community engagement	Widespread and with further potential	A key feature of ICM programmes and project but questions over sustainability

Barcelona Convention was adopted in 2010, ICZM attains a limited status. Generally it is confined to a non-statutory, somewhat peripheral activity, with the longevity of even the most acclaimed ICZM programmes frequently questioned (Shipman and Stojanovic 2007).

This may reflect the perceived 'environmental' agenda of ICZM by many stakeholders which can make it difficult for industry to take it on board. Equally, ICZM may seem overly complex, peripheral and challenging to traditional sectors of decision-making. As a result, as budgets tighten so 'coastal policy squeeze' occurs as it competes with other policy areas (op. cit., Christie 2005). Subsequent resource issues, associated with limited staffing, skills and funding, make it difficult for many local ICZM programmes to look beyond the short-term (Shipman and Stojanovic 2007) and to adequately address the real challenges of consensus-building (Poitras et al. 2003). Confined by their resource base and limited status, ICZM programmes often then have focused on soft, less challenging issues, such as recreation and education, rather than taking on fully fledged integrated planning and management *per se*. Increasingly too academics have questioned the wisdom of too much decentralisation of ICZM activity (Lowry 2002), fearing that this can lead to appropriate local agendas dominating as local power struggles come into play.

20.5 Conclusions

Traditional, technocentric approaches to coastal risk management are under strain as the coastal zone becomes under increasing pressure from population growth and development amid concerns over climate change predictions. Even the UK's Institution of Civil Engineers and a think tank of the Royal Institute of British Architects have recently suggested UK should establish it long-term position as to whether or not major coastal and estuarine cities, such as Portsmouth, should be defended and allowed to continue to expand or indeed, should retreat from the shore (Institution of Civil Engineers 2010). Whilst such monumental decisions may be some time away, governments are already having to face up to a wide range of adaptation options and contemplate abandonment of 'hold the line' in many locations where the cost benefit of improving or even maintaining current defences is not convincing. Many new approaches require involvement of multiple actors, challenging the engineering dominance of centuries. Given the complexities of potential impacts, the costs and benefits to a wide range of coastal interests and the ever increasing scrutiny of decisions, traditional governance structures are also being questioned.

In this context and drawing on a range of examples, particularly from North West European experiences, this chapter has evaluated the extent to which ICZM can contribute to the management of coastal risk. Whilst not a panacea, it concludes that ICZM may be able to facilitate the development of more adaptable and palatable approaches for local communities, much needed in the context of coastal climate change impacts. The ICZM principles are certainly of potential importance. However, there remain specific questions about the current underperformance of ICZM, particularly its floundering status within the European Union, where it is somewhat overshadowed by strategic debate on the future of marine planning and management. Indeed, it could be argued that such discussions may lead to further fragmentation rather than integration of governance.

Whatever the shape of ICZM or indeed general governance structures for the management of coastal risk in future, there are several essential elements without which coastal risk decision-making will be destined to fail. These include mechanisms to ensure better scientific underpinning of decisions, improved public understanding and community engagement in decisions, and better understanding of how perceptions influence individual, community and government behaviour related to coastal risk. There, no doubt, will be challenges ahead for the management of coastal risk, but we need to ensure that communities can understand and make reasoned choices between difficult trade-offs (Tompkins et al. 2008; Alexander et al. 2012). Such decisions must also not preclude options or create public liability nightmares for future generations. With seemingly so many intractable and complex issues involved, it is, therefore, likely that coastal governance will remain 'on the edge' for some time to come.

References

Abel, N., R. Gorddard, B. Harman, A. Leitch, J. Langridge, A. Ryan, and S. Heyenga. 2011. Sea level rise, coastal development and planned retreat: Analytical framework, governance principles and an Australian case study. *Environmental Science & Policy* 14(3): 279–288.

Agency, Environment. 2010. *The national flood and coastal erosion risk management strategy for England*. London: Environment Agency.

Alexander, K.S., A. Ryan, and T.G. Measham. 2012. Managed retreat of coastal communities: Understanding responses to projected sea level rise. *Journal of Environmental Planning and Management* 55(4): 409–433.

Ballinger, R.C. 1999. The evolving organisational framework for Integrated Coastal Management in England and Wales. *Marine Policy* 23(4/5): 479–500.

Ballinger, R.C., and T.S. Stojanovic. 2010. Policy development and the estuary environment: A Severn Estuary case study. *Marine Pollution Bulletin* 61(1–3): 132–145.

Ballinger, R.C., J. Taussik, and J.S. Potts. 2002. Managing coastal risk: Making the shared coastal responsibility work. In *Proceedings of the 37th DEFRA, flood and coastal management conference,* Keele, September 2002.

Ballinger, R., V. Cummins, A.M. O'Hagan, and M. Phillips. 2008. *The point of COREPOINT: Improving capacity for Integrated Coastal Zone Management in North West Europe, COREPOINT report*. Available from: http://corepoint.ucc.ie/index.htm.

Ballinger, R., A. Pickaver, G. Lymbery, and M. Ferreira. 2010. An evaluation of the implementation of the European ICZM principles. *Ocean and Coastal Management* 53: 738–749.

Billé, R. 2007. A dual-level framework for evaluating integrated coastal management beyond labels. *Ocean and Coastal Management* 50(10): 796–807.

Bucx, T. 2010. *Assessment of European coastal erosion policies in relation to ICZM, Concepts and Science for Coastal Erosion Management (CONSCIENCE) project report*. Available from: http://www.conscience-eu.net/documents/deliverable11-assessment.pdf.

Charlier, R.H. 1989. Coastal zone: Occupance, management and economic competitiveness. *Ocean and Coastal Management* 12: 383–402.

Cheong, S.-M. 2011. Policy solutions in the U.S. special issue: Coastal adaptation. *Climatic Change* 106(1): 57–70.

Christie, P. 2005. Is integrated coastal management sustainable? *Ocean and Coastal Management* 48: 208–232.

Cicin-Sain, B., and R.W. Knecht. 1998. *Integrated coastal and ocean management. Concepts and practices*. Washington, DC: Island Press.

Cooper, J.A.G., and J. McKenna. 2008. Social justice in coastal erosion management: The temporal and spatial dimensions. *Geoforum* 39(1): 294–306.

Creel, L. 2003. *Ripple effects: Population and coastal regions*. Washington, DC: Population Reference Bureau. Available at: http://www.prb.org/pdf/RippleEffects_Eng.pdf.

Deboudt, P. 2010. Towards coastal risk management in France. *Ocean and Coastal Management* 53(7): 366–378.

Defra (Department for Environment, Food and Rural Affairs). 2009. *Appraisal of flood and coastal erosion risk management. A Defra policy statement*. London: Defra.

Defra (Department for Environment, Food and Rural Affairs). 2012. *Coastal change pathfinder review final report.* January 2012, PB 13720. London: Defra.

Dodds, W., J.A.G. Cooper, and J. McKenna. 2010. Flood and coastal erosion risk management policy evolution in Northern Ireland: "Incremental or leapfrogging?". *Ocean and Coastal Management* 53(12): 779–786.

Dolan, A.H., and I.J. Walker. 2006. Understanding vulnerability of coastal communities to climate change related risks. *Journal of Coastal Research*, Special Issue No. 39, Proceedings of the 8th International Coastal Symposium, 1316–1323.

Doody, J.P. 2004. 'Coastal squeeze' – An historical perspective. *Journal of Coastal Conservation* 10: 129–138.

Environment Agency. 2012. *Shoreline management plan policies: What do they mean?* Available from: http://www.environment-agency.gov.uk/homeandleisure/134834.aspx. Accessed Sept 2012.

European Commission. 2012. Communication from the Commission to the European Parliament, The Council, The European Economic and Social Committee and the Committee of the Regions – *A blueprint to safeguard Europe's water resources, Brussels*, 14.11.2012, COM(2012)673 final.

European Parliament and Council. 2002. *Recommendation of the European Parliament and of the Council of 30 May 2002 concerning the implementation of Integrated Coastal Zone Management in Europe,* 2002/413/EC. Available from: http://eurlex.europa.eu/LexUriServ/LexUriServ.do?uri=OJ:L:2002. Accessed July 2012.

EUROSION. 2004. Living with coastal erosion in Europe: Sediment and space for sustainability, *PART I – Major findings and Policy Recommendations of the EUROSION project,* Service contract B4-3301/2001/329175/MAR/B3 "Coastal erosion – Evaluation of the need for action" Directorate General Environment, European Commission, 10 May 2004.

Gibson, J. 2011a. UK Coastal zone law notes, available from: http://www.coastlaw.uct.ac.za/iczm/notes/note4.htm#sect6. Last updated on: 1/3/11.

Gibson, J. 2011b. *Legal liabilities for coastal erosion and flooding in the United Kingdom due to climate change, Institute of marine and Environmental law*. Powerpoint presentation, available from: www.iucnael.org/en/documents/doc…/1011-gibson-john.html. Accessed July 2012.

Gilbert, J., and P. Vellinga. 2005. Coastal Zone Management. In *Climate change: The IPCC response strategies*, 133–159. Available from: http://www.ipcc.ch/ipccreports/far/wg_III/ipcc_far_wg_III_chapter_05.pdf. Accessed Aug 2012.

Greiving, S., M. Fleischhauer, and S. Wanczura. 2006. Management of natural hazards in Europe: The role of spatial planning in selected EU member states. *Journal of Environmental Planning and Management* 49(5): 739–757.

Grimmond, C.S.B. 2011. Urban climate in the coastal zone: Resilience to weather extremes. In Pelling, M. (ed.). *Megacities and the coast: Transformation for resilience,* A preliminary review of knowledge, practice and future Research, LOICZ, June 2011, 37–42. Abingdon: Routledge.

Hadley, D. 2009. Land use and the coastal zone. *Land Use Policy* 26(Supplement 1): S198–S203.

Hillen, M.M., S.N. Jonkman, W. Kanning, M. Kok, M.A. Geldenhuys, and J.F. Stive. 2010. *Coastal defence cost estimates case study of the Netherlands, New Orleans and Vietnam,* Delft University of Technology, in cooperation with Royal Haskoning, April 2010.

House of Commons Select Committee on Agriculture. 1998. *Report on flood and coastal defence.* Fifth special report, session 1997–98. House of Commons, London. http://www.publications.parliament.uk/pa/cm199798/cmselect/cmagric/1117/111701.htm

Institution of Civil Engineers. 2010. *Facing up to sea level risk: Retreat? Defend? Attack? The Future of our Coastal and Estuarine Cities*. London: Institution of Civil Engineers.

Klijn, F., P. Samuels, and A. van Os. 2008. Towards flood risk management in the EU: State of affairs with examples from various European countries. *International Journal of River Basin Management* 6(4): 307–321.

Li, H. 2003. Management of coastal mega-cities—A new challenge in the 21st century. *Marine Policy* 27: 333–337.

Lowry, K. 2002. Decentralized coastal management. *International Newsletter of Coastal Management* 42: 1–42.

McFadden, L. 2007. Governing coastal spaces: The case of disappearing science in integrated coastal zone management. *Coastal Management* 35(4): 429–443.

McGranahan, G., D. Balk, and B. Anderson. 2007. The rising risks of climate change: Urban population distribution and characteristics in low elevation coastal zones. *Environment and Urbanization* 19: 17–37.

McKenna, J., and J.A. Cooper. 2006. Sacred cows in coastal management: the need for a 'cheap and transitory' model. *Area* 38(4): 421–431.

Mee, L. 2010. Between the devil and the deep blue sea: The coastal zone in an era of globalisation. *Estuarine, Coastal and Shelf Science* 96: 1–8.

MESSINA. 2006. *Integrating the shoreline into spatial policies, Practical Guide*. Project part-financed by the European Union, (European Regional Development Fund) within the INTERREG IIIC Programme. Available from: http://www.interreg-messina.org/documents/CoastalToolkit/MESSINA%20-%20Practical%20Guide%20-%20Integrating%20the%20Shoreline.pdf.

Milligan, J., and T. O'Riordan. 2007. Governance for sustainable coastal futures. *Coastal Management* 35(4): 499–509.

Morris, R.K.A. 2008. English nature's estuaries initiative: A review of its contribution to ICZM. *Ocean and Coastal Management* 51: 25–42.

Moser, S.C., J. Williams, and D.F. Boesch. 2012. Wicked challenges at land's end: Managing coastal vulnerability under climate change. *Annual Review of Environmental Resources* 37. Available from: http://www.susannemoser.com/publications.adaptation.php.

National Institute for Coastal and Marine Management of the Netherlands. 2004. *A guide to coastal zone erosion practices in Europe: Lessons learnt*. Report to Directorate General Environment European Commission as part of the EUROSION project. Available from: http://www.eurosion.org/shoreline/lessons_learned.pdf.

Nicholls, R. 1995. Coastal megacities and climate change. *GeoJournal* 37(3): 369–379.

Nicholls, R.J., and A. Cazenave. 2010. Sea-level rise and its impact on coastal zones. *Science* 328(5985): 1517–1520.

O'Connor, M.C., J.A.G. Cooper, J. McKenna, and D.W.T. Jackson. 2010. Shoreline management in a policy vacuum: A local authority perspective. *Ocean and Coastal Management* 53(12): 769–778.

O'Hagan, A.M., and R.C. Ballinger. 2009. Coastal governance in NW Europe: An assessment of approaches to the European Stocktake. *Marine Policy* 33(60): 912–922.

O'Hagan, A.M., and R. Ballinger. 2010. Implementing Integrated Coastal Zone Management in a national policy vacuum: Local case studies from Ireland. *Ocean and Coastal Management* 53: 750–759.

O'Riordan, T., S.A. Nicholson-Cole, and J. Milligan. 2008. Designing sustainable coastal futures. *Twenty-First Century Society: Journal of the Academy of Social Sciences* 3(2): 145–157.

Pethick, J. 2001. Coastal management and sea-level rise. *Catena* 42: 307–322.

Pettit, S.J. 1999. The statutory approach to coastal defence in England and Wales. *Marine Policy* 23(4/5): 465–477.

Phillips, M.R., and S. Crisp. 2010. Sea level trends and NAO influences: The Bristol Channel/Severn Estuary. *Global and Planetary Change* 73(3–4): 211–218.

Poitras, J., R. Bowen, and J. Wiggin. 2003. Challenges to the use of consensus building in integrated coastal management. *Ocean and Coastal Management* 46: 391–405.

Reis, J., T. Stojanovic, and H. Smith. 2014. Relevance of systems approaches for implementing Integrated Coastal Zone Management principles in Europe. *Marine Policy* 43: 3–12.

Severn Estuary Partnership. 2011. *State of the Severn Estuary Report*. Available from: http://viewer.zmags.com/publication/ad5b93bd#/ad5b93bd/1.

Shipman, B., and T. Stojanovic. 2007. Facts, fictions and failures of Integrated Coastal Zone Management in Europe. *Coastal Management* 35: 375–398.

Small, C., and R.J. Nicholls. 2003. A global analysis of human settlement in coastal zones. *Journal of Coastal Research* 19(3): 584–599.

Sorensen, J. 1993. The international proliferation of integrated coastal zone management efforts. *Ocean and Coastal Management* 21(1–3): 45–80.

Sorensen, Jens. 2000. *Baseline 2000 background report: The status of integrated coastal management as an international practice (Second iteration).* Urban Harbors Institute Publications, Paper 31. http://scholarworks.umb.edu/uhi_pubs/31

Stallworthy, M. 2006. Sustainability, coastal erosion and climate change: An environmental justice analysis. *Journal of Environmental Law* 18(3): 357–373.

Stojanovic, T., and N. Barker. 2008. Improving governance through local coastal partnerships in the UK. *Geographical Journal* 174: 344–360.

Taussik, J. 2000. The role of town and country planning in implementing shoreline management strategies. In *Proceedings of 35th MAFF conference of river and coastal engineers.* Keele University, 5–7 July 2000. London: Ministry of Agriculture, Fisheries and Food.

Thia-Eng, C. 1993. Essential elements of integrated coastal zone management. *Ocean and Coastal Management* 21(1): 81–108.

Tompkins, E.L., R. Few, and K. Brown. 2008. Scenario-based stakeholder engagement: Incorporating stakeholders preferences into coastal planning for climate change. *Journal of Environmental Management* 88(4): 1580–1592.

Transport and Water Management Inspectorate. 2006. *Assessment of primary flood defences in The Netherlands,* National Report on 2006. Available from: http://www.safecoast.nl/editor/databank/File/Assessment%20primary%20flood%20defences%20in%20NL.pdf.

Turner, R.K. 2000. Integrating natural and socio-economic science in coastal management. *Journal of Marine Systems* 25(3–4): 447–460.

van Aalst, M.K., T. Cannon, and I. Burton. 2008. Community level adaptation to climate change: The potential role of participatory community risk assessment. *Global Environmental Change* 18(1): 165–179.

van der Weide, J. 1993. A systems view of ICM. *Ocean and Coastal Management* 21: 129–148.

Vellinga, P., and R.J.T. Klein. 1993. Climate change, sea level rise and integrated coastal zone management: An IPCC approach. *Ocean and Coastal Management* 21(1–3): 245–268.

Wales Audit Office. 2009. *Report on coastal erosion and tidal flooding risks in Wales.* Cardiff: Wales Audit Office.

Wetzel, F.T., W.D. Kissling, H. Beissmann, and D.J. Penn. 2012. Future climate change driven sea-level rise: Secondary consequences from human displacement for island biodiversity. *Global Change Biology* 18(9): 2707–2719.

WWF and IUCN. 1998. *Creating a sea change: The WWF/IUCN Marine Policy.* Available from: http://www.iucn.org/about/work/programmes/marine/marine_resources/?1605/Creating-a-Sea-Change-The-WWF-IUCN-Marine-Policy.

Rhoda Ballinger is Senior Lecturer in the School of Earth and Ocean Sciences, Cardiff University. She undertakes teaching in marine geography and environmental geoscience. Her research interests include integrated coastal management, coastal policy and coastal risk management. She has undertaken a wide number of research projects both within the UK and Europe and published many articles in this field. She also takes a practical interest in coastal and environmental management, particularly through her involvement with the Severn Estuary Partnership and Flood Risk Management Wales. Over the last decade, she has engaged in a quest for model institutional and policy frameworks to deliver Integrated Coastal Management (ICM).

Chapter 21
Ocean Governance and Risk Management

Geoffrey Wescott

21.1 Introduction

The high seas have always engendered a range of emotions and reactions from humans. Curiosity, fear, even terror, of this great expanse of ocean which cover 70 % of Earth – the blue planet. Yet the sheer size of the oceans and the difficulty of transporting across them meant the 'high seas' were largely ignored by the vast majority of humans for centuries. Humans were largely confined to land with the only interest in the seas being as trade routes and the defence of the land. In fact all the way up to the last quarter of the twentieth century a nation's territorial sea extended only three nautical miles off shore – the distance that a cannon ball could be fired.

This almost casual relationship to the oceans changed dramatically in the 1960s and 1970s as technology played an ever increasing role in the exploitation of the natural resources of the seas. Fishing was made far easier by being able to use sophisticated sonar systems to detect the fish and by advanced nets and vessels. But it was probably the technological ability to first find and then extract oil and gas off shore on continental shelfs, and at increasing depths, which stimulated interest in exploiting marine resources. Dreams of other deep sea mineral resources (e.g. manganese nodules) simply fuelled interest in the oceans, not to mention some of the pharmaceuticals that were being discovered.

But there was an issue: the "high seas" – those areas beyond the three nautical mile territorial zones of nations were a "commons" – they were not owned, controlled or policed by any nation state. So how were resource allocation decisions, ownership regulation and control to be exercised? The answer lay in removing vast areas of exploitable oceans from the "commons" by allocating areas of seas to

G. Wescott (✉)
School of Life and Environmental Sciences, Deakin University,
Melbourne Campus, 221 Burwood Highway, Burwood 3125, VIC, Australia
e-mail: wescott@deakin.edu.au

U. Fra.Paleo (ed.), *Risk Governance*, DOI 10.1007/978-94-017-9328-5_21

neighboring nations. This was to be done by a United Nations Convention on what became known as the "Law of the Sea". Now referred to as the Law of the Sea Convention (LOSC) it lies at the centre of international and national ocean governance in the twenty first century and is the major focus of this chapter. In this chapter we will start with a little background on the marine environment followed by a brief history of how the LOSC came to become the basis of world's ocean governance system. This will be followed by a brief description of how different nations have attempted to implement the LOSC and the responsibilities to manage and protect their new marine domain which comes with being a signature and ratifier to the Convention. Following this broad based description a brief case study of how Australia's attempts at meeting its responsibilities will be described and discussed. Australia is chosen for the case study because it an early implementer of ocean governance policies, because it has one of the largest Exclusive Economic Zones declared under LOSC in the world, because there is considerable literature on Australia's trials and tribulations in implementation and finally as the author was directly involved in this implementation a few unpublished and reflective insights may also assist the reader.

From these descriptions a brief overview, or conclusion, on the global state of ocean governance will be offered. The theme of the book will then be tested on this area of governance: What can risk management offer global ocean governance? In fact given the problematic state of ocean governance – which will be seen has no single clear cut approach operating – is risk management a possible unifying approach to ocean governance? These questions will be tackled in the concluding sections of the chapter.

21.2 The Marine Environment

Cicin-Sain et al. (2011) provide an overview of the marine environment (oceans) and its importance to humans. They point out that oceans are the quintessential sustainable development issue. That is that ocean management exemplifies the need to consider economic development, social development and environmental protection in parallel. Some of the salient points highlighted in this report include:

- Oceans generate one half of all the oxygen on the planet,
- Oceans regulate climate and temperature and absorb most of the carbon dioxide produced,
- 90 % of global trade is carried by ship,
- Fish provide more than 4.2 billion people with over 15 % of their annual protein intake,
- 50 % of the world's fisheries occur in the 7 % of the oceans that are coastal waters – the areas are most affected by land based sources of pollution,
- There are 183 coastal countries in the world including 52 small island developing states (SIDS),

- Over 50 % of the world's population live in coastal areas. (Author's note: the largest proportion and largest number (3.5 billion) of people ever in terms of human existence. This approximately a 700 fold increase in coastal zone numbers since 5,000 BC).
- Marine and coastal tourism is a major global industry and is more prevalent in developing countries

21.3 Development of Ocean Governance and the Law of the Sea Convention

Whilst there are a number of summaries available of the development of LOSC, this brief introduction uses that of Haward and Vince (2008) as its primary source.

Public awareness of the oceans was stimulated in a number of ways in the 1950s and 1960s including by the underwater cinematography of Jacque Cousteau and the famous 1969 iconic photograph of the "blue planet" from an Apollo spacecraft. Following a number of significant marine pollution events the UN General Assembly in 1970 passed a resolution establishing an international conference to revive discussions on the law of the sea. The conference was formally titled the Third United Nations Conference on the Law of the Sea (UNCLOSIII) and covered 1974–1982 resulting in the Law of the Sea Convention (LOSC). In 1992 the United Nations Conference on Environment and Development (UNCED) resulted in a global action plan – called Agenda 21 – with an entire chapter (Chap. 17) devoted to oceans and coasts, with a strong focus on integration. The importance of the LOSC cannot be overstated – it has given stability to international negotiations and is possibly the most significant re-drawing of national boundaries we have ever witnessed.

Since then as technological exploitation of the oceans has increased rapidly there have been considerable efforts to develop institutions and processes (governance frameworks) to attempt to manage and protect vast areas of oceans now brought under national jurisdiction as Exclusive Economic Zones (EEZs) by nations which have ratified the LOSC. Regular reports now are produced by the UN Secretary General on "Oceans and the Law of the Sea" and there are a range of regional and national responses to the implementation of the LOSC. These advances are briefly covered in the next section. Apart from the LOSC, the other critical governance framework comes through the International Maritime Organisation (IMO) and its various conventions (see below).

21.3.1 Current State of Law of the Sea Convention

Regionally there have been a variety of attempts to introduce ocean governance frameworks (Haward and Vince 2008). The Asia Pacific Economic Cooperation (APEC) forum has a Marine Resources Conservation Working Group and the "Partnerships in Environmental Management for the Seas of East Asia" (PEMSEA) has operated now for over a decade. The Pacific Islands Forum has a regional oceans policy whilst in Europe "An Integrated Maritime Policy for the European Union" was produced in 2007.

Meanwhile nations also have been very active. Canada possess an Oceans Act (1996), Australia has an Oceans Policy (1998, see Sect. 21.3.3. below) and USA has an Oceans Act 2000 with a "Commission on Ocean Policy" with a "parallel" non-government "Pew Ocean Commission" attempting to encourage debate on ocean governance in the world's most powerful maritime nation – which incidentally, and importantly, is yet to ratify the LOSC.

In all, as of June 2011, 162 nations have ratified the Law of the Sea Convention with 78 nations agreeing to implement the provisions relating to the conservation and management of straddling fish stocks and highly migratory fish stocks one of the more difficult areas of ocean governance. Land-locked nations can sign on to the LOSC as a way of guaranteeing an access route through a neighboring country to a sea port, providing that nation is also a signature to the LOSC. Nations which ratify the LOSC agree to recognise other ratifiers' rights and responsibilities under the LOSC (including the recognition of Exclusive Economic Zones) and hence it is a mutual recognition of these national claims and responsibilities. A difficulty arises when a nation does not ratify the LOSC but never the less attempts to claim the EEZ e.g. the USA.

21.3.2 Current National Governance Approaches to Implementing the Law of the Sea Convention

Ecosystem Based Management (EBM) is the most often quoted proposed management approach to deliver the objectives of integrated ocean planning and management seen as an essential base for sound ocean governance. The concept underlying EBM is to go beyond traditional approaches based on single species (e.g. in fisheries) and single sectors to emphasise ecological processes which connect all these organisms and their environment, i.e. to include all elements of an ecosystem in management – including humans (Espinosa-Romero et al. 2011). Whilst it is beyond the scope of this short chapter to review the progress of this evolving approach, the reader who wishes to delve into the literature on the topic and its current status, in terms of the practical development of a largely theoretical concept, should consult Espinosa-Romero et al. (2011) and Ruckelhaus et al. (2008). The latter gives four practical continuing attempts to implement EBM and concludes: "The (Great Barrier Reef Marine Park Authority) GBRMPA is the current gold standard for EBM in the oceans, and its success thus far in applying EBM principles is in large part because of its equal attention to both the human and natural systems parts of ecosystem management." EBM is clearly a rapidly developing field but as the author above emphasises it still requires considerable testing before it will be easily incorporated into ocean governance.

A number of books have also attempted to bring together material associated with ocean governance. The 'watershed' book of Cicin-Sain and Knecht (1998) brought together much of the material on the then rapidly emerging concept of integrated coastal and ocean management. Later books have examined how the implementation of ocean governance systems have begun to work in practice including the notable volumes of Rothwell and VanderZwaag (2006) and Ebbin et al. (2005). These independent (i.e. non-government) reviews are invaluable contributions to progressing ocean governance. McLeod and Leslie (2009) have explored what ecosystem based management may actually mean in terms of ocean management (particularly in North America) and a very recent book focusing on marine management in Australia, but not exclusively so, is a welcome addition to the literature on ocean governance (Gullett et al. 2011).

Whilst there have been a considerable number of other publications on ocean governance development and implementation over the last decade and a half, in particular there have been a number of attempts at collating these experiences are available for the reader who wishes to consider these in detail. In 2008 the "International Journal of Marine and Coastal Law" published a special issue (23) devoted to "International Ocean Governance in the 21st Century" – see the editor's introduction for a summary (VanderZwaag and Oral 2008). These included regional reviews on the Black Sea, the Caribbean, the Arctic, the "Convention on the Conservation of Antarctic Marine Living Resources (CCAMLR)" and the Mediterranean Sea. These articles are preceded by important overviews on attempts at integrated management of oceans beyond national jurisdictions by Rayfuse and Warner (2008). The recent book of Haward and Vince (2008) combines a series of reviews of institutional arrangements and initially concludes (Haward and Vince 2008; p. 199):

> In all cases oceans governance arrangements are very effective at the sectoral level, with these arrangements providing stability and consistency. There is less evidence of success in relation to the attempt to implement integrated arrangements across sectors and across jurisdictions, simply because issues of horizontal and vertical governance have not been satisfactorily resolved. Challenges remain, as we have noted, in addressing the demands of integrated management (Haward 2003, p. 49). One immediate challenge is providing frameworks and processes that can accommodate, and resolve, conflicts between the vast range of oceans-related interests and values.

They also comment on how "sustainable development" underpins most of these efforts and how there is a trend towards using economic market based approaches and collaborative efforts rather than purely state based regulatory frameworks (although they comment this does not mean removing all regulatory approaches). They repeatedly comment on the difficulty at all levels (international, regional and national) to deliver integrated planning and management despite efforts of governments to coordinate entrenched sectorally based approaches.

In Europe the "Integrated Maritime Policy" brings together many of the recurring theories of ocean governance (Evans 2011) that other early implementers of ocean governance systems have noted. Included in these themes are:

- The need for "data conjunction" i.e. combined data bases where all information is stored and where it is easily available and accessible to all stakeholders. (Author's note: data hoarding seems to be a characteristic of sectoral management regimes).

- Shared enforcement (this is also connected to a focus on regional strategies below). If large areas of EEZ are to be covered not only should enforcement be shared amongst agencies within a nation (e.g. the coastguard, navy, border security, customs, etc.) but also potentially – although more difficult and fraught – across enforcement agencies from different nations within one regional area.
- Ecosystems Based Management (EBM) is seen to be the basis of management approaches adopted. Although EBM is far from a well-defined technique the overall principle/concept of managing an entire ecosystem in a coordinated "working within nature" basis has strong intrinsic appeal.
- Marine spatial planning. The basis of planning to be a spatial/zoning system with designated objectives and uses consistent with these objectives prescribed for each spatial area. In essence this could be visualized as taking classic land use planning to the sea i.e. the identification of zones on a map and then an accompanying table designated what uses are a permitted and not permitted.
- Regional strategies. Particularly relevant to areas of sea where there are many nation's EEZs in close proximity but also relevant to other areas regional strategies combining and coordinating the governance approaches of the nation states are highly desirable and efficient.

Whilst Haward and Vince (2008) and the other books mentioned above have given recent critiques of existing approaches, Juda (2003) had some valuable and still relevant insights into the three nations (Australia, Canada and the USA) which have invested possibly the most effort into attempting to implement, in reality, the theoretical elements on integrated ocean planning, management and governance. Juda (2003) highlights that one of the un-stated but historical human uses of the oceans is as a dumping ground for human wastes of all kinds. He argues that sectoral historic approach to oceans management is "dysfunctional because of generated externalities and mutual interference of different users." (p. 162). He believes one of the stumbling blocks to effective integrated management is that governments are 'problem oriented not place oriented' (p. 166) and this leads to an emphasis on short term electoral cycle driven decision-making which is easier to carry out on traditional sectoral lines. He concludes from his comparison that there has been an increasing recognition of the sectoral problem, the need to consider externalities and to broaden the factors considered beyond a single sector. But there is still a lack of institutional arrangements available to move towards the stated desired aim of integration of planning, management and governance. So the conclusions from Juda (2003) are reinforced in later work while his optimism for the Australian and Canadian systems have to be now tempered by a swing back to sectoral groups yielding more influence over ocean planning and governance frameworks than before in these key "experimenting" nations.

The concept of marine spatial planning (ocean zoning) and how it may progress integrated ocean management has been excellently summarized by Crowder a et al. (2006). They highlight that the failure to integrate oceans management in the USA can be assigned to a failure in the ocean governance arrangements (rather than in any characteristics of the marine environment for example). The USA has over 20 federal agencies and 140 federal statutes that are ocean-related but no clear single authority to co-ordinate, if not integrate, these institutional arrangements to deliver integrated oceans management. Crowder et al. (2006) argued strongly for ecosystems based management (rather than sector based management) as the solution and that this can be delivered through using ocean zoning techniques. They argue that this is the best way to succeed because it matches the scale of the ecosystem to the scale of the governance framework. In effect issues span vast territorial areas (e.g. fish stocks) and certainly cross jurisdictional boundaries such as national- state waters and EEZ – high seas boundaries. Hence the institutional arrangements need to be constructed to be of similar scale through the marine spatial planning schemes crossing these boundaries as they are based on ecosystems characteristics not human use issues. As well as these 'spatial mismatches' they also recognise 'temporal mismatches' – where the short term electoral cycles of usually 2–4 years do not match either the long term cycles of oceanic phenomena or sometimes the short term impacts such as the sudden crash of a long established fishery. They believe the Great Barrier Reef marine planning approach is a sound well established model to overcome these issues.

This concludes an overview of how national states (and some regions) have attempted to establish ocean governance systems covering their EEZ. But nearly 50 % of the planet is still "high seas", i.e. areas outside the control of national jurisdiction under the LOSC (Freestone 2009). Although there are some international protocols covering these areas (e.g. 1972 London Convention and its 1996 Protocol on Ocean Dumping) the lack of clarity over governance of these areas was recognised in the establishment, by the UN General Assembly, in 2004 of an "Ad hoc Open-Ended Informal Working Group to study issues relating to the conservation and sustainable use of marine biological diversity beyond areas of national jurisdiction" (Freestone 2009). This group along with the IUCN 4th World Conservation Congress has derived a draft of ten working principles for high sea governance which are listed here (Freestone 2009):

- Conditional freedom of the seas (navigation, overflight, fishing, etc.)
- Protection and preservation of the marine environment (Freestone 2009, lists some of the regional sea agreements which assist here)
- International cooperation
- Science based approach to management
- The precautionary approach
- Ecosystem approach
- Sustainable and equitable use
- Public availability of information
- Transport and open decision making processes
- Responsibility of states as stewards of the global marine environment

These principles are clearly an extension of national jurisdictional LOSC principles, but are an excellent summary of the current state of development of the critical components of this ocean governance framework and as such provide an excellent conclusion to this section. The themes emerging from this literature can begin now to be summarized.

There appears to be a clear recognition in both the technical literature and high level policy discussions at international, regional and national levels that, if the problems now confronting the oceans are to satisfactorily addressed , if not solved in the short term, then there must be a move to integrated ocean planning and management (i.e. consideration of the whole system and all its human uses and the interaction and impact of these uses) which entails a move away from the sector by sector historically, culturally and politically based approach which still dominates in this realm. In essence there needs to be a decisive move towards a systems based approach.

A consensus of what this integrated approach would entail appears to have a number of elements. These include recognition that large ecosystems in the oceans should be the fundamental 'unit' (or focus) of planning and management. Another key element is to deliver ecosystems based management (EBM) probably through a marine spatial planning (ocean zoning) scheme.

The governance framework, and in particular the institutional arrangements, by which to then implement such an approach, is more problematic (and experimental) not least because of differences in cultural and political systems between different countries. As well it needs to be acknowledged by all stakeholders that the concepts fundamental to delivering good ocean governance – marine spatial planning and ecosystems based management – are in their infancy i.e. in an early stage of development – with few examples of practical implementation to draw on globally.

Politically it appears that this lack of maturity in the development of these concepts is seized upon, or at least used to their advantage, by sectoral (development/exploitation based) interests to exert pressure on politicians (often behind closed doors) to fall back to historical sectoral based (silo) approaches which clearly favour the powerful economically-based vested interests and disenfranchise the general community and future generations.

In conclusion these authors state while the high level political commitment to integration, use of marine spatial planning and ecosystems based management has been growing over the last decade there appears to be an even more urgent need for ocean governance arrangements that have strong institutional leadership (i.e. leadership and coordination within government agencies) and a clear and transparent open consultation process for implementation.

Many writers (Juda 2003; Crowder et al. 2006) suggest that the best model currently available as an approach to planning, management (and governance?) of the Great Barrier Reef (Marine Park) by the Australian Government through the national agency the Great Barrier Reef Marine Park Authority (GBRMPA). They suggest that it is the world's "stand out" example of integrated oceans planning and management. We will return to this point in the next section after considering the case study of Australian ocean governance, the framework in which the GBRMPA model resides.

21.3.3 Ocean Governance Implementation in Australia 1996–2011: A Case Study

Australia has been chosen as a brief case study because it was an early adopter of a policy response to the UN Convention on the Law of Sea (given the potential benefits of its enormous EEZ), the development of its policy responses and how these have fared over time (1996–2011) are well documented and because the author can give some insights into the machinations associated with these processes as he was a member of the National Oceans Advisory Group and its predecessor the Ministerial Advisory Group on Oceans Policy.

A chronological approach will be used here using a series of references including several contributions from the book "Marine Resources Management" (Gullett et al. 2011). Haward and Vince (2008, 2009), Vince (2003, 2008), Rothwell and VanderZwaag (2006), Reichelt and Wescott (2005), National Oceans Office (NOO 2004), Bateman and Bergin (2009) and Wescott (2000), are source references for this material. Wescott (2000) describes the early history of implementing the LOSC within Australia's waters and the challenges that nation faced at the turn of the century. Australia ratified the LOSC in late 1994 several months after the Convention came into force and in so doing assumed responsibility for over 15 million square kilometres of Exclusive Economic Zones (EEZ) one of the largest areas in the world. This EEZ stretches from sub-Antarctic regions through all of the temperate zones to sub-tropical and tropical ecosystems. Soon after ratification Australia decided to meet its obligations under the LOSC by developing an oceans policy.

In late 1997 the then Minister for Environment and leader of the government in the Australian Senate (upper house of parliament) committed Australia to a comprehensive and integrated oceans policy with the aims of establishing Australia's sovereign rights over to EEZ and to ensure ecologically sustainable development (ESD) of the ocean's resources for both wealth creation and environmental protection (Wescott 2000). Through 1996 most of the work has been carried out by bureaucrats but pressure from NGOs and community based groups such as the Marine and Coastal Community Network (MCCN) resulted in the Minister "opening up" the process to wider consultation including his establishment of a Ministerial Advisory Group on Oceans Policy (MAGOP) which included independent scholars, NGOs and industry representatives.

An earlier "consultation paper" (March 1997) launched by the Prime Minister had failed to stimulate much discussion in the wider community (63 submissions only) but the initiation of MAGOP (September 1997) and the production of a series of discussion papers in late 1997 stimulated interest, with the assistance of the 10,000 plus participants of the MCCN (Binkley et al. 2006). This was supported by a major forum which emphasised the 'national', comprehensive and integrated components of the forthcoming policy. 'National' refers to the fact that in Australia whilst the Federal government controls off shore waters the States control those within 3 nautical miles of shore and hence a 'national' (as distinct from 'federal') policy would cover both waters.

Hence Australia was committing itself to a national, comprehensive, integrated whole of government oceans policy – that is one in which sectoral and inter-government and inter-departmental interests would be subordinate to the greater national interest and to an integrated governance system. This was a very ambitious objective given that Commonwealth-State Government environmental relations in Australia have been very acrimonious in the past (see Wescott 2006), sectoral (particular exploitative) interests have dominated resource allocation decision making and inter-departmental rivalries and power struggles are the dominant feature of the federal bureaucracy. In fact, as we will see below, this commitment was over ambitious and without legislative backing was problematic from the beginning. In March 1998 (the International Year of the Ocean) the Ministerial advisory body delivered its recommendations (MAGOP 1998). There was consensus on the need for a comprehensive all-encompassing policy and that it should be based on ESD principles. But there was disagreement on whether there should be an ecosystems based management approach combined with integrated regional planning and management. Sectoral groups – again particularly the exploitative industries – wanted a continuation of sectoral based approaches unified solely by the common objectives of ESD whilst other members favoured genuine integration of planning and management based on ecosystems based management.

An "Issues Paper" published for consultation in May 1998. Whilst comprehensively covering the scope of an oceans policy the paper did not propose any specific implementation process for such a policy. A theme was emerging here even in the early stages of preparing Australia's Ocean Policy (AOP) – the policy was indeed to be "comprehensive" but there was no such agreement on an integrated implementation process or how binding the policy would be over various sectors. The final AOP (Commonwealth of Australia 1998a, b) did little to resolve this dilemma. It was certainly comprehensive and it did establish a clear set of institutional arrangements by which the policy would be delivered. This involved the establishment of a National Oceans Ministerial Board (NOMB) chaired by the Minister for Environment and containing the Ministers covering primary industry, shipping and science. To aid the NOMB a National Oceans Advisory Group (NOAG) was established with 16 representatives of groups with interests in marine affairs. Both bodies were to be serviced by a National Oceans Office (NOO). Most of the recommendations in the Ocean Policy were to be implemented through a regional marine planning process (based on marine spatial zoning) which focused on a series of regions which were essentially Large Marine Ecosystems (LMEs). The first plan was to be for the most populous area of Australia – the south east.

Unfortunately, the State Governments refused to cooperate with implementing the AOP, which meant it was not going to be a national plan and was not to be integrated across the boundary of state verse commonwealth waters. As Wescott (2000) pointed out, the implementation phase, in the absence of any legislation and with the exclusion of state and territories, was always going to be difficult – particularly as the most powerful groups (oil, gas, fisheries, shipping) were never committed to moving away from individual sector based historical approaches. Other groups pointed to the rhetorical commitment to integration in the AOP and the apparently integrated regional marine planning process proposed. Both groups could not be right!

The first RMP demonstrated how difficult resolving this conflict was going to be. In total it took nearly 4 years to finish the South-east Regional Marine Plan (Vince 2011; NOO 2004) and even then it lacked the precise boundaries of the Marine Protected Areas (MPAs) at the heart of the zoning system. This MPA system was finally released in 2006 and the MPAs declared in 2007 i.e. 7 years after the planning process commenced. Given the entire AOP had been produced in less than 2 years, 7 years to produce one RMP did not auger well for implementation.

A review of the process (TFG International 2002) reinforced concerns about the lack of State Government cooperation but also suggested moving NO Office from Hobart (Tasmania) to Canberra. Whilst this would clearly place NOO close to real political power in the nation's capital it was ironic that an oceans office was to be based hundreds of kilometres from the sea – and in the federal government's capital – when a lack of state cooperation was a primary issue.

A series of institutional changes followed. NO Office was taken into the Department of Environment and Heritage (2004). The NO Ministerial Board was disbanded (2005) and the preparation of RMPs were converted to Marine Bioregional Plans under sections of the federal Environment Protection and Biodiversity Conservation Act 1999.

Whilst the above changes were probably well intentioned as a means of giving legislative backing to the regional marine planning process the appearance to external observers was that the "whole of government" approach in the "integrated" Australian Ocean Policy had been supplanted by a more restricted view that this was an environmental policy implemented through the environment department under environmental legislation i.e. apparently a clear sectorally based implementation process. This was the perception to be sure, if not the reality. The author was a member of NO Advisory Group throughout this time and the view that was expressed 'in house' was that this was an environmental "takeover". If it was an "environmental takeover", it certainly did not originate with environmentalists who were still arguing for an Oceans Act and total integration of ocean planning and management (e.g. Smyth et al. 2003).

These changes have increased the rate of production of Marine Bioregional Plans with most being published as drafts in late 2011 into 2012.

One of the features though of these Marine Bioregional Plans are that in essence the end result is largely a series of Marine Protected Areas (MPAs) are identified with internal zoning. Multiple Use Zones dominate and whilst there are a series of recommended approaches to marine management in the rest of the area, without any zoning identified for these at all it is unclear how these are to be implemented. In effect these 'plans' could be more honestly presented as being a plan with 'general use' zones with minimal regulation present covering the majority of the area and a few 'protected areas' (some highly protected 'no-take' reserves) with more detailed zoning within. If presented as such they would more closely resemble a genuine attempt at marine spatial planning, rather than MPA identification.

The Great Barrier Reef (Marine Park) is a very special case within these Commonwealth waters rather than the general case. Here the whole region is designated a 'park' under federal legislation and then is zoned throughout its entire area with over 20 % in highly protected 'no-take' national park zones. Most of the 'park' is zoned 'general use' with few restrictions (all the park is protected from mining which is the major 'purpose' of designating it a park). Large vessels navigate through these areas and commercial and recreational fishing is usually permitted (with some restrictions).

The park has its own management authority – the Great Barrier Reef Marine Park Authority and its own legislation (Great Barrier Reef Marine Park Act 1975). Management throughout is based on EBM and obviously marine spatial planning based on considerable public participation are the basis for park management. These mechanisms certainly qualify the park as meeting the criteria of integrated oceans management and hence the praise it draws from many quarters as a model for ocean governance is entirely justified. There is close cooperation in management between state and federal agencies and possibly the only area in which it is challenged is in controlling land based sources of pollution to the reef.

In conclusion – what does the Australian experience inform us about ocean governance?

First, it illustrates the difficulty of developing governance mechanisms in a new domain – the oceans. Secondly, this ocean governance system was being imposed 'over the top' of well established historically based sectoral planning and management arrangements. Change is never easy and the perceived "threat" of a new system, especially one involving groups and sectors you have not had to deal with before, is challenging to sectoral groups – especially groups that have mutually evolved with cooperative, supportive and like-minded government agencies.

Thirdly the Australian experience highlights how difficult implementing the central tenet of "integration" is going to be in ocean governance and demonstrates that whilst concepts of ecosystems based management through marine spatial planning are intuitively and scientifically attractive their implementation in practice is going to be a long and adaptive learning process.

Finally, the major focus on Marine Protected Areas as the most high profile outcome of regional marine planning is problematic. Given that in the Australian context historically a small proportion of the area within large MPAs (notably excluding the GBRMP) are highly protected marine resources (often called "no take", a term which can lead to confusion. It may well have been better to use the various categories of the IUCN that define high protection – categories 1 and 2) there has been a considerable "blurring" of the boundaries between general marine spatial planning and zoning within MPA boundaries. In effect the long term Australian experience is that the very large "general use" zones within MPAs differ very little, if any, from the general use zone outside an MPA. This suggests that the large MPAs, delivered and proposed, in the Australian system may have been more about political perception than marine environmental protection and hence this "blurring" is a distraction from a rational derivation of an ocean governance system based on an objective of integrated marine and coastal planning and management (see also Cicin-Sain and Belfiore 2005).

21.4 Concluding Remarks on Progress with Ocean Governance: Integration as a Unifying Theme

The common theme throughout the literature on ocean governance and management is "integration". The challenge though is that most ocean governance remains mired in historical sectoral management processes. More pointed is that in nations where genuine attempts have been made towards integration (e.g. Canada and Australia) after a period of attempted integration the approaches have fallen back into sectoral management practices over time. This does beg the question as to 'why is integration so hard to achieve and then maintain?'

The answer probably lies in the power relationships between the various stakeholders and how stakeholders perceive a governance system should operate to achieve their specific aims. Hence the stakeholders exploiting natural resources (and usually not on a sustainable basis over time) most notably commercial fisheries and the oil and gas mining industries, prefer sectoral governance approaches because in general the agencies meant to regulate them are often far too close to the industry and do not appear at times to be acting in the wider community interest much less with the aim of conservation of natural environments. The power of the economic arguments presented by these sectors will often overpower ecological arguments offered by the sectors favouring integration. Hence this 'begs the question' as to whether a risk management based approach may be more appropriate.

21.5 Is Risk Management an Option for Ocean Governance?

Before attempting to answer this question a brief introduction of what 'risk management' might be taken to mean in the context of ocean governance will be introduced. Following on from this introduction a very brief risk appraisal of the marine environment is performed to yield the current major "risks" confronting ocean environments and hence ocean governance.

In this chapter the definition and model for risk management and governance is based on the work of the International Risk Governance Council (IRGC 2005, 2008). Risk Governance is defined as (IRGC 2008, p. 4):

> Risk governance deals with the identification, assessment, management and communication of risks in a broad context. It includes the totality of actors, rules, conventions, processes and mechanisms and is concerned with how relevant risk information is collected, analysed and communicated, and how management decisions are taken. It applies the principles of good governance that include transparency, effectiveness and efficiency, accountability, strategic focus, sustainability, equity and fairness, respect for the rule of law and the need for the chosen solution to be politically and legally feasible as well as ethically and publicly acceptable.

The IRGC (2008, p. 4) also states:

> The challenge of better risk governance lies here: to enable societies to benefit from change while minimising the negative consequences of the associated risks.

The IRGC (2008) proposes a risk governance framework comprising five linked (cyclic) phases, in order:

- Pre-assessment
- Appraisal
- Characterisation and evaluation
- Management
- Communication

At all stages of the framework communication with all stakeholders is regarded as crucial with in effect the first half of the cycle focusing on the understanding of the risks present and the second half on decision making (phase 3, characterisation and evaluation bridge both).

The IRGC approach (IRGC, 2005, 2008) differs from earlier risk management approaches through the inclusion of both a societal context and through the categorisation of risk related knowledge (into simple, complex, uncertain or ambiguous) assessing to what extent, and how, cause and effect are understood in the specific context being studied (IRGC 2005). As such it unequivocally includes and embraces three major valued based premises and assumptions (IRGC 2005):

- Both "factual" (technical/data input if you prefer) and socio-cultural dimensions must be considered.
- "Inclusiveness" in the governance process is a necessary element. This refers to ensuring an "early and meaningful" involvement of all stakeholders and, in particular, civil society (IRGC 2005, p. 12).
- That "values" are implicit in the principles of "good governance" including:
  - A commitment (point 2 above) to participation; and
  - Transparency,
  - Effectiveness and efficiency,
  - Accountability,
  - Strategic focus,
  - Sustainability,
  - Equity and fairness,
  - Respect for the law, and
  - A solution which is politically and legally realistic (practical) and publicly acceptable.

This is a "tall order" in anyone's language but an admirable, comprehensive and well considered framework. Would such a framework be a suitable framework for ocean governance?

In previous sections the well agreed and well documented threats to the marine environment have been mentioned. But just to collate them together and to place them in a "risk appraisal" framework they are briefly listed here again:

- Marine pollution of all types and sources from the high profile oil spillages from off shore platforms and ships running aground through to the 70 % of marine pollution that comes from land based activities.

- Severely declining fish stocks. The global decrease in fishery resources both within coastal state waters through lack of regulation, compliance and enforcement through to the massive threat of IUU (illegal, unreported and unregulated) fishing of the high seas.
- Decline in Marine Biodiversity. Through a multitude of causes (including excessive bycatch in legal and IUU fishing and climate change impacts, see below) marine biodiversity is decreasing even before we have described, probably most, of the species the seas. Included here could also be the threat to the ecological processes that drive marine ecosystems and provide ecosystem services to humans.
- The impacts of climate change on the marine environment. Ocean acidification, coral bleaching, changes to the distribution of marine species and of ocean currents and sea level rise are impacts resulting from increasing sea water temperatures and increased concentrations of dissolved carbon dioxide in sea water. These impacts whilst critical in their own right become even more dangerous when combined with the other impacts listed here.
- The translocation and introduction of pest species (largely in ship ballast water) from one side of the world to the other hence threatening local species and local ecological processes.

Finally a combined threat/risk could be identified from this review as well: the failure to, as yet, develop a comprehensive and integrated global ocean governance system.

There is insufficient space and time in this chapter to do a comprehensive testing of the IRGC's (2008) framework as a model for ocean governance. But nevertheless a preliminary brief assessment will be attempted with the strong recommendation that full assessment would be a very worthwhile exercise for a government or regional ocean group to attempt.

The risks confronting the ocean environment appraised above would certainly be worthy of "characterisation and evaluation" under the IRGC framework and without doubt the inclusion of the "societal context" and in particular all the elements of "values" listed above would enhance – indeed should be a compulsory component of – any ocean governance framework. The difficulty may arise in the complexity – both in the natural environment and the socio-political environment – of ocean systems, e.g. risks to underlying ecological processes and for the determination; and assessment of the "driving forces" behind Large Marine Ecosystems may be difficult to unravel in the framework but it certainly would greatly assist in identifying the areas of inadequate knowledge and the "flash points" between the very powerful vested interest stakeholders in ocean affairs (oil and gas, fishing and the shipping industry) and the wider general community's and environmental interests.

As such the IRGC framework for risk governance goes far beyond currently used, and limited, risk management processes which are too weighted towards technical (engineering based) appraisals of quantifiable known areas of knowledge and based on information kept and guarded usually by the industry themselves and not available to wider society.

In conclusion there is little doubt that the application of a risk governance framework to the varied ocean governance approaches currently being tested across the world would undoubtedly yield very valuable information on the failings of these current ocean governance systems and hence, if applied rigorously, would be expected to greatly improve these currently imperfect approaches.

The bigger question of whether a risk governance framework could replace the current approaches would await the outcome of the review proposal above but it may find that the lack of knowledge and understanding of the functioning of marine environmental systems combined with the lack of transparency, participation and other "value" based components in global ocean governance practices is a "bridge too far" at least at this time, even for an approach as innovative and hopeful as risk governance.

21.6 Conclusion

There are common themes that emerge from any analysis of ocean governance frameworks across the world – a commitment to integrated ocean management, encouragement to cooperate regionally with neighboring LOSC nations, marine spatial planning as the preferred planning approach and Ecosystems Based Management as the basis for management. The precise manner in which these components fit together is less clear and some countries are notably more advanced in their implementation of these still largely theoretical concepts. The other more disappointing trend is a return, under political pressure, to sectoral based approaches which are favored by large corporations exploiting the marine environment. These approaches often are not transparent and disenfranchise the general community from decision making.. This struggle between integration and sectoral or system management will continue for some time.

But is risk management and governance the answer to break through these barriers? Probably not in a holistic sense but it is certainly a very worthwhile analytical tool for enhancing integrated oceans management and strong continuing ocean governance mechanisms.

Acknowledgments As always all errors are those of the author but I would like to thank Marcus Haward and Larry Juda for providing valuable references for this paper and Heather Andrews for assisting with its preparation.

References

Bateman, S., and A. Bergin. 2009. *Sea change: Advancing Australia's ocean interests*. Canberra: The Australian Strategic Policy Institute.

Binkley, M., A. Gill, P. Saunders, and G. Wescott. 2006. Community involvement in marine and coastal management in Australia and Canada. In *Towards principled ocean governance*, ed. D. Rothwell and D.L. VanderZwaag, 249–279. London/New York: Routledge.

Cicin-Sain, B., and S. Belfiore. 2005. Linking marine protected areas to integrated coastal and ocean management: A review of theory and practice. *Ocean and Coastal Management* 48: 847–868.

Cicin-Sain, B., and R. Knecht. 1998. *Integrated coastal and ocean management. Concepts and practices*. Washington, DC: Island Press.

Cicin-Sain, B., M. Balgos, J. Appiott, K. Wowk, and G. Hamon. 2011. Oceans at Rio+20. *How well are we doing in meeting the commitments from the 1992 Earth Summit and the 2002 World Summit on Sustainable Development? Summary for decision makers*. Newark: Global Ocean Forum.

Commonwealth of Australia. 1998a. *Australia's ocean policy*, vol. 1. Canberra: Commonwealth of Australia.

Commonwealth of Australia. 1998b. *Australia's ocean policy*, vol. 2. Canberra: Commonwealth of Australia.

Crowder, L.B., G. Osherenko, O.R. Young, E.A. Airame, E.A. Norse, N. Baron, J.C. Day, F. Douvere, C.N. Ehler, B.S. Halpern, S.J. Langdon, K.L. McLeod, J.C. Ogden, R.E. Peach, A.A. Rosenburg, and J.A. Wilson. 2006. Resolving mismatches in U.S. ocean governance. *Science* 313: 617–618.

Ebbin, S.A., A.H. Hoel, and A.K. Sydnes (eds.). 2005. *A sea change: The Exclusive Economic Zone and governance institutions for living marine resources*. Dordrecht: Springer.

Espinosa-Romero, M.J., K.M.A. Chan, T. McDaniels, and D.M. Dalmer. 2011. Structuring decision-making for ecosystem-based management. *Marine Policy* 35: 575–583.

Evans, L. 2011. *The integrated maritime policy: One right answer to a difficult economy*. Aarhus: Speech to the CPMR general Assembly, 29 September 2011.

Freestone, D. 2009. Modern principles of high seas governance. The legal underpinnings. *Environmental Policy and Law* 39(1): 44–49.

Gullett, W., C. Schofield, and J. Vince (eds.). 2011. *Marine resources management*. Sydney: LexisNexis Butterworths.

Haward, M., and J. Vince. 2008. *Oceans governance in the twenty-first century: Managing the blue planet*. Cheltenham: Edward Elgar.

Haward, M., and J. Vince. 2009. Australian ocean governance: Initiatives and challenges. *Coastal Management* 37: 1–6.

International Risk Governance Council. 2005. *White paper on risk governance. Towards an integrative approach*. Geneva: IRGC.

International Risk Governance Council. 2008. *An introduction to the IRGC risk governance framework*. Geneva: IRGC.

Juda, L. 2003. Changing national approaches to ocean governance: The United States, Canada and Australia. *Ocean Development and International Law* 34: 161–187.

McLeod, K., and H. Leslie. 2009. *Ecosystems based management for the oceans*. Washington, DC: Island Press.

Ministerial Advisory Group on Oceans Policy. 1998. *Australia's oceans policy: A report of the Ministerial Advisory Group on oceans policy*. Canberra: Environment Australia.

National Oceans Office. 2004. *South-east Regional Marine Plan: Implementing Australia's oceans policy in the South-east Marine Region*. Hobart: National Oceans Office.

Rayfuse, R., and R. Warner. 2008. Securing a sustainable future for the oceans beyond national jurisdiction: The legal basis for an integrated cross-sectoral regime for high seas governance for the 21st century. *International Journal of Marine and Coastal Law* 23: 399–421.

Reichelt, R., and G. Wescott. 2005. Integrated oceans management and the institutional performance of exclusive Economic Zones: The Australian case. In *A sea change: The Exclusive Economic Zone and governance institutions for living marine resources*, ed. S.A. Ebbin, A.H. Hoel, and A.K. Sydnes, 64–77. Dordrecht: Springer.

Rothwell, D., and D.L. VanderZwaag (eds.). 2006. *Towards principled ocean governance*. London/New York: Routledge.

Ruckelhaus, M., T. Klinger, N. Knowlton, and D.P. DeMaster. 2008. Marine ecosystem-based management in practice: Scientific and governance challenges. *BioScience* 58(1): 53–63.

Smyth, C., M. Prideaux, K. Davey, and M. Grady. 2003. *Oceans eleven*. Melbourne: Australian Conservation Foundation.

TFG International. 2002. *Review of the implementation of oceans policy*. Hobart: TFG International.

VanderZwaag, D.L., and N. Oral. 2008. Editor's introduction. International ocean governance in the 21st century. *International Journal of Marine and Coastal Law* 23: 393–397.

Vince, J. 2003. Australia's oceans policy: Five years of integration across sectors and jurisdictions? *Maritime Studies* 133: 1–13.

Vince, J. 2008. Policy transfer in oceans Governance: Learning lessons from Australia's oceans policy process. *Ocean Yearbook* 22: 159–181.

Vince, J. 2011. Managing Australia's oceans through the oceans policy process. In *Marine resources management*, ed. W. Gullett, C. Schofield, and J. Vince, 77–90. Sydney: LexisNexis Butterworths.

Wescott, G. 2000. Development and initial implementation of Australia's 'integrated and comprehensive' Oceans Policy. *Coastal Management* 43: 853–878.

Wescott, G. 2006. Conservation policy in Australia. In *Ecology: An Australian perspective*, 2nd ed, ed. P. Attiwell and B. Wilson, 433–445. Melbourne: Oxford University Press.

Geoffrey Wescott is the Associate Professor of Environment at the School of Life and Environmental Sciences, Deakin University, Australia. He is a Fellow of the Environment Institute of Australia and New Zealand. Geoff has published extensively on environmental policy, planning and management and is a current member of the Zoos Victoria Board and Vice President of the Australian Coastal Society. He has been a member of the National Oceans Advisory Group, the Board of Parks Victoria, the Victorian Coastal Council, the Marine and Coastal Community Network and the National Parks Advisory Council in Australia among many other community and professional bodies.

Chapter 22
Governance of Megacity Disaster Risks: Confronting the Contradictions

James K. (Ken) Mitchell

...cities rather than states are becoming the islands of governance on which the future world order will be built. ... The defining feature of this new urban age will be megalopolises whose populations are measured in the tens of millions, with jagged skylines that stretch as far as the eye can see.

(Khanna 2010)

...most of today's largest cities are in areas where earthquakes, floods, landslides and other disasters are most likely to happen.

(Lewis and Mioch 2005)

22.1 Introduction

When the first scholarly book devoted exclusively to natural hazards in megacities[1] was published (Mitchell 1999), the term "risk governance" had not yet been coined. The effectiveness of different governmental policies for reducing the societal burden of hazards and disasters was then the main focus of attention but the concept of governance itself was unproblematic. Now it has become commonplace to recognize that inadequate governance is both part of the problem of megacity risk management and also that new forms of governance can be part of the solution. As a result, existing arrangements for addressing issues of environmental risk are undergoing fundamental restructuring.

[1] Megacities are massive spatially discrete urbanized areas. The U.N. Commission on Human Settlements (United Nations Human Settlements Program 2008) has set the lower population threshold for megacities at ten million. About 45 urban areas have already reached this size or are likely to do so within the next decade. They include: Bangkok, Bangalore, Beijing, Bogota, Buenos Aires, Cairo, Chennai, Chicago, Chongquing, Delhi, Dhaka, Dongguan, Essen, Guangzhou, Ho Chi Minh City, Hong Kong, Hyderabad, Istanbul, Jakarta, Johannesburg, Karachi, Kinshasa, Kolkata, Lagos, Lahore, Lima, London, Los Angeles, Manila, Mexico City, Moscow, Mumbai, Nagoya, New York, Osaka, Paris, Rio de Janeiro, Sao Paulo, Seoul, Shanghai, Shenzhen, Taipei, Tehran, Tianjin, Tokyo.

J.K. Mitchell (✉)
Department of Geography, Rutgers University,
54 Joyce Kilmer, Piscataway, NJ 08854-8045, USA
e-mail: jmitchel@rci.rutgers.edu

U. Fra.Paleo (ed.), *Risk Governance*, DOI 10.1007/978-94-017-9328-5_22

No megacity has achieved unqualified success in the governance of disaster risks but some are actively confronting present risks and moving ahead to engage new ones while others are either not engaging existing risks or are failing in the effort. The great majority of published studies focus on large, poor and otherwise disadvantaged megacities outside Europe, Japan and the USA because such places are perceived as highly vulnerable and vulnerability is widely considered the most important factor that contributes to disaster, as well as the one most within human ability to affect. From the perspective of actors like the World Bank, UN-Habitat (United Nations Human Settlements Program), UNDP (United Nations Development Programme) and the ISDR (International Strategy for Disaster Reduction) the premier goal is to close vulnerability gaps by improving the security of the disadvantaged to a point where they have caught up with their better-provisioned contemporaries. This is no bad thing so long as it does not blind us to the existence of additional processes that construct hazard differently in megacities and that probably require a different suite of solutions.

Climate change is an unprecedented problem that comprehensively burdens megacities and challenges all of their inhabitants, not just the currently disadvantaged. Climate change risks prompt us to be concerned about their systemic effects on the infrastructure that makes great cities work and the wellbeing of interconnected economic systems that generate material goods and much of the world's wealth. This kind of problem requires an approach that foregrounds new kinds of risk and new risk habitats, views avoidance of catastrophic surprises as the master task of governance and looks toward creation of self-regulating systems of risk management as the best means for addressing emerging challenges of a highly uncertain twenty-first century. The most prominent institutional exponents of this latter approach have been actors in the global reinsurance industry as well as civic and business leaders in major cities but it is also present among global change researchers involved in the work of the Intergovernmental Panel on Climate Change (IPCC) and other bodies that focus on the intersection of technological and environmental risks, e.g. International Risk Governance Council (IRGC).

The future of risk governance will be strongly affected by the success of researchers and policy makers in addressing the aforementioned contradiction between vulnerable, mainly poor, megacities struggling to cope with existing natural risks and better-provisioned affluent ones that face the added burden of new kinds of threats and new kinds of vulnerability associated with the functions they perform in the interconnected global urban-economic system. This is one of four contradictions that will be examined herein. The other three are not specific to megacities but affect all realms of risk governance, albeit most complexly and ambiguously in megacities (Table 22.1).

Second is the contradiction between risk governance traditions that have evolved mainly in response to failures of human-devised technologies and those that owe

Table 22.1 Contradictions of megacity risk governance

Locus	More developed city	Less developed city
Dominant risks	Human created	Natural
Decision style	Technocratic	Political
Ethics	Value-neutral	Value-charged

much to experience with natural disasters; with caution these might be labeled European versus North American perspectives. Third is the contradiction between so-called "technocratic" conceptions of risk governance viewed as a task for expert management on the one hand and as an arena of political action, sometimes geopolitical and sometimes bio-political, on the other. Finally, there is a contradiction between the utilitarian tradition of risk governance (i.e. effectiveness getting the job done) from the ethically charged concept of "good governance". All four of these contradictions are ripe for investigation by researchers and attention by public policy makers, as well as an increasingly empowered lay public.

22.2 Megacity Risks

Humans who live, visit, or work in megacities are presently exposed to a wide range of potentially catastrophic risks. Among others these include earthquakes, flooding and tropical cyclones. Likely future risks that are associated with anticipated climate changes include sea level rise and changes in temperature and rainfall regimes that may exacerbate the existing risks. On a per event basis, earthquakes today inflict the heaviest death tolls and the most physical destruction. In the aggregate, floods are the most ubiquitous and costly (Jha et al. 2011; Adikari et al. 2010). Tropical cyclones combine the most complex array of loss-causing mechanisms (winds, storm surges, rain fueled floods, lightning, embedded tornadoes) and are particularly threatening to low-lying deltas that shelter vast human populations (Seto 2011). Together with terrorism, an entirely human-created risk, these three are not only threats that have elicited great public concern but also ones that have fueled the most active efforts at avoidance, prevention and reduction. Systematic information about drought susceptibility is not available but it is suspected that almost all megacities are pressing against the limits of available water supplies (World Wildlife Fund-Germany 2011). Other risks that are significant in certain cities include: landslides, volcanism, dust storms, blizzards, tornadoes, subsidence and wildfires. While a dozen or so of the 45 megacities reviewed herein are exposed to one or two kinds of risks, approximately half (23) are exposed to four or five. The latter include; Chennai, Chongqing, Dhaka, Dongguan, Guangzhou, Ho Chi Minh City, Hong Kong, Istanbul, Jakarta, Karachi, Kolkata, Lima, Los Angeles, Manila, Mumbai, Nagoya, Osaka, Seoul, Shanghai, Shenzhen, Taipei, Tianjin, Tokyo. All but three of these megacities are located in Asia.

22.2.1 The Growing Importance of Economic Impacts

The public policy salience of economic losses in large cities of the developed world is growing relative to the tolls of deaths and injuries (United Nations Radio 2012). This is certainly true of the United States. With the signal exception of hurricane Katrina in 2005, the long-term trend of fatalities from most types of natural disaster in the United States has been going unmistakably down for around a century and it is now so low that efforts to achieve further reductions may be reaching the point of diminishing returns. At the same time the trend of economic losses is rapidly moving upwards. This means that the reduction of economic impacts may well become the most pressing problem that faces researchers, policymakers and managers in megacities of the USA and other More Developed Countries that have also seen their disaster death rates diminish. New York City is the first contemporary U.S. megacity to experience the direct impact of a major natural disaster (see below). Its economic effects are still being tallied at the time of writing but some estimates suggest they may exceed $50 Billion, making it the second worst disaster in U.S. history after hurricane Katrina (2005). In the absence of a full accounting of Sandy, data from disasters in smaller cities are indicative. For example, the shift to concern about economic impacts is detectable in New Orleans, the site of the most deadly U.S. disaster in more than three-quarters of a century. Five years after hurricane Katrina the Brookings Institution published a comprehensive assessment of the city's recovery (Brookings Institute 2010). Therein economic aspects of recovery received the lion's share of attention. In this authoritative accounting, economic recovery was the criterion by which long-term recovery as a whole is judged. The shift towards privileging economic recovery has profound implications for other aspects of recovery and for the humanitarian-inflected emergency response policies that characterize international efforts to reduce the disaster burdens of Less Developed Countries (McDonald 2011).

Economic loss estimation procedures have also become much more salient in recent years. Some of these were pioneered because of concern by banks and lending agencies about potentially catastrophic global economic consequences of a major earthquake in Tokyo (Lewis 1989; Committee on Earthquake Engineering 1992; Shah 1994; Hadfield 1995; Davidson and Shah 1997). Others grew out of similar worries about U.S. cities by the Federal Emergency Management Agency (U.S. Federal Emergency Management Agency 2004; Rose 2007; New York City Consortium for Earthquake Loss Mitigation 2005). Yet others reflected the concern of reinsurance companies about excessive levels of financial exposure in such places (Munich Re Research Group 2007).

However, a shift to economics-led assessments moves policy debate into ambiguous territory. This is because experts are divided about whether to regard disasters (natural and otherwise) as intrinsic, incidental and continuing attributes of capitalist economies, that can be beneficial (Schumpeter 1942; Hirschleifer 1993; Lindner 2005; Tierney 2007; Klein 2007; Cuaresma et al. 2008) – or as dire events whose importance is expanding, forced by globalization disruptions and unprecedented risks of climate change (Intergovernmental Panel on Climate Change 2007; Stern 2007; Leichenko and O'Brien 2008).

In light of the constraints on risk assessments and loss estimations, perhaps we can gain a better appreciation of the evolving hazard burden of contemporary megacities by examining recent disaster impacts as reported by the mass media and disaster assistance organizations during 2011. These are of two contrasting types (a) places that mainly suffer disruptions of economic functions and human mobility, and (b) places that incur substantial loss of life and material destruction.

22.2.2 *Economic Impacts in Privileged Megacities*

In 2011 several megacities that are important pivots of the global economic system were affected by damaging events. Record monsoon flooding in the vicinity of Bangkok severely dislocated worldwide manufacturing and distribution supply chains for computers while the massive undersea Tohoku earthquake and subsequent tsunamis drastically reduced the production of automobiles from companies based in greater Tokyo. In the USA, New York City suspended daily public services and conducted large-scale evacuations of exposed populations in advance of the onrushing Hurricane Irene. In none of these cases was there significant loss of life within the megacity and only limited physical destruction.

Nor are these isolated examples. In April 2010 international travel was suspended for several days when airports in Paris, London and smaller European cities shut down during the eruptions of Iceland's Eyjafjallajökull volcano. Also in 2010, Jakarta's international airport closed owing to heavy concentrations of volcanic ash from Mt. Merapi. During the peak travel season of January 2008 heavy snow paralyzed rail travel and impaired the movement of workers and materials that serve Guangzhou, Dongguan, Shenzhen and the booming export-based urban megaregion of China's Pearl River delta (Ding 2008; Shi et al. 2009). Monsoon-related floods in India's premier global city during 2005 brought about the closure of Mumbai's international and domestic airports, forced the shutdown of trading on its stock exchange and disrupted banking within the country and beyond (Government of Maharashtra 2005). As these examples demonstrate, the entire system of interconnected megacities is sensitive to extreme events that affect any one of its members, illustrating a previously overlooked aspect of what has come to be labeled the "globalization" of disasters (Alexander 2006).

In summary, the subsequent effects of local disasters may ripple far afield. For a growing number of megacities that are closely coupled with the global economic system their future risk governance agendas will include avoidance of infrastructure disruption and the dislocation of businesses flows dependent on transportation systems as well as the accommodation of displaced populations.

22.2.3 *Deaths, Displacement and Destruction in Disadvantaged Megacities*

Poor and otherwise marginalized urban populations are, without doubt, the most heavily penalized victims of disaster in megacities. This is particularly the case in the megacities of Latin America, Africa and South Asia where occasional major disasters are typically interspersed with frequent sub-catastrophic losses from lesser extremes.

Within the past decade there has been a dismal parade of such reports from places like Sao Paulo, Lagos, Karachi, Delhi, Dhaka and Manila. For example, in 2011 Karachi's shorefront fishing colonies were evacuated ahead of a threatened cyclone, adding to an already heavy burden of poor refugees who had come into the city fleeing catastrophic rural floods that affected the Indus river system to the east. In 2009 tropical storm Ketsana submerged more than 80 % of Metro Manila, killing almost 250 people and displacing 280,000–300,000 more (World Bank 2010a). Between November 2009 and March 2010 semi-continuous flooding made more than 20,000 people homeless in poor neighborhoods of Sao Paulo and killed 78 (World Bank 2010b, p. 82). In February 2007, 70 people died and 340,000 were displaced from their homes by floods in Jakarta (Jha et al. 2011). Typhoon-generated floods inundated more than 200,000 houses in Shanghai on August 7, 1991 (Zhong and Chen 1999). Central and southwest Dhaka suffered especially damaging floods in 1988, 1998 and 2004, some of which lasted for months and directly involved over 60 % of the city's population (Alam and Rabbani 2009, p. 86). After being displaced many victims of megacity disasters return to face the same problems. Thus, floods in 1995 destroyed shantytowns along the Yamuna River in Delhi but 15,000 people who were rendered homeless re-established themselves on these floodplain sites within 2 months (Delhi Disaster Management Authority 2014). These are just a small selection of the more or less routine disasters that affect poorer megacities. There is clear potential for much greater losses.

22.2.4 Risk in the Megacity Writ Larger

Most writings about megacity risks view them mainly as external physical agents that periodically inflict damage on humans and are therefore to be avoided, prevented, controlled or reduced. Alternatively, they might be viewed from a more expansive perspective that regards risk as additionally produced by everyday human routines and sees hazard reduction measures as potentially disruptive of urban functions other than those that are connected with the protection of human life and property. Though this view does not repeal an independent role for natural risks, it is consistent with scientific findings about the increasing role of humans as environmental modifiers and with characterizations of cities as multifunctional entities (Mitchell 2006).

An influential group of mainly European sociologists have taken the notion of risk as socially created and placed it at the center of a conception of society that is animated by experiences of environmental disaster as well as processes of modernity, globalization, and technological change. Among others they include: Anthony Giddens (1990), Ulrich Beck (1992); Juergen Habermas (1999); Charles Perrow (1999, 2007); Lee Clarke (1999, 2006); Zygmunt Bauman (2000) and Niklas Luhmann (2008). Beck, in particular, has responded to disasters involving flawed technologies (e.g. Chernobyl, BSE) not just with heightened awareness of the phenomena themselves but also by encouraging a re-conceptualization of the entire corpus of national and international governance as a risk management project (i.e. the "risk society") (Beck 1992). For many of these scholars increased public preoccupation with risks and hazards reflects a massive cultural shift in vernacular and elite worldviews that has affected agendas of governance and much more.

Insofar as megacities are often thought of as exemplars of modernity, the elevation of risk to a kind of master trope of urban life has clear implications for risk governance. Theories of urban political geography that focus on issues of safety, terrorism, crime, fear, surveillance and the loss of public space in cities also employ risk as a conceptual pivot although, with few exceptions (Davis 1998), like the risk-centered sociological theories they are more often prompted by technological failures and heavy-handed hazards adjustments to natural hazards than to natural risks (Mitchell 2003; Swyngedouw 2007; Rossi and Vanolo 2012).

22.2.5 Contrasting Trajectories of Risk: Mumbai and New York

A detailed analysis of differences in risk governance between all privileged and disadvantaged megacities is beyond the scope of this chapter but some of the contrasts are worth indicating in brief. Governance of flood risks in Mumbai and New York might stand as general exemplars (Ranger et al. 2011). India's largest city Mumbai is exposed to a variety of natural risks, most notably flooding during annual monsoons. Though the city's internationally connected economy has been disrupted by floods and with it the lives of Mumbai's better off residents as well as others to whom they are linked internationally, the chief burdens of risk are borne by poor marginalized shantytown dwellers who account for at least half of the city's population. Despite the fact that they provide much of the labor that sustains a huge informal economy, these predominantly recent in-migrants have no official standing as residents and are ignored in the provision of municipal infrastructure and public services. As well as emergency management and disaster relief assistance these include trash disposal and drainage programs that are essential for the prevention of flooding in the low-lying lands that make up most of the city. During emergencies shantytown dwellers are forced to rely on charitable assistance and local voluntary loss-sharing systems largely of their own making. Moreover the land on which they (illegally) reside occupies a large fraction of city space and is much in demand for redevelopment by entrepreneurs who are members of the currently buoyant formal economy. The process of redevelopment often exacerbates flooding problems by infilling floodplains and displacing stream channels into adjacent shantytown communities. Accelerated sea level rise poses a looming threat to almost the entire city.

From a governance perspective, Mumbai presents a variety of challenges. Flooding needs to be recognized as a joint product of natural and human causes that require the deployment of mutually reinforcing technocratic and political solutions. Scientific knowledge about climate change as a driving force of environmental risks might play a major role in shaping a more resilient city while good governance reforms could give all Mumbai residents a stake in hazard mitigation and avenues for facilitating it. The central task for reformers is to foster a comprehensive integrated understanding of the many factors that contribute to Mumbai's flood risk while changing the status of potential victims into that of stakeholders who find common purposes in the process hazard management and at the same time redesigning the institutions of governance in support of that broadly shared goal.

For some time it has been known that the New York metropolitan area also faced formidable natural risks including flooding associated with coastal storms (New York City Panel on Climate Change 2010) but the arrival of Superstorm Sandy has thrown the city's vulnerability into high relief. Here, in what is arguably the world's richest megacity and one of its largest, existing twentieth century governance arrangements have provided significant security against natural threats but the protective system is now showing severe strains in the face of rising climate risks and increasing human exposures and vulnerabilities.

The municipality of New York dominates the entire megacity and it is clear that it suffered heavily during Sandy. However, the governance problems of communities on the megacity's expanding edge are perhaps more revealing of contemporary hazard dynamics. Here one finds a patchwork of small municipalities that are being incorporated into the continuously urbanized area without benefit of the governance skills, equipment, facilities and budgets that would be necessary to discharge their responsibilities for effective hazard management. Charged with maintaining criss-crossing infrastructure networks that are vital to the support of the entire metropolis some of these municipalities are also being hollowed out by loss of businesses and services to external "Greenfield" sites and burdened by the growth of poor minority groups some of which seek to remain invisible to disaster-assistance organizations because of their status as undocumented immigrants. Others are summer beach resorts with and without significant year round populations, some of which are affluent and others of modest means. When floods arrive they wreak havoc on municipal facilities like police stations, libraries and government offices that are often located in publicly owned floodplains and on homes and retail businesses that provide the bulk of real estate tax revenues on which local governments depend. During floods the sprawling megacity is subject to a range of effects that vary dramatically from place to place thereby posing serious difficulties for state-level emergency agencies, river basin authorities, watershed associations and other regional hazard management organizations. As revealed by Sandy and other recent coastal storms (e.g. Hurricane Irene 2011) impact patterns are strongly spatially differentiated.

The degree to which risk governance arrangements helped or hindered the public response to Superstorm Sandy is not yet known but, compared with Katrina, vertical coordination across the tiered set of federal, state and local government units appears to have been effective, lateral mutual aid agreements among different emergency institutions seem to have worked well and public-private partnerships have not experienced any conspicuous failures. Whether these initial impressions will stand the test of time remains to be seen but, if they do, they will testify to the potential for effective adaptive action against burgeoning climate risks that reside in affluent megacities. The extent to which that knowledge and those institutional arrangements might be modified and transferred to less privileged cities is considerably less certain.

The experience of Sandy in the greater New York area illustrates that even privileged megacities can sustain heavy disaster losses, especially in pockets of relative deprivation and increased vulnerability within them. But it also throws light on the contradictions that were identified at the beginning of this chapter. The disaster has been predominantly viewed as the product of an extreme natural phenomenon, though the connection with anthropogenic climate change is also being heard increasingly in public discourse and with it willingness to question existing hazard management policies that originated in an era when worsening climate risks were rarely taken into account. The beginnings of a tussle between advocates of high technology engineering solutions for hazard management and ecologically sensitive ones is apparent in calls for Netherlands-style flood gates on New York harbor on the one hand and proposals for withdrawal of humans from exposed coasts and revitalization of natural systems. Dispassionate technocratic styles of decision-making have been initially dominant but community leaders are also being forced to recognize that emotional attachments of victims to damaged places and public sensitivity to human suffering also influence perceptions of acceptable public responses. Value neutrality in the selection of appropriate responses by governmental agencies remains the orthodoxy but the value-loaded tenets of good governance are also making inroads on customary practice. However, it is too early to be certain how these trends will eventually develop.

Risk governance in the New York megacity is not only about making the city's safety bureaucracies pay attention to new threats posed by climate change – a task that for which they have shown considerable aptitude – but also about adapting an antiquated system of local "home rule" governance to protect residents against worsening conventional flood vulnerabilities that manifest themselves in a bewildering array of effects with which today's hazard-management systems were not designed to cope. Ironically, if properly supported, a system of diffuse local capabilities for risk management like the kind that is being stressed in Greater New York might serve Mumbai's presently inadequate unified and centralized risk governmental apparatus quite well while at the same time needing to be reinvented on its home territory in the United States. An adaptation that is failing in one place may – with appropriate modifications to local specifics – suffice quite well in another; this is one of the paradoxes of contemporary megacity risk governance.

As we shall see in the next section improvements in governance are viewed as necessary responses to burgeoning risks. But first it is important to set the context in which risk governance has become perceived as a concept whose application might rescue megacities from disaster.

22.3 Risk Governance

Risk governance refers to public arrangements for the prevention, avoidance and reduction of technological and natural risks; it usually connotes a comprehensive and systematic approach that employs the tools of science, and is coordinated with other public policies that are environmentally and economically sustainable. Although this term is new in the lexicon of public decision-making it is rooted in studies of human responses to catastrophic environmental risks that began in the early twentieth century. These developed first in the United States, where there was initially more attention to natural hazards (Burton et al. 1978, 1993; Kates 1978; Kasperson et al. 1988; Douglas and Wildavsky 1982; Slovic 1992, 1999; Rayner 1992; Wisner et al. 2004), and later in Europe where the emphasis has been on human-produced risks (van Asselt and Renn 2011; Lofstedt et al. 2011; Boholm et al. 2012). Now, scholars of risk science and governance everywhere are giving special attention to unprecedented catastrophic risks that push against the limits of available management strategies (Integrated Risk Governance Project 2010, 2011). The field is expanding to address global governance of climate change risks (Intergovernmental Panel on Climate Change 2012; Deere-Birkbeck 2009). However, several challenges remain. These include: puzzles of ambiguity; tensions between the (pragmatic) forms of governance that are currently practiced and (ethically grounded) alternatives that are promoted under the insurgent concept of "good governance"; and public misgivings about the primacy of scientific knowledge.

22.3.1 Ambiguity

Complexity, uncertainty and ambiguity are central attributes of environmental risks (Renn 2008). Uncertainty and complexity can be successfully addressed by scientific means and there has been much progress in this respect. But, from the perspective of this writer, ambiguity is more opaque, more context-driven and more resistant to rational analysis, perhaps insolubly so.

Human engagement with risk is shot through with ambiguities that hamper rationality-based governance systems (Menzel 2009). For example: humans often simultaneously seek incompatible goals; the longer the world remains safe the more vulnerable it may be to the next disaster; and, hazardous places are frequently among the most desirable. Ambiguous circumstances come in many forms that include contradictions, paradoxes, contested interpretations, inconsistent expectations and surprises, among others. Under conditions of ambiguity the rules of choice are unstable – customary guides to choice are either missing, non-functioning, undependable or so deeply conflicted that decision making is effectively paralyzed. These are circumstances where frames of reference shift, evaluative criteria are fluid and choices are fuzzy. Large crowded cities undergoing rapid change are particularly ambiguous places, especially when widely shared taken-for-granted assumptions about institutions, policies and values are being challenged and renegotiated therein.

22.3.2 Good Governance

Good governance is a morally charged concept that undergirds a program of bureaucratic and societal reform, mostly targeted at poor developing countries and those that lack stable governments. The concept has separate but related roots in several different initiatives. These include: the anti-corruption efforts of the World Bank; the anti-poverty campaigns of the Bank, humanitarian organizations and the UN system (World Bank 1997; United Nations 2012); and advocacy of human rights by many of these same groups (U.N. General Assembly 2000; Office of the United Nations High Commissioner for Human Rights 2007). According to the definition employed by the U.N. High Commissioner for Human Rights good governance is "participatory, consensus oriented, accountable, transparent, responsive, effective and efficient, equitable and inclusive and follows the rule of law." UN-Habitat prefers a slightly different but overlapping list of criteria: sustainability, equity, efficiency, transparency and accountability, security, civic engagement and citizenship (United Nations Human Settlements Program 1999).

Advocates of good governance generally encourage: (1) decentralization of political and economic power; (2) strengthening of local autonomy, frequently by leveraging social capital that already exists among grass-roots communities and by nurturing the growth of capacity in municipal level institutions; (3) cultivation of socially inclusive and transparent decision-making processes; and (4) requiring accountability of those who execute policy to those who are affected by it. Van Doeveren (2011) identifies accountability, effectiveness and efficiency as the most widely employed criteria, followed in order by transparency, participation and the rule of law, anti-corruption and equity and, finally, coherence, responsiveness, consensus, regulatory quality, and absence of violence. Some of these criteria are recognized as being integral to procedures of risk governance that are emerging from the risk management community (e.g. participation/inclusion) but others are not yet included within that purview (van Asselt and Renn 2011).

Science-based risk governance is challenged by a gap that separates the values of science from broader public values (Boseman and Sarewitz 2011). Thus Meyer (2011) offers a general caution about conflating scientific progress with public good. He suggests that programs of risk governance that are underpinned by science may be blind to the scientific community's own self-interests and argues implicitly for the application of good governance standards to science applications that enter the arena of public policy.

Not everyone regards "good governance" as a self-evident positive contribution to public welfare. Following the "governmentality" rationale put forward by Michael Foucault (Foucault 2009), some scholars have pointed out that governments can shape citizenship to fit their needs just as citizens can shape governance. This can lead to certain governance arrangements becoming regarded as "good" in and of themselves rather than as reflections of particular (and contestable) political worldviews. Zeiderman has argued that coupling the vision of megacities as dystopias with the privileging of risk management as a public policy concern reflects a neo-liberal conception of governance rather than a disinterested application of knowledge (Zeiderman 2008). In a case study of Bogota he suggests that officially designated landslide risk zones "come to inhabit the territory of the poor" (p. 5) through the work of risk professionals whose judgments about vulnerability are made in the course of face to face interactions with underprivileged populations that give rise to subjective, as well as objective, assessments (Zeiderman 2012).

Rather than questioning the intrinsic value of good governance, other researchers have challenged the validity of its central premises. For example, Bhide (2006) questions the assumption that the kind of smooth dialog between top-level decision makers and grass roots constituencies on which good governance depends, can be assured by managerial means. In India, responsibilities for facilitating such a dialog are increasingly out-sourced by governments to NGOs and CBOs (Community Based Organizations) that are not representative of the urban poor and function instead as agents of conflict containment (Bhide 2006).

The reduction of natural hazards and disasters is today often included as a goal of good governance that contributes to better human security by improving decisions about development. A marriage between the principles of good governance and ongoing programs of disaster risk reduction and climate change adaptation would be a major step toward truly integrated risk governance.

A heterogeneous bundle of approaches and ideas cluster under the heading of risk management. Although a common commitment to orderly, human-centered, science-based principles of decision-making is generally shared within the risk management community, it is a broad church in which the adherents continue to differ about a range of issues including, the appropriate framing of environmental risks within broader policy contexts (e.g. safety, security, sustainability, social justice), the relative importance of humans and nature in the construction of risk, and the degree to which management of risk has become the master rubric of modern governance.

22.4 Governance of Disaster Risks in Megacities

During the past couple of decades programs to reduce worldwide disaster risks have been initiated and led by international science and public policy organizations (e.g. U.N. agencies). These have usually focused on improving the capacities of national government units that have formal responsibilities for risks but there have also been programs directed at improving global or local governance.

Some organizations have prioritized urban disaster risks but without singling out megacities for special attention; only a few have made megacity risks an explicit focus. Among groups that promote the reduction of disaster risks in general and/or urban risks in particular are the United National Development Programme (UNDP) (2004), United Nations Human Settlements Program (UN-Habitat) (2008, 2011), the Intergovernmental Program on Climate Change (IPCC) (2007, 2012), and the International Federation of Red Cross and Red Crescent Societies (IFRC) (2010). The United Nations International Strategy for Disaster Reduction (UNISDR) mentions megacities in its "Making Cities Resilient" campaign (http://www.unisdr.org/campaign/resilientcities) but the biggest cities have received more focused attention from a mixed set of non-governmental organizations, research institutes, business groups and metropolitan mayors. These include: the World Wildlife Fund (2009), the 'International Council for Local Environmental Initiatives' (ICLEI 2012), the Earthquake and Megacities Initiative (EMI) (2012), the United Nations University (Bohle and Warner 2008), the International Human Dimensions Program (IHDP) (Pelling 2011), the International Institute of Applied Systems Analysis (IIASA) (Berse and Reyes 2007), commercial insurance bodies (Munich Re 2007), other business research support groups (Siemens 2007; Maplecroft 2012), and a consortium of big city mayors (C40 Cities Climate Leadership Group 2011). Many of these groups are partnered with others, for which megacity risk reduction may not be a central continuing concern but one that they are willing to lend occasional support to at international conferences and other venues.

22.4.1 Institutional Responses to Megacity Complexity

The preferred cure for ills that beset megacities has usually been increased attention to scientific planning and management. Government leaders are advised to: lengthen planning horizons; consolidate fragmented territory; avoid information and reporting "stovepipes"; harness underemployed and/or dispersed resources; focus on a few priority matters; rely on scientific information, expert advice and skilled professional managers. Consolidation of territory and/or of administrative functions is a popular curative prescription. This follows from the widespread observation that the action space of many environmental risks transcends the boundaries of municipalities and departments and cannot be adequately managed without raising thorny issues of jurisdiction or authority (Gillen 2005). In the past situations like this have usually prompted calls for the creation of unified metropolitan governments or recognition that area-wide organizations, crafted for purposes of managing regional risks, that provide foundations for broad metropolitan governance initiatives leading to the creation of formal unified structures (Desfor and Kiel 2000; Bourdreau et al. 2006). More recently, this line of reasoning has been joined by another view that suggests cooperative agreements among existing governments offer alternative frameworks for addressing the complexities of large urbanized regions (Innes et al. 2011; Nelles 2012). Most of these debates have focused on cities in affluent countries (e.g. Toronto, Montreal, Los Angeles, Sydney) whose risk management capabilities would usually be considered among the world's best but it is clear that here too existing governance is being severely stressed.

There is no doubt that metropolitan risk governance could be significantly improved by the adoption of some rational science-based management procedures where these are presently missing. Types of adjustment to risk that typically require large amounts of scientific or technical knowledge if they are to perform optimally include: hazard prediction and warning systems; structural and infrastructural engineering works; and flood, storms, drought or earthquake insurance schemes. Well-resourced burgeoning megacities like Shanghai have invested heavily in such measures (Shanghai Meteorological Bureau 2010) but local critics often allege that they do not offer comprehensive protection to populations at risk (Shanghai Institute for Disaster Prevention and Relief 2012).

If one preferred risk governance strategy is to plan, consolidate and manage based on the application of accumulated scientific knowledge, an alternative evolving strategy is to improvise, decentralize and adapt using a combination of expert and vernacular knowledge. In this case improvisation does not mean withdrawing from science-based approaches but rather focuses on a particular phase of the scientific process namely, discovery through experimentation with emergent and provisional alternatives that are added to the roster of useful governance mechanisms as they prove their worth. A crucial part of these experiments involves empowering and expanding the number of stakeholders who play a role in formal decision-making. However, proof of worth tends to be equated with acceptance by the societal groups that previously had been thwarted in the achievement of their own goals, but capable of frustrating the goals of governing elites; there are as yet few proofs of worth as to the long term effectiveness of these novel governance arrangements in achieving resilience in the face of worsening risks.

Some scholars have labeled this alternative approach to risks adaptive governance, others consensual governance. It offers clear advantages to two very different constituencies among the critics of present risk governance. It addresses a need, long felt by existing public administrators, for a new framework that can accommodate the multiplicity of horizontal and vertical linkages among formal and informal groups that typically exist in large polities like those of megacities, but are not well integrated in conventional arrangements. Integration, coordination and networking are bywords in this discourse. It also offers new hope to campaigners for social justice that those who are most vulnerable to risks will have a greater say in shaping the means by which they are protected and will thereby help to strengthen their effectiveness as well as enriching the social fabric for all. With these preliminary observations in mind let us make a closer inspection of the current research oeuvre.

22.4.2 A Synopsis of the Published Literature

There is as yet no discrete field of study that focuses on the governance of megacity risks. Instead contributions come from a variety of sources that touch on different aspects of the topic. Among others, these include scholarship on governance, urbanization, the perception, analysis and communication of risks, vulnerability, sustainability and climate adaptation. It is helpful to identify some of the main features of this heterogeneous body of work.

With some notable exceptions (Inam 2005; Vale and Campanella 2005; Mitchell 1999) most publications on urban risk governance focus on small to medium-sized places (e.g. Pelling 2003; Pelling and Wisner 2008; Satterthwaite et al. 2011). Very few of the intergovernmental organizations that promote improved disaster risk reduction and the reform of risk governance single out megacities for special attention. Moreover, the great bulk of published literature on megacities addresses climate risks; although earthquakes and other geo-hazards have prompted important risk governance innovations, especially under the auspices of the Earthquakes and

Megacities Initiative, research and management of geo-risks is not well integrated with the much larger field of climate risk governance.

Early work on the hazardousness of megacities tended to focus on the lack of sound information about risk and vulnerability-forcing processes. Many assessments of risks and hazards in megacities have since been published although coverage is not complete for all cities nor is the data always comparable. What is available provides an acceptable basis for drawing up risk management plans and putting them into effect but there is general agreement that implementation has been patchy, slow and sometimes entirely absent.

Now researchers tend to lay the blame for much of what continues to ail megacities at the door of poor governance. Writing in 2006 researchers associated with the Earthquake and Megacities Initiative, who analyzed data from Manila, Mumbai, Kathmandu, Bogota, Quito, Tehran and Istanbul, commented: "Risk analysis and evaluation, particularly of earthquake risk, seems not to be the problem in the studied megacities. All of the cities … have a good understanding of the hazards they face, and major social and structural vulnerabilities have been identified; consequently, their associated risks are pretty much known and have been accurately mapped. However, serious limitations related to governance and knowledge management surfaced in most of the cities. Appropriate legal and regulatory frameworks and strong institutions and coordination need to be examined" (Fernandez et al. 2006, pp. 13–14). Another analyst observed that "the growing problems of Asian megacities in general and Dhaka in particular …(shows)… how governance has developed in a sectoral and national way rather than being place oriented (Talukder 2006, p. iii). More recently other investigators have drawn on a comparative analysis of climate change adaptation efforts in nine large cities (including the megacities of London and New York) to call for "a paradigm shift to move from the dominant focus on the adjustment of physical structures towards the improvement of planning tools and governance processes and structures themselves" (Birkmann et al. 2010). Many published studies identify and dissect the failings of megacity governance and there is no shortage of advice about needed reforms and best practices that are suitable for widespread adoption. Most advocate mainstreaming hazard management and climate adaptation throughout existing organs of government at all levels and encourage grass roots initiatives that would increase resilience among vulnerable groups.

The preceding information provides a contemporary overview of megacity hazards based on scientific sources that gather data about the occurrence of physical risks as well as humanitarian organizations and mass media that report immediate human emergencies. While certainly useful this body of knowledge does not tell us much about the dynamics of exposure, vulnerability, response and resilience in the megacities. These matters are mainly addressed in the scholarly research literature and data about them are only beginning to be collected by hazard management agencies. Given the spate of findings about megacity hazards that is now pouring forth, it is important to be aware of the research literature's strengths and limitations. In this respect three features stand out: (1) the profusion of case studies; (2) the salience of consensus-seeking collaboration and researchers; and (3) the prevalence of judgments that there is no standardized approach to governance; no one size that fits all.

22.4.2.1 A Profusion of Case Studies

Most of what is known about megacity risks is derived from case studies. There are many analyses of single kinds of risks in specific neighborhoods of particular megacities, sometimes from the perspective of especially vulnerable groups (Rashid 2000; Zoleta-Nantes 2002; Kelly 2003; Mgcuba and Vogel 2004; Maantay and Maroko 2009; Adelekan 2010; Chatterjee 2011). There are also all-hazards analyses of single at-risk cities like Los Angeles (Davis 1998) or Santiago de Chile (Heinrichs et al. 2012) as well as studies that employ a comparative approach across a range of cities (Mitchell 1999). Typically, the specialized comparative analyses focus on a single type of hazard (Orsi 2004; Varis 2006) or a specific type of vulnerability (Kovats and Akhtar 2008), or a particular component of risk response (e.g. reconstruction – Inam 2005; Vale and Campanella 2005; Gencer 2007; Edgington 2010), or one kind of megacity habitat (Nicholls et al. 2007; Hanson et al. 2011; Seto 2011). Others address the megacities of a particular continent, sub-region or country (Shaw et al. 2010; Hardoy and Lankao 2011; Hochrainer and Melcher 2011), or confine attention to the cluster of related hazards that is forced by climate change (De Sherbinin et al. 2007; Tanner et al. 2009; Mills et al. 2010). Some explore applications of methodological tools (e.g. remote sensing) (Beckel 2001; Taubenboch et al. 2011, 2012), or assess attitudes about risks, or institutional capacities for managing different sorts of risks – either separately or as part of integrated programs nested within broader strategies of sustainability, security and the like (Brauch 2003). In other words, researchers have minutely subdivided the universe of megacity risks and their findings are often quite narrowly specific.

In risk research it is common to find that a second round of comparative investigations follows an initial round of single case studies. The comparative work provides a foundation for broader generalizations about risks, vulnerabilities and adjustments that can act as guides to public policy. We now seem to be in an early phase of the second round of megacity risk studies. Most of these compare constructions of risk and vulnerability in different places but some also address choices of adjustments. Most frequently the latter are commentaries on possible future adjustments that might be suitable for adoption (Jha et al. 2011; Hanson et al. 2011), or anecdotal evidence about selected ongoing projects (Douglas et al. 2008) or cost-effectiveness studies that confine attention mostly to economic measures (World Bank 2010a, b). Almost all of the comparative work relies on collective analyses of single research case studies or policy planning documents for different cities that have already been published (Hunt and Watkiss 2011; Birkmann et al. 2010), or on rapid assessment surveys carried out by visiting international experts sometimes in conjunction with specialists from the targeted megacities (Tanner et al. 2009). A few employ opinion surveys of multiple experts in each city, including city mayors and their support staff (C40 Cities Climate Leadership Group 2011). However, detailed comparative investigations of risk perceptions and actions taken by local lay populations at risk, across a range of megacities, are very rare (Wisner 2003). Insofar as these types of informants are usually the targets of good governance initiatives there is a clear information gap between top-down and bottom-up perspectives. Taken together with the spottiness of single case studies, the uneven coverage of comparative analyses strongly suggests that a thorough appraisal of megacity risk as a whole is still lacking.

22.4.2.2 Consensus-Seeking Collaboration

Consensus is one of the core principles of good governance, though not one that has been successfully codified and put into practice in most public policy venues; it is also a tenet of much collaborative research on environmental risks. There has been an upsurge in the number of studies that are produced by very large teams of researchers and writers working under the auspices of organizations whose methods and goals are often standardized and programmatic (United Nations Development Programme 2004; UN-Habitat 2011; IPCC 2012). Moreover, multiple co-authors are common for single scholarly papers on risk, sometimes exceeding several dozens (Baklanov et al. 2010).

Publications like these are doubtless motivated by desires to make order out of the vast and heterogeneous scientific literature on environmental risk and the contributors' involvement is reinforced by long-term trends in the commissioning and funding of research that favor collaboration across disciplinary boundaries. But they are also a bid to gain the kind of legitimacy and authority that attaches to pronouncements on publically controversial topics by large numbers of collaborating experts. The resulting products are usually admirable in their scope and ambitions for synthesis, but the striving for consensus that animates them, and the editorial procedures that reinforce the quest, leave little room for discussion of contradictions and ambiguities that are also part of the picture, especially in megacities. Insofar as ambiguity is one of the distinguishing characteristics of risk, the accommodation of ambiguities poses great difficulties for management and governance. It is also striking that consensus among scientists has not yet proven to be a compelling rationale for political actions that might regulate human contributions to the risks of climate change (Donner 2011; Holmes 2012). The EU's phasing out of risk regulation-making based on consensus-style procedures among a limited number of privileged constituencies in favor of a more open, but potentially adversarial, process that involves a much wider range of stakeholders, signals uncertainty about the boundaries within which consensus might be sought (Lofstedt et al. 2011).

Despite the difficulties of maintaining consensus as a principle of governance the trend towards more intricate forms of collaboration in risk governance is likely to continue as the full capabilities of electronic information systems are tapped to permit inputs from an enormous array of potential lay informants. Since the Haiti earthquake of 2010 crowd sourcing of information for purposes of emergency management has vastly accelerated and it is already clear that it will become necessary to develop new procedures for verifying, sorting and organizing inputs from people who are reporting from their own local communities (Heinzelman and Waters 2010; Heipke 2010).

Consensus-seeking as a style of collaboration in research and governance has obvious merits but it is not a panacea for managing risks. It may be appropriate to seek consensus about some matters but not others. Scale magnifies the difficulties of achieving consensus and megacities are at the upper end of the scale of community complexity. The technical problems of handling vast quantities and new kinds of information within a risk decision-making system are formidable, and the existence of ambiguity poses especially tough problems. There is also the question of whether consensus-seeking tends to marginalize scientific ideas that are disapproved but not disproved. In addition, it has also been argued that U.S. engagement with some mega-problems (e.g. climate change) has benefitted from the separation of science from policy-making because of the difficulties of simultaneously achieving consensus between the membership of both groups (Clark and Dickson 2001, p. 277). This suggests that efforts to recombine these two aspects of risk management in the service of risk governance may face similar problems. Agents of risk governance confront knotty information problems and may have to work very hard to avoid putting science and policy on a collision course.

22.4.2.3 Tailor Made Solutions

A third frequent feature of research on megacity risks is the prevalence of findings that indicate each place requires tailor made management and governance arrangements (Kotter and Friesecke 2008; Nuissl et al. 2012). At least one seasoned observer has opined that the best examples of megacity risk governance are probably not replicable elsewhere (Corfee-Merlot et al. 2011). Although real world applications often require modification to fit local circumstances, the lack of robust generic findings suggests that research on risk governance in megacities is still at an early stage. In time a menu of standardized governance arrangements may become available from which cities might choose (and doubtless modify) those that are best suited to their needs.

22.5 Conclusions

Megacities are recent, fast developing, social formations that both generate, and are affected by, new kinds of risks and vulnerabilities as well as different modifications and permutations of existing risks and vulnerabilities. They pose particular challenges for public policy, planning and management because the high-pressure context in which public decision-making takes place requires rapid application of research knowledge to problems that are growing worse while at the same time reinventing the institutions that are charged with carrying out that task. In short, the structures, instruments and substance of megacity risk governance are simultaneously in flux.

Research on the governance of megacity risks is proceeding rapidly and risk management institutions are receptive to inputs of scientific knowledge as well as other types of (non-scientific) information. However the field is still at a youthful stage of development that exhibits a variety of problems and limitations.

Social scientific investigations of megacity risk governance are heavily reliant on case studies that make for a rich array of findings but also create difficulties in establishing generic conclusions. The awkward fit between idiomatic case study-based research on risks in specific urban places and generic simulation model-based research on risks associated with atmospheric change at global and sub-global scales is a case in point. The continuing divergence of risk experience between privileged and disadvantaged megacities also makes it difficult to foster a single integrated body of research findings about megacity risks. Technocratic interpretations of risk governance vie with political interpretations; the dominant existing pragmatic approaches to governance are challenged by ethically charged alternatives. Finally, the prevalence of studies that conclude by endorsing a case-by-case approach to risk management signals a failure to devise convincing general explanations that would have broad applicability.

Discourses of risk governance have performed a valuable service by querying long held assumptions about the immutability of existing institutional arrangements for choosing responses to environmental threats. They have encouraged us to think about new alternatives that may be more robust, not just in the face of new risks and worsening familiar ones but also, and more importantly, in societies that are remaking themselves and their vulnerabilities in unprecedented ways. Risk scientists and risk managers who once might have operated within taken-for-granted institutional settings now face the additional task of discovering new roles for themselves in the unfamiliar societal circumstances of the twenty-first century. Nowhere is this engagement more appropriate and more needed than in the world's burgeoning megacities.

Research on the governance of megacity risks has accelerated in recent years and the knowledge base has grown correspondingly. Researchers are now moving beyond case studies into an era of systematic comparative investigations across many different megacities. These are beginning to promote conversations among the different research communities (Birkmann et al. 2010; Solecki et al. 2011). Calls for new kinds of research on hazards and disasters are appearing with increasing frequency, signaling the existence of barriers to further progress within existing strategies (Burton 2010; Moser 2010; Siembieda 2012).

The present assessment of megacity risk studies has sought to call attention to the barriers and the need for new departures. It suggests that a particular set of comparative research initiatives would now be timely. These should address enduring contradictions that encumber the field and have been so identified in this chapter. Such contradictions can be stated as a series of research questions. Are privileged megacities and disadvantaged ones on converging, parallel or divergent trajectories of risk? How salient might economically grounded rationales for risk management become in cities that are increasingly organized to function as points of wealth accumulation and exchange in a global system and how can those functions best be reconciled with humanitarian and other values? To what extent do regionally different experiences of risk and disaster influence the formulation of risk governance theories and methods that are candidates for global application? In what ways might political perspectives on megacity risks be incorporated into theories and methods of risk governance? Finally, how shall the joint application of criteria for achieving sustainability (an ongoing project of reform in existing risk institutions) and good governance (a evolutionary new concept) be arranged in a megacity risk governance system? Although these questions are posed here in the context of megacity risks they have broad applicability within the corpus of risk governance research.

References

Adelekan, I. 2010. Vulnerability of poor urban coastal communities to flooding in Lagos, Nigeria. *Environment and Urbanization* 22(2): 433–450.

Adikari, Y., R. Osti, and T. Noro. 2010. Flood-related disaster vulnerability: An impending crisis of megacities in Asia. *Journal of Flood Risk Management* 3(3): 185–191.

Alam, M., and G. Rabbani. 2009. Vulnerabilities and responses to climatic change for Dhaka. In *Adapting cities to climate change: Understanding and addressing the development challenges*, ed. J. Bicknell, D. Dodman, and D. Satterthwaite, 93–110. London: Earthscan.

Alexander, D. 2006. Globalization of disaster: Trends, problems and dilemmas. *Journal of International Affairs* 59(2): 1–22.

Baklanov, A., M. Lawrence, S. Pandis, A. Mahura, S. Finardi, N. Moussiopoulos, M. Beekmann, P. Laj, L. Gomes, J. Jaffrezo, A. Borbon, I. Coll, V. Gros, J. Sciare, J. Kukkonen, S. Galmarini, F. Giorgi, S. Grimmond, I. Esau, A. Stohl, B. Denby, T. Wagner, T. Butler, U. Baltensperger, P. Builtjes, D. van den Hout, H. van der Gon, B. Collins, H. Schluenzen, M. Kulmala, S. Zilitinkevich, R. Sokhi, R. Friedrich, J. Theloke, U. Kummer, L. Jalkinen, T. Halenka, A. Wiedensholer, J. Pyle, and W. Rossow. 2010. MEGAPOLI: Concept of multi-scale modelling of megacity impact on air quality and climate. *Advances in Science and Research* 4: 115–120.

Bauman, Z. 2000. *Liquid modernity*. New York: Polity Press.

Beck, U. 1992. *Risk society: Towards a new modernity*. New Delhi: Sage.

Beckel, L. (ed.). 2001. *Megacities: The European Space Agency's contribution to a better understanding of a global challenge*. Salzburg: Geospace-Verlag.

Berse, K., and M. Reyes. 2007. Enhancing megacities disaster resilience: The 3cd program in Metro Manila and Kathmandu. Presentation at 7th *IIASA-DPRI Forum on integrated disaster risk management*, Stresa, Italy, September 19–21. http://www.iiasa.ac.at/Research/RAV/conf/IDRiM07/Papers/Berse.pdf. Accessed 1 Oct 2012.

Bhide, A. 2006. Governance, civil society, and partnership: A view from the periphery. *Social Development Issues* 28(1): 53–68.

Birkmann, J., M. Garschagen, F. Kraas, and N. Quang. 2010. Adaptive urban governance: New challenges for the second generation of urban adaptation strategies to climate change. *Sustainability Science* 5: 185–206.

Boholm, A., H. Corvellec, and M. Karlsson. 2012. The practice of risk governance: Lessons from the field. *Journal of Risk Research* 15(1): 1–20.

Bohle, H.-G., and K. Warner. 2008. *Megacities: Resilience and social vulnerability*, SOURCE, no. 10. Bonn: United Nations University, Institute for Environment and Human Security, Munich Re Foundation.

Boseman, B., and D. Sarewitz. 2011. Public value mapping and science policy evaluation. *Minerva* 49(1): 1–23.

Bourdreau, J.-A., P. Hamel, B. Jouve, and R. Keil. 2006. Comparing metropolitan governance: The cases of Montreal and Toronto. *Progress in Planning* 66: 7–59. http://aitec.reseau-ipam.org/IMG/pdf/progress_in_planning.pdf. Accessed 23 Sept 2014.

Brauch, H. 2003. Natural disasters in the Mediterranean (1900–2001): From disaster response to disaster preparedness. In *Security and environment in the Mediterranean: Conceptualizing security and environmental conflicts*, ed. H. Brauch, H. Liotta, A. Marquina, P. Rogers, and R. Selim, 863–906. Berlin: Springer.

Brookings Institution. 2010. *The New Orleans index at five*, ed. A. Liu and A. Plyer. http://www.brookings.edu/reports/2007/08neworleansindex.aspx.

Burton, I. 2010. Forensic disaster investigations in depth: A new case study model. *Environment* 52(5): 36–51.

Burton, I., R. Kates, and G. White. 1978. *The environment as hazard*. New York: Oxford University Press (2nd ed., 1993. New York: Guilford Press).

C40 Cities Climate Leadership Group. 2011. *Climate action in megacities: C40 cities baseline and opportunities* Version 1.0. ARUP. http://c40citieslive.squarespace.com/storage/ARUP%20C40%20Basline%20Report.pdf. Accessed 1 Oct 2012.

Chatterjee, M. 2011. Flood loss redistribution in third world mega city: The case of Mumbai. In *Coping with global environmental change, disasters and security threats, challenges, vulnerabilities and risks*, ed. H.G. Brauch et al., 603–612. Berlin: Springer.

Clark, W., and N. Dickson. 2001. Civic science: America's encounter with global environmental risks. In *Learning to manage global environmental risks: A comparative history of social responses to climate change, ozone depletion and acid rain*, ed. W.C. Clark, N.C. Dickson, J. Jaeger, and J. van Eindhoven, 259–294. Cambridge: MIT Press.

Clarke, L. 1999. *Mission improbable: Using fantasy documents to tame disaster*. Chicago: University of Chicago Press.

Clarke, L. 2006. *Worst cases: Terror and catastrophe in the popular imagination*. Chicago: University of Chicago Press.

Committee on Earthquake, Engineering, National Research Council. 1992. *The economic consequences of a catastrophic earthquake: Proceedings of a forum*. Washington, DC. http://www.nap.edu/catalog.php?record_id=2027. Accessed 1 Oct 2012.

Corfee-Merlot, J., I. Cochran, S. Hallegatte, and P. Teasdale. 2011. Multilevel risk governance and urban policy. *Climatic Change – Special Issue: Understanding Climate Change Impacts, Vulnerability and Adaptation at City Scale* 104(1): 169–197.

Cuaresma, J., J. Hlouskova, and M. Obersteiner. 2008. Natural disasters as creative destruction? Evidence from developing countries. *Economic Inquiry* 46(2): 214–226.

Davidson, R., and H. Shah. 1997. *An urban earthquake disaster risk index*. Report no. 121. John A. Blume Earthquake Engineering Research Center, Stanford University. https://blume.stanford.edu/sites/default/files/TR121_Davidson.pdf. Accessed 1 Oct 2012.

Davis, M. 1998. *Ecology of fear: Los Angeles and the imagination of disaster*. New York: Metropolitan Books.

De Sherbinin, A., A. Schiller, and A. Pulsipher. 2007. The vulnerability of global cities to climate hazards. *Environment & Urbanization* 19(1) April. http://www.ciesin.org/documents/vulofglob_contactshtml.pdf. Accessed 1 Oct 2012.

Deere-Birkbeck, C. 2009. Global governance in the context of climate change: The challenges of increasingly complex risk parameters. *International Affairs* 85(6): 1173–1194.

Delhi Disaster Management Authority. 2014. *Floods*. Government of the National Capital Territory of Delhi. http://delhi.gov.in/wps/wcm/connect/DOIT_DM/dm/home/vulnerabilities/hazards/floods. Accessed 23 Sept 2014.

Desfor, G., and R. Kiel. 2000. Every river tells a story: The Don River (Toronto) and the Los Angeles River (Los Angeles) as articulating landscapes. *Journal of Environmental Policy and Planning* 2(1): 5–23.

Ding, Y. 2008. *Handling emergency in mega-cities: The case of South China under snow storm in 2008*. Unpublished paper, Institute for Social and Economic Geography, Leuven. http://www.irmgard-coninx-stiftung.de/fileadmin/user_upload/pdf/urbanplanet/identities/ws1/072%20Ding. Accessed 1 Oct 2012.

Donner, S. 2011. Making the climate a part of the human world. *Bulletin of the American Meteorological Society* 92(10): 1297–1302.

Douglas, M., and A. Wildavsky. 1982. *Risk and culture: An essay on the selection of technological and environmental dangers*. Berkeley: University of California Press.

Douglas, I., K. Alam, M. Maghenda, Y. McDonnell, L. Mclean, and J. Campbell. 2008. Unjust waters: Climate change, flooding and the urban poor in Africa. *Environment and Urbanization* 20(1): 187–205.

Earthquake and Megacities Initiative. 2012. *Partnering with cities to reduce urban risk in a changing climate*. http://www.emi-megacities.org/home/. Accessed 1 Oct 2012.

Edgington, D. 2010. *Reconstructing Kobe: The geography of crisis and opportunity*. Vancouver: University of British Columbia Press.

Fernandez, J., F. Bendimerad, S. Mattingly, and J. Buika. 2006. Comparative analysis of disaster risk management practices in seven megacities. Paper presented at the *2nd Asian conference on earthquake engineering*. Manila, March 10–11. http://emi.pdc.org/DRMlibrary/General/Comparative-analysis-DRM-in-7-megacities.pdf. Accessed 1 Oct 2012.

Foucault, M. 2009. *Security, territory, population: Lectures at the College de France 1977–1978*, ed. M. Foucault, M. Sennelart, and F. Ewald. London: Macmillan.

Gencer, E.A. 2007. *Vulnerability in hazard-prone megacities: An overview of global trends and the case of the Istanbul metropolitan area*. Draft from summer institute on social vulnerability. http://www.ehs.unu.edu/file/get/4034. Accessed 1 Oct 2012.

Giddens, A. 1990. *The consequences of modernity*. Cambridge: Polity.

Gillen, M. 2005. Urban governance and vulnerability: Exploring the tensions and contradictions in Sydney's response to bushfire threat. *Cities* 22(1): 55–64.

Government of Maharashtra. 2005. *Maharashtra floods 2005*. Department of Relief and Rehabilitation. http://mdmu.maharashtra.gov.in/pdf/Flood/statusreport.pdf. Accessed 1 Oct 2012

Habermas, J. 1999. Learning by disaster: A diagnostic look back on the short 20th century. *Constellations* 5(3): 307–320.

Hadfield, P. 1995. *The coming Tokyo earthquake: Sixty seconds that will change the world*. Boston: C. E. Tuttle Company.

Hanson, S., R. Nicholls, N. Ranger, S. Hallegatte, J. Corfee-Morlot, C. Herweller, and J. Chateau. 2011. A global ranking of port cities with high exposure to climate extremes. *Climatic Change* 104: 89–111.

Hardoy, J., and P. Lankao. 2011. Latin American cities and climate change: Challenges and options to mitigation and adaptation responses. *Current Opinion in Environmental Sustainability* 3(1): 158–163.

Heinrichs, D., K. Krellenberg, B. Hansjurgens, and F. Martinez (eds.). 2012. *Risk habitat megacity*. Berlin/Heidelberg: Springer.

Heinzelman, J., and C. Waters. 2010. *Crowdsourcing crisis information in disaster-affected Haiti*. Special report 252. Washington DC: U.S. Institute of Peace. http://www.usip.org/files/resources/SR252%20-%20Crowdsourcing%20Crisis%20Information%20in%20Disaster-Affected%20Haiti.pdf. Accessed 1 Oct 2012.

Heipke, C. 2010. Crowdsourcing geospatial data. *SPRS Journal of Photogrammetry and Remote Sensing* 65(6): 550–557.

Hirschleifer, J. 1993. Disaster and recovery. In *The fortune encyclopedia of economics*, ed. D.R. Henderson. Warner Books. Reprinted in the Concise Encyclopedia of Economics. (2007). http://www.econlib.org/library/Enc1/DisasterandRecovery.html.

Hochrainer, S., and R. Mechler. 2011. Natural disaster risk in Asian megacities: A case for risk pooling? *Cities* 28(1): 53–61.

Holmes, B. 2012. Planting seeds of climate doubt. *New Scientist* 213(2): 6–7.

Hunt, A., and P. Watkiss. 2011. Climate change impacts and adaptation in cities: A review of the literature. *Climatic Change* 104(1): 13–49.

ICLEI – Local Governments for Sustainability. 2012. *Third global forum on urban resilience and adaptation*. Bonn, Germany, May 12–15. http://resilient-cities.iclei.org/bonn2012/home/.

Inam, A. 2005. *Planning for the unplanned: Recovering from crises in megacities*. New York: Routledge.

Innes, J., D. Booher, and S. Di Vittorio. 2011. Strategies for megaregion governance. *Journal of the American Planning Association* 77(1): 55–67.

Integrated Risk Governance Project. 2010. *Science plan. Potsdam and Beijing: International human dimensions programme*. http://irg-project.org. Accessed 1 Oct 2012.

Integrated Risk Governance Project. 2011. *Integrated risk governance project: Annual report 2011. Potsdam and Beijing: International human dimensions programme*. http://irg-project.org. Accessed 1 Oct 2012.

Intergovernmental Panel of Climate Change. 2007. *Climate change 2007: Impacts, adaptation and vulnerability. Working group II contribution to the fourth assessment report*. New York: Cambridge University Press. http://www.ipcc-wg2.gov/publications/AR4/index.html. Accessed 1 Oct 2012.

Intergovernmental Panel on Climate Change. 2012. *Managing the risks of extreme events and disasters to advance climate change adaptation*, ed. C.B. Field, V. Barros, T.F. Stocker, D. Qin, D.J. Dokken, K.L. Ebi, M.D. Mastrandrea, K.J. Mach, G.-K. Plattner, S.K. Allen, M. Tignor, and P.M. Midgley. A special report of working groups I and II of the intergovernmental panel on climate change. Cambridge: Cambridge University Press.

International Federation of Red Cross and Red Crescent Societies. 2010. *World disasters report 2010: Focus on urban risk*. Geneva. www.ifrc.org/Global/Publications/disasters/WDR/WDR2010-full.pdf. Accessed 1 Oct 2012.

Jha, A., R. Bloch, and J. Lamond (eds.). 2011. *Cities and flooding: A guide to integrated flood risk management for the 21st century*. Washington, DC: World Bank and Global Facility for Disaster Reduction and Recovery. http://reliefweb.int/sites/reliefweb.int/files/resources/Full_Report_3518.pdf. Accessed 1 Oct 2012.

Kasperson, R., O. Renn, P. Slovic, H. Brown, J. Emel, R. Gobel, J. Kasperson, and S. Ratick. 1988. The social amplification of risk: A conceptual framework. *Risk Analysis* 8(2): 177–187.

Kates, R. 1978. *Risk assessment of environmental hazard*. New York: Wiley.

Kelly, C. 2003. *Acute food insecurity in mega-cities: Issues and assistance options*, Working paper 7. London: Benfield Hazard Research Centre.

Khanna, P. 2010. Beyond city limits. *Foreign Policy* 181(September/October): 120–128.

Klein, N. 2007. *The shock doctrine: The rise of disaster capitalism*. New York: Metropolitan Books.

Kotter, T., and F. Friesecke. 2008. *Developing urban indicators for managing Mega Cities*. Department of Urban Planning and Real Estate Management. Institute of Geodesy and Geoinformation, University of Bonn.

Kovats, S., and R. Akhtar. 2008. Climate, climate change and human health in Asian cities. *Environment and Urbanization* 20(1): 165–175.

Leichenko, R., and K. O'Brien. 2008. *Environmental change and globalization: Double exposures*. New York: Oxford University Press.

Lewis, M. 1989. *How a Tokyo earthquake could devastate Wall Street and the world economy, Manhattan*. Reprinted in Vanity Fair.com, March 18, 2011. http://www.vanityfair.com/business/features/2011/03/michael-lewis-japan-201103. Reprinted as a book by W. W. Norton and Company, New York, 1991 and 2011.

Lewis, D., and J. Mioch. 2005. Urban vulnerability and good governance. *Journal of Contingencies and Crisis Management* 13(2): 50–53.

Lindner, C. 2005. Willa Cather, Daniel Libeskind, and the creative destruction of Manhattan. *The Journal of American Culture* 28(1): 114–121.

Lofstedt, R., F. Bouder, J. Wardman, and S. Chakraborty. 2011. The changing nature of communication and regulation of risk in Europe. *Journal of Risk Research* 14(4): 409–429.

Luhmann, N. 2008. *Risk: A sociological theory* (with a new introduction by Nico Stehr and Gotthard Bechmann). New Brunswick: Transaction Publishers.

Maantay, J., and A. Maroko. 2009. Mapping urban risk: Flood hazards, race, & environmental justice in New York. *Applied Geography* 29(1): 111–124.

Maplecroft. 2012. *Climate change and environment risk atlas*, 4th ed. http://maplecroft.com/about/news/ccvi_2012.html. Accessed 1 Oct 2012.

McDonald, F. 2011. Afterword. *Environmental Hazards* 10(1): 93–98.

Menzel, B. 2009. Vulnerability and resilience in natural disasters: A marketing and public policy perspective. *Journal of Public Policy and Marketing* 28(1): 114–123 (Special issue).

Meyer, R. 2011. The public values failures of climate science in the US. *Minerva: A Review of Science, Learning and Policy* 49(1): 47–70.

Mgcuba, S.K., and C. Vogel. 2004. Living with environmental risks and change in Alexandra Township. *South African Geographical Journal* 86(1): 30–39.

Mills, G., H. Cleugh, R. Emmanuel, W. Endlicher, E. Erell, G. McGranahan, E. Ng, A. Nickson, J. Rosenthal, and K. Steemer. 2010. Climate information for improved planning and management of mega cities (Needs Perspective). *Procedia Environmental Sciences* 1: 228–246.

Mitchell, J. (ed.). 1999. *Crucibles of hazard: Megacities and disasters in transition*. Tokyo: United Nations University Press.

Mitchell, D. 2003. *The right to the city: Social justice and the fight for public space*. New York: Guilford Press.

Mitchell, J. 2006. The primacy of partnership: Scoping a new national disaster recovery policy. In Shelter from the storm: Repairing the national emergency management system after Katrina, ed. W.L. Waugh. *Annals of the American Academy of Political and Social Science* 604(1): 228–255.

Moser, S. 2010. Now more than ever: The need for more societally relevant research on vulnerability and adaptation to climate change. *Applied Geography* 30(4): 464–474.

Munich Re Research Group. 2007. *Natural hazard risk index for megacities*. Munich. Reprinted as a poster in Social Vulnerability, UNU Summer Academy 2007 Megacities as hotspots of risk. http://www.munichre-foundation.org/NR/rdonlyres/5FC116A6-B33F-4370-A047-71AE170339B3/0/PosterLoewNaturalHazardRiskIndex.pdf. Accessed 1 Oct 2012.

Nelles, J. 2012. Cooperation and capacity? Exploring the sources and limits of city-region governance partnerships. *International Journal of Urban and Regional Research*. http://onlinelibrary.wiley.com/doi/10.1111/j.1468-2427.2012.01112.x/pdf.

New York City Consortium for Earthquake Loss Mitigation. 2005. *Full report: Earthquake risk mitigation in the New York, New Jersey and Connecticut Region*. http://www.nycem.org/techdocs/FinalReport/tocnew.asp. Accessed 1 Oct 2012.

New York City Panel on Climate Change. 2010. *Climate change adaptation in New York City: Building a risk management response*. Report no. 1196. New York: Annals of the New York Academy of Sciences. http://www.nyas.org/publications/annals/Detail.aspx?cid=ab9d0f9f-1cb1-4f21-b0c8-7607daa5dfcc. Accessed 1 Oct 2012.

Nicholls, R.J., S. Hanson, C. Herweijer, N. Patmore, S. Hallegatte, J. Corfee-Morlot, J. Chateau, and R. Muir-Wood. 2007. *Ranking of the world's cities most exposed to coastal flooding today and in the future*, OECD environment working paper no. 1, 1. Paris: OECD.

Nuissl, H., C. Hohnke, M. Lukas, G. Duran, and C.R. Seeger. 2012. Megacity governance: Concepts and challenges. In *Risk habitat megacity*, ed. D. Heinrichs, K. Krellenberg, B. Hansjurgens, and F. Martinez, 87–108. Heidelberg: Springer.

Office of the United Nations High Commissioner for Human Rights. 2007. *Good governance practices for the protection of human rights*. HR/PUB/07/4. New York/Geneva. http://www.ohchr.org/Documents/Publications/GoodGovernance.pdf. Accessed 1 Oct 2012.

Orsi, J. 2004. *Hazardous metropolis: Flooding and urban ecology in Los Angeles*. Berkeley: University of California Press.

Pelling, M. 2003. *The vulnerability of cities: Natural disasters and social resilience*. London: Earthscan.

Pelling, M (ed.). 2011. *Megacities and the coast: Transformation for resilience. A preliminary review of knowledge, practice and future research. Land-ocean interactions in the coastal zone, a joint IGBP-IDHP core project*. Geesthacht, Germany. http://www.loicz.org/imperia/md/content/loicz/hotspots/urbanization/Megacities_and_the_coast_report_4_6_2011.pdf. Accessed 1 Oct 2012.

Pelling, M., and B. Wisner (eds.). 2008. *Disaster risk reduction: Cases from urban Africa*. London: Earthscan.

Perrow, C. 1999. *Normal accidents: Living with high-risk technologies*. Princeton: Princeton University Press.

Perrow, C. 2007. *The next catastrophe: Reducing our vulnerabilities to natural, industrial, and terrorist disasters*. Princeton: Princeton University Press.

Ranger, N., S. Hallegatte, S. Bhattacharya, M. Bachu, S. Priya, K. Dhore, F. Rafique, P. Mathur, N. Naville, F. Henriet, C. Herweijer, S. Pohit, and J. Corfee-Morlot. 2011. An assessment of the potential impact of climate change on flood risk in Mumbai. *Climatic Change* 104(1): 139–167.

Rashid, S. 2000. The urban poor in Dhaka City: Their struggles and coping strategies during the floods of 1998. *Disasters* 24(3): 240–254.

Rayner, S. 1992. Cultural theory and risk analysis. In *Social theories of risk*, ed. S. Krimsky and D. Golding, 27–64. Westport: Praeger.

Renn, O. 2008. *Risk governance: Coping with uncertainty in a complex world*. London: Earthscan.

Rose, A. 2007. Economic resilience to natural and man-made disasters: Multidisciplinary origins and contextual dimensions. *Environmental Hazards* 7(4): 383–398.

Rossi, U., and A. Vanolo. 2012. *Urban political geographies: A global perspective*. Los Angeles: Sage.

Satterthwaite, D. et al. 2011. Environment and Urbanization 2(2). Special issue on disasters. http://eau.sagepub.com.proxy.libraries.rutgers.edu/content/23/2.toc. Accessed 23 Sept 2014.

Schumpeter, J.A. 1942. *Capitalism, socialism and democracy*. New York: Harper and Brothers.

Seto, K. 2011. Exploring the dynamics and migration to mega-delta cities in Asia and Africa: Contemporary divers and future scenarios. *Global Environmental Change* 21S: S94–S107.

Shah, H. 1994. *Casualty, damage estimates of great quakes updated*. Press Release. Stanford University. http://news.stanford.edu/pr/94/940815Arc4158.html Accessed 1 Oct 2012.

Shanghai Institute for Disaster Prevention and Relief. 2012. *General introduction of SIDPR*. http://www.idpr.sh.cn/en/. Accessed 1 Oct 2012.

Shanghai Meteorological Bureau. 2010. *Overview of Shanghai multi-hazard early warning system and the role of meteorological services*. Powerpoint presentation, March 23, Shanghai. http://www.wmo.int/pages/prog/drr/events/MHEWSCostaRica/Docs/Session%202/Shanghai/Shanghai_MHEWS_CostaRica.pdf. Accessed 1 Oct 2010.

Shaw, R., J. Pulhin, and J. Pereria (eds.). 2010. *Climate change adaptation and disaster risk reduction: An Asian perspective*. Bingley: Emerald.

Shi, P., W. Fang, W. Dong, N. Li, W. Xu, and S. Chen. 2009. *Experiences and lessons of large-scale disaster governance in China: Perspective to the response of freezing rain and snowstorm disaster in South China, Early 2008, Beijing*. Integrated risk governance report no. 2, catastrophe governance case analysis II, IDHP 2009 open meeting, April 26–30, Bonn. http://www.irisknet.cn/riskdb/globalrisk/juzai/7.1.pdf. Accessed 1 Oct 2012.

Siembieda, W. 2012. Transactions and friction as concepts to guide disaster recovery policy. *International Journal of Disaster Risk Science* 3(1): 38–44.

Siemens, A.G. 2007. *Megacity challenges: A stakeholder perspective. A research project conducted by GlobeScan and MRC McLean Hazel, sponsored by Siemens*. Munich. http://www.siemens.com/entry/cc/features/.../study_megacities_en.pdf. Accessed 1 Oct 2012.

Slovic, P. 1992. Perceptions of risk: Reflections of the psychometric paradigm. In *Social theories of risk*, ed. S. Krimsky and D. Golding, 117–152. Westport: Praeger.

Slovic, P. 1999. Trust, emotion, sex, politics and science: Surveying the risk-assessment battlefield. *Risk Analysis* 19(4): 689–701.

Solecki, W., R. Leichenko, and K. O'Brien. 2011. Climate change adaptation strategies and disaster risk reduction in cities: Connections, contentions, and synergies. *Current Opinion in Environmental Sustainability* 3: 135–141.

Stern, N. 2007. *The economics of global climate change*. Cambridge: Cambridge University Press.

Swyngedouw, E. 2007. Impossible 'sustainability' and the post-political condition. In *The sustainable development paradox: Urban political economy in the United States and Europe*, ed. R. Kruger and D. Gibbs, 255–274. London: Belhaven Press.

Talukder, S. 2006. *Managing megacities: A case study of metropolitan regional governance for Dhaka*. Ph.D. thesis, Murdoch University. http://researchrepository.murdoch.edu.au/330/. Accessed 1 Oct 2012.

Tanner, T., T. Mitchell, E. Polack, and B. Guenther. 2009. *Urban governance for adaptation: Assessing climate change resilience in ten Asian cities*, Working paper no. 315. London: Institute for Development Studies.

Taubenboch, H., M. Wurm, M. Netzband, H. Zwenzner, A. Roth, A. Rahman, and S. Dech. 2011. Flood risks in urbanized areas – Multi-sensoral approaches using remotely sensed data for risk assessment. *Natural Hazards and Earth System Sciences* 11: 431–444.

Taubenbock, H., T. Esch, A. Felbier, M. Wiesner, A. Roth, and S. Dech. 2012. Monitoring urbanization in mega cities from space. *Remote Sensing of the Environment* 117: 162–176.
Tierney, K.J. 2007. From the margins to the mainstream? Disaster research at the crossroads. *Annual Review of Sociology* 33: 503–525.
U.S. Federal Emergency Management Agency. 2004. *Earthquake loss estimation methodology (HAZUS)*. Washington, DC: National Institute of Building Sciences.
United Nations. 2012. *We can end poverty 2015: Millennium development goals*. New York. http://www.un.org/millenniumgoals/bkgd.shtml. Accessed 1 Oct 2012.
United Nations Development Programme. 2004. *Reducing disaster risk: A challenge for development – A global report*. New York: Bureau for Crisis Prevention and recovery. http://www.preventionweb.net/files/1096_rdrenglish.pdf. Accessed 1 Oct 2012.
United Nations General Assembly. 2000. *Millennium declaration*. A/Res/55/2. September 18. http://www.un.org/millennium/declaration/ares552e.pdf. Accessed 23 Sept 2014.
United Nations Human Settlements Program (UN-Habitat). 1999. *Global campaign on urban governance*. Nairobi, Kenya. http://www.unhabitat.org/content.asp?typeid=19&catid=25&cid=2097 Accessed 1 Oct 2012.
United Nations Human Settlements Program (UN-Habitat). 2008. *State of the world's cities 2010–2011: Bridging the urban divide*. London: Earthscan. http://www.unhabitat.org/pmss/listItemDetails.aspx?publicationID=3016. Accessed 1 Oct 2012.
United Nations Human Settlements Program (UN-Habitat). 2011. *Cities and climate change: Global report on human settlements*. London: Earthscan. http://www.unhabitat.org/content.asp?typeid=19&catid=555&cid=9272.
United Nations Radio. 2012. *2011 peak year of disaster losses in the world*. Comments by Margareta Wahlström, Special representative of the Secretary-General for Disaster Risk Reduction. http://www.unmultimedia.org/radio/english/2012/03/2011-peak-year-of-disaster-losses-in-the-world/. Accessed 10 Oct 2012.
Vale, L., and T. Campanella. 2005. *The resilient city: How modern cities recover from disaster*. New York: Oxford University Press.
Van Asselt, M., and O. Renn. 2011. Risk governance. *Journal of Risk Research* 14(4): 431–449.
Van Doeveren, V. 2011. Rethinking good governance: Identifying common principles. *Public Integrity* 13(4): 301–318.
Varis, O. 2006. Megacities, development and water. *Water Resources Development* 22(2): 199–225.
Wisner, B. 2003. Disaster risk reduction in megacities: making the most of human and social capital. In *Building safer cities: The future of disaster risk*, Disaster risk management series no. 3, ed. A. Kreimer, M. Arnold, and A. Carlin. Washington, DC: World Bank. http://www.bvsde.paho.org/bvsacd/cd46/cap13-risk.pdf. Accessed October 1, 2012.
Wisner, B., P. Blaikie, I. Davis, and T. Cannon. 2004. *At risk: Natural hazards, people's vulnerability and disasters*, 2nd ed. London: Routledge.
World Bank. 1997. *World development report: The state in a changing world*. Washington, DC: Oxford University Press for the World Bank. http://www-wds.worldbank.org/external/default/WDSContentServer/IW3P/IB/1997/06/01/000009265_3980217141148/Rendered/PDF/multi0page.pdf. Accessed 1 Oct 2012.
World Bank. 2010a. *Climate risks and adaptation in Asian coastal megacities: A synthesis report*. Washington, DC. http://siteresources.worldbank.org/EASTASIAPACIFICEXT/Resources/226300-1287600424406/coastal_megacities_fullreport.pdf. Accessed 1 Oct 2012.
World Bank. 2010b. *São Paulo case study overview. Climate change, disaster risk, and the urban poor: Cities building resilience for a changing world*. http://siteresources.worldbank.org/INTURBANDEVELOPMENT/Resources/336387-1306291319853/CS_Sao_Paulo.pdf. Accessed 11 Oct 2012.
World Wildlife Fund. 2009. *Mega-stress for mega-cities: A climate vulnerability ranking of major coastal cities in Asia*. Gland, Switzerland. http://assets.wwf.org.uk/downloads/mega_stress_cities_report.pdf. Accessed 1 Oct 2012.
World Wildlife Fund-Germany. 2011. *Big cities, big water, big challenges: Water in an urbanizing world*. Berlin. http://www.wwf.se/source.php/1390895/Big%20Cities_Big%20Water_Big%20Challenges_2011.pdf. Accessed 1 Oct 2012.

Zeiderman, A. 2008. Cities of the future?: Megacities and the space/time of urban modernity. *Critical Planning: UCLA Journal of Urban Planning* 15: 23–39. http://www.spa.ucla.edu/critplan/past/volume015/Zeiderman.pdf. Accessed 1 Oct 2012.

Zeiderman, A. 2012. On shaky ground: The making of risk in Bogota. *Environment and Planning A* (forthcoming). http://stanford.academia.edu/AustinZeiderman/Papers/1002215/On_Shaky_Ground_The_Making_of_Risk_in_Bogota. Accessed 1 Oct 2012.

Zhong, Y., and X. Chen. 1999. Typhoon hazards in the Shanghai area. *Disasters* 23(1): 66–80.

Zoleta-Nantes, D. 2002. Differential impacts of flood hazards among the street children, the urban poor and residents of wealthy neighborhoods in Metro Manila, Philippines. *Mitigation and Adaptation Strategies for Global Change* 7(3): 239–266.

James K. (Ken) Mitchell is Professor of Geography at Rutgers, The State University of New Jersey. Born in Northern Ireland, he is a Fellow of the American academy of the Advancement of Sciences and a member of the International Research Committee on Disasters, who founded the international journal *Global Environmental Change* and has authored more than 140 publications on human dimensions of natural hazards, technological risks and terrorism. He has served on science and public policy advisory bodies for hazards, coastal and offshore marine environments at state and federal levels in the United States and the United Nations system. He edited the volume *Crucibles of hazard: Megacities and disasters in transition* (1999).

Chapter 23
Natech Disaster Risk Reduction: Can Integrated Risk Governance Help?

Ana Maria Cruz, Yoshio Kajitani, and Hirokazu Tatano

23.1 Introduction

Natech risk refers to risk originating from conjoint natural and technological hazards. In this chapter, we are concerned with risk governance of natechs involving technological hazards arising from the processing, handling and/or storage of hazardous materials (hazmats), as well as the transportation of oil and gas by pipeline. Generally speaking, natechs can occur anywhere in the world where industrial facilities and other infrastructure housing hazardous materials are located in areas exposed to natural hazards.

Hurricanes Katrina and Rita which impacted the United States Gulf Coast in 2005 serve as an example. The hurricanes triggered more than 200 hazardous materials (hazmat) releases from fixed industrial facilities and over 400 oil and gas releases from offshore platforms with harmful impacts to the surrounding communities and the environment (Cruz and Krausmann 2009; Santella et al. 2009). In one case, more than 1,800 homes were affected and resulted in a class action settlement for US$330 million (US District Court 2007). Figure 23.1 shows an affected home after flood waters receded. Although the hazmat releases in these two disaster events received less attention given the dramatic emergency management problems that resulted in their aftermath, they none the less represented an additional burden on emergency responders, disaster victims in neighboring residential areas, and government officials. Furthermore, the damage inflicted by Hurricanes Katrina and Rita

A.M. Cruz (✉) • H. Tatano
Disaster Prevention Research Institute, Kyoto University,
Gokasho, Uji, Kyoto 611-0011, Japan
e-mail: anamaria@drs.dpri.kyoto-u.ac.jp; tatano@imdr.dpri.kyoto-u.ac.jp

Y. Kajitani
Central Research Institute of Electric Power Industry,
1-6-1 Ohtemachi, Chiyodaku, Tokyo 1008126, Japan
e-mail: y-kaji@creipi.denken.or.jp

U. Fra.Paleo (ed.), *Risk Governance*, DOI 10.1007/978-94-017-9328-5_23

Fig. 23.1 One of more than 1,800 homes in Chalmette, Lousiana, affected by floods and an oil release from a neighboring refinery following Hurricane Katrina in 2005

resulted in huge losses to the oil and gas industry, shortages of much needed fuel for emergency response and clean up and economic repercussions that were felt around the globe (Cruz and Krausmann 2008, 2009).

The images of raging fires at an oil refinery in Chiba Prefecture following the Great East Japan magnitude 9.0 earthquake on 11 March 2011 serves as another example of a natech accident. The earthquake forces initiated a chain of events at the Cosmo oil refinery that resulted in multiple fires and explosions which completely destroyed 17 liquefied petroleum gas (LPG) storage tanks, caused damages to neighboring industrial plants, warehouses, residential buildings and vehicles.

The Great East Japan earthquake and the tsunami it triggered caused catastrophic consequences in a large area, including severe damage and disruption to industry leading to multiple natech accidents with offsite consequences. The economic losses to industry and the cost of reconstruction and clean up due to the natech accidents were probably high, although there is only limited information available. As an example, Krausmann and Cruz (2013) found data of economic loss for restoration, destruction of assets and running costs for three of JX Holdings companies in Japan, including US$1.2 billion for the JX Refinery in Sendai, US$251 million for the Kashima Oil refinery in Kashima, and US$100 million for the JX Nippon Mining and Metal's plant in Isohara. These losses were incurred in the Japanese fiscal year 2010 which ended on 31 March 2011 (Krausmann and Cruz 2013).

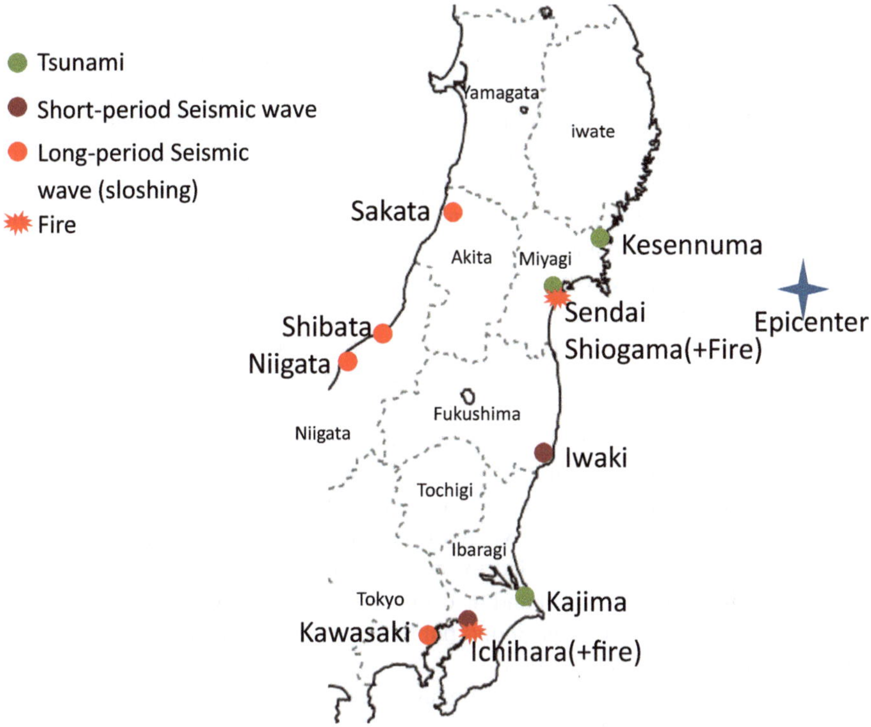

Fig. 23.2 Map of North East Japan showing the location of damaged oil refineries following the 9.0 magnitude earthquake and subsequent tsunami of 11 March 2011. The figure indicates whether the damage was mainly caused by the earthquake or the tsunami (Source: Adapted from NRIFD 2012)

In the aftermath of the disaster, six oil refineries (about 30 % of Japanese refining capacity) had to be shut down for a week. Figure 23.2 shows areas where damage to oil refining was observed, caused by the earthquake or the tsunami. Furthermore, damage to the facilities and hazmat releases compounded by earthquake aftershocks, tsunami alerts, evacuation of personnel, lack of utilities and damage to infrastructures resulted in a 34 % loss of production capacity in the Tōhoku area (Tatano and Kajitani 2012). Overall, the economic losses to industry due to property and production capacity losses following the Great East Japan earthquake and tsunami were huge and have had repercussions worldwide through supply chain effects.

The Great East Japan earthquake and subsequent tsunami also triggered one of the worst nuclear power accidents since Chernobyl. Water as high as 15 m caused by the tsunami inundated the Dai-ichi nuclear power plant (NPP) in Fukushima causing loss of power which led to disruption of controls and cooling system failure shortly after the earthquake. Venting of gases, hydrogen explosions, and a fire in a spent fuel pond in one of the units resulted in the release of radionuclide contaminants into air and water with a relatively high atmospheric release rate through March 24 (Buesseler et al. 2011).

The Fukushima nuclear power accident triggered by the earthquake and tsunami is an extreme case of a natech which has had huge social, environmental and economic impacts for the people of Japan but also the world. The catastrophic impact of the tsunami in the Tōhoku area and the Fukushima disaster have led scientists, professional organizations, public authorities and the public in general to question the design basis of hazardous facilities as well as structural protection measures against natural hazards (e.g., tsunami, tropical cyclones). One of the main concerns has been how to address and prepare for 'beyond-design-basis' events, for example the earthquake-tsunami combination that overwhelmed Fukushima (Lochbaum and Lyman 2012).

The Fukushima disaster has also had profound impacts on risk governance of nuclear power plants in Japan including a decision to set up a new independent nuclear regulatory agency under the Ministry of the Environment. This reorganization will create an entity responsible for regulating nuclear power generation, which is separate from the entity that is promoting it. Furthermore, the Fukushima accident had prompted an immediate review of the safety of nuclear energy in most countries with nuclear programs. Many of these countries announced comprehensive safety reviews, which have led to regulatory changes that may slow or even eliminate plans for expansion of and investment in nuclear power. The most drastic public policy changes due to profound public reactions included Japan, Germany, Italy, and Switzerland (WEC 2011).

Natech disasters have occurred in many other parts of the world. Krausmann et al. (2010) report that during the devastating Wenchuan earthquake on 12 May, 2008 in Sichuan Province in China hundreds of industrial facilities were completely destroyed or severely damaged. This resulted in multiple hazardous materials releases including ammonia and sulfuric acid from severely damaged industrial establishments in Honbai, Yinghua, and Shifang City. Leakage of approximately 150 t of liquid ammonia and about 1,000 t of sulfuric acid from damaged tanks were reported at a completely destroyed facility in Honbai serve as examples (Krausmann et al. 2010).

Risk governance of natechs presents particular challenges. Natech risk originates from the overlapping of natural, environmental and technological causes, making it predominantly complex and uncertain. Growing urban populations, industrialization, and globalization have resulted in more people and property at risk from natural hazards and secondary effects such as major natech accidents.

As the world becomes more and more interconnected, disaster impacts on industry in one part of the world can result in impacts in other areas through supply chain effects resulting also in high economic losses as the examples above illustrate. The floods in Thailand in 2011 serve as another example. Flood damage and losses have been estimated at US$45.7 billion (World Bank 2011); most of these losses have been attributed to inundation of seven major industrial manufacturing facilities (Mydans 2011). The disruption of the manufacturing supply chains affected many industrial sectors including the computer industry which caused global shortage of hard disk drives (Fuller 2011; Mydans 2011). Furthermore, in Bangkok alone, the number of people getting ill from dangerous substances caused by flooding of industrial areas and farmland reached around 2,000 patients in December 2011, mostly from industrial and agricultural chemicals (Labboun 2011).

Awareness of the need to better address natech risks has been increasing. Prominent examples include a natech workshop organized by the Organization for Economic Cooperation and Development (OECD) held in Dresden, Germany in May 2012 to investigate elements of natech risk management which are not part of the national (OECD member states) chemical accidents programs in order to make recommendations for good practices with respect to natech risk reduction. In the aftermath of the March 11, 2011, Tōhoku earthquake and tsunami, the authors, at the Disaster Prevention Research Institute, and researchers from the Joint Research Centre of the European Commission are studying the multiple natech accidents triggered by the Japanese disaster (Krausmann and Cruz 2013) in an effort to prepare recommendations for policy makers for improved natech risk management. While concern over natechs has been on the rise among researchers and government officials, a general framework for the governance of natech risk is lacking in most countries. This chapter provides an overview of natech hazards, their characteristics and the problems associated with natech risk governance. We use the IRGC risk governance framework for guidance.

23.2 Natechs and Their Characteristics

Natech hazard risk exists where natural hazards and hazardous industrial activities co-exist. Thus, it is not surprising that the highest numbers of natech accidents in the United States every year (US) occur in the states of Louisiana, Texas, and California which are densely populated, heavily industrialized and are subject to earthquake, hurricane and flood hazards (Sengul et al. 2012). Unfortunately, many of the world's largest cities are located in areas subject to natural hazard risk, are densely populated, and have large or rapidly growing industrial sectors. Furthermore, many of these areas are already suffering environmental changes due to population pressures and climate change (e.g., soil erosion, loss of marshland, changes in run off due to rapid urbanization, pollution) (Cruz and Krausmann 2013) which could result in increased risk of natechs.

23.2.1 Incidence of Natechs

Natech accidents have been documented both in the developed and developing world (Lindell and Perry 1997; Malhotra 2001; ABAG 1990, 1991; Steinberg and Cruz 2004; Cruz and Steinberg 2005; Cruz and Krausmann 2008, 2009; Krausmann et al. 2010); although stricter rules regarding reporting of accidental hazmat releases in countries like the United States and the European Union provide better databases for natech analysis.

The incidence of natechs in countries like the United States is increasing. For example, the number of natech problems in the Northridge earthquake in California in 1994, tripled those of the Loma Prieta earthquake in 1989 (Lindell and Perry 1997). More recently, Sengul et al. (2012) found that on average, weather- and storm-related hazmat releases reported to the US Federal National Response Center (NRC) between 1990 and 2008 had significantly increased by approximately 8 and 5 % per year, respectively; whereas events due to tornadoes had significantly increased by 5 % per year.

23.2.1.1 Weather-Related Natechs

In fact, weather related natechs in the US accounted for over 80 % of all natech accidents reported to the NRC. Natechs represent about 1–7 % of all NRC reported releases per year representing approximately 16,600 hazardous material (hazmat) releases between 1990 and 2008 (Sengul et al. 2012). The authors found a higher number of hazmat releases triggered by heavy rain (26 % of the total), followed by hurricanes (20 % of the total) and storms, winds and other unspecified types of weather related phenomena (25 % of the total). Similar results were found by Rasmussen (1995) who studied data from US and European accident databases. The author reported between 1 and 5 % of hazmat releases were caused by natural causes, 80 % of these were due to various weather related events.

Santella et al. (2011) quantified the conditional probabilities of natechs at industrial plants regulated by various US Federal regulatory requirements including the Toxic Release Inventory (TRI), the Risk Management Program (RMP), and the Standard Industrial Code (SIC) 1311 identified from the US EPA (Environmental Protection Agency)'s Facility Registry System (FRS). The authors found that during hurricanes, a higher probability of releases was observed due to storm surge (7.3 releases per 100 TRI/RMP facilities exposed vs. 6.2 for SIC 1311 facilities), as compared to category 1–2 hurricane winds (5.6 TRI, 2.6 SIC 1311). Areas inundated during flood events had a probability of 1.1 releases per 100 facilities but demonstrated widely varying natech occurrence during individual events, indicating that factors not quantified in their study such as flood depth and speed were important for predicting flood natechs. A changing climate and increased urbanization, particularly in coastal areas, may result in a higher frequency of extreme weather events, hence the potential for more natechs. Recent research has shown that the oil and gas industry, which handles large volumes of hazmats, located in low-lying coastal areas and areas subject to severe weather will be most vulnerable (Cruz and Krausmann 2013).

23.2.1.2 Earthquake-Related Natechs

In addition to weather related natechs, major earthquakes (magnitude 7.0 and higher) in urbanized areas in the last two decades have increasingly resulted in hazmat releases, some of them with severe consequences -Kocaeli Earthquake,

Turkey, 1999 (Cruz and Steinberg 2005); Gujarat earthquake in India in 2001 (Malhotra 2001), Great Indian Ocean earthquake and tsunami 2004 (UNEP 2005), Wenchuan earthquake in China in 2008 (Krausmann et al. 2010), and the Great East Japan Earthquake in 2011 (Krausmann and Cruz 2013) serve as examples.

Understanding and estimating the likelihood of natechs during earthquakes has been an important research topic as it would allow for improved impact assessment and emergency response planning. Lindell and Perry (1997) found that hazmat releases during the Northridge earthquake in Los Angeles County in 1994 had likely occurred from 19 % of the industrial facilities in the county (where the Modified Mercalli Index (MMI) values were VIII -IX). Similar results were reported by Santella et al. (2011). The authors found that the probability of natechs at TRI/RMP facilities during earthquakes increased from 0.1 releases per 100 facilities at MMI V to 21.4 releases per 100 facilities at MMI IX.

In another study, Cruz and Steinberg (2005) found that hazmat releases during the Kocaeli earthquake of 1999, occurred in 8 % of the industrial facilities that handled hazardous chemicals. Although the percentage of natech events reported for the Turkey earthquake was lower than that reported by Lindell and Perry (1997) for the Northridge earthquake, the magnitude of the events, and the overall effects on public health and emergency response in Turkey were much greater. Furthermore, Cruz and Steinberg (2005) found that larger industrial plants, as well as older facilities that handle hazmats were significantly more likely to suffer damage and result in hazmat releases during the Kocaeli earthquake.

23.2.2 Prevalence of Natech Accidents

Overall, the likelihood of experiencing a natech accident will not only depend on the type of natural hazard trigger and its magnitude or severity, but also on other factors including the extent of exposure, the type of chemical, quantity, storage tank and storage conditions (pressure and temperature), structural integrity of the vessel/structure containing the material, its design, age, maintenance, or proximity to other structures, among others. Atmospheric storage tanks, pressurized tanks and pipelines were the units most frequently damaged by past earthquakes as well as past flood events (Antonioni et al. 2009).

Previous research indicates that the likelihood of simultaneous releases from one or more sources may be higher during a natural disaster because the natural hazard forces impact large areas and similar structures simultaneously often causing common-cause failures (Steinberg and Cruz 2004; Cruz et al. 2006; Steinberg et al. 2008). For example, four fires in a naphtha tank farm were started simultaneously when metal-to-metal vibration of the tanks' floating-roofs against the tank shells created sparks that ignited all four tanks during the Kocaeli earthquake. Elephant foot buckling, liquid sloshing, or floating roof damage to several tanks simultaneously have been documented during past earthquakes; floating off of partially empty oil tanks, and damage to pipelines and flange connections -among other impacts- were commonly observed in several tanks during floods (Antonioni et al. 2009). In fact Antonioni et al. (2009) found that individual risk increased by an order of magnitude when probabilistic risk assessment of accidental scenarios (for storage tanks) triggered by earthquakes were considered as compared to accidental scenarios triggered by internal system failures. The authors found similar effects for flood triggered accidents: both frequency and magnitude of the expected events increased.

23.2.3 Natech Accidents Versus Other Types of Industrial Accidents

Natechs are characterized by several aspects which are important in understanding their significance versus other types of industrial accidents (Steinberg et al. 2008): (a) multiple releases may occur simultaneously; (b) safety and mitigation measures may not work properly due to the natural event; (c) emergency response personnel and resources may not be available; (d) emergency response to the chemical release may be hampered by the natural disaster, or the natural disaster may exacerbate its effects, and; (e) recovery from the hazmat release may be significantly slowed by impacts from the natural disaster or vice-versa, recovery from the natural disaster may be slowed by the hazmat release.

During a natural disaster, many industrial facilities may be affected simultaneously resulting in a high number of hazmat releases. During a natural disaster the area impacted is often quite large, e.g., the impact zone of a hurricane, earthquake or flood can be hundreds of square kilometers. Thus, local responders working for the local and regional government may be overwhelmed, and any mutual aid agreements among facilities may be unavailable if each facility has been affected or needs to respond to its own releases (Steinberg et al. 2008).

Mitigation measures may not work properly as the natural hazard forces may have damaged them or rendered them inoperable due to damage to other equipment in particular electrical power, sensitive electronic equipment, heating and cooling, and air. For example, an earthquake may cause damage to containment dike walls, or cause breaking of pipelines carrying water for foam-suppression, floods may damage low-lying equipment such as pumps and motors or cause salt-water intrusion of electric control panels. Tsunami wave and floating debris can cause extensive damage to emergency vehicles and fire trucks (see Fig. 23.3). The possibility of cascading disasters also exists, as mitigation measures fail again and again and one release triggers another. Explosions and fires are particularly likely to trigger domino effects, as the force of an exploding tank can damage the structural integrity of others, and a fire in one tank can impinge on another causing additional fires or explosions. At the Cosmo oil refinery, a fire broke out due to a gas release from an LPG storage tank. The LPG leakage was started when the LPG tank collapsed during an aftershock after its legs were damaged by the Great East Japan Earthquake. Fire impingement on other LPG tanks lead to the explosion of other tanks in a chain of events that destroyed all 17 LPG tanks at the refinery.

Fig. 23.3 Fire truck damaged by the Great East Japan tsunami

Response personnel, equipment and emergency materials such as water and foam for firefighting may not be available due to damage caused by the natural disaster, or because the disaster has overwhelmed response capacity. For example the natural disaster may cause the collapse of buildings housing fire trucks, and flood waters may inundate fire stations or access roads impeding passage of response personnel. If several hazmat accidents have been triggered, there may not be sufficient personnel readily available to respond. In addition to the possible need to respond to a large number of simultaneous releases, response personnel and their equipment may be called to respond to the natural disaster-caused problems, especially search and rescue operations of natural disaster victims. Other personnel may be unavailable as they may wish to stay close to their families, or because they themselves have been hurt in the natural disaster.

The natural disaster may hinder response to the chemical release or exacerbate its effects. Roadways may be blocked or impassable to response vehicles and flooding may prevent responders from reaching the release site, or they may flood areas where response equipment is stored. People may be trapped in buildings or in other areas from which they cannot escape, subjecting themselves to the effects of the released chemicals. Conflicts may arise wherein it is safer to 'shelter-in-place' from the chemical release, but it is better to leave the building to escape aftershocks. Alternatively, it may be safer to shelter-in-place, but building structural integrity may have been compromised by the natural disaster (Steinberg et al. 2008).

Recovery from the release may be significantly slowed by impacts from the natural disaster, and vice-versa, recovery from the natural disaster may be hampered due to contamination or pollution from the release. The repair and rebuilding of the damaged equipment, the clean-up of the contaminated natural environment or man-made structures, and the overall ability of the industrial facility to resume operation may be significantly hampered by lack of personnel, housing, and other infrastructure. Repair and rebuilding of neighborhoods affected by the release may take longer depending on the type of chemicals released, the availability of personnel and equipment, ¡ and clean-up methods available given the natural disaster conditions. The economic and social recovery of the devastated area may make the economic justification for repairing the facility damage questionable; industrial or governmental money to address environmental contamination may be in short supply; local construction costs for material and labor may significantly rise or may be unavailable; and other area-wide influences may inhibit recovery from the accident (Steinberg et al. 2008).

These characteristics of natech accidents imply that they require special attention in terms of risk governance as natech risk reduction will involve many players and stakeholders including the industrial facility owners/operators and its contractors or suppliers, but also government officials in charge of chemical accident prevention, first responders, neighboring industrial facilities, and residents, among others.

23.2.4 Risk Management Practices

There are a variety of ways in which the risk to people and the environment from chemical accidents triggered by natural disasters can be reduced. These may include design codes and standards, chemical process safeguards, combined natural hazard and chemical process safeguards, land-use planning and disaster mitigation, and response planning (Steinberg et al. 2008). For new facilities an environmental impact assessment is usually required to identify and minimize potential environmental effects of proposed industrial projects. Risk management should not only include a risk assessment of potential impacts of natural hazards to an industrial facility, but should also consider the potential impacts of natural hazard forces on neighboring industrial plants and other infrastructure such as utilities, roads, and nearby communities. To ensure the best results, risk and risk-reduction alternatives should be evaluated and adopted during the entire life cycle of a plant.

In most countries industrial risk management practices for chemical accident prevention fall short when it comes to natech accidents (Cruz et al. 2006; Cruz and Okada 2008; Krausmann and Baranzini 2009; Krausmann et al. 2011) particularly because most regulatory frameworks do not require the analysis of natech risk in a territory. Most risk management rules and regulations around the world concern individual facilities, with the exception of Japan which has some requirements concerning earthquake risk at certain industrial parks or complexes. Furthermore, there are limited risk assessment methods and tools for natech risk evaluation, and only limited guidance is available on what industry and government authorities can do to assess natech risk.

Industrial risk management generally calls for hazard identification and risk analysis in order to quantify the probability of occurrence and expected consequences of the identified hazards; a prevention program including adoption of safety and mitigation measures; and an emergency response program which should include emergency response procedures and training.

23.2.4.1 Risk Assessment

Natech risk assessment differs from standard industrial risk assessment in that it requires a detailed characterization of the triggering natural hazard or disaster and analysis of the final accident scenarios (Krausmann et al. 2011). Krausmann et al. (2011) summarize the following steps for natech risk assessment at an industrial facility: (a) characterization of the natural event in terms of frequency and severity; (b) identification of target equipment; (c) identification of damage states and reference scenarios and development of event trees; (d) estimation of damage probability; (e) consequence evaluation of the reference scenarios; (f) identification of credible combinations of events; (g) frequency/probability calculation for each combination; and (h) consequence calculation for each combination and overall vulnerability mapping. Krausmann et al. (2011) point out the need for multidisciplinary efforts in order to characterize the natural event at the site (e.g., for earthquakes through probabilistic seismic hazard analysis), to analyze vulnerability of the equipment to the natural hazard loads, and to assess the consequences of equipment damage on hazardous industrial processes.

Unfortunately, many country regulatory frameworks do not explicitly require industrial facilities to assess natech risks. A few exceptions include the California Accidental Release Prevention (CalARP) program in the US, which specifically calls for a risk assessment of potential releases due to an earthquake (Steinberg and Cruz 2004); and the Law on the Prevention of Disasters in Petroleum Industrial Complexes and Other Petroleum Facilities (PDPC) amended after the Tokaichi-oki earthquake in Hokkaido, Japan, in September 2003, which triggered a large fire at a refinery (Cruz and Okada 2008). The PDPC and related regulations include anti-earthquake safety measures for floating roof tanks and fire-fighting tactics in case a full-scale fire occurs due to an earthquake (Krausmann and Cruz 2013). At the European Union level, the Seveso II Directive requires industrial establishments to consider external hazards in the hazard analysis. However, the Directive does not specify methodologies or actions that can be taken to achieve these requirements leading to uneven levels of preparedness among countries (Cruz et al. 2006; Krausmann and Baranzini 2012).

Compliance with the Seveso II Directive requirement of taking into consideration external hazards is often dealt with by insuring that buildings are constructed to building codes. However, several studies have shown that this may not be sufficient if the particular characteristics of natech hazards have not been properly considered (Steinberg and Cruz 2004; Krausmann et al. 2011). For example, under CalARP a seismic risk assessment is required, but only processes that contain certain regulated chemicals must be analyzed. Thus, structures adjacent to the covered processes whose structural failure or excessive displacement could result in the failure of systems which contain hazmat materials are not covered under the CalARP rule. The damage to the LPG storage tanks at the Cosmo oil refinery during the Great East Japan earthquake occurred because at the time of the earthquake the tank was filled with water, which is much heavier than LPG, for a maintenance check. The damaged storage tank was designed to resist earthquakes while containing LPG, not water. Authorities believe the additional weight of the water caused the leg braces to break (Cruz and Krausmann 2013). Simple mitigation measures such as the use of anchoring bolts to protect tanks from floating off during flooding or to avoid tank displacement and buckling during an earthquake may not be considered without a full natural hazard risk assessment.

Nonetheless, there is evidence of growing interest in natechs as an emerging risk issue (Krausmann and Baranzini 2012) and countries are taking a second look at their risk management practices. France has revised its Environmental Code to reflect the lessons learned from the floods of 2002 (Cruz et al. 2006). The French Law No. 2003-699 on the 'prevention of technological and natural risks and the compensation of the damages they cause' (Law of July 30, 2003) and the (Decree No. 2005-1466 of 28 November 2005) require that the national insurance program for natural hazards be extended to cover major accident hazards involving regulated hazardous materials. In 2010, a new regulation (Decrees 2010-1254 and 2010-1255 of October 22, 2010) introduced a new zoning for seismic activity in France where industrial establishments may be considered under 'normal risk' or 'special risk'. According to the Ministerial Order of 24th January 2011, 'special risk' classified sites are pieces of equipment in low and upper-tier Seveso facilities that may lead, in case of an earthquake, to one or more dangerous phenomena with lethal offsite consequences (Köppke 2012).

Concerned over a possible increase of hazards by precipitation and floods due to expected climate change, the German Federal Ministry for the Environment, Nature Conservation and Nuclear Safety passed the new Technical Rule for Plant Safety 310 in 2012. This Technical Rule requires industrial establishments with major chemical accident potential to assess the risk of flood triggered accidents at their installations, including the consideration of potential impacts due to climate change, and requires establishments to take necessary risk reduction measures.

Concern about a powerful earthquake (Richter scale 8.0 or more) with a 95 % probability in the next 50 years in the region off the Pacific coast in central and western Japan, mainly the Tokai, To-nankai, and Nankai regions, has resulted in the passing of the Large-Scale Earthquake Countermeasures Special (LSECS) Act. This act has prompted national, regional, and local governments in these potentially affected areas to take special disaster-prevention measures. The LSECS Act has also resulted in amendments to the Japanese High Pressure Gas Safety (HPGS) Law specifically requiring industrial establishments to take any additional measures necessary to reduce the risk of accidents, to protect its workers and the public from any accidental release caused by severe ground motion or earthquake triggered tsunami (High Pressure Gas Safety Law). Japan also uses performance-based building standards where the building is required to satisfy performance criteria (e.g., remain operational) with regard to materials, equipment, and structural methods (Japan External Trade Organization 2005). Thus, industrial facilities that handle certain hazardous chemicals for example, will be subject to strict building design codes that would permit withstanding the 1 in 800–1,500 year event, depending on particular fault characteristics, such as distance to the fault or soil type (Yoshimura 2005). The relatively low damage to buildings and industry due to the 9.0 magnitude Great East Japan earthquake pays tribute to the effectiveness of Japan's earthquake mitigation efforts (Mori et al. 2012).

Japan's PDPC law stipulates detailed restrictions for petrochemical industrial zones regarding the layout of processing facilities with regards to earthquake risk. Thus, the PDPC law in Japan requires that industrial facilities located together within a territory be treated as an industrial complex. For example, the PDPC law requires the establishment of united collective hazard mitigation and emergency response systems to prevent disasters in petroleum and chemical complexes due to earthquakes (Cruz and Okada 2008).

Many challenges for natech risk management remain, including lack of data on past accidents and lessons learned; limited availability of industrial equipment vulnerability relationships for natural hazards -only some data exists for earthquakes and limited data for floods-, and overconfidence of industrial personnel and government officials in risk management practices for day-to-day chemical accident prevention which they believe will provide sufficient protection against natural hazards.

23.2.4.2 Prevention and Mitigation Measures

Industrial facilities and other infrastructure (e.g., chemical warehouses, chemical storage farms and ports, oil and gas pipelines) housing hazardous chemicals located in areas subject to natural hazards may be at risk of damage or disruption if appropriate measures are not taken to prevent or prepare for such events. The adoption of prevention and mitigation measures to reduce the risk of chemical accidents triggered by natural hazards will vary depending on the type of hazards present in a territory and their potential magnitude or severity and frequency.

Land use regulations should be used to limit the construction of hazardous installations in areas subject to high natural hazard risk. Prohibiting construction in river flood plains and restricting certain type of structures in the 50- and 100-year flood zone serve as examples. Prohibiting construction on or at a certain distance from a known earthquake fault is another example. Unfortunately, land use restrictions are often difficult to establish, particularly for existing structures.

Industrial facilities and other establishments housing hazmats can be protected against natural hazards through the adoption, monitoring and enforcement of design codes and standards that account for natural hazard loads (e.g. seismic, tsunami, wind, and flood loads) on buildings, steel support structures for processing equipment and storage tanks, and other structures. Industrial risk management practices for the prevention of accidental chemical releases which specifically address the potential impacts of natural hazards on safety and mitigation measures, plant utilities and external lifeline systems are needed. Additional risk management actions that can be promoted to make plants less vulnerable to natural hazards include the use of redundant safety systems, passive mitigation measures that do not rely on external power sources or human intervention, natural hazard-resistant design for protective structures (e.g., earthquake resistant tank walls), the provision of guidelines to inform industry about how to plan for natural hazards, the implementation of safety distances between certain types of equipment and passages for emergency response, and requiring the strategic placement of hazardous substances in less vulnerable areas (Cruz 2011).

23.2.5 Emergency Response for Natechs

Natech accidents will require special planning in terms of emergency management because the natural disaster may impact safety and mitigation measures and emergency response capacity to deal with the natech accidents. Furthermore, a natural disaster can contribute to the escalation of a chemical accident, often resulting in more severe consequences and complicating emergency response (Steinberg and Cruz 2004; Antonioni et al. 2009).

As discussed in the previous section, the natural disaster may impact large areas triggering multiple and simultaneous chemical releases. Most industrial emergency response plans are conceived based on normal day-to-day plant operation which foresee a single hazardous event. Thus, emergency response may be insufficient. In addition, safety and mitigation measures may not be available due to impacts from the natural disaster, and standard emergency operation procedures may be inadequate under the natural disaster conditions. Furthermore, external emergency responders (e.g., public fire departments) may be unavailable as they may be busy attending to natural disaster victims or residential fires.

Natech accidents may also exacerbate impacts or hamper emergency response to the natural disaster victims. Following the Wenchuan earthquake in China on 12 May 2008, a large number of industrial facilities suffered damage resulting in hazardous chemical releases in Honbai, Shifang City, and Yinghua, among others. Krausmann et al. (2010) report that in one town at least 6,000 earthquake victims had to be evacuated due to ammonia and other chemical releases, and that in another town an ammonia cloud was reported to have drifted down a valley engulfing survivors of the earthquake, some still trapped under the debris in a village and killing some of them.

Similar impacts on earthquake victims were documented during the Kocaeli earthquake. The earthquake triggered over 21 hazardous chemical releases and multiple fires with offsite consequences on people, property and the environment. Toxic releases and the threat of explosion forced government authorities in two different municipalities to order the evacuation of residents and emergency response personnel less than 12 h after the earthquake. The evacuation order resulted in the abandonment of search and rescue activities by squads, as well as family and relatives of victims, who were forced to leave victims behind still trapped in collapsed buildings and debris (Steinberg and Cruz 2004).

These examples point out the need for careful consideration of conflicting emergency-management objectives during a natech accident. They also highlight the need to carefully evaluate and plan for threats involving the impact of natural hazards on chemical industry located in industrialized, urbanized areas that can result in potentially high death tolls, damage to property and environmental pollution. Because natural hazards may impact large areas, thus exposing a high number of facilities and communities, the need to address Natech risk reduction as a risk governance issue is imperative.

23.2.6 Industrial Infrastructure Interdependencies and Natechs

Natural disasters can cause major disruptions to industrial areas and other infrastructure systems (e.g., oil and petrochemical plants, electric power generation, communications, transportation, emergency services). The huge losses inflicted by the Great East Japan earthquake, hurricanes Katrina and Rita, and recently the floods in Thailand have pointed out the need to better understand infrastructure failure interdependencies and their societal significance. Interdependencies of industrial and other critical infrastructures can lead to cascading failures which may cross regional, national and international boundaries (Cruz 2011). The impact of a natural disaster on various systems including industrial facilities, lifelines, hospitals and emergency response facilities, residential buildings, and people have shown that couple effects and interconnections may induce secondary effects in each of these systems (e.g. gas leakages, toxic releases, toxic release from hospital), which lead to effects in other subsystems such as the economic, emergency services, and social systems (Menoni 2001). Menoni noted the need to incorporate both parameters of the physical environment such as lifelines and building stock as well as organizational, social and systemic factors into the analysis of natural hazard risk.

There have been efforts to better understand the interdependencies and possible consequences of major infrastructure failures (Santella et al. 2009). Efforts to model complex and interlinked infrastructures systems, and their interdependencies have been proposed including the Critical Infrastructure Protection Decision Support System (CIPDSS) developed between Los Alamos, Sandia, and Argonne National Laboratories, sponsored by the US Department of Homeland Security (DHS) (Santella et al. 2009). The University of British Columbia promotes research and disseminates knowledge that is needed to prioritize investments for fostering disaster-resilient infrastructures, and thus, more disaster-resilient communities

(UBC 2010). The European Community has set out non-binding guidelines for the identification and designation of European critical infrastructure, and their risk assessment in order to improve their protection (Bouchon et al. 2008).

Recently, joint work between the Disaster Prevention Research Institute (DPRI) and the International Risk Governance Council (IRGC) led to analysis of gaps in risk governance of global maritime critical infrastructure (MGCI) systems (IRGC 2011). The Straits of Malacca and Singapore, and the Port of Singapore were selected as a MGCI case study, and an oil refinery fire and explosion in the Singapore Strait was identified as an initiating major disruptive event that could lead to total closure of the Straits with huge social, economic and environmental consequences. The disruption scenario was identified through several stakeholder workshops (IRGC 2011). The results from this work point out the potential high consequences of major chemical accidents. If a concurrent natural disaster is included in the scenario, one can imagine the devastating consequences this could have.

23.3 Environmental Liability

Environmental liability regarding natech accidents is an area that requires further development. The Polluter-Pays Principle, as defined by the OECD's 1989 recommendation (OECD 1989), states that the "operator of a hazardous installation should bear the cost of reasonable measures to prevent and control accidental pollution from that installation which are introduced by public authorities in member countries in conformity with domestic law prior to the occurrence of an accident in order to protect human health or the environment." The OECD guidance includes measures to improve safety, development of emergency plans, and prompts action to respond to accidents as well as clean up operations to minimize environmental damage. However, with regard to natural disasters not reasonably foreseeable, the OECD guidance states that "the polluter pays principle should not apply but the public purse" (Larsson 1999).

At the European Union level, the Directive 2004/35/CE of the European Parliament and of the Council of 21 April 2004 on environmental liability with regard to the prevention and remedying of environmental damage says that establishments must apply the polluter-pays principle: "an operator causing environmental damage or creating an imminent threat of such damage should, in principle, bear the cost of the necessary preventive or remedial measures." As with the OECD guidelines, Article 4 of this Directive excludes environmental damage or an imminent threat of such damage caused by a natural phenomenon of exceptional, inevitable and irresistible character. Nonetheless, Article 174 (2) of the European Community Treaty provides that Community policy on the environment shall be based on the precautionary principle: "risk prevention in the face of scientific uncertainty." The precautionary principle aims to prevent harm before a hazard has come into existence. The precautionary principle could be used to justify polluter's liability.

In the United States the Comprehensive Environmental Response, Compensation, and Liability Act (CERCLA) of 1980, commonly known as Superfund, the Superfund Amendments and Reauthorization Act (SARA) of 1986, and the Oil Pollution Liability (OPL) Act provide for liability of persons/operators responsible for releases of hazardous substances. Similar to the EU Directive, SARA excludes natural disasters: "A responsible party is not liable for removal costs or damages under section 1002 if the responsible party establishes, by a preponderance of the evidence, that the discharge or substantial threat of a discharge of oil and the resulting damages or removal costs were caused solely by an act of God." In reality, many responsible parties have been held liable for pollution incidents following natural disasters in the United States. The US$330 million paid by Murphy Oil to affected residents following the oil spill triggered by Hurricane Katrina's damage to one of the company's oil storage tanks serves as an example (US District Court 2007).

Environmental liability for natech accidents will become more and more of an issue in the face of climate change and known impacts of certain natural hazards such as earthquakes, tropical storms and floods on hazardous establishments. The need to better address liability issues for natech risk management was evident at the OECD Workshop on Natech Risk Management in Dresden mentioned above, including a dedicated session on "Natechs and the polluter-pays principle."

As awareness for the need to address natech risk increases, as countries make efforts to reduce losses from natural disasters, they will be less tolerant with facilities that are not making the minimum efforts to reduce their vulnerability to natural hazards and the risk they pose to human health, property and the environment due to accidental hazardous materials releases.

23.4 Stakeholder Participation in Natech Risk Governance

Risk governance of chemical accident hazards has generally been carried out through industrial risk management regulatory frameworks and practices, engineering and safety codes and equipment standards, environmental regulations, and land use planning. The risk from natural hazards has been addressed through building codes (mostly to insure life safety), land use planning, structural and non-structural mitigation measures, and emergency preparedness and response planning. However, government agencies in charge of natural disaster prevention, preparedness and response generally work separately from those in charge of chemical accident prevention resulting in risk governance gaps.

Natech risks differ from other types of industrial accident risk in that natech risk is a complex risk which requires a broader, more comprehensive analysis. Because natural disasters have the potential to impact large areas and trigger multiple hazmat releases simultaneously, the assessment, prevention, preparedness for and response to natechs requires special governance arrangements, and coordination and planning to insure that all possible hazards and consequences in a territory are identified and adequately addressed. Thus, only by bringing all players and stakeholders together, can the full picture of possible interactions and failure modes be foreseen, and their impacts reduced or mitigated. Governance of natech risks requires bringing together industry owners/operators, professionals from differing backgrounds including engineers, flood hazard specialists, hydrologists, meteorologists, earthquake engineers, industrial equipment manufacturers, officials from government agencies in charge of industrial risk management, public health and environmental pollution control, regional and city planners, civil protection officials and first responders, and policy-makers as well as potentially affected communities.

Most importantly, when addressing the protection of hazardous industrial facilities and their associated systems, it is necessary to analyze them not as independent entities, but as a part of a much larger, connected system. Disaster consequences may be greatly reduced with a collective effort to understand and prevent ripple effects from natech type failures (Cruz 2011).

23.5 Final Remarks and Conclusions

The human, economic, and environmental cost of natech accidents can be very high. Cascade failures and ripple effects are also common because of the interdependence of industry, lifeline systems and community infrastructure (both hard and soft systems). A comprehensive approach to natech risk governance that integrates structural and nonstructural risk reduction measures, strengthens the capacity of local communities to make their own informed natech risk management choices, and promotes the participation of all stakeholders, in particular natural hazard specialists, disaster managers, land use planners, engineers, industry and professional associations, community groups, non-governmental organizations (NGO) and municipal governments, in all stages of natech risk reduction should be adopted.

Current practices for industrial risk management have worked fairly well in the wealthier nations with professional engineers and builders, well trained inspectors, and well educated users. Nonetheless, as the examples of Hurricane Katrina and the Great East Japan earthquake have shown there have been failures even in the rich developed world. These events demonstrate the need for a more careful assessment of potential impacts of natural hazards on hazardous facilities and their consequences on public health and the environment. They also show that the potential impacts of natechs may be so far reaching as the natural disaster itself that they require an integrated risk governance approach.

The economic impacts of natech accidents may be high, particularly for affected regions as many industrial facilities are likely to be affected simultaneously. Furthermore, the costs of clean-up and remediation of contaminated areas may be high and fall on the public purse, particularly during floods and tsunami because chemicals get dispersed and it is often impossible to determine the sources of the releases. Reporting of natech accidents by industrial facility operators/owners including information on economic losses and costs of clean-up and remediation should be encouraged by public authorities. This would permit a better estimate of the cost of natechs and allow for better cost-benefit analysis concerning investments in natech risk reduction measures.

Natech accidents may have profound impacts on risk perception and tolerability, and may force authorities to make changes to regulations which could result in additional technical and economic burden to industry. In the face of a possible increase in the number of severe weather related events due to climate change, and growing urban populations and industrialization in areas subject to high natural hazard risk, governments should take heed in addressing the risk of natechs not only to avoid losses caused by natural disasters in industry, but also to avoid or reduce possible health, environmental, and economic impacts.

However, there is good news. International as well as national efforts are being made to better understand, prevent, and/or prepare for natechs (Cruz et al. 2004; Krausmann and Baranzini 2012; OECD 2012). Recently at the OECD workshop on natech risk management in Dresden, Germany, representatives of member states and international experts worked together to identify gaps and challenges for improved natech risk reduction. The workshop concluded with the identification of priority areas for future work including the development of guidance on natech risk management, the development/improvement of natech risk analysis methodologies, and the development of dedicated natech risk maps (OECD 2012).

Given the complex and interconnected nature of natechs, a more comprehensive risk governance framework is needed that brings together major players and stakeholders in order to adequately capture the full range of issues and alternative solutions.

References

ABAG. 1990. *Hazardous materials problems in earthquakes: A guide to their cause and mitigation*. Oakland: Association of Bay Area Governments.

ABAG. 1991. *Sources of toxic gas releases in earthquakes*. Oakland: Association of Bay Area Governments.

Antonioni, G., S. Bonvicini, G. Spadoni, and V. Cozzani. 2009. Development of a general framework for the risk assessment of natech accidents. *Reliability Engineering & System Safety* 94: 1442–1450.

Bouchon, S., C. di Mauro, C. Logtmeijer, J.P. Nordvik, R. Pride, B. Schupp, and M. Thornton. 2008. *Non-binding guidelines for application of the Council Directive on the identification and designation of European critical infrastructure and the assessment of the need to improve their protection*. European Commission, Joint Research Centre, Ispra, Italy: EUR 23665 EN – 2008.

Buesseler, K., M. Aoyama, and M. Fukasawa. 2011. Impacts of the Fukushima nuclear power plants on marine radioactivity. *Environmental Science & Technology* 45: 9931–9935.

Cruz, A.M. 2011. Protecting infrastructure. In *Handbook of hazards and disaster risk reduction*, ed. I. Kelman, B. Wisner, and J.C. Gaillard. London: Routledge.

Cruz, A.M., and E. Krausmann. 2008. Damage to offshore oil and gas facilities following Hurricanes Katrina and Rita: An overview. *Journal of Loss Prevention in the Process Industries* 21(6): 620–626.

Cruz, A.M., and E. Krausmann. 2009. Hazardous materials releases from the offshore oil and natural gas facilities following hurricanes Katrina and Rita. *Journal of Loss Prevention in the Process Industries* 22(1): 59–65.

Cruz, A.M., and E. Krausmann. 2013. Vulnerability of the oil and gas sector to climate change and extreme weather events. *Climate Change* 121(1): 41–53.

Cruz, A.M., and N. Okada. 2008. Consideration of natural hazards in the design and risk management of chemical industrial facilities. *Journal of Natural Hazards* 44(2): 213–227.

Cruz, A.M., and L.J. Steinberg. 2005. Industry preparedness for earthquakes and earthquake-triggered hazmat accidents during the Kocaeli Earthquake in 1999: A survey. *Earthquake Spectra* 21(2): 285–303.

Cruz, A.M., L.J. Steinberg, A.L. Vetere-Arellano, J.P. Nordvik, and F. Pisano. 2004. *State of the art in natech (Natural Hazard Triggering Technological Disasters). Risk Assessment in Europe*. EUR Report 21292 EN, DG Joint Research Centre, EC and UN/ISDR, Ispra, Italy.

Cruz, A.M., L.J. Steinberg, and A.L. Vetere-Arellano. 2006. Emerging issues for natech disaster risk management in Europe. *Journal of Risk Research* 9(5): 1–19.

Fuller, T. 2011. Thailand flooding cripples hard-drive suppliers. *The New York Times*, Global Business, 6 November 2011. http://www.nytimes.com/2011/11/07/business/global/07iht-floods07.html?pagewanted=all. Accessed 12 June 2012.

International Risk Governance Council (IRGC). 2011. *Risk governance of maritime global critical infrastructure: The example of the Straits of Malacca and Singapore*. Geneva: International Risk Governance Council (IRGC).

Japan External Trade Organization. 2005. *Amended Building Standard Law*. Japan.

Köppke, K.E. 2012. *Discussion document*. OECD workshop on Natech Risk Management (Natural-hazard triggered technological accidents), 12 May 2012, Dresden, Germany. http://www.oecd.org/chemicalsafety/riskmanagementofinstallationsandchemicals/50337868.pdf.

Krausmann, E., and D. Baranzini. 2009. *Natech risk reduction in OECD Member Countries: Results of a questionnaire survey*. Report JRC 54120, European Communities, 2009, Ispra, Italy.

Krausmann, E., and D. Baranzini. 2012. Natech risk reduction in the European Union. *Journal of Risk Research* 15(8): 1027–1047.

Krausmann, E., and A.M. Cruz. 2013. Impact of the 11 March, 2011, Great East Japan earthquake and tsunami on the chemical industry. *Journal of Loss Prevention* 121(1): 41–53.

Krausmann, E., A.M. Cruz, and B. Affeltranger. 2010. The impact of the 12 May 2008 Wenchuan earthquake on industrial facilities. *Journal of Loss Prevention in the Process Industries* 23(2): 242–248.

Krausmann, E., V. Cozzani, E. Salzano, and E. Renni. 2011. Industrial accidents triggered by natural hazards: An emerging risk issue. *Natural Hazards and Earth System Sciences* 11: 921–929.

Labboun, S. 2011. Thai floods impact on environment. *Press TV*, Bangkok, 29 December 2011. http://www.presstv.ir/detail/218272.html. Accessed 17 Sept 2012.

Larsson, Marie-Louise. 1999. *The law of environmental damage: Liability and reparation*, Stockholm studies in law. Stockholm: Kluwer Law International.

Lindell, M.K., and R.W. Perry. 1997. Hazardous materials releases in the Northridge earthuqke: Implications for seismic risk assessment. *Risk Analysis* 17(2): 147–156.

Lochbaum, D., and E. Lyman. 2012. *U.S. nuclear power safety: One year after Fukushima*. Union of Concerned Scientists. http://www.ucsusa.org/nuclear_power. Accessed 27 Feb 2013.

Malhotra, P.K. 2001. *Industrial facilities: Bhuj, India earthquake*. Earthquake Engineering Abstracts. Earthquake Engineering Research Institute, Oakland. http://www.eeri.org/earthquakes/Reconn/bhuj_India/indusfac.pdf.

Menoni, S. 2001. Chains of damages and failures in a metropolitan environment: Some observations on the Kobe Earthquake in 1995. *Journal of Hazardous Materials* 86: 101–119.

Mori, N., T. Takahashi, and The 2011 Tōhoku Earthquake Tsunami Joint Survey Group. 2012. Nationwide post event survey and analysis of the 2011 Tōhoku earthquake tsunami. *Coastal Engineering Journal* 54(1): 1250001-1-27.

Mydans, S. 2011. Thai Prime Minister to take command of flood control efforts. *The New York Times*, Asia Pacific, 21 October 2011. http://www.nytimes.com/2011/10/22/world/asia/bangkok-opens-floodgates-as-government-response-is-criticized.html?_r=2&partner=rss&emc=rss&. Accessed 12 June 2012.

National Research Institute of Fire and Disaster (NRIFD). 2012. *Report on damage to industry from the Great East Japan Earthquake and Tsunami*. National Research Institute of Fire and Disaster, Japan. http://www.fri.go.jp/cgi-bin/hp/index.cgi?ac1=JNN2&ac2=&ac3=740&Page=hpd_view. Accessed 27 Nov 2012.

Organization for Economic Cooperation and Development (OECD). 1989. *The application of the polluter-pays principle to accidental pollution*. OECD Rec C(89)99 (Final), July 7, Paris, France.

Organization for Economic Cooperation and Development (OECD). 2012. *OECD Workshop on Natech Risk Management (Natural-hazard triggered technological accidents)*. 23–25 May 2012, Dresden, Germany. http://www.oecd.org/env/chemicalsafetyandbiosafety/riskmanagementofinstallationsandchemicals/oecdworkshoponnatechriskmanagementnatural-hazardtriggeredtechnologicalaccidents23-25may2012dresdengermany.htm. Accessed 9 July 2012.

Rasmussen, K. 1995. Natural events and accidents with hazardous materials. *Journal of Hazardous Materials* 40(1): 43–54.

Santella, N., L.J. Steinberg, and K. Parks. 2009. Decision making for extreme events: Modeling critical infrastructure interdependencies to aid mitigation and response planning. *Review of Policy Research* 26(4): 409–422.

Santella, N., L.J. Steinberg, and G.A. Aguirre. 2011. Empirical estimation of the conditional probability of natech events within the United States. *Risk Analysis* 31(6): 951–968.

Sengul, H., N. Santella, L.J. Steinberg, and A.M. Cruz. 2012. Analysis of hazardous material releases due to natural hazards in the United States. *Disasters* 36(4): 723–743.

Steinberg, L.J., and A.M. Cruz. 2004. When natural and technological disasters collide: Lessons from the Turkey earthquake of August 17, 1999. *Natural Hazards Review* 5(3): 121–130.

Steinberg, L.J., H. Sengul, and A.M. Cruz. 2008. Natech risk and management: An assessment of the state of the art. *Journal of Natural Hazards* 46(2): 143–152.

Tatano, H., and Y. Kajitani. 2012. *Production capacity losses due to the 311 disaster—Facility damage and lifeline disruption impacts*. International disaster and risk conference, Davos, Switzerland, 26–30 August 2012.

United Nations Environment Programme (UNEP). 2005. *After the tsunami: Rapid environmental assessment*. Asian Tsunami Disaster Task Force, United Nations Environmental Program, Geneva, Switzerland.

United States District Court (Eastern District of Louisiana). 2007. *Class action against Murphy Oil. Order & reasons approving class action settlement and award of common-benefit fees and expenses*. Case 2:05-cv-04206-EEF-JCW, Document 1072, Filed 01/30/2007.

University of British Columbia (UBC). 2010. *Analyzing infrastructures for disaster-resilient communities project*. http://www.chs.ubc.ca/dprc_koa. Accessed 4 June 2012.

World Bank. 2011. *The World Bank supports Thailand's post-floods recovery effort*. World Bank News and Views, 3 December 2011. http://www.worldbank.org/en/news/2011/12/13/world-bank-supports-thailands-post-floods-recovery-effort. Accessed 12 June 2012.

World Energy Council (WEC). 2011. *World energy perspective: Nuclear energy one year after Fukushima*. London: World Energy Council. http://www.worldenergy.org/documents/world_energy_perspective__nuclear_energy_one_year_after_fukushima_world_energy_council_march_2012_1.pdf. Accessed 19 Sept 2012.

Yoshimura, N. 2005. *Personal communication*. Tokyo: High Pressure Gas Safety Institute of Japan, October 2005.

Ana Maria Cruz is Professor at the Disaster Prevention Research Institute (DPRI) of Kyoto University. Ana Maria is an international expert in industrial risk management and emergency planning for natech. She has a unique background having work experience as a chemical engineer in industry. Her research interests include risk analysis of flooding, storm, earthquake, tsunami and climate change -induced impacts on infrastructure systems, particularly the oil and gas industry; investigation of natech accidents and consequence analysis; and natech disaster prevention/preparedness policy development and implementation. She has published over 40 peer-reviewed articles.

Yoshio Kajitani is a senior researcher at the Central Research Institute of the Electric Power Industry in Tokyo, Japan. He received his Ph.D. in Civil Engineering from Kyoto University in 2004. His thesis work concerned *Spatial-temporal exposure assessment and loss estimation during disasters*. Kajitani obtained a master's degree (1999) in statistics from the University of Western Ontario in 1999. His major is disaster risk management, focusing on socio-economic impacts of natural disasters. His interests are mainly in the applications of statistics, economics and transportation and utility engineering to the field of disaster prevention, reduction and recovery.

Hirokazu Tatano is Professor at the Disaster Prevention Research Institute, Kyoto University, Japan. He serves as an academic advisor to national, state and local government's disaster risk governance and management policy formulation, implementation and impact evaluation. Hirokazu Tatano is an expert on disaster economic analysis and economic loss modeling, and disaster risk management, infrastructure planning and theory of risk governance. He leads the socio-economic assessment survey team of JSCE (Japan Society of Civil Engineering) for the Great East Japan Earthquake and is the Head of the Business Humanitarian Logistics Planning for Osaka Bay Area in the Committee for Business Continuity Planning Operations of the Osaka Bay System.

Part III
Directions for Further Advancement in Risk Governance

Chapter 24
From Risk Society to Security Society

Simin Davoudi

In the pre-modern time, natural hazards were viewed as divine retributions decreed from on high by divine forces such as Fortuna -the Roman goddess of fate and translated into English as fortune. People were passively exposed to these 'strokes of fate' and believed that they were unable to change them. The Enlightenment project sought to bring such fates under human control by moulding the world to their purposes. The interventions meant that natural hazards which were previously seen as external to and beyond the social realm became increasingly intertwined with it. Modernity and its "technoscience turned what was nonhuman Nature into something contingent and coincident with human society" (Luke 1999: 10), and by doing so it transformed hazards into risks. The distinction between the two lies in the role of human intervention in nature. Whereas hazard refers to a natural event, risk refers to an event whose occurrence is directly or indirectly linked to human action. "Risks are made, hazards naturally occur", as Ulrich Beck (2012: 13–15) put it. It is this understanding of risk which is at the heart of Beck's 'risk society' and the hallmark of what he calls 'reflexive modernity' (Beck 1996). He argues that, the present ecological crisis, along with other social transformations, signifies the emergence of a new form of societal arrangement which he describes as 'risk society'. Risk society represents a new phase in the process of modernisation in which the 'production of *wealth* is systematically accompanied by the social production of *risks*' (Beck 1992: 19). He suggests that "the more modern a society becomes, the more unintended consequences it produces, and as these become known and acknowledged, they call the foundations of industrial modernisation into question" (Beck 1998: 91). He describes this contemporary social experience as 'reflexive modernisation' referring to an era "when modernity is dealing with problems

This note draws on Davoudi, S. (2014). Climate change, securitisation of nature, and resilient urbanism. *Environment and Planning C: Government and Policy* 32(2): 360–375.

S. Davoudi (✉)
Environmental Policy and Planning, School of Architecture, Planning and Landscape,
Newcastle University, Claremont Tower, Newcastle upon Tyne NE1 7RU, UK
e-mail: simin.davoudi@ncl.ac.uk

U. Fra.Paleo (ed.), *Risk Governance*, DOI 10.1007/978-94-017-9328-5_24

literally of its own making" (Dalby 2008: 445). Concerns about risk usher in deep anxieties about security. The more we have 'fabricated uncertainties' (Beck 1996) the greater our sense of insecurity. In the contemporary 'ecologies of fear' (Davis 1999) the risk society becomes intertwined with the security society. Like risk, security is socially produced but, "whereas risk threatens, security promises" (Zedner 2003: 176). It gives people both a *sense* of being safe and the *means* to achieve that. It promises a condition in which risk is non-existent, neutralised or avoided (ibid). While non-existence of risk is utopian, the desire for neutralization or avoidance of risk provides the rationale for relentless pursuit of security (Davoudi 2012). Risk and security, therefore, feed from one another in the sense that keeping up the demand for security requires maintaining a heightened sense of risk. Attraction of such circularity has led to the recasting of many social and environmental problems as security measures. Furthermore, security is not just a means to an end (i.e. protection from risk), but is an end in itself (i.e. a positive good). It is "sold as a desirable product in and of its own right" (Zedner 2003: 160). The pursuit of security is as much about security providers seeking *raison d'etres* for their operations as it is about risk prevention. As a commodity with a price, security becomes factored into both private suppliers' and urban governance's strategic decisions and calculations with profound distributional implications and potential for political exploitation. For urban governance security is becoming a highly sought-after commodity which competes with other commodities in terms of economic and social costs. As Sassen (2011) suggests, security is increasingly urbanized, and cities are increasingly in competition with one another in positioning themselves on the world's league tables of 'safe places'. Emphasis is shifting from urban sustainability to risk and security. Together, risk and security provoke strong emotions and legitimise extraordinary exercise of power. They renounce or displace social conflicts and lead to practices which may otherwise seem indefensible. They create imaginaries of fear which renounce social conflict, foreclose politics, and crowd out descending voices. They squeeze out the arenas in which questions about justice, fairness and conflicts can be raised. Thus, 'the hallmark of the reflexive modernity has become not just the risk society, as Beck suggests, but also the security society' (Davoudi 2014: 371). The recasting of social and environmental problems as security problems reflects and reinforces securitisation as the hegemonic discourse of the twenty-first century.

References

Beck, U. 1992. *Risk society: Towards a new modernity*. London: Sage.

Beck, U. 1996. World risk society as a cosmopolitan society? Ecological questions in a framework of manufactured uncertainty. *Theory Culture and Society* 13: 1–32.

Beck, U. 1998. *Democracy without enemies*. Cambridge: Polity Press.

Beck, U. 2012. *The social and political dynamics of the world at risk: The cosmopolitan challenge*. Paper presented at the 26th annual congress of the Association of European Schools of Planning, July 11–15, Ankara.

Dalby, S. 2008. Green geopolitics. In *A companion to political geography*, ed. J. Agnew, K. Mitchell, and G. Toal, 440–454. Oxford: Blackwell.
Davis, M. 1999. *Ecology of fear. Los Angeles and the imagination of disaster*. New York: Vintage.
Davoudi, S. 2012. Climate risk and security: New meanings of 'the environment' in the English planning system. *European Planning Studies* 20(1): 49–69.
Davoudi, S. 2014. Climate change, securitisation of nature, and resilient urbanism. *Environment and Planning C: Government and Policy* 32(2): 360–375.
Luke, T.W. 1999. *Placing ecocritique in context: Technology, democracy and capitalism as environment*. http://www.vtonline.vt.edu/tim. Accessed on 20 Aug 2012.
Sassen, S. 2011. Against all odds: The urbanizing of human security? In *Bottom politics: An agency-centred approach to globalisation*, ed. D. Kostovicova and D. Marlies, 2016–2226. London: Palgrave.
Zedner, L. 2003. The concept of security: An agenda for comparative analysis. *Legal Studies* 23: 153–176.

Simin Davoudi is Professor of Environmental Policy and Planning at the School of Architecture, Planning and Landscape and Associate Director of the Institute for Sustainability at Newcastle University. She is past President of the Association of European Schools of Planning (AESOP) and one of the editors for Research on Sustainability, at Newcastle University. Simin has led the UK Office of Deputy Prime Minister's Planning Research Network. She is one of the editors of *Journal of Environmental Planning and Management*, founding member of the editorial team of *Journal of the Academy of Social Sciences*, and member of editorial boards of ten international journals. Her research focuses largely on planning and environmental governance. Her recent books include: *Conceptions of space and place in strategic spatial planning* (Routledge 2009), *Planning for Climate change* (Earthscan 2009) and *Reconsidering Localism* (Routledge in press).

Dalby, S. 2003. Green geopolitics. In *A companion to political geography*, ed. J. Agnew, K. Mitchell, and G. Toal, 440–454. Oxford: Blackwell.

[illegible]

Chapter 25
Governing Risk Tolerability

Frederic Bouder

The recurring nature of systemic failures and disasters that intrinsically link technology to the vulnerability of the environment and ecosystems -from the Bhopal or Chernobyl tragedies to the more recent BP oil spill of 2010 and the 2011 Tsunami/ Fukushima Daiichi disaster- raises the fundamental question of risk acceptance. The benefit from technology that we enjoy at an individual level seems to be accompanied by growing scepticism about science. This is combined with increasing fear about the potential downside of technological innovations for health and the environment. How safe is safe enough? Under what conditions are risks to be accepted or refused? Who decides and according to what criteria? In a democratic society, the politics of risk acceptance is particularly puzzling. Born of utilitarian considerations, probabilistic expert-based models have often been viewed as the most rational tool for risk decisions. A strict elitist-technocratic approach to risk decisions raises ethical and political concerns (Bijker et al. 1987). It may not create acceptable risk-benefits tradeoffs for each member of society (Fischhoff 1994).

As risk decisions become less straightforward, the need to formalise democratic risk management procedures becomes even more pressing. Specific methodologies have been devised to organise active engagement in risk situations (NRC 1983; IRGC 2005). Yet, too little has been said about how the new relationship between engagement, expertise and democracy may re-shape the procedures that govern decisions about the risks that we may collectively accept without question, tolerate under specific conditions, or even refuse. For instance, how can twenty-first century 'post-trust' societies (Löfstedt 2005) envisage a reasoned and democratic way of dealing with risk when relatively minor mistakes tend to jeopardise social acceptability?

F. Bouder (✉)
Assistant Professor, Department of Technology and Society Studies,
Maastricht University, Grote Gracht 76, 6211 SZ Maastricht, The Netherlands
e-mail: f.bouder@maastrichtuniversity.nl

U. Fra.Paleo (ed.), *Risk Governance*, DOI 10.1007/978-94-017-9328-5_25

Risk research has used the notion of 'tolerable risk' to describe activities considered worthwhile for the added value or the benefits they provide but sufficiently uncertain to require specific measures to diminish and limit their likely adverse consequences. In practice, simple decision heuristics can offer a valuable conceptual help; in distinct policy fields – i.e. nuclear safety, occupational health and safety – formalized tolerability of risk (ToR) models have been successfully developed. ToR models tend to combine technical probabilistic estimates about the magnitude and harm of a risk with a societal criterion that integrates the perceptions of the non-experts (HSE 1988, 2001). In order to achieve a result that is acceptable to society, stakeholders involved in the bargaining process should be carefully selected to represent the major forces in society.

When Fairman (2007) conducted an institutional analysis of the UK health and safety tolerability model, she came to the conclusion that such models can be established when the objectives of all sides are similar at heart and all sides win by being part of the process of decision-making. Fairman also suggested that this approach to acceptability/tolerability is only possible when its objectives are modest and clearly in the interests of each organisation. Other critical factors for success include the ability of organizations to speak with a single voice, and the ability or those sitting at the negotiation table to sell their decisions to their constituents.

Arguably, the allocation of risks into 'acceptable', 'tolerable' and 'unacceptable' regions is more difficult to achieve in highly confrontational situations, especially when key stakeholders -for example industry and prominent NGOs- fundamentally disagree about how a risk should be handled. Controversies about whether Genetically Modified Organisms (GMO) should be pursued or banned offer a topical example. In the future, more efforts need to be made to formalise risk acceptance procedures in such problematic areas. Failing to agree on risk tolerability procedures may not only lead to further conflicts, it may also result in poor risk-benefit and risk-risk judgments. The consequence may be more harm to health and the environment, which would undermine public support and trust in the fairness of the allocation of risks and benefits.

How to move forward and improve the governance of risk tolerability? Faced with these challenges, the contours of the new 'collaborative discourse' (Renn 1999) remain to be defined. If anything, the formalisation of tolerability decision procedures, especially when conflict arise, will require, in the future, to pay much more attention to cognitive and perception factors. In sensitive areas, the first step should be to launch an honest two-way non-persuasive dialogue between experts, government and non-experts.

References

Bijker, W., T.P. Hughes, and T.J. Pinch (eds.). 1987. *The social construction of technological systems. New directions in the sociology and history of technology*. Cambridge, MA/London: The MIT Press.

Fairman, R. 2007. What makes tolerability of risk work. In *The tolerability of risk: A new framework for risk management*, ed. F. Bouder, D. Slavin, and R. Lofstedt. London/Sterling: Earthscan.

Fischhoff, B. 1994. Acceptable risk: A conceptual proposal. *Risk: Health, Safety & Environment* 1: 1–28.

Health and Safety Executive (HSE). 1988. *The tolerability of risks from nuclear power stations*. Sudbury: HSE Books.

HSE. 2001 (revised 2002). *Reducing risks, protecting people*. Sudbury: HSE Books.

International Risk Governance Council (IRGC). 2005. *Risk governance. Towards an integrative approach*, White paper no. 1. Geneva: IRGC.

Löfstedt, R.E. 2005. *Risk management in post-trust societies*. Basingstoke: Palgrave.

National Research Council (NRC). 1983. *Risk assessment in the federal government: Managing the process*. Washington, DC: National Academy Press.

Renn, O. 1999. A model for an analytic deliberative process in risk management. *Environmental Science and Technology* 33(18): 3049–3055.

Frederic Bouder is an Assistant Professor at the Faculty of Arts and Social Sciences, Maastricht University, in the Department of Technology and Society Studies. His research interests concentrate on developing innovative models of risk regulation and communication, and the crucial concepts of acceptability and tolerability He co-edited the volume The tolerability of risk: A new framework for risk management (2007). For several years Bouder was an Administrator in the Public Governance and Territorial Development Directorate of the OECD, a Senior Analyst for the British Sustainable Development Commission (SDC), Department for Environment, Food and Rural Affairs (DEFRA), United Kingdom and a National Expert at the European Medicines Agency.

Chapter 26
Risk and Adaptive Planning for Coastal Cities

Timothy Beatley

Coastal cities around the world will face tremendous shocks and challenges in the years ahead—sea level rise, cyclones and coastal storms, on top of a variety of climate change-exacerbated health challenges including heat waves and drought, among others. These are tough times for cities, especially those in low-lying coastal settings, suggesting the importance of even more rapidly developing the capabilities and capacities to function in ways that will make them profoundly more resilient in the future.

Is there a silver lining here? Amongst the concerns and understandable fear of human, economic and ecological disruption are there also reasons to be optimistic? In some important ways the answer is 'yes'. For one, the shift towards resilience will bring with it opportunities to re-think and redesign the ways in which cities function, the kinds of infrastructures they invest in, the ways services are provided, often with the chance to enhance quality of life and livability.

The city of Rotterdam, for instance, has developed an extensive climate adaptation program -they speak of the goal of 'climate-proofing' their city by 2025.[1] For Rotterdam one of the key challenges is finding creative new ways to expand the capacity of the city to absorb and hold rainwater on-site. The city has an ambitious goal for installing new green rooftops (40,000 m^2 per year), providing a subsidy for their installation. And it has been developing new, creative ways to retain rainwater, for instance through the concept of 'water plazas'—public square plazas designed to hold storm water some of the time, while most of the time serving as important gathering spaces.

In Rotterdam the impetus to adapt to climate change may help it to become a greener, more biophilic city, with all of the other value contact with nature holds for

[1] Rotterdam Climate Initiative, Rotterdam Climate Proof: Adaptation Programme 2010, City of Rotterdam, Netherlands, 2009.

T. Beatley (✉)
School of Architecture, University of Virginia,
P.O. Box 400122, Campbell Hall, Charlottesville, VA 22904-4122, USA
e-mail: beatley@virginia.edu

U. Fra.Paleo (ed.), *Risk Governance*, DOI 10.1007/978-94-017-9328-5_26

urbanites. In an interview with the head of Rotterdam's adaptation program, Arnoud Molenaar,[2] he described their emphasis on 'no regrets' measures -policies, programs, actions that the city might want to do anyway, and that whatever happens with climate change and weather will deliver important benefits and value. That seems a sensible approach for other coastal cities as well.

In Rotterdam, many of these adaptation measures are viewed as steps towards making the city a more attractive and desirable place to live in the long run. And the city views their growing knowledge-edge and expertise in the area of coastal adaptation as a potential area of economic growth and competitive advantage as a city. As Malenaar says: "If we are able in Rotterdam to keep dry feet…the knowledge that we have developed and techniques that we have implemented can be of much use for other delta cities in the world. Investing in climate adaptation is also in our opinion investing in a promising market."

The Rotterdam vision includes exploring new opportunities to enliven and strengthen the city while also adapting to climate change. The city, for instance, imagines that an entire portion of its harbor might serve as a floating neighborhood, moving up and down with sea levels and floodwaters. Mostly an idea for the future, the city has funded the design and building of one quite interesting floating structure in this area -a visually striking structure of three connected domes, called the Floating Pavilion. Now serving as a climate and energy center, the dome demonstrates the use of newer, lighter materials, and is helping to advance green building as well as climate adaptation.

This idea of floating homes and neighborhoods is not so odd in a country with a love of boats and water and with cities like Amsterdam already with a rich history of house boats. New and creative design expressions of the floating home have made their way into Amsterdam's new IJburg district, west of the city. Here several floating suburbs have emerged. Most of these homes are not especially small, with all the amenities of a more conventional single family house, including a floating deck for outdoor time and in several cases even floating gardens, providing an element of nature and greenery in an otherwise watery setting.

This new approach to thinking about housing in the city's harbor, outside the protection of its dikes, is indicative of the many new ways coastal cities will need to re-imagine shoreline edges and spaces around them. Structural reinforcement in the form of floodwalls and dikes and gates will have their role, but increasingly new more dynamic views of these edges will be necessary. Softer edges where human uses may be more ephemeral and multifunctional, and where opportunities to enhance and restore ecological values abound, may carry the day.

Similarly, mitigating risk will present positive opportunities to advance a larger agenda of sustainability, and this is being demonstrated in a number of other coastal cities. In Houston, Texas, for instance, a shift towards distributive systems of renewable energy is argued as necessary for that city to be resilient in the face of future storm activity, and in New York City, similarly, more decentralized and sustainable

[2] Interview with Arnoud Molenaar, July, 2012, Rotterdam.

water systems have been advocated. Climate resilience has the potential to deliver many other sustainability benefits, such as reducing water and energy consumption.

Urban resilience is a goal that once achieved could provide many other positive benefits. The challenges faced by coastal cities in mitigating future risk are immense and unprecedented, yet they might as well provide an unprecedented opportunity to lay the foundations for a more sustainable, resilient urban culture.

Timothy Beatley is the Teresa Heinz Professor of Sustainable Communities in the School of Architecture at the University of Virginia, where he has taught for the last twenty-five years. Much of Beatley's work focuses on the subject of sustainable communities, and creative strategies by which cities and towns can fundamentally reduce their ecological footprints, while at the same time becoming more livable and equitable places. Beatley is the author of a number of books, including *Green urbanism: Learning from European cities* (2000), *Native to nowhere: Sustaining home and community in a global age* (2004), and most recently *Biophilic cities: Integrating nature into urban design and planning* (2010).

water systems have been advocated. Climate [illegible] the potential to deliver many other sustainability benefits, such [illegible]

[illegible]

[illegible]

Chapter 27
Risk Governance and Development

Andrew E. Collins

Reducing risk is a question of governance in decisions and how people make their decisions. As risk assessment and risk management comprise an ongoing process of decision making it is reasonable to conceptualise a risk governance cycle whereby different governance influences have positive, negative or benign impacts on prevention, coping and recovery from disruptive events. Environmental and human systems are proven to be inherently unpredictable, so that progress in reducing the risk of disaster includes subjective risk assessment and decision-making under conditions of uncertainty. The complexes of decision-making influences in the risk governance cycle relate to *knowledge, power, culture* and *environment*. Capacity to address environmental and technological hazards through improved risk governance is therefore a development issue. The remainder of this short paper outlines some key factors for dealing with risk governance in a developmental way.

Risk assessment is influenced by measurement, knowledge and the understanding and perception of risks. Decisions made on the basis of risk assessments are influenced not only by remaining uncertainty and the politics of the stakeholders involved, but by the effectiveness of knowledge integration and communication. Risk management is influenced by capacity to engage risk reduction, risk transfer, risk ownership and responsibility. The developmental process in the cycle of assessment and management, particularly following a major emergency, is then also dependent on institutional learning, investment in further research and in legislation. An improved awareness of the facilitating and limiting aspects of this process can be elaborated through personalization of risk governance, where closer analyses at the individual level inform likely wider societal and institutional challenges. It is timely to revisit the *knowledge, power, culture* and *environmental* aspects at this localized level in more detail.

A.E. Collins (✉)
Department of Geography, Northumbria University,
B201, Ellison Building, Newcastle upon Tyne NE18ST, UK
e-mail: andrew.collins@northumbria.ac.uk

U. Fra.Paleo (ed.), *Risk Governance*, DOI 10.1007/978-94-017-9328-5_27

Individual, or personal level risk governance (with implication for local and wider interpretations) based on *knowledge, power, culture* and *environment* involves the knowledge base for identifying reactions to risk, the presence or not of constraints to being able to react, rationales to choice, motivation or belief, and the nature of place specific interactions with risks. However, the composition and context of risks are subject to intervening influences such as a person's experience (age, activity and education), structural contexts (the economy, politics and society), social origins (agency, tradition and faith factors), activity and physical environmental change (systemic and chaotic hazards and the changing nature of places). For example, disaster and development related research can show risk related impacts through a lack of experience or education, uneven development, inappropriate technology, access and rights. Though less well covered by systematic studies in high risk environments the role of personality, altruism, kinship, gender, beliefs in immortality, faith and non-faith based traditions also mediate people's reactions to risks. These intervening influences apply together with environmental change aspects in influencing how people are vulnerable and resilient. An applied understanding of risk governance therefore demands an integrated approach as hazards are driven and negotiated by environmental, social and economic change. Ultimately, *knowledge, power, culture* and *environment* aspects of risk governance invoke a need for *awareness*, with *rights* and *responsibilities* for both disaster reduction and development.

In considering risk governance and development applications and implications, resultant resilience achieved is not only in terms of environmental risk reduction, but related objectives of health risk reduction, community strengthening, cost effectiveness and sustainable development. Furthermore, preparedness, early warning and early action for disaster avoidance require that people and communities are motivated to engage in risk reduction for improved wellbeing, which is a condition for any of the above to function. However, questions remain as to the manner in which risk and resilience can be addressed through organized local governance or more as adaptation based on human agency already latent within communities. In relation to this brief synopsis it is suggested for consideration that risk and resilience governance involves:

- Disaster risk reduction as a governance and development process
- Empowering people through capacity and rights that support locally grounded risk reduction strategies
- Improving risk communication, awareness and action
- Delineating the responsibilities of state and citizens regarding risk reduction
- Legislating for rights to the means of risk reduction, for individuals and in terms of the responsibilities of private and public development initiatives
- Facilitating increased scientific and technical support
- Sensitising institutions through all of the above

It is suggested that characteristic 'good disaster risk governance' is where it is:

- Informed by ongoing real and perceived threats of the governed

- Practitioner orientated -guided by perpetual interpretation and review
- Proactively engaged -with hazards, vulnerability, and coping to facilitate resilience
- Based on lessons learnt using evaluation before, during and after risk reduction activities
- Relative to localized knowledge through grounded research
- People centred -driven and motivated by disaster assessment that is multidisciplinary, integrated and perpetual
- In receipt of adequate prioritisation and investment

Ultimately good risk governance not only offsets disaster risk but invests in local wellbeing and its sustainability.

Andrew E. Collins is Professor of Disaster and Development Studies at the Department of Geography and Environment and Leader of the Disaster and Development Network (DDN) at Northumbria University. He has worked for extended periods in the UK, Africa, Asia and Europe. His work involves research, consultancy and institutional development around the rationale that integrated disaster and development studies illuminate options for people centred disaster reduction and future sustainability. He has elaborated this in most detail with respect to health. His publications include titles such as Environment, health and population displacement (1998), Disaster and development (2009) and Hazards, risks and disasters in society (2014).

Chapter 28
Profiling Risk Governance in Natural Hazards Contexts

Gordon Walker and Fiona Tweed

Risks are always managed within a broader context of relationships between government, citizens, civil society and business. To talk of risk governance is therefore to tie the management of threats, including those posed by natural hazards, to on-going changes in how societies achieve their goals. The various elements of what has been termed 'new governance' include the emergence of multi-level structures and processes and the 'hollowing out' of the nation state; moves away from the exercise of centralised authority towards the collaboration of a multiplicity of actors; the application of new forms of authority and control; and a changing distribution of responsibilities between the state and other actors.

Whilst such governance characteristics can be discussed in sweeping terms, in practice there are considerable differences between countries and regions in the extent to which these trends of change have taken place. In the case of natural hazards it is possible to identify instances of 'new governance' characteristics; for example in new forms of involvement of private, NGO and community stakeholders and in new modes of management, which emphasize those at risk taking greater responsibility for their own protection. However, these are not universal features of the ways in which risks are now being governed and much variation exists across the diverse political contexts, for example, of the European region.

Through work in the EU-funded CapHaz-Net project (Walker et al. 2010) we have developed a framework for profiling some of the key dimensions of natural hazard governance. The aim was to capture the variability and dynamism of governance practice through a simple structure that enables any chosen national,

G. Walker (✉)
Lancaster Environment Centre, Lancaster University, Lancaster LA1 4YD, UK
e-mail: g.p.walker@lancaster.ac.uk

F. Tweed
Geography and Environment, Science Centre, Staffordshire University,
Leek Road, Stoke-on-Trent ST4 2DF, UK
e-mail: f.s.tweed@staffs.ac.uk

U. Fra.Paleo (ed.), *Risk Governance*, DOI 10.1007/978-94-017-9328-5_28

regional or local natural hazard governance context to be profiled against a set of eight governance characteristics. These characteristics address:

- governance scale and its distribution between national (1), regional (2) and local levels (3), with a spectrum from weak to strong in each case,
- how much those at risk are expected to be responsible for protecting themselves, compared to how much responsibility rests with government (4),
- the extent and culture of stakeholder participation in the governance system (5),
- the type of insurance provision in place, in terms of how much this is marketized and segmented according to level of risk (6),
- the extent of communication with the public about risks (7), and
- the degree of balance between governance tasks and the availability of resources for such tasks to be carried out (8)

Two examples of the application of this framework, as completed by the authors, are provided in Fig. 28.1, the first for flood risk governance in the UK, the second for volcanic hazard governance in Iceland. In each case and for each characteristic, the current position is indicated along a spectrum with extremes at either end. Arrows towards the current position indicate the direction and extent of change (if any) over the last 5 years. Arrows away from the current position indicate any shifts expected in the future, again over a 5 year timespan.

The result is a visualised governance profile that is deliberately qualitative and subjective in character reflecting the value that we see in the use of the frame-work. We can envisage it being used with a group of stakeholders each producing their own version and then discussing their similar or contrasting perspectives. It could also be used to compare governance profiles for the same hazard across different contexts (e.g. for earthquake hazards in different national systems), or between hazards in the same context (e.g. for the range of natural hazards in Italy), thereby generating questions about why similarities or differences exist. In such ways the profiling of key risk governance characteristics could promote debate about the different ways of governing risks from natural hazards, and greater understanding of both the complexities and choices involved.

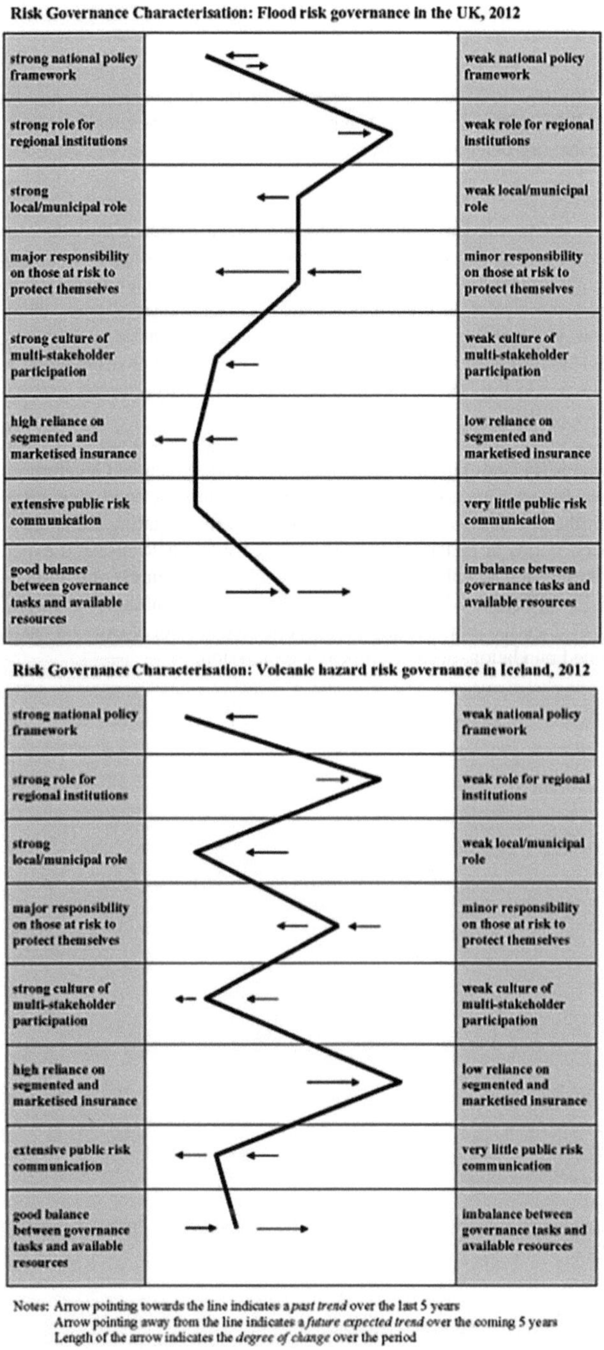

Fig. 28.1 Two examples of risk governance profiles

Reference

Walker, G., R. Whittle, W. Medd, and N. Watson. 2010. *Risk governance and natural hazards. CapHaz-Net WP2 report.* Lancaster Environment Centre, Lancaster University. Available at: http://caphaz-net.org/outcomes-results.

Gordon Walker is Chair in Environment, Risk and Social Justice at the Lancaster Environment Centre at Lancaster University, and has a distinctive profile of research on the social and spatial dimensions of environment, risk and sustainability issues. He has undertaken UK and EU research projects focused on a range of different technological and natural risks issues, examining issues of public perception, vulnerability, risk governance, risk communication and land use planning. He has also been a leading figure in the development of research on environmental inequalities and environmental justice in the UK. His most recent book is *Environmental justice: Concepts, evidence and politics* (2012).

Fiona Tweed is Professor of physical geography in the Department of Geography at Staffordshire University. Fiona Tweed is a recognised expert on glacier outburst floods, ice/volcano interactions and natural hazards. She has over 20 years' experience of investigating the processes and impacts of glacial outburst floods in Iceland, where she has led many field research expeditions. Fiona's research also focuses on risk, vulnerability and resilience in the context of natural hazards, particularly in Europe. Fiona has served on the governing bodies of international research projects, on the editorial boards of scientific journals and is an experienced advisor to external agencies such as the European Science Foundation, or the European Commission.

Chapter 29
Risk Governance and the Social Amplification of Risk: A Commentary

Roger E. Kasperson

One of the most perplexing problems in risk governance is why some relatively minor risks or risk events, as assessed by technical experts, often elicit strong public concerns and result in substantial effects on the governance. Social amplification is an analytic framework that seeks to link systemically the technical assessment of risk with psychological, sociological, and cultural perspectives of risk perception and risk-related behavior (Kasperson et al. 1988; Pidgeon et al. 2003). The main thesis is that hazards interact with the governance process that may shape extensively the public responses to the risk or risk event. Risk amplification in this framework occurs at two stages: in the societal diffusion of information about the risk, and in the decision process of society. Involved are the scientists who communicate risk assessments, the news media, cultural groups, and interpersonal networks that interpret and reframe the risk and locate it in governance priorities. One important result is secondary impacts or what we term 'ripple effects'. One implication is that governance typically is 'risk-informed' rather than 'risk-based.'

The starting-point in the risk governance process is 'risk events,' which include accidents and incidents or even new reports on existing risk. They remain limited to the domain of science, unless human beings observe and communicate them to others. A key part of governance, therefore, is the communication process. Depictions of risks are portrayed through various risk signals which in turn interact with a wide range of psychological, social, institutional, or political processes in ways that intensify or attenuate assessments of risk and their severity. The experience of risk, therefore, is not only an experience of physical harm but the result of communication processes by which groups and individuals learn to acquire or create interpretations of risk and risk frames. Social amplification may qualitatively and quantitatively increase or diminish the risk and its consequences. This was quite apparent in the Fukushima nuclear accident in Japan as Germany restructured its energy system in

R.E. Kasperson (✉)
George Perkins Marsh Institute, Clark University,
950 Main Street, Worcester, MA 01610, USA
e-mail: rkasperson@clarku.edu

U. Fra.Paleo (ed.), *Risk Governance*, DOI 10.1007/978-94-017-9328-5_29

a ripple effect, the U.S. launched a major review of its nuclear plants, and France and Japan, two leading nuclear states, pondered potential phase of nuclear power.

Perhaps no problem is more complex in the social amplification of risk than uncertainty. Uncertainty is an inescapable ingredient of life. Even for familiar situations—such as crossing a street—some level of uncertainty inevitably exists. Past experience is relevant for decision involving the future but contexts change and new elements affecting risk may unexpectedly appear. Usually this residual uncertainty remains within reasonable bounds and humans make their way in an uncertain and changing world where existing knowledge and experience suffice as a guide to future expectations. But where highly complex systems with extensive connectivity and interactions exist, or where novel problems or technology limit experience as a resource, governance must proceed and decisions must be taken even under conditions of high uncertainty. Many decisions cannot wait. It is not surprising that in a world of complex systems involving rapid technological change, highly coupled human and natural systems, and a kaleidoscope of social, economic, and political institutions, high levels of uncertainty challenge existing assessment methods, and public consideration and communication of risk decisions.

While there is little question about how challenging risk uncertainty and social amplification are for the scientific community, they are not issues for the scientists alone. Uncertainty and social amplification have much to do with the differential pattern of vulnerability, to nature and to human communities, and to the ambiguities surrounding effective governance and potential risk interventions. To take a prominent example, risk assessment assumes that sufficient knowledge and quantification can be achieved so that command-and-control strategies and regulation can be employed. Much depends, however, on how large the residual uncertainties and amplification are, whether they can be reduced significantly, and how they affect the social acceptability of the risk. Where large uncertainties and social amplification exist, other governance approaches may be called for. 'Adaptive governance,' for example, recognizes that large uncertainties may still exist and that risk knowledge is co-evolving with the risk, so an alternative approach to governance -going with the flow and making mid-course corrections- may be required. Much depends upon both the risk and the governance system, however, and the extent to which mid-course corrections can be made in technologies, institutions, public values, and communication networks, and how extensively amplification affects decision options.

References

Kasperson, R.E., O. Renn, P. Slovic, H.S. Brown, J. Emel, R. Goble, J.X. Kasperson, and S.J. Ratick. 1988. The social amplification of risk: A conceptual framework. *Risk Analysis* 8(2): 177–187.

Pidgeon, N., R. Kasperson, and P. Slovic (eds.). 2003. *The social amplification of risk.* Cambridge: Cambridge University Press.

Roger E. Kasperson is Research Professor and Distinguished Scientist at the George Perkins Marsh Institute at Clark University. He has taught at Clark University, the University of Connecticut, and Michigan State University. His expertise is in risk analysis, global environmental change, and environmental policy. He has served on EPA's Science Advisory Board, and on numerous committees of the National Research Council. He is a member of the Executive Steering Committee of the START Programme of the IGBH. He is an elected member of the National Academy of Sciences and the American Academy of Arts and Sciences. From 2000 to 2004, Kasperson was Executive Director of the Stockholm Environment Institute in Sweden. He has been honored by the AAG for his hazards research.

Chapter 30
Help or Hindrance? The Contribution of the Resilience Approach to Risk Governance

Georg Frerks

30.1 Introduction

The notion of resilience has rapidly gained popularity in the field of disaster studies and emergency management. Improving the resilience of individuals, communities and societies is thought to be an effective and efficient way to reduce prevailing vulnerabilities and thereby the risk of disaster. The advantage of strengthening resilience is that it is an all-hazard approach killing several birds with one stone. Resilience has been embraced as the new catchword for the decade to come and at present there is an avalanche of initiatives, workshops and publications on the subject, very much like happened to the notion of vulnerability that dominated the disaster discourse in the 1990s. In a recent ODI Background Note Tom Mitchell discusses various options for including disaster resilience in post-2015 development goals, including a 'standalone goal on disaster resilience' or a 'mainstreaming approach' incorporating the theme in other sector-oriented goals (2012: 2).

On the other hand, the ascendancy of resilience has also attracted serious criticisms. For example, Ben Aguirre and Eric Best (2015) consider the current widespread usage of the concept of resilience just a 'fad', and in fact redundant when applied to research and management of disasters, as the strengthening of the institutions of society faced with disasters has been already ongoing practice for half a century, they state. While taking a much less radical stance, Kathleen Tierney also observed several fundamental weaknesses with regard to the resilience

G. Frerks (✉)
Conflict Prevention and Conflict Management, Centre for Conflict Studies, Utrecht University, Drift 6, 3512 BS Utrecht, The Netherlands

International Security Studies, Faculty of Military Sciences, Netherlands Defense Academy, P.O. Box 90004, 3509 AA, MPC 55A Utrecht, The Netherlands

Disaster Studies, Sociology of Development and Change, Social Sciences Group, Wageningen University, Hollandseweg 1, 6706 KN Wageningen, The Netherlands
e-mail: g.frerks@uu.nl; ge.frerks@nlda.nl

U. Fra.Paleo (ed.), *Risk Governance*, DOI 10.1007/978-94-017-9328-5_30

approach in a recent keynote delivered on the subject.[1] While acknowledging its stimulating force in policy, she wondered whether the concept was really innovative or rather 'old wine in new wineskins' and whether it comprised a sufficiently deep analysis of root causes. She, among others, further critiqued its under-theorization of power and claimed that there was a need to focus much more on the 'pathologies of power' that generated wide-spread vulnerabilities in society, including different forms of policy denial and denigration of initiatives in the face of an unsustainable future.

So what to think of the strengths and weaknesses of resilience? Below I give first an overview of the resilience concept and approach and then discuss its significance in terms of policy and politics. Summing up my arguments at the end, I try to conclude what the resilience approach can contribute to risk governance.

30.2 The Ascendancy of Resilience in Disaster Studies and Its Definition

Since the 1990s the field of Disaster Studies has taken on board some ideas from environmental systems analysis, resilience being one of them. The ecologist Holling defined resilience as "the ability of a system to maintain its structure and patterns of behavior in the face of disturbance" (Holling 1986: 296). In disaster research, the definition of resilience means the ability to survive and cope with a disaster with minimum impact and damage. However, it was further expanded to include social and institutional aspects. Harrald and Veldhuis (2010) provide an overview of the recent debate on resilience in the US and include a series of definitions currently in use by US departments and in academic literature. Box 30.1 gives three of them showing an increasing complexity.

These definitions emphasize the capacity or ability to anticipate risk or disturbance, absorb or limit impact, and bounce back after a crisis but -more importantly- they include adaptive community capacity, and processes of change, as evidenced in definitions of Cutter et al. (2008) and Norris et al. (2008). It must be stressed that these capacities and abilities mentioned are not some mysteriously in-built systemic property of individuals or organizations, but are based on interactive and contingent community-level and societal processes involving change, entrepreneurship, learning and increased competence. Hence, these definitions move far beyond the ecologists' traditional equilibrium thinking. In that sense resilience does not need to be a return to a previous equilibrium. The current debate about disaster rehabilitation asserts that rather than 'building back' we should be 'building back better', giving disaster survivors more capabilities, options and flexibility in their coping with

[1] Keynote Kathleen Tierney, 3rd Conference on Community Resilience, organised by the Center for Community Security and Resilience, Virginia Tech, Arlington, USA, in collaboration with the Metropolitan Institute, Congress Center, Davos, Switzerland, 24–25 August 2012.

Box 30.1: Definitions of Resilience

Community resilience "is defined as the sustained ability of communities to withstand and recover – in both the short and the long terms – from adversity" (U.S. Department of Health and Human Services 2009: 5).

"Resilience refers to the ability of human systems to respond and to recover. It includes those inherent conditions that allow the system to absorb impacts and cope with the event, as well as post-event adaptive processes that facilitate the ability of the systems to recognize, change and learn in response to the event" (Cutter et al. 2008).

Resilience is "a process linking a set of adaptive capacities to a positive trajectory of functioning and adaptation after a disturbance. Community resilience emerges from four primary sets of adaptive capacities – Economic Development, Social Capital, Information and Communication and Community Competence" (Norris et al. 2008).

(Derived from Harrald and Veldhuis 2010: 9–10)

future adversity, and also making progress by structural vulnerability reduction and the increase of institutional capabilities.

In this connection, the strength of the resilience approach is that it is human-centered and community-focused, but simultaneously situated in a larger macro-setting of environmental, macro-economic and policy processes and cognizant of global-local dynamics. It is also interdisciplinary and multi-layered, requiring new forms of stakeholders' engagement and public-private partnerships. One critique on the earlier vulnerability approach in disaster management pointed out that it victimized and disempowered people. It would engender a fatalistic and passive outlook and take away the agency from people, thereby creating external dependency. In fact, vulnerability was and still is often externally attributed to groups of people, who rarely label themselves as vulnerable. Anderson and Woodrow (1989) highlighted already two decades ago that people have important physical, social and motivational capacities that can offset their vulnerabilities. The thinking on local disaster capacities has sociologically been further influenced by debates on actor-orientation and the role of agency. Actor-orientation is a constructivist perspective focusing on the making and remaking of society through the self-transforming actions and perceptions of a diverse and interlocked world of actors (Long 2001). Actor-oriented approaches form a counter-balance to approaches that basically see human behaviour as externally determined. In relation to earlier paradigms in disaster studies the resilience approach moves beyond the vulnerability and victimization discourse towards agency and capacity, and from short-term coping towards longer-term adaptation and innovation. It focuses on process rather than being a static state of affairs.

30.3 Resilience as a Policy Approach

Turning to the policy world, it seems to make sense to invest in resilience in view of its merits outlined above. On the other hand, there is as yet fairly little insight in how to translate resilience into a workable concept and policy approach. We need more substantive work on the operationalization of the concept and its use in policy practice. The resilience approach is associated with a clear shift in responsibilities and roles in public disaster policy and with regard to the composition of the actor alliances involved. In the field of disaster management collaboration between authorities and citizens was already promoted in the 1994 Yokohama and 2005 Hyogo frameworks. It is however necessary to ascertain the impacts of such policy shifts on the anticipation and prevention of, and recovery after shocks. As grass-roots or community-based perspectives have often been welcomed merely on ideological grounds or 'feel-good' sentiments, it is essential to provide for a critical and evidence-based framework to inform policy and practice on resilience initiatives and enhance their effectiveness. Such a framework should include: (a) a further definitional delineation and conceptual elaboration of resilience, building on the ample literature that exists today; (b) define descriptive-analytical benchmarks or indicators for resilience (also here much work is ongoing already); (c) collect empirical evidence on the application of the resilience approach in practice or work with pilot cases (this evidence is still weak); (d) analyze the larger policy and political context and its impact (see my remarks below) and (e) propose policy measures to enhance resilience.

Though such steps can help and promote community and societal resilience in disaster-prone areas, there still remains a need to critically approach the resilience paradigm. Whether or not such interventions may have a beneficial impact in terms of risk governance and the target population also depends on the broader political and economic context, as already mentioned by Tierney in her keynote referred to above. Therefore I suggest that alongside the policy work outlined above, a more politically informed analysis takes place that looks at and deconstructs the resilience discourse as a political project.

30.4 Resilience as a Political Project

What are in effect the political underpinnings of the resilience approach? It can -in my view- be considered as part of the larger neo-liberal project that is taking hold of contemporary society. In terms of (risk) governance it relates to a model that includes parliamentary democracy, a liberalized economy with a retreating state, and western model of security provision based on the securitization of certain external threats. Some authors have claimed that this neo-liberal ordering of the world has led on the one hand to an interventionist attempt to govern and control parts of the globe, implying the erosion of civil rights and liberties, while on the other hand it is excluding and marginalizing those people deemed useless, who have been called the 'insecured' or 'surplus life' (Duffield 2007) or 'wasted lives' (Bauman 2004).

The emphasis on resilience indeed seems to be the product of a political discourse that seeks to shift the responsibility for mediating the impact of disasters from the state to the society and therefore may engender the same problems and feelings of disenchantment as the neo-liberal project creates in other societal domains and the economy at large.

Reid (2010) suggests that "the resilient subject is a subject which must permanently struggle to accommodate itself to the world". By doing so resilience backgrounds the political, the imagining of alternatives and foregrounds adaptivity, accepting "the imperative not to resist or secure themselves from the difficulties they are faced with". Coaffee and Rogers (2008) claim that the notion of social resilience has been instrumentalized, leading to a new governance and policy structure exerting domination and causing inequality. They talk in this connection about a 'dark side' to resilience planning. Though those warnings help us to focus on potential risks and dangers resulting from the political context, the ultimate test of the resilience approach lies in what it achieves in practice. As I said above, evidence is still largely absent or patchy and hence, the jury is still out.

30.5 Conclusion

In recent years, resilience has rapidly become a mainstream notion as a useful addition to hazard and vulnerability. The concept of social resilience focuses our minds on the social capacities available well beyond the capacities of the formal disaster management sector, and is also redressing the victimizing and disempowering effects of the vulnerability notion. While having a number of strong points, the resilience project also carries risks to society. Whether promoting resilience reduces people's vulnerability to disaster is highly dependent on a person's socio-economic standing. Here, a more differentiated approach is called for than the current generalized one to promoting resilience implies. In this connection, we should be critical about the fiction promoted by the retreating neo-liberal state that everyone can be equally resilient. We have to study the potential negative political effects the neo-liberal project inheres in order to fully gauge its impact on vulnerable disaster-stricken individuals and communities, and how it may affect the governance of risk ultimately.

References

Aguirre, B.E., and E. Best. 2015. How not to learn: Resilience in the study of disaster. In *Learning and calamities. Practices, interpretations, patterns*, ed. H. Egner, M. Schorch, and M. Voss. London: Routledge.

Anderson, M., and P. Woodrow. 1989. *Rising from the ashes, development strategies in times of disaster*. Boulder/San Francisco/Paris: Westview Press/UNESCO.
Bauman, Z. 2004. *Wasted lives. Modernity and its outcasts*. Cambridge: Polity Press.
Coaffee, J., and P. Rogers. 2008. Rebordering the city for new security challenges: From counter terrorism to community resilience. *Space and Polity* 12(2): 101–118.
Cutter, S.L., L. Barnes, M. Berry, C. Burton, E. Evans, E. Tate, and J. Webb. 2008. A place-based model for understanding community resilience to natural disasters. *Global Environmental Change* 18(4): 598–606.
Duffield, M. 2007. *Development, security and unending war. Governing the world of peoples*. Cambridge: Polity Press.
Harrald, J.R., and P. Veldhuis. 2010. *Resilience as a strategic concept, a next step in disaster management US-NL*. Blacksburg, VA: Virginia Tech (exploratory research paper).
Holling, C.S. 1986. The resilience of terrestrial ecosystems; Local surprise and global change. In *Sustainable development of the biosphere*, ed. W.C. Clark and R.E. Munn, 292–317. Cambridge: Cambridge University Press.
Long, N. 2001. *Development sociology: Actor perspectives*. London/New York: Routledge.
Mitchell, T. 2012. *Options for including disaster resilience in post-2015 development goals*. ODI background note, September 2012. London: ODI.
Norris, F., S. Stevens, B. Pfefferbaum, K. Wyche, and R. Pfefferbaum. 2008. Community resilience as a metaphor, theory, set of capacities, and strategy for disaster readiness. *American Journal of Community Psychology* 41(1): 127–150.
Reid, J. 2010. *The disastrous and politically debased subject of resilience*. Paper presented at the symposium on the biopolitics of development: Life, welfare, and unruly populations, 9–10 September 2010. http://www.mcrg.ac.in/Development/draft_Symposium/Julian.pdf.
U.S. Department of Health and Human Services. 2009. *National health security strategy*. http://www.hhs.gov/aspr/opsp/nhhs/nhhs0912.pdf.

Georg Frerks holds a Chair in Conflict Prevention and Conflict Management at Utrecht University, The Netherlands as well as a Chair in International Security Studies at the Netherlands Defence Academy. Till mid 2014 he held the Chair of Disaster Studies at Wageningen University, The Netherlands. As a sociologist and policy analyst he focuses on disaster and conflict-induced vulnerabilities and local responses as well as on policies and interventions implemented at international and national levels. Frerks has done extensive work on the Asian tsunami and is presently working on issues of local resilience and community involvement in disaster preparedness and response. He participated in the Netherlands US Crisis Research Network NUWCReN. He co-edited with Greg Bankoff and Dorothea Hilhorst the volume *Mapping vulnerability: Disasters, development & people* (2004).

Chapter 31
Risk Mitigation: We Are All Going to Die

Graham A. Tobin

We are all going to die! If we believe the media, we would not venture outside for fear of mayhem and destruction. The popular media seem often to construct a picture of a world gone wild, where life is inherently unsafe, yet clear information about genuine risks and how to prepare is lacking. Perhaps this is not surprising, because risk is indeed difficult to define, as demonstrated in the chapters in this book. Academic disciplines, government agencies, private organizations, and insurance companies all conceptualize risk in different ways; they use terms that appear synonymous, but have intrinsically different meanings - presenting critical communication difficulties that hinder effective planning.

Natural hazards' risks are often portrayed as the product of the probability of an event's occurrence and its magnitude. We can see the utility of this from a governance perspective; the probability of a specific size event occurring in a given period informs us as to a level of risk. This measure has scientific credibility in that fairly accurate projections of recurrence intervals at specific locations can be calculated. However, the equation does not incorporate critical human elements that must be integrated to account for societal factors. Four components are pertinent here to governance and risk awareness: spatial and temporal scales, communication, and human vulnerability.

For example, the USA National Flood Insurance Program identifies flood risk based on the 100 year floodplain, using analyses of flood records; this generally results in reasonable spatial and temporal estimates of flood frequencies. It should be stressed, however, that while the 100 year floodplain demarcates the probability of a 100 year event, in reality the floodplain transitions from frequent to less frequent flooding probabilities. Risk then varies spatially, so location is crucial and thus the appropriate spatial scale of analysis for planning purposes needs to be determined.

G.A. Tobin (✉)
School of Geosciences, University of South Florida,
4202 E. Fowler Ave (NES 107), Tampa, FL 33620, USA
e-mail: gtobin@usf.edu

U. Fra.Paleo (ed.), *Risk Governance*, DOI 10.1007/978-94-017-9328-5_31

Furthermore, the hydrological system is dynamic, responding to natural and human-induced perturbations in different ways but rarely reaching any long-term equilibrium, so flood recurrence probabilities are ever fluctuating. Global climate change, for instance, will increase flood frequency in some areas and reduce it in others, but it is (usually) not possible to predict specific outcomes given the complexity of dynamic natural and human systems. In this context, what should the appropriate temporal scale for governance projections be?

The picture is further complicated by the human dimension, with risk determined by degree of exposure to extreme geophysical events. A hurricane making landfall in Miami, Florida, will create greater impacts than one in less densely populated areas. At the same time, differential human vulnerabilities create variations in risk that must be examined if mitigation strategies are to be effective. It is well established that extreme events do not affect everyone equally but are determined by a web of social, economic and political forces that can exacerbate or ameliorate impacts. The poor and marginalized are invariably negatively affected in disasters and thus experience higher risk than other members of society.

Effective communication of these risks also presents real challenges. The concept of the 100-year event is confusing and it is not unusual to hear flood victims who have experienced such an event remark that they will not experience another, since it will be a century away. From their perspective, the risk has disappeared and they may be ill-prepared when another 100-year flood occurs. Good communication in clear concise language is critical to instill political will and individual motivation to undertake preventative action.

Understanding the hazardousness of place, therefore, presents challenges, especially for those tasked to mitigate risk. To determine risk for governance purposes, it is essential to consider the following: (1) *Geophysical research*: Studies must continue to assess the ever-changing probabilities and recurrence intervals of geophysical events; (2) *Vulnerability metrics*: Further research is necessary to ascertain how differential vulnerabilities affect risk perceptions and responses to enhance the predictive value of vulnerability models; (3) *Behavioral concerns*: It is still not known exactly how individuals will react to disaster. Personal experience can have both positive and negative impacts, as can communication and education, and risk taking is not fully understood in this context; (4) *Risk levels*: Acceptable levels of risk need to be determined in association with governance strategies; (5) *Hazardousness of place*: A focus on context of place is required, which also raises issues as to the appropriate scale of analysis for planning; (6) *Dynamic systems*: The dynamism of both the human and natural systems must be recognized and flexibility be incorporated into the planning process; (7) *Personal and community responsibility*: Even when extensive mitigation strategies are implemented, there remains a residual risk for which individuals and communities must take responsibility.

In essence, the various dimensions of risk are difficult to encapsulate. Certainly, we are all going to die but our risk of doing so in a specific disaster can be mitigated by taking account of decades of hazards research that informs the most appropriate remedial action.

Acknowledgments This brief commentary has been shaped by the vast literature on risk undertaken by many researchers from diverse academic disciplines.

Graham A. Tobin is a Professor in the School of Geosciences the University of South Florida. Dr. Tobin's research interests in natural hazards, water resources policy, and environmental contamination focus on sustainability concerns of human vulnerability, community resilience, social networks, and health conditions in hazardous environments. His current research is being conducted in communities around two active volcanoes in Ecuador and Mexico, and in hurricane prone areas of the United States.

Chapter 32
Confronting the Risk of Large Disasters in Nature

John B. Rundle and Donald L. Turcotte

Risk is usually defined to be the product of (hazard) × (vulnerability), or alternately, (hazard) × (exposure). Ameliorating risk can be accomplished either by reducing the hazard, or reducing the exposure to the hazard. For natural hazards, it is usually not possible to prevent the hazard. Earthquakes and hurricanes are examples, although events such as human-induced earthquakes in areas of ground water injection may be exceptions. One strategy is to simply move away from hazardous areas, thereby reducing the exposure to the hazard. Another method is to alter and strengthen physical structures and social processes so that they are less susceptible to disruption and destruction.

Wildfires and floods offer interesting examples of hazards that might be greatly minimized through preventive measures. In areas where vegetation is systematically removed by controlled burns or other means, wildfires tend to occur in smaller clusters, rather than as a large conflagration. Case studies of interest involve southern California and northern Baja California (Mexico) (Minnich and Chou 1997). South of the border, wildfires are often allowed to burn themselves out, thereby removing combustible fuel at regular intervals. The result is a series of typically small clusters of fires that burn frequently during the summer months. North of the border, US law requires that fires be rapidly extinguished as they occur. The result is the buildup of large quantities of combustible fuel load that, over decades, may produce large conflagrations consuming hundreds of residential structures within a few days.

Floods can be diminished if the growth of vegetation is encouraged on slopes prone to heavy rainfall. Denuded slopes can occur following mudslides and wildfires. Other sources of flooding are related to man-made activities, such as the levees

J.B. Rundle (✉)
Department of Physics, University of California, One Shields Ave, Davis, CA 95616, USA
e-mail: jbrundle@ucdavis.edu

D.L. Turcotte
Department of Geology, University of California, One Shields Ave, Davis, CA 95616, USA
e-mail: dlturcotte@ucdavis.edu

U. Fra.Paleo (ed.), *Risk Governance*, DOI 10.1007/978-94-017-9328-5_32

around New Orleans, LA, and Sacramento, CA. Failure of the levees in the New Orleans area following hurricane Katrina led to the inundation of the city and its Ninth Ward (Zaninetti and Colton 2012).

Statistics of natural hazards such as these are often characterized by fat-tailed (power-law) distributions. An example is the Gutenberg-Richter law of seismology, which can be written in the form (Scholz 1990):

$$N = AM^{-\beta} \tag{32.1}$$

N is the cumulative number of events larger than seismic moment M, A is a measure of the total number of events, and β is a scaling exponent. For earthquakes, M is related to the more familiar magnitude m via the relation $m = (2/3)\log_{10} M - 6.0$. For earthquakes, the exponent $\beta \approx 2/3$. Wildfires (Malamud et al. 1998) and floods (Malamud and Turcotte 2006) can also be described by power law statistics.

Equation (32.1) is an example of fat-tailed statistics. The frequency of occurrence of many such natural phenomena are described by equations (Sachs et al. 2012) similar to (32.1). Questions of research interest include whether Eq. (32.1) uniformly describes the frequency of occurrence of all sizes of events, or whether there are several power laws or scaling regimes with different exponents. Other questions relate to whether the largest events occur more frequently than as described by Eq. (32.1), i.e. whether all events fall on the scaling line (32.1), or whether the largest earthquakes are 'characteristic earthquakes' or 'nucleation events' that occur disproportionately often.

If the occurrence of natural hazards such as large wildfires cannot be prevented, an alternative strategy is to develop forecast methods. Such methods are highly visible for weather or climate hazards, but are much less so for earthquakes. Most seismically active countries do have official earthquake forecasts or probabilistic hazard maps, primarily to set insurance rates and to determine building construction standards and codes for the reduction of vulnerability or exposure. In California, the official forecast has been published for over 20 years by several Working Groups on California Earthquake Probabilities (Field et al. 2009), which have been associations of federal and state agency scientists, together with groups of academic scientists.

Improving methods for forecasting earthquake and other natural hazards is currently an active subject of research. Validation and verification methods are critical, as is the use of these methods for assimilating data into forecasts (Joliffe and Stephenson 2003). New methods for earthquake forecasting are under development, including the Natural Time Weibull (NTW) method (Rundle et al. 2012), which is based on the idea that the largest and most critical events 'fill in' the fat-tailed distribution (32.1) at infrequent intervals. For the case of earthquakes, it is found from Eq. (32.1) that for every magnitude $m \geq 6$ earthquake, there are approximately 1,000 earthquakes having $3 \leq m < 6$. So if 1,000 $m \geq 3$ earthquakes have occurred in a region since the last earthquake, it can be concluded that a large earthquake is highly probable in the near future (Rundle et al. 2012). Using the Weibull probability law to convert this statement to a quantitative probability results in the NTW model for earthquake probabilities.

Disseminating information about natural hazard occurrence and risk has in past years been the province of federal and state agencies. Most members of the public have been completely unaware of the risk they face from these events. Recently, however, the World Wide Web has enabled new modes of information distribution. Examples of effective use of this information infrastructure include the QuakeSim (www.quakesim.org) and OpenHazards (www.openhazards.com) web sites. These sites provide tools and information to researchers and the public. It is likely that this type of direct and open access information and sharing will grow rapidly in the future.

Traditional approaches to risk management rely on a top down process, in which risk information is provided to governments and large corporations by groups of scientists and engineers working in government agencies or in the private sector. By contrast, the new emergent paradigm is directed at using the world wide web to provide information and tools on risk directly to the public, allowing them to decide on courses of action to remediate their risk. The dichotomy between the old and new approaches is best seen in the area of weather forecasting and probability. Examples of groups providing weather forecasts include the NOAA National Weather Service, the companies www.weather.com, www.wunderground.com, and www.climate.com, and for weather catastrophes, www.rms.com and www.eqecat.com.

Quality and reliability of forecasts is always an issue, both for government as well as commercial organizations. Some type of formal backtesting and historical analysis of forecast success is required. Unfortunately, this is not possible for a common approach to earthquake forecasting, which relies on 'expert elicitation' more usually known as 'expert opinion'. In order to backtest forecast models, a completely automated forecast method is required. Backtesting of this type is now used routinely in weather forecasting (Joliffe and Stephenson 2003).

References

Field, E.H., T.E. Dawson, K.R. Felzer, A.D. Frankel, V. Gupta, T.H. Jordan, T. Parsons, M.D. Petersen, R.S. Stein, R.J. Weldon II, and C.J. Wills. 2009. Uniform California earthquake rupture forecast, Version 2 (UCERF 2). *Bulletin of the Seismological Society of America* 99: 2053–2107.

Joliffe, I.T., and D.B. Stephenson. 2003. *Forecast verification. A practitioners guide in atmospheric science*. New York: Wiley.

Malamud, B.D., and D.L. Turcotte. 2006. The applicability of power-law frequency statistics to floods. *Journal of Hydrology* 322: 168–180.

Malamud, B.D., G. Morein, and D.L. Turcotte. 1998. Forest fires: An example of self-organized critical behavior. *Science* 281: 1840–1842.

Minnich, R.A., and Y. Chou. 1997. Wildland fire patch dynamics in the chaparral of southern California and northern Baja California. *International Journal of Wildland Fire* 7: 221–248.

Rundle, J.B., J.R. Holliday, W.R. Graves, D.L. Turcotte, K.F. Tiampo, and W. Klein. 2012. Probabilities of large events in driven threshold systems. *Physical Review E* 86(2–1): 021106.

Sachs, M.K., M.R. Yoder, D.L. Turcotte, J.B. Rundle, and B.D. Malamud. 2012. Black swans, power laws, and dragon-kings: Earthquakes, volcanic eruptions, landslides, wildfires, floods, and SOC models. *The European Physical Journal Special Topics* 205: 167–182.

Scholz, C.H. 1990. *The mechanics of earthquakes and faulting*. Cambridge: Cambridge University Press.

Zaninetti, J.M., and C.E. Colton. 2012. Shrinking New Orleans: Post-Katrina population adjustments. *Urban Geography* 33: 675–699.

John B. Rundle is a Distinguished Professor of physics and geology at the Departments of Physics and Geology at University of California, Davis. He is also Director of the California Institute for Hazard Research of the University of California, and Executive Director of the APEC Cooperation for Earthquake Simulation (ACES), The Santa Fe Institute. His research is focused on understanding the dynamics of earthquakes through numerical simulations, pattern analysis of complex systems, dynamics of driven nonlinear earth systems, and instabilities in financial markets. Rundle is a Fellow of the American Physical Society and a Fellow of the American Geophysical Union. He was recently named a co-winner of the NASA Software of the Year Award in 2012.

Donald L. Turcotte is a Distinguished Professor of geology at the Department of Geology at University of California, Davis. His research is focused on applying dynamical systems to geological problems, including crustal deformation, seismicity, topography, and forest fires. Turcotte has a Ph.D. from the California Institute of Technology. He is a Fellow of the American Geophysical Union and the Geological Society of America. He has been elected to the National Academy of Sciences (USA) and the American Academy of Arts and Sciences. He has also won the Whitten medal and recipient of the William Bowie Award of the American Geophysical Union.

Chapter 33
Transitions Into and Out of a Crisis Mode of Socio-ecological Systems

Armin Haas, Qian Ye, Peijun Shi, and Carlo C. Jaeger

33.1 Socio-ecological Systems

The basic idea of *socio-ecological systems* (SES) has been developed in different communities using different labels and highlighting different aspects of them. Gallopin (1991) and Young et al. (2006) labeled them SES, whereas, for example, Berkes and Folke (1998) used the notion *social-ecological systems*, Turner et al. (2003) prefer the notion of *coupled human-environment systems*, while Jaeger (1994) opts for the notion of *human ecological systems*. SES can be found at all scales, from the local household with its surroundings, to the society of nations inhabiting planet earth. In any SES, human and ecological (environmental, natural, or biophysical) subsystems interact.

Often, an SES is conceptualized as being composed of four subsystems in interaction: a social, an economic, an ecological, and an institutional one (Fig. 33.1). The analysis of the interrelationships within the system creates the foundation for managing risks. Hereby three kinds of interactions are always considered: (i) interactions between different risks; (ii) interactions between risks and subsystems; and (iii) interactions between subsystems themselves. In this way, a better understanding can be developed of the risks under consideration in a specific SES of interest.

This can be sketched using the example of the European heat wave in 2003. According to current knowledge, this caused 70,000 deaths and was thus to date the largest civil catastrophe in Europe since the influenza outbreak after the First World War (Robine et al. 2008; Lass et al. 2011). It would certainly be erroneous to ascribe

A. Haas
Global Contract for Sustainability, Institute for Advanced Sustainability Studies, Berliner Str. 130, 14467 Potsdam, Germany
e-mail: armin.haas@iass-potsdam.de

Q. Ye • P. Shi • C.C. Jaeger (✉)
State Key Laboratory of Earth Surface Processes and Resource Ecology, Beijing Normal University, 19 Xinjiekouwai Street, Beijing 100875, China
e-mail: qianye@yahoo.com; spj@bnu.edu.cn; carlo.jaeger@globalclimateforum.org

U. Fra.Paleo (ed.), *Risk Governance*, DOI 10.1007/978-94-017-9328-5_33

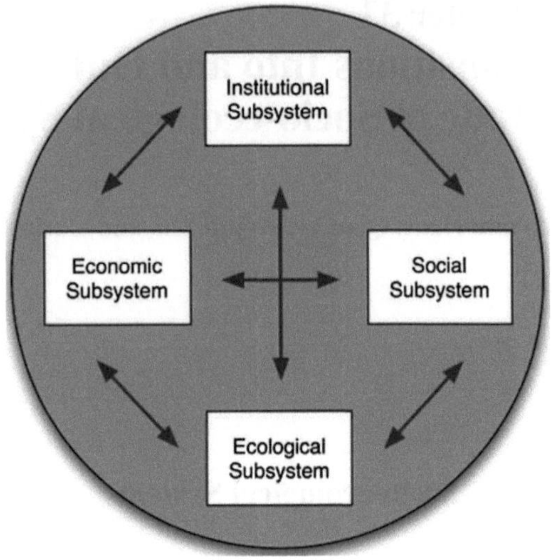

Fig. 33.1 Socio-ecological system

this catastrophe solely to the weather. Many of the elderly people died from lack of access to water, however, not because there was not enough water. Instead, they more likely died because the people themselves, as well as the institutions, were not capable of coping with such an unprecedented situation. Moreover, many buildings did not offer sufficient protection from the high temperatures. Thus the reasons for the high number of deaths are not solely to be found in the ecological subsystem.

This example shows that for each concrete instance of socio-ecological systems the interplay of its subsystems must be understood and specific risks be evaluated and analyzed, in order to establish how, and whether, the interplay of the subsystems can spawn certain risks. This is also important for the reason that a well-intentioned intervention in a particular subsystem can lead to unintended effects in other subsystems. Likewise, this kind of interrelationship can be used to achieve effects in one subsystem through an intervention in another – a classic policy tool, which requires, however, a sufficient understanding of the system under scrutiny.

33.2 Transitions into and Out of a Crisis Mode of SES

Building on the concept of SES, we can define *transitions into and out of a crisis mode of SES*. Our focus on entry and exit transitions builds on previous work on the social amplification of risk (Kasperson et al. 1988; Kasperson and Kasperson 1996; Lofstedt and Renn 1997; Pidgeon et al. 2003). An SES transits into crisis mode when a risk realizes, or is expected to realize.

A specific event under consideration can be a sudden onset event like an earthquake, or a slow onset event like sea level rise. Figure 33.2 illustrates our concept of

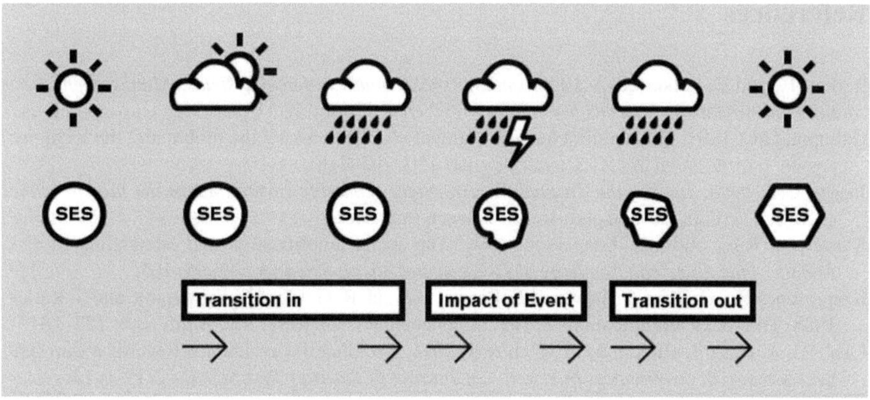

Fig. 33.2 Entry and exit transition of a SES

entry and exit transitions in a simplified manner. Depending on whether the event is a sudden or slow onset event, the time span for the transition into the crisis mode may range from minutes to decades, or even centuries. In Fig. 33.2, the realization of a specific risk is broadly sketched. Cloudiness and heavy rain stand for increasingly threatening signals that can help humans anticipate the realization of a specific risk. The very realization is depicted as *impact of event*.

We define *entry transition* as the sequence of changes in the decision-making processes, including the deployment and reorganization of actions, actors and resources, to cope with the perceived risk. Depending on the specific risk, this may imply something from the whole disaster risk management spectrum, starting from orderly resorting to emergency plans readily available, to spontaneously jumping into chaotic emergency mode.

An *exit transition* is the sequence of changes in the decision-making processes that takes place after the risk has realized. If an SES is sufficiently *structurally stable*, it may return to its pre-event state. Depending on the academic domain of the discourse, such a structural stability is termed *resilience*, or *human coping capacity*. An event may, however, exceed the human coping capacity of an SES, which is permanently transformed by the event during the transition out of the crisis mode. Judged from specific normative standpoints, such a transformation could lead to a better or worse performance of the SES.

When the realization of the risk and its impact lead to a weakened human coping capacity, forgetting the lessons learned, etc., it may decrease the resilience of the SES and make the system more vulnerable to future risks.

The transformation could, however, also trigger processes of *social learning* that lead to a considerably *higher* level of resilience of the SES. The 2008 winter icestorms in China triggered, for example, a conscious process of social learning that led to several far-reaching changes in how high risks are approached administratively and intellectually in China (Shi et al. 2013). These social learning processes are considered as enhancing China's resilience as SES.

References

Berkes, F., and C. Folke (eds.). 1998. *Linking social and ecological systems: Management practices and social mechanisms for building resilience*. Cambridge: Cambridge University Press.

Gallopín, G.C. 1991. Human dimensions of global change: Linking the global and the local processes. *International Social Science Journal* 130: 707–718.

Jaeger, C.C. 1994. *Taming the dragon. Transforming economic institutions in the face of global change*. Yverdon: Gordon and Breach Science Publishers.

Kasperson, R.E., and J.X. Kasperson. 1996. The social amplification and attenuation of risk. *Annals of the American Academy of Political and Social Sciences* 545: 95–105.

Kasperson, R.E., O. Renn, P. Slovic, H.S. Brown, J. Emel, R. Goble, J.X. Kasperson, and S. Ratick. 1988. The social amplifications of risk. A conceptual framework. *Risk Analysis* 8: 177–187.

Lass, W., A. Haas, J. Hinkel, and C.C. Jaeger. 2011. Avoiding the avoidable: Towards a European heat waves risk governance. *International Journal of Disaster Risk Science* 2(1): 1–14.

Lofstedt, R., and O. Renn. 1997. The Brent Spar controversy: An example of risk communication gone wrong. *Risk Analysis* 17: 131–136.

Pidgeon, N., R.K. Kasperson, and P. Slovic (eds.). 2003. *The social amplification of risk*. Cambridge: Cambridge University Press.

Robine, J., S. Cheung, S. Le Roy, H. van Oyen, C. Griffiths, J. Michel, and F. Herrmann. 2008. Death toll exceeded 70,000 in Europe during the summer of 2003. *Comptes Rendus Biologies* 331(2): 171–178.

Shi, P., W. Fang, W. Dong, N. Li, W. Xu, and S. Chen. 2013. Experience, lessons and recommendations of China's response to sleet & snow disaster in south. In *Integrated risk governance: Science plan and case studies of large-scale disasters*, ed. P. Shi, C.C. Jaeger, and C. Ye, 89–118. Beijing/Berlin: Springer.

Turner II, B.L., R.E. Kasperson, P.A. Matson, J.J. McCarthy, R.W. Corell, L. Christensen, N. Eckley, J.X. Kasperson, A. Luers, M.L. Martello, C. Polsky, A. Pulsipher, and A. Schiller. 2003. A framework for vulnerability analysis in sustainability science. *Proceedings of the National Academy of Sciences of the U S A* 100: 8074–8079.

Young, O.R., F. Berkhout, G.C. Gallopin, M.A. Janssen, E. Ostrom, and S. Sander van der Leeuw. 2006. The globalization of socio-ecological systems: An agenda for scientific research. *Global Environmental Change* 16: 304–316.

Armin Haas is Senior Researcher in the Research Domain Transdisciplinary Concepts and Methods at the Potsdam Institute for Climate Impact Research (PIK), Senior Fellow in the Research Cluster Global Contract for Sustainability of the Institute for Advanced Sustainability Studies (IASS) Potsdam, and the leader of the research process Integrated Risk Governance of the Global Climate Forum (GCF). His current research interests are the interaction of financial markets and climate change, and innovative approaches for the management and governance of large-scale uncertainties. Armin is co-founder of the Munich Climate Insurance Initiative (MCII), and of The Compensators, a non-profit organisation giving citizens low-cost access to the European emission trading system, and co-inventor (together with Antonella Battaglini) of the SuperSmart Grid.

Qian Ye is Professor of climate change impacts at Beijing Normal University, and the leading scientist of the research group on risk governance and sustainability at the State Key Laboratory of Earth Surface Processes and Resource Ecology. He is Deputy General Secretary of the Chinese National Committee for IHDP and member of the urban meteorological committee of the China Meteorological Society. He is Executive Director of the IHDP core project Integrated Risk Governance Project, and member of its Scientific Steering Committee. His current research interest is focused on the role of social and ecological dynamic systems in dealing with climate change and its impacts, meteorological economics, climate change risks, green growth, and risk governance issues.

Peijun Shi is the Executive Vice-president of Beijing Normal University. He is the Vice Chairman of the Expert Committee of the National Disaster Reduction Commission of China. He is also a member of OECD's High Level Advisory Board on Financial Management of Large-Scale Catastrophes. Shi's research focuses on natural disaster theories and risk governance. He has been the principal investigator of many national and ministerial research programs and has conducted numerous publications. He co-chairs the Integrated Risk Governance (IRG) project, a core science project of the International Human Dimensions Programme on Global Environmental Change (IHDP), whose mission is to improve the governance of new risks that exceed current human coping capacities.

Carlo C. Jaeger is an economist working on global systems, in particular on the role of financial markets in addressing -or exacerbating- the problem of climate change. He is co-founder and Chairman of the Global Climate Forum, Professor at Potsdam University, Germany, and the Beijing Normal University (BNU). Jaeger's current research is focused on problems of global coordination, especially in view of sustainable development. Until March 2012 Carlo C. Jaeger was Head of the Research Domain Transdisciplinary Concepts and Methods at the Potsdam Institute for Climate Impact Research. He is a member of the Scientific and Technical Council of the International Risk Governance Council, and has served on the boards of various scientific organizations.

Index

U. Fra.Paleo (ed.), *Risk Governance*, DOI 10.1007/978-94-017-9328-5

D

MIX
Papier aus verantwortungsvollen Quellen
Paper from responsible sources
FSC® C105338

If you have any concerns about our products,
you can contact us on
ProductSafety@springernature.com

In case Publisher is established outside the EU,
the EU authorized representative is:
Springer Nature Customer Service Center GmbH
Europaplatz 3, 69115 Heidelberg, Germany

Printed by Libri Plureos GmbH
in Hamburg, Germany